AP®
ENVIRONMENTAL SCIENCE
PREMIUM PREP

19th Edition

The Staff of The Princeton Review

PrincetonReview.com

The Princeton Review
110 East 42nd St, 7th Floor
New York, NY 10017

Published in the United States by Penguin Random House LLC,
New York.

ISBN: 978-0-593-51765-9
eBook ISBN: 978-0-593-51766-6
ISSN: 2690-540X

Editor: Laura Rose
Production Editors: Lyssa Mandel and Kathy Carter
Production Artist: Amanda Shurgin
Content Contributor: Eliz Markowitz, EdD

Printed in the United States of America.

10 9 8 7 6 5 4 3 2 1

19th Edition

The Princeton Review Publishing Team
Rob Franek, Editor-in-Chief
David Soto, Senior Director, Data Operations
Stephen Koch, Senior Manager, Data Operations
Deborah Weber, Director of Production
Jason Ullmeyer, Production Design Manager
Jennifer Chapman, Senior Production Artist
Selena Coppock, Director of Editorial
Orion McBean, Senior Editor
Aaron Riccio, Senior Editor
Meave Shelton, Senior Editor
Chris Chimera, Editor
Patricia Murphy, Editor
Laura Rose, Editor
Isabelle Appleton, Editorial Assistant

Penguin Random House Publishing Team
Tom Russell, VP, Publisher
Alison Stoltzfus, Senior Director, Publishing
Emily Hoffman, Associate Managing Editor
Patty Collins, Executive Director of Production
Mary Ellen Owens, Assistant Director of Production
Alice Rahaeuser, Associate Production Manager
Maggie Gibson, Associate Production Manager
Suzanne Lee, Senior Designer
Eugenia Lo, Publishing Assistant

Acknowledgments

The Princeton Review would like to give a special thanks
to Eliz Markowitz for her work on this year's edition.

We would also like to extend our thanks to Amanda
Shurgin, Lyssa Mandel, and Kathy Carter for their time
and attention while producing this title.

For customer service, please contact
editorialsupport@review.com,
and be sure to include:

- full title of the book

- ISBN

- page number

Contents

Get More (Free) Content
at **PrincetonReview.com/prep**

As easy as 1 • 2 • 3

1 Go to PrincetonReview.com/prep or scan the **QR code** and enter the following ISBN to register your book:

9780593517659

2 Answer a few simple questions to set up an exclusive Princeton Review account. *(If you already have one, you can just log in.)*

3 Enjoy access to your **FREE** content!

Once you've registered, you can...

- Access an additional practice test online

- Take a full-length practice SAT and/or ACT

- Get valuable advice about the college application process, including tips for writing a great essay and where to apply for financial aid

- Use our searchable rankings of *The Best 390 Colleges* to find out more information about your dream school.

- Access comprehensive study guides, digital flashcards to review core content, and a variety of printable resources, including bubble answer sheets and Key Terms lists

- Check to see whether there have been any corrections or updates to this edition

LET'S GO MOBILE! Access free, additional resources by downloading the new Princeton Review app at **www.princetonreview.com/mobile/apps/highschool** or scan the QR Code to the right.

Need to report a potential **content** issue?

Contact **EditorialSupport@review.com** and include:
- full title of the book
- ISBN
- page number

Need to report a **technical** issue?

Contact **TPRStudentTech@review.com** and provide:
- your full name
- email address used to register the book
- full book title and ISBN
- Operating system (Mac/PC) and browser (Chrome, Firefox, Safari, etc.)

Look For These Icons Throughout The Book

 ONLINE ARTICLES

 PROVEN TECHNIQUES

 APPLIED STRATEGIES

 ONLINE VIDEO TUTORIALS

 MORE GREAT BOOKS

 STUDY BREAK

Part I
Using This Book to Improve Your AP Score

- Preview: Your Knowledge, Your Expectations
- Your Guide to Using This Book
- How to Begin

PREVIEW: YOUR KNOWLEDGE, YOUR EXPECTATIONS

Your route to a high score on the AP Environmental Science Exam depends a lot on how you plan to use this book. Respond to the following questions:

1. Rate your level of confidence about your knowledge of the content tested by the AP Environmental Science Exam.
    A.   Very confident—I know it all
    B.   I'm pretty confident, but there are topics for which I could use help
    C.   Not confident—I need quite a bit of support
    D.   I'm not sure

2. Circle your goal score for the AP Environmental Science Exam.
    5   4   3   2   1   I'm not sure yet

3. What do you expect to learn from this book? Circle all that apply to you.
    A.   A general overview of the test and what to expect
    B.   Strategies for how to approach the test
    C.   The content tested by this exam
    D.   I'm not sure yet

YOUR GUIDE TO USING THIS BOOK

This book is organized to provide as much—or as little—support as you need, so you can use this book in whatever way will be most helpful to improving your score on the AP Environmental Science Exam. Please note that this book should be used in conjunction with your AP course textbook.

*   The remainder of **Part I** will provide guidance on how to use this book and help you determine your strengths and weaknesses.

*   **Part II** of this book contains your first practice test, the Diagnostic Answer Key, answers and explanations, and a scoring guide. (A bubble sheet can be found in the back of the book or you can download them from your Student Tools.) This is where you should begin your test preparation in order to realistically determine:
    o   your starting point right now
    o   which question types you're ready for and which you might need to practice
    o   which content topics you are familiar with and which you will want to carefully review

    Once you have nailed down your strengths and weaknesses with regard to this exam, you can focus your test preparation, build a study plan, and be efficient with your time. Our Diagnostic Answer Key will assist you with this process.

*   **Part III** of this book will:
    o   provide information about the structure, scoring, and content of the AP Environmental Science Exam
    o   help you make a study plan
    o   point you toward additional resources

- **Part IV** of this book will explore the following strategies:
  - o how to handle multiple-choice questions
  - o how to approach free-response questions
  - o how to manage your time to maximize the number of points available to you

- **Part V** of this book covers the content you need to know for the AP Environmental Science Exam.

- **Part VI** of this book contains Practice Test 2 and Practice Test 3, with answers and explanations and a scoring guide. (A bubble sheet can be found in the very back of the book for easy tear-out.) If you skipped Practice Test 1, we recommend that you do all three tests (with at least a day or two between them) so that you can compare your progress between the three. Additionally, this will help to identify any external issues. If you get a certain type of question wrong all three times, you probably need to review it. If you only got it wrong once, you may have run out of time or been distracted by something. In either case, this will allow you to focus on the factors that caused the discrepancy in scores and to be as prepared as possible on the day of the test.

> **Want More?**
> Check out your online Student Tools for some great ways to extend your prep, including study guides. Scan the code below and register your book!
>
>

HOW TO BEGIN

1. **Take a Test**

   Before you can decide how to use this book, you need to take a practice test. Doing so will give you insight into your strengths and weaknesses and help you make an effective study plan. If you're feeling test-phobic, remind yourself that a practice test is a tool for diagnosing yourself—it's not how well you do that matters, but how you use information gleaned from your performance to guide your preparation.

   So, before you read further, take Practice Test 1 starting on page 9 of this book. Be sure to do so in one sitting, following the instructions that appear before the test.

2. **Check Your Answers**

   Using the Diagnostic Answer Key on page 27, follow our three-step process to identify your strengths and weaknesses with regard to the tested topics. This will help you determine which content review chapters to prioritize when studying this book. Don't worry about the explanations for now, and don't worry about missed questions. We'll get to that soon.

3. **Reflect on the Test**

   After you take your first test, respond to the following questions:
   - How much time did you spend on the multiple-choice questions?
   - How much time did you spend on each free-response question?
   - How many multiple-choice questions did you miss?
   - Do you feel you had the knowledge to address the subject matter of the free-response questions?

Bonus Tips and Tricks...

Check out our test-taking tips and must-know strategies at youtube.com/c/ThePrincetonReview/videos

- Check the content areas that were most challenging for you and draw a line through the ones in which you felt confident/did well.
  o Geological eras
  o Atmosphere
  o Soil dynamics
  o Population
  o Pollution (specifically solid waste, acid rain, ozone depletion, and thermal pollution)
  o Water conservation
  o Conservation and economic impact
  o Interpreting charts and graphs
  o Interpreting experimental data
  o Analyzing experiments and experimental reasoning
  o Ecosystems and biodiversity
  o Biogeochemical cycles
  o Global climate change
  o Environmental regulation and international treaties
  o Energy resources and consumption
  o Energy science
  o Mining

4. **Read Part III of this Book and Complete the Self-Evaluation**

   As discussed in the Guide section above, Part III will provide information on how the test is structured and scored. It will also outline areas of content that are tested.

   As you read Part III, reevaluate your answers to the questions you just answered. At the end of Part III, you will revisit and refine your answers to the previous questions. You will then be able to make a study plan, based on your needs and available time, that will allow you to use this book most effectively.

5. **Engage with Parts IV and V as Needed**

   Notice the word *engage*. You'll get more out of this book if you use it intentionally than if you read it passively, hoping for an improved score through osmosis.

   Strategy chapters in Part IV will help you think about your approach to the question types on this exam. Part IV will open with a reminder to think about how you approach questions now, and then close with a reflection section asking you to think about how/whether you will change your approach in the future.

   Content chapters in Part V are designed to provide a review of the content tested on the AP Environmental Science Exam, including the level of detail you need to know and how the content is tested. You will have the opportunity to assess your mastery of the content of each chapter through test-appropriate questions and a reflection section.

6. **Take Practice Tests 2 and 3. Assess Your Performance**
Once you feel you have developed the strategies you need and gained the knowledge you lacked, you should take Practice Test 2 and then Practice Test 3. Take each test in one sitting, with maybe a day or two in between.

When you are done, check your answers to the multiple-choice sections. See if a teacher will read your responses to the free-response questions and provide feedback.

Once you have taken the test, reflect on what areas you still need to work on, and revisit the chapters in this book that address those deficiencies. Through this type of reflection and engagement, you will continue to improve.

7. **Keep Working**
As Part III will discuss, there are other resources available to you, including a wealth of information on AP Students. You can continue to explore areas in which you can stand to improve and engage in those areas right up to the day of the test.

AP Students
Check out the College Board's homepage for more information about the updated exam at apstudents.collegeboard.org/courses/ap-environmental-science

Part II
Practice Test 1

- Practice Test 1
- Practice Test 1: Diagnostic Answer Key and Explanations
- How to Score Practice Test 1

Practice Test 1

AP® Environmental Science Exam

SECTION I: Multiple-Choice Questions

DO NOT OPEN THIS BOOKLET UNTIL YOU ARE TOLD TO DO SO.

Instructions

At a Glance

Total Time
1 hour and 30 minutes
Number of Questions
80
Percent of Total Grade
60%
Writing Instrument
Pencil required

Section I of this examination contains 80 multiple-choice questions. Fill in only the ovals for numbers 1 through 80 on your answer sheet (located at the back of this book).

Indicate all of your answers to the multiple-choice questions on the answer sheet. No credit will be given for anything written in this exam booklet, but you may use the booklet for notes or scratch work. After you have decided which of the suggested answers is best, completely fill in the corresponding oval on the answer sheet. Give only one answer to each question. If you change an answer, be sure that the previous mark is erased completely. Here is a sample question and answer.

Sample Question   Sample Answer

Chicago is a

    (A) state

    (B) city

    (C) country

    (D) continent

Use your time effectively, working as quickly as you can without losing accuracy. Do not spend too much time on any one question. Go on to other questions and come back to the ones you have not answered if you have time. It is not expected that everyone will know the answers to all the multiple-choice questions.

About Guessing

Many candidates wonder whether or not to guess the answers to questions about which they are not certain. Multiple-choice scores are based on the number of questions answered correctly. Points are not deducted for incorrect answers, and no points are awarded for unanswered questions. Because points are not deducted for incorrect answers, you are encouraged to answer all multiple-choice questions. On any questions you do not know the answer to, you should eliminate as many choices as you can, and then select the best answer among the remaining choices.

GO ON TO THE NEXT PAGE.

ENVIRONMENTAL SCIENCE
Section I
Time—1 hour and 30 minutes
80 Questions

Directions: Each of the questions or incomplete statements below is followed by four suggested answers or completions. Select the one that is best in each case and then fill in the corresponding oval on the answer sheet.

1. Which of the following concepts best describes a situation where individuals, acting in their self-interest, deplete shared resources and, in turn, cause environmental degradation?

   (A) The Tragedy of the Commons

   (B) Slash-and-burn

   (C) The Green Revolution

   ✓(D) The Tragedy of Free Access

2. Which of the following is a primary greenhouse gas responsible for trapping heat in the Earth's atmosphere?

   (A) Nitrogen

   ✓(B) Carbon Dioxide

   (C) Oxygen

   (D) Hydrogen

Questions 3–4 refer to the following graph.

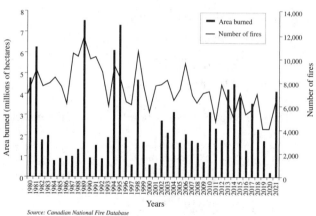

Source: Canadian National Fire Database

3. Based on the graph provided, Canada experienced the fewest fires in

   ✓(A) 2020

   (B) 2011

   (C) 1992

   (D) 1989

4. Which of the following statements regarding the information in the provided graph is true?

   ✓(A) More hectares of land were burned in 1995 than in any other year.

   (B) In 2017, Canada experienced approximately 4,000 fires.

   (C) The number of hectares burned in 1985 was approximately twice that burned in 1984.

   (D) More hectares of land were burned in 1989 than in any other year.

5. All of the following statements regarding climate change are true EXCEPT

   (A) water vapor does not contribute significantly to global climate change, despite being a greenhouse gas

   ✓(B) chlorofluorocarbons have the highest global warming potential

   (C) an increase in the global temperature will decrease the frequency and duration of storms, decrease the number of hot days, and increase the number of cold days

   (D) the increase in average global temperature since the mid-1900s is likely due to increases in anthropogenic greenhouse gas concentrations

6. In general, r-selected organisms

   (A) have populations above the carrying capacity of the environment

   (B) tend to have a large body size

   ✓(C) reproduce late in life

   (D) have short lifespans

GO ON TO THE NEXT PAGE.

7. Which of the following type of volcanoes have broad bases and are tall with gentle slopes?

(A) Shield volcanoes

(B) Composite volcanoes

✓(C) Cinder volcanoes

(D) Lava domes

Questions 8–10 refer to the following graph.

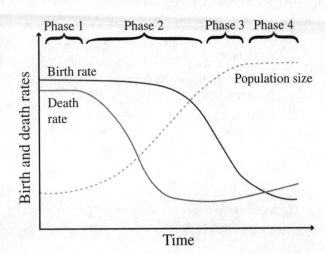

8. The graph above best describes the

(A) boom-and-bust cycle

(B) survivorship curve

(C) logistic growth curve

/(D) demographic transition model

9. Which phase is referred to as the transitional state, during which birth rates are high, but death rates are lower because of improved access to food, water, and healthcare?

(A) Phase I

✓(B) Phase II

(C) Phase III

(D) Phase IV

10. During which phase does the population approach and reach a zero-growth rate?

(A) Preindustrial state

✓(B) Transitional state

(C) Industrial state

(D) Postindustrial state

11. Lowland areas that are saturated with moisture, such as marshes and swamps, and are the natural habitat of wildlife are referred to as

✓(A) limnetic zones

(B) taiga

(C) wetlands

(D) estuaries

Questions 12–13 refer to the following graph.

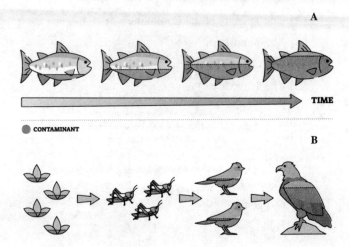

12. The portion of the graph labeled A above represents the concept of

(A) biological oxygen demand

(B) bioaccumulation

✓(C) biomass augmentation

(D) biomagnification

13. The portion of the graph labeled B above represents the concept of

(A) biological oxygen demand

(B) bioaccumulation

✓(C) biomass augmentation

(D) biomagnification

GO ON TO THE NEXT PAGE.

14. Also known as the zone of illuviation, which of the following horizons is where organic matter, soil, clay, and minerals that are washed out of the upper horizons accumulate?

    (A) A horizon

    (B) B horizon

    (C) C horizon

    √(D) O horizon

15. Jets streams

    (A) are areas of relatively still air because the air is constantly rising rather than blowing

    (B) are caused by the surface currents of the Hadley cells and were named for their ability to quickly propel trading ships across the ocean

    (C) are caused by the surface currents of Ferrel cells and are a result of the Coriolis effect

    √(D) are high-speed currents of wind that occur in the tropopause and influence local weather patterns

16. The rate of population change over time can be found by

    (A) subtracting the sum of the death rate and emigration rate from the sum of the birth rate and immigration rate and dividing by ten

    (B) subtracting the sum of the death rate and immigration rate from the sum of the birth rate and emigration rate and dividing by ten

    √(C) subtracting the sum of the birth rate and immigration rate from the sum of the death rate and emigration rate and multiplying by ten

    (D) subtracting the sum of the death rate and emigration rate from the sum of the birth rate and immigration rate and multiplying by ten

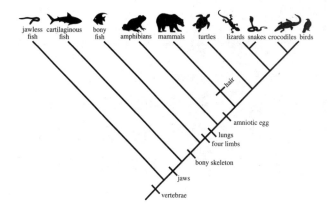

17. The figure above is known as a

    (A) food chain

    (B) energy pyramid

    √(C) phylogenic tree

    (D) food web

18. All of the following are potential safety issues with nuclear energy EXCEPT

    (A) thermal pollution

    (B) highly radioactive waste

    (C) nuclear weapons development

    √(D) greenhouse gas emissions

19. Which of the following is NOT a problem that arises from genetically modified organisms (GMOs)?

    √(A) They discourage biodiversity.

    (B) They can decrease antibiotic resistance.

    (C) They could encourage the rise of pesticide-resistant pests.

    (D) They can pose new allergen risks.

20. All of the following are examples of point-source pollution EXCEPT

    √(A) runoff from agricultural fields

    (B) an oil spill from a tanker accident

    (C) industrial discharge from a factory pipe

    (D) effluent from a wastewater treatment plant

GO ON TO THE NEXT PAGE.

21. Which of the following weather phenomenon is depicted in the image below?

❶ Land is warmer than ocean

Warm air rises

❷ Moist air blows from above ocean to land. Moisture condenses and becomes rain.

Cool air sinks

Ocean is cooler than land

(A) Rain shadow effect

(B) Lake effect

✓ (C) Monsoon formation

(D) El Niño

22. All of the following are advantages of wind power EXCEPT

(A) lower land and water use requirements than fossil fuels

(B) increased resource availability

(C) lower greenhouse gas emissions than fossil fuels

✓ (D) intermittent supply

23. Salinization refers to

✓ (A) a significant accumulation of salts on the soil's surface that results in land unusable for agricultural purposes

(B) the cutting of furrows between crop rows and filling them with water

(C) the practice of cutting down and burning an area of vegetation prior to sowing crops

(D) the combination of environmentally sensitive methods used to keep the pest population down to an economically viable level

24. Unwanted or unanticipated consequences of using resources are known as

(A) marginal costs

(B) externalities

(C) tangible properties

(D) intangible properties

25. The purest type of coal, which is almost pure carbon, is known as

(A) bituminous

(B) subbituminous

✓ (C) lignite

(D) anthracite

26. Silviculture is

(A) the management of forest plantations for the purpose of harvesting timber

(B) the removal of select trees in an area

✓ (C) the creation of a mutualistic symbiotic relationship by planting trees and crops together

(D) the removal of all trees in an area

Questions 27–29 refer to the following chart.

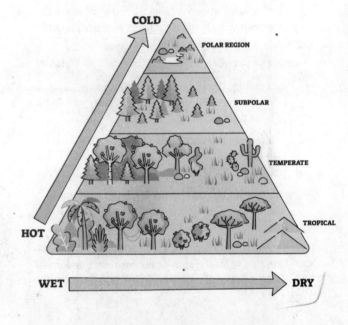

COLD

POLAR REGION

SUBPOLAR

TEMPERATE

TROPICAL

HOT

WET   DRY

27. The types of ecosystems that are based on land, as depicted in the provided chart, are referred to as

(A) aquatic life zones

✓ (B) biomes

(C) ecotones

(D) ecozones

GO ON TO THE NEXT PAGE.

28. The type of ecosystem that one would most likely find in the polar region is referred to as

   (A) tundra

   (B) taiga

   (C) a temperate rainforest

   (D) a tropical rainforest

29. The type of ecosystem that has hardwood trees as its major vegetation, and is found in regions that are both temperate and tropical, is known as

   (A) a tropical rainforest

   (B) chapparal

   (C) a deciduous forest

   (D) a coniferous rainforest

Questions 30–31 refer to the following figure.

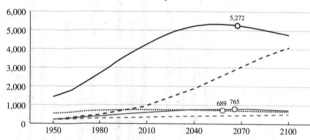

World Population Trend
Estimated development of the population (in millions)

30. Based on the figure above, the populations of which regions will be declining in 2100?

   (A) Africa and North America

   (B) Europe, North America, Latin America, and the Caribbean

   (C) Asia, Oceania, Latin America, the Caribbean, and Europe

   (D) Asia, Oceania, and Africa

31. Which of the following statements regarding the figure is valid?

   (A) The population of Europe is expected to reach 5,272 million people before beginning to shrink.

   (B) North America is expected to experience a greater percent increase in population size between 1950-2100 than is Africa.

   (C) The population sizes of Africa and Europe were roughly equal in 1995.

   (D) Asia and Oceania will experience population decline sooner than any other geographic area shown.

32. The Dust Bowl led to the passage of the

   (A) Food Security Act of 1985

   (B) Soil Conservation Act of 1935

   (C) Water Salinization Act of 1938

   (D) Soil and Water Conservation Act of 1977

33. Which of the following is NOT a primary goal of green taxes?

   (A) To generate revenue to correct past pollution damage and reduce future pollution

   (B) To change consumption behaviors

   (C) To increase the market share of green technologies

   (D) To use the funds received from pollution taxes for restoration

34. A species that is under a very high risk of extinction is known as a(n)

   (A) endangered species

   (B) extinct species

   (C) critically endangered species

   (D) vulnerable species

35. Which of the following is the leading cause of coral bleaching?

   (A) Herbicides

   (B) Ocean acidification

   (C) Oil and chemical spills

   (D) Rising water temperatures

GO ON TO THE NEXT PAGE.

36. Radioactive waste can be categorized as either low- or high-level based on the amount of ionizing radiation produced. Furthermore, the Environmental Protection Agency (EPA) categorizes radioactive waste according to its place of origin. Which of the following does NOT pair the EPA category of radioactive waste with the type of ionizing radiation produced?

(A) Nuclear reactor waste; high-level radiation

(B) Waste from the reprocessing of spent nuclear fuel; low-level radiation

(C) Waste from the mining and processing of uranium ore; high-level radiation

(D) Radioactive waste from industrial or research industries; low-level radiation

Questions 37–39 refer to the following figure.

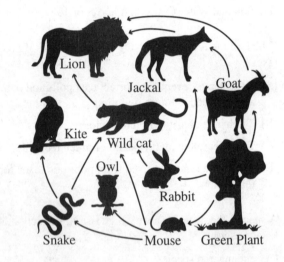

37. The figure above is an example of a

(A) food chain

(B) energy pyramid

(C) phylogenic tree

(D) food web

38. Which of the following is a food source for kites?

(A) Mice

(B) Snakes

(C) Green plants

(D) Owls

39. A rabbit is an example of a

(A) primary consumer

(B) secondary consumer

(C) tertiary consumer

(D) detritivore

40. Which of the following cut the emissions of CFCs that damage the ozone layer and was later amended to include other key ozone-depleting chemicals?

(A) Kyoto Protocol

(B) National Environmental Policy Act

(C) Montreal Protocol

(D) Pollution Prevention Act

Questions 41–43 refer to the following graph.

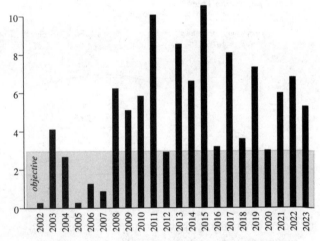

Western Lake Erie Bloom Severity

Source: NOAA

41. According to the provided graph, the objective is to have a severity level of algal blooms in Western Lake Erie of

(A) 0

(B) less than or equal to 3

(C) greater than or equal to 3

(D) 10

GO ON TO THE NEXT PAGE.

54. Which of the following is NOT a human factor that can cause extinction and decrease biodiversity?

(A) Overexploitation

(B) The introduction of invasive species

(C) Pollution

(D) Habitation reunification

55. Which of the following descriptions best explains the purpose of the Pollution Prevention Act (PPA)?

(A) The PPA cut the emissions of chlorofluorocarbons that damage the ozone layer.

(B) The PPA bans the capture, exportation, or sale of endangered and threatened species.

(C) The PPA was designed to promote source reduction.

(D) The PPA mandates that federal agencies prepare environmental impact statements.

56. Citizens can exert political influence and guide their leaders towards policies for sustainability by doing which of the following activities?

(A) Voting

(B) Donating to politicians

(C) Posting on social media

(D) Spending less on unsustainable products

57. The image above shows an elephant population that exhibits which type of population dispersion?

(A) Uniform

(B) Clumped

(C) Random

(D) Linear

58. Riparian rights refer to

(A) water rights given to those who have historically used the natural water resources in a certain area

(B) rights provided to certain countries that allow the capture of a certain number of whales annually

(C) private water-use rights tied to the ownership of land bordering a natural river

(D) the extension of each country's jurisdiction to 200 miles from shore for the purpose of offshore fisheries

59. Which of the following is NOT an environmental concern related to mining?

(A) Mining can disrupt the ecosystem and harm the surrounding land.

(B) Coal mining can result in radioactive waste that is highly dangerous to an ecosystem and human health.

(C) Mineral extraction can leave pollutants resulting from the surface exposure of underground minerals.

(D) Mining can result in acid mine drainage that severely harms local water systems.

GO ON TO THE NEXT PAGE.

Questions 60 and 61 refer to the following chart.

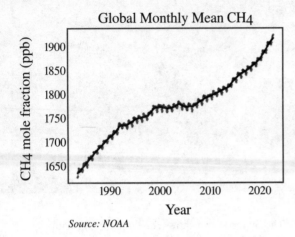

Global Monthly Mean CH₄

Source: NOAA

60. The global monthly mean of CH_4 experienced the greatest percent increase between

    (A) 2020–2023

    (B) 2010–2014

    (C) 2000–2004

    (D) 1990–1994

61. All of the following statements regarding the provided chart are valid EXCEPT that

    (A) in 1984, the global monthly mean of CH_4 was approximately 1,635 ppb

    (B) the global monthly mean of CH_4 experienced an increase between 2004–2006

    (C) the global monthly mean of CH_4 is over 250 ppb greater than it was when measurements were first recorded

    (D) in 2012, the global monthly mean of CH_4 was approximately 175 ppb greater than it was in 1994

62. Federal legislation that was created to oversee and control prospecting and mining for economic minerals is known as the

    (A) Prospectors Regulation Act of 1870

    (B) Mineral Lands Act of 1866

    (C) Homestead Act of 1862

    (D) General Mining Act of 1872

63. All of the following factors directly impact a population's birth rate EXCEPT

    (A) a population's religious beliefs, culture, and traditions

    (B) the demand for children in the labor force

    (C) the availability of nutritious food

    (D) a base level of education for women

64. Which of the following is NOT one of the four physical spheres that make up our planet and regulate life on Earth?

    (A) Troposphere

    (B) Pedosphere

    (C) Atmosphere

    (D) Hydrosphere

65. Species of fish, mammals, and birds that are caught during fishing operations, but are not the target fish, are caught with

    (A) drift nets

    (B) long lining

    (C) capture fisheries

    (D) bottom trawling

GO ON TO THE NEXT PAGE.

Questions 66 and 67 refer to the following graphs.

Number of Species of Organisms Going Extinct

Conservation Status

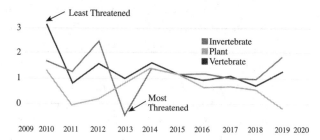

66. Which of the following statements regarding the graphs is TRUE?

    (A) In 2016, the number of species of organisms that were going extinct was higher than in any other year.

    (B) The number of species of organisms going extinct has decreased over the recorded period.

    (C) The number of organisms that were going extinct in 2019 was 300% higher than those in 2015.

    (D) In 2012, the number of threatened plant species was lower than in any other recorded year.

67. In 2015, which organisms were facing the greatest threat?

    (A) Invertebrates

    (B) Vertebrates

    (C) Plants

    (D) Invertebrates, vertebrates, and plants were equally threatened.

68. The only form of energy that can travel through empty space is referred to as

    (A) kinetic energy

    (B) radiant energy

    (C) potential energy

    (D) elastic energy

69. All of the following are units of energy EXCEPT

    (A) joules (J)

    (B) calories (cal)

    (C) kilowatt hours (kWh)

    (D) watts (W)

70. Which of the following statements regarding renewable and nonrenewable energy is true?

    (A) Nuclear energy provides over 50% of the world's electricity.

    (B) Wind power is the most abundant nonrenewable energy source in the United States.

    (C) Fracking, or hydraulic fracturing, is used to extract natural gas and oil that lies deep underground.

    (D) China consumes more oil, in millions of barrels per day, than any other developed country.

71. The IPAT model is a

    (A) mathematical model to describe the impact that humans have on the environment

    (B) graphical model that is used to model evolution

    (C) mathematical model that says that the time it takes for a population to double can be approximated by dividing 70 by the current growth rate of the population

    (D) graphical model used to predict population trends based on the birth and death rates of a population

72. Urban areas that contain abandoned factories or former residential sites and are unlikely to experience redevelopment due to possible soil and water contamination are known as

    (A) brown zones

    (B) tailings

    (C) brownfields

    (D) greenbelts

GO ON TO THE NEXT PAGE.

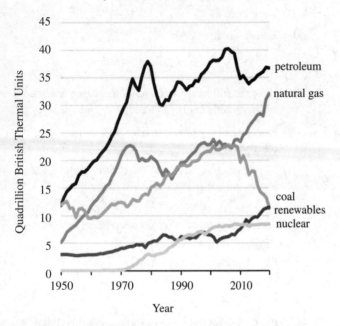

U.S. primary energy consumption by source (1950–2020)

Year

Questions 73–75 refer to the above graph.

73. Based on the graph above, which of the following statements is valid?

   (A) The U.S. consumption of nuclear energy has decreased since 1970.

   (B) Renewable energy comprises approximately 20% of U.S. primary energy consumption.

   (C) The U.S. consumption of petroleum has decreased since 1950.

   (D) Fossil fuels comprised approximately 80% of U.S. primary energy consumption in 2020.

74. U.S. consumption of coal was highest in

   (A) 2020

   (B) 2005

   (C) 1990

   (D) 1960

75. Which energy source had approximately the same rate of consumption in both 1950 and 2020?

   (A) Nuclear

   (B) Natural gas

   (C) Petroleum

   (D) Coal

76. The reaction that frees chlorine from chlorine monoxide, which destroys the ozone, is

   (A) $Cl + O_3 \rightarrow ClO + O_2$

   (B) $ClO + O \rightarrow Cl + O_2$

   (C) $O_2 + UV \rightarrow O + O + O + O_2 \rightarrow O_3$

   (D) $6CO_2 + 6H_2O \rightarrow C_6H_{12}O_6 + 6O_2$

77. All of the following statements regarding climate change are true EXCEPT

   (A) human activities have led to increased emissions of greenhouse gases

   (B) ocean warming will likely cause changes in coastlines, ocean currents, and sea surface temperatures

   (C) there will be increased crop yields in warm environments, which will likely be offset by the loss of croplands in other areas

   (D) increased ocean acidification will damage corals by decreasing their ability to calcify and form shells

78. A biodiversity hot spot, which is a highly diverse region that faces severe threats, has already lost what percent of its original natural habitat by area?

   (A) 10%

   (B) 25%

   (C) 70%

   (D) 90%

GO ON TO THE NEXT PAGE.

79. What book did Rachel Carson write that raised the environmental consciousness of Americans?

    (A) *The Population Bomb*

    (B) *Silent Spring*

    (C) *Walden*

    (D) *Man and Nature*

80. Which of the following was NOT a provision of the 1976 Resource Conservation and Recovery Act?

    (A) The establishment of a trust fund to provide for cleanup when no responsible party could be identified

    (B) The creation of the underground storage tank program to regulate underground storage tanks containing hazardous substances and petroleum products

    (C) The establishment of a system for controlling hazardous waste from the time it is generated until its ultimate disposal

    (D) The development of comprehensive plans to manage nonhazardous industrial solid waste and municipal solid waste, set criteria for municipal solid waste landfills, and prohibit open dumping of solid waste

END OF SECTION I

ENVIRONMENTAL SCIENCE
SECTION II
Time—1 hour and 10 minutes
3 Questions

Directions: Answer all three questions, which are weighted equally; the suggested time is about 23 minutes for answering each question. Where calculations are required, clearly show how you arrived at your answer. Where explanation or discussion is required, support your answers with relevant information and/or specific examples.

1. Human activity has led to rapidly increasing levels of carbon dioxide worldwide.

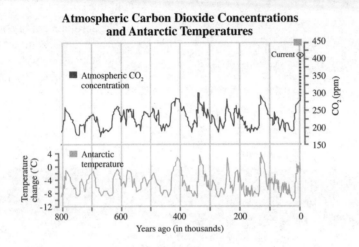

(a) The graphs show the analyses of ice core data extending back 800,000 years that show both atmospheric CO_2 concentration and Antarctic temperature.

   i   Based on the data in the graph, **identify** the highest carbon dioxide concentration.

   ii   Based on the data in the graph, **identify** the dependent variables in each graph.

   iii   Based on the data in the graph, **describe** the relationship between atmospheric CO_2 concentration and Antarctic temperature.

   iv   **Explain** one negative environmental effect of increased atmospheric CO_2 concentration and increases in Antarctic temperature.

(b) Climate scientists claim that the global warming trend observed since the mid-20th century is due to increased human activity that has exacerbated the greenhouse effect.

   i   **Describe** one human activity that has led to the global warming trend described by the scientists.

   ii   **Identify** one of the six criteria pollutants that contribute to global warming.

   iii   **Identify** one way that increased levels of atmospheric CO_2 and increased global temperatures can negatively affect human health.

(c) In order to stabilize global temperature at present levels, the world would need to either eliminate all emissions of heat-trapping gases or achieve a carbon-neutral society, wherein individuals remove as much carbon from the atmosphere as they emit.

   i   **Explain** TWO ways that countries can reduce their greenhouse gas emissions.

   ii   **Explain** TWO ways that individuals can reduce greenhouse gas emissions.

GO ON TO THE NEXT PAGE.

2. In agriculture, genetic engineering is the process of improving plants by adding genes from one species to another to encourage desirable characteristics.

   (a) Genetic engineering has advantages, disadvantages, and unintended consequences.
      i   **Identify** one benefit of genetically engineered plants.
      ii **Explain** one unintended consequence of genetically engineered plants.

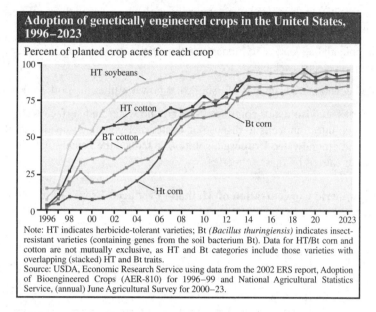

Adoption of genetically engineered crops in the United States, 1996–2023

Percent of planted crop acres for each crop

Note: HT indicates herbicide-tolerant varieties; Bt *(Bacillus thuringiensis)* indicates insect-resistant varieties (containing genes from the soil bacterium Bt). Data for HT/Bt corn and cotton are not mutually exclusive, as HT and Bt categories include those varieties with overlapping (stacked) HT and Bt traits.
Source: USDA, Economic Research Service using data from the 2002 ERS report, Adoption of Bioengineered Crops (AER-810) for 1996–99 and National Agricultural Statistics Service, (annual) June Agricultural Survey for 2000–23.

Genetically Engineered (GE) seeds were commercially introduced in the United States for major field crops in 1996, with adoption rates increasing rapidly in the years that followed. Currently, over 90 percent of U.S. corn, upland cotton, and soybeans are produced using GE varieties. GE crops are broadly classified in this data product as herbicide-tolerant (HT), insect-resistant (Bt), or stacked varieties that are a combination of both HT and Bt traits.

   (b) Use the data in the graph above to answer the following questions.
      i.   **Identify** the crop that had the fastest increase in percent of planted crop acres between 1996 and 2023.
      ii. **Describe** the relationship in the percentages of planted crop acres for herbicide-tolerant corn and insect-resistant corn between 1996 and 2023.
      iii **Explain** how improvements in the development of herbicide-tolerant and insect-resistant varieties influenced the adoption of genetically engineered crops in the United States, based on the data in the graph.
      iv **Describe** TWO disadvantages of genetically modified crops, such as those in the graph provided.

   (c) There are alternatives to the use of insect-resistant varieties to reduce pest populations in agricultural production.
      i   Integrated Pest Management (IPM) is often used to control the insect population in crops. **Define** Integrated Pest Management.
      ii **Identify** TWO methods used in IPM.
      iii **Propose** one reasonable method, other than IPM and genetic engineering, to reduce crop pest populations and maintain a high crop yield.

GO ON TO THE NEXT PAGE.

3. There are approximately 129 operable petroleum refineries in the United States that break crude oil down through the fractional distillation to produce gasoline. One problem with using such gasoline is that it contributes to global warming through the emission of greenhouse gases. Additionally, concerns exist regarding rising fuel costs, so many car companies are working to develop hybrid vehicles or full electric vehicles.

(a) **Describe** why petroleum is considered a nonrenewable energy source.

(b) **Describe** one potential environmental advantage of replacing a car with an oil-based internal combustion engine with an electric vehicle.

(c) **Describe** one potential advantage to human health of replacing a car with an oil-based internal combustion engine with an electric vehicle.

(d) **Describe** TWO economic advantages of using solar power, rather than oil, to produce electricity.

(e) Some oil refineries are also major contributors to ground water and surface water contamination, as some refineries use deep-injection wells to dispose of wastewater generated inside the plants. These wastes end up in aquifers and groundwater. **Propose** a solution to reduce the contamination levels of ground water and surface water caused by these refineries.

Atmospheric Concentration of Methane Produced From Oil in 2000 and 2016

Year	Average Annual Global Methane Emissions (Mt)
2000	46.6
2022	45.6

(f) **Calculate** the percent change in the average annual global methane emissions from 2000 to 2022. **Show** your work.

(g) Zeolite, an abundant and inexpensive type of clay, treated with copper has been found to be extremely effective at absorbing methane from the air. If 1,000 kg of zeolites decreases the average annual global methane emissions by 0.8 Mt, **calculate** the average annual global methane emissions if 10,000 kg of zeolites had been introduced in 2022. **Show** your work.

END OF SECTION II

Practice Test 1: Diagnostic Answer Key and Explanations

PRACTICE TEST 1: DIAGNOSTIC ANSWER KEY

Let's take a look at how you did on Practice Test 1. Follow the three-step process in the diagnostic answer key below to identify your strengths and weaknesses with regard to the tested topics. This will help you determine which content review chapters to prioritize when studying this book. In addition, read the explanations for any questions you struggled with or got wrong.

STEP 1 ≫ Check your answers and mark any correct answers with a ✔ in the appropriate column.

Section I—Multiple Choice							
Q #	Ans.	✔	Chapter #, Section Title	Q #	Ans.	✔	Chapter #, Section Title
1	A		7, Share and Share Alike?	27	B		4, The World's Ecosystems
2	B		6, The Atmosphere	28	A		4, The World's Ecosystems
3	A		7, Forest Resources	29	C		4, The World's Ecosystems
4	D		7, Forest Resources	30	C		5, Human Populations
5	C		10, Climate Change	31	C		5, Human Populations
6	D		5, Population Growth	32	B		7, Agriculture
7	A		6, Welcome to Planet Earth	33	C		10, Avenues to Environmental Progress
8	D		5, Human Populations	34	C		10, Human Impacts on Biodiversity
9	B		5, Human Populations	35	D		10, Climate Change
10	D		5, Human Populations	36	B		9, Hazardous Waste
11	C		4, The World's Ecosystems	37	D		4, Food Chains and Food Webs
12	B		9, Water Pollution	38	B		4, Food Chains and Food Webs
13	D		9, Water Pollution	39	A		4, Food Chains and Food Webs
14	B		6, Soil	40	C		10, What Can We Do About It?
15	D		6, The Atmosphere	41	B		9, Water Pollution
16	A		5, Human Populations	42	B		9, Water Pollution
17	C		4, Evolution	43	D		9, Water Pollution
18	D		8, Nonrenewable Energy	44	A		4, How Ecosystems Change
19	B		7, Agriculture	45	A		4, Food Chains and Food Webs
20	A		9, Air Pollution	46	D		9, Air Pollution
21	C		6, The Atmosphere	47	C		9, Toxicity and Health
22	D		8, Renewable Energy	48	C		9, Toxicity and Health
23	A		7, Agriculture	49	D		5, Human Populations
24	B		7, Economics and Resource Utilization	50	A		5, Human Populations
25	D		8, Nonrenewable Energy	51	B		5, Human Populations
26	A		7, Forest Resources	52	B		4, Evolution

Q #	Ans.	✔	Chapter #, Section Title	Q #	Ans.	✔	Chapter #, Section Title
\multicolumn{8}{c}{Section I—Multiple Choice (Continued)}

Q #	Ans.	✔	Chapter #, Section Title	Q #	Ans.	✔	Chapter #, Section Title
53	A		**9,** Water Pollution	67	D		**10,** Human Impacts on Biodiversity
54	D		**10,** Human Impacts on Biodiversity	68	B		**8,** Energy Resources and Consumption
55	C		**10,** What Can We Do About It?	69	D		**8,** Units of Energy
56	A		**7,** The Importance of Being Sustainable	70	C		**8,** Nonrenewable Energy
57	B		**5,** Terms Used to Describe Populations	71	A		**7,** Housing and Community
58	C		**7,** Water Use	72	C		**7,** Housing and Community
59	B		**7,** Mining	73	D		**8,** Nonrenewable Energy
60	A		**8,** Nonrenewable Energy	74	B		**8,** Nonrenewable Energy
61	B		**8,** Nonrenewable Energy	75	D		**8,** Nonrenewable Energy
62	D		**10,** What Can We Do About It?	76	A		**10,** The Greenhouse Effect
63	C		**5,** Factors Influencing Population Growth	77	C		**10,** Climate Change
64	A		**6,** The Atmosphere	78	A		**10,** Human Impacts on Biodiversity
65	C		**7,** Ocean Resources	79	B		**10,** Avenues to Environmental Progress
66	C		**10,** Human Impacts on Biodiversity	80	A		**9,** Hazardous Waste

Section II—Free Response			
Q #	**Ans.**	**✔**	**Chapter #, Section Title**
1(a)(i)	See Explanation		**6,** The Atmosphere; **10,** Climate Change
1(a)(ii)	See Explanation		**6,** The Atmosphere; **10,** Climate Change
1(a)(iii)	See Explanation		**6,** The Atmosphere; **10,** Climate Change
1(a)(iv)	See Explanation		**6,** The Atmosphere; **10,** Climate Change
1(b)(i)	See Explanation		**6,** Soil; **7,** Forest Resources; **8,** Nonrenewable Energy; **9,** Water Pollution; **10,** Human Impacts on Biodiversity; **10,** Climate Change; **10,** The Greenhouse Effect
1(b)(ii)	See Explanation		**9,** Air Pollution
1(b)(iii)	See Explanation		**9,** Toxicity and Health; **9,** Air Pollution
1(c)(i)	See Explanation		**7,** The Importance of Being Sustainable; **7,** Forest Resources; **8,** Renewable Energy; **8,** Energy Conservation: A Final Note; **10,** What Can We Do About It?
1(c)(ii)	See Explanation		**7,** The Importance of Being Sustainable; **7,** Forest Resources; **8,** Renewable Energy; **8,** Energy Conservation: A Final Note; **10,** What Can We Do About It?
2(a)(i)	See Explanation		**7,** Agriculture
2(a)(ii)	See Explanation		**4,** Evolution; **7,** Agriculture
2(b)(i)	See Explanation		**7,** Agriculture
2(b)(ii)	See Explanation		**7,** Agriculture
2(b)(iii)	See Explanation		**4,** Biodiversity; **4,** Evolution; **7,** Agriculture
2(b)(iv)	See Explanation		**4,** Biodiversity; **4,** Evolution; **7,** Agriculture
2(c)(i)	See Explanation		**7,** Agriculture
2(c)(ii)	See Explanation		**7,** Agriculture
2(c)(iii)	See Explanation		**7,** Agriculture
3(a)	See Explanation		**8,** Nonrenewable Energy
3(b)	See Explanation		**8,** Energy Conservation: A Final Note; **9,** Air Pollution
3(c)	See Explanation		**8,** Energy Conservation: A Final Note; **9,** Air Pollution
3(d)	See Explanation		**8,** Nonrenewable Energy; **8,** Renewable Energy
3(e)	See Explanation		**9,** Water Pollution
3(f)	See Explanation		**9,** Air Pollution
3(g)	See Explanation		**9,** Air Pollution

 Tally your correct answers from Step 1 by chapter. For each chapter, write the number of correct answers in the appropriate box. Then, divide your correct answers by the number of total questions (which we've provided) to get your percent correct.

CHAPTER 4 TEST SELF-EVALUATION

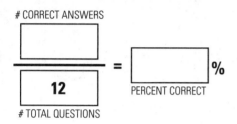

CHAPTER 8 TEST SELF-EVALUATION

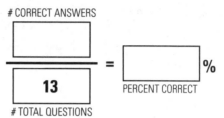

CHAPTER 5 TEST SELF-EVALUATION

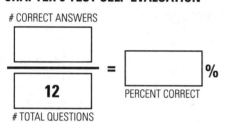

CHAPTER 9 TEST SELF-EVALUATION

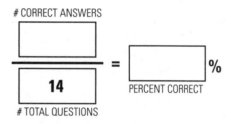

CHAPTER 6 TEST SELF-EVALUATION

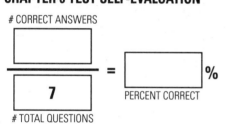

CHAPTER 10 TEST SELF-EVALUATION

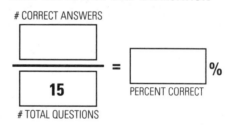

CHAPTER 7 TEST SELF-EVALUATION

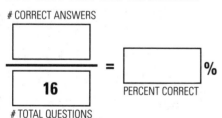

 Use the results above to customize your study plan. You may want to start with, or give more attention to, the chapters with the lowest percents correct.

HOW TO SCORE PRACTICE TEST 1

Section I: Multiple-Choice

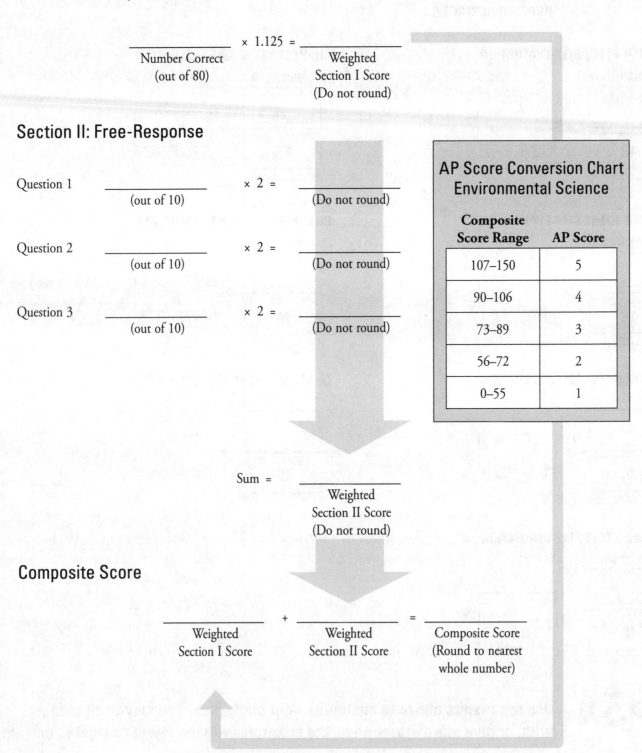

_____ × 1.125 = _____
Number Correct   Weighted
(out of 80)   Section I Score
                                  (Do not round)

Section II: Free-Response

Question 1   _____   × 2 =   _____
                 (out of 10)   (Do not round)

Question 2   _____   × 2 =   _____
                 (out of 10)   (Do not round)

Question 3   _____   × 2 =   _____
                 (out of 10)   (Do not round)

**AP Score Conversion Chart
Environmental Science**

Composite Score Range	AP Score
107–150	5
90–106	4
73–89	3
56–72	2
0–55	1

Sum =   _____
              Weighted
              Section II Score
              (Do not round)

Composite Score

_____ + _____ = _____
    Weighted   Weighted   Composite Score
  Section I Score   Section II Score   (Round to nearest
                                            whole number)

Note: This score sheet is to help you estimate your approximate score for the official exam, not your actual score.

PRACTICE TEST 1: ANSWERS AND EXPLANATIONS

Section I—Multiple-Choice Questions

1. **A** The Tragedy of the Commons refers to the concept in which a piece of open land, or a commons, was used collectively by the townspeople for grazing livestock; as each individual added livestock to the land, the land became overgrazed and, in turn, unusable. Keep choice (A). Slash-and-burn is an agricultural practice that involves cutting down and burning vegetation before planting new crops; eliminate (B). The Green Revolution, which occurred in the 1950s and 1960s, refers to a post-Industrial Revolution time when farming became mechanized and crop yields in industrialized nations increased dramatically; eliminate (C). The Tragedy of Free Access refers to the depletion of worldwide marine fisheries resulting from the United Nations authorization that member nations could expand their limits of jurisdictions to 200 miles from shore; eliminate (D). The correct answer is (A).

2. **B** The troposphere contains the air we breathe, which is composed of 78% nitrogen and 21% oxygen, with the remaining 1% including greenhouse gases; eliminate (A) and (C) as they are not considered greenhouse gases. Despite a low proportion of *greenhouse gases* in the troposphere, they have a tremendous impact on Earth, with water vapor, carbon dioxide, and methane absorbing a lot of solar radiation and, in turn, trapping heat in the Earth's atmosphere. Accordingly, (B), carbon dioxide, is the correct answer.

3. **A** Make sure that you know how to read and interpret charts and graphs for this exam! Here, you need to find the year that Canada experienced the fewest fires. According to the legend, the bars represent the number of hectares burned, which are shown on the left y-axis, and the line represents the number of fires that occurred, which are shown on the right y-axis. Since you need to find the year that Canada experienced the fewest fires, you need to find the lowest point on the line graph. According to the graph, Canada experienced the fewest fires in 2020, so the correct answer is (A).

4. **D** Here you need to interpret the information on the graph and use POE to find the correct answer. According to the legend, the bars represent the number of hectares burned, which are shown on the left y-axis, and the line represents the number of fires that occurred, which are shown on the right y-axis. Based on the graph, more hectares of land were burned in *1989* than in any other year; eliminate (A). Based on the graph, Canada experienced approximately *6,000* fires in 2017; eliminate (B). The graph shows that the number of hectares burned in 1985 was approximately *10% greater than* that burned in 1984; eliminate (C). According to the graph, more hectares of land were burned in 1989 than in any other year; this is true, so the correct answer is (D).

5. **C** Here, you need to assess each answer choice to determine which one is false. Choice (A) is true: water vapor does not contribute significantly to global climate change, despite being a greenhouse gas; eliminate (A). Choice (B) states that chlorofluorocarbons have the highest global warming potential; this is true, so eliminate (B). Choice (C) states that an increase in the global temperature will decrease the frequency and duration of storms, decrease the number of hot days, and increase the number of cold days; this is

incorrect, as an increase in the global temperature will *increase* the frequency and duration of storms, *increase* the number of hot days, and *decrease* the number of cold days; keep (C). Choice (D) states that the increase in average global temperature since the mid-1900s is likely due to increases in anthropogenic greenhouse gas concentrations; this is true, so eliminate (D). Thus, the correct answer is (C).

6. **D**   This question requires you to distinguish between *r*-selected and *K*-selected organisms. Here, you are asked about the characteristics of *r*-selected organisms. In general, *r*-selected organisms have populations *below* the carrying capacity of the environment, tend to have a *small* body size, reproduce *early* in life, and have short lifespans. Thus, (A), (B), and (C) are incorrect, and the correct answer is (D).

7. **A**   In order to answer this question, you need to know the characteristics of the four different types of volcanoes. Shield volcanoes have broad bases and are tall with gentle slopes; keep (A). Composite volcanoes have broad bases and are tall, but have much steeper slopes than shield volcanoes; eliminate (B). Cinder volcanoes are small, short, and have steeply sloped cones; eliminate (C). Lava domes are small and short with steep slopes and rounded tops; eliminate (D). Accordingly, the correct answer is (A).

8. **D**   Here, you need to identify the type of graph provided. The boom-and-bust cycle shows the population growth over time among *r*-strategists, in which there is a rapid increase in the population followed by an equally rapid decline in the population; eliminate (A). The survivorship curve, which varies depending on a given species, represents the number of individuals in a population born at a given time that remain alive as time continues; eliminate (B). The logistic growth curve models population growth, indicating that a population that is below the carrying capacity of a given region will grow exponentially, but as the population approaches the carrying capacity, the resource base of the population shrinks relative to the population itself; while the line representing population size on the graph mirrors the logistic growth curve, this graph provides more information and (C) can be eliminated. The demographic transition model (DMT) is used to predict population trends based on the birth and death rates of a population as a population transitions through various stages of growth; the four phases of the DMT are the preindustrial state, transitional state, industrial state, and postindustrial state. Thus, the graph provided shows the demographic transition model, and the correct answer is D).

9. **B**   This question requires you to identify the transitional state, during which birth rates are high, but death rates are lower because of improved access to food, water, and healthcare. Looking at the graph, you can see that birth rates are still high, but death rates are lower during Phase II, which is known as the transitional state. Therefore, the correct answer is (B).

10. **D**   In order to answer this question, you must know the names associated with each phase of the demographic transition model (DMT). In the DMT, Phase I is the preindustrial state, Phase II is the transitional state, Phase III is the industrial state, and Phase IV is the postindustrial state. In the preindustrial state, the population exhibits a slow rate of growth and has both high birth and death rates due to poor living conditions; eliminate (A). During the transitional state, the birth rates are high, but death rates are lower due to improved living conditions; eliminate (B). During the industrial state, population growth is still relatively high, but the birth rate lowers to become similar to the death rate; eliminate (C). Finally, during the postindustrial state, the population approaches and reaches a zero-growth rate. Thus, the correct answer is (D).

11. **C**   Here you need to know the characteristics of different aquatic ecosystems. Limnetic zones refer to the surface of open water where sunlight can penetrate, often hosting organisms like photosynthesizing phytoplankton; eliminate (A). Taiga is a biome characterized by moderate precipitation, acidic soil, and coniferous trees; eliminate (B). Wetlands are lowland areas, including marshes and swamps, that are saturated with moisture, and regarded as the natural habitat of much wildlife; keep (C). Estuaries are part of the wide lower course of a river where its current is met by the tides; eliminate (D). Thus, the correct answer is (C).

12. **B**   This question requires you to analyze the given graph and identify the different processes that are occurring in each portion of the graph—here you need to identify the process occurring in the section of the graph labeled *A*. Biological oxygen demand is a measure of the rate at which bacteria absorb oxygen from the water; this is not demonstrated in the section of the graph labeled *A*, so eliminate (A). Bioaccumulation is the absorption of pollutants from the environment that have entered the food chain—a process that is demonstrated in the portion of the graph labeled *A*; keep (B). Biomass augmentation is not a scientific concept, so eliminate (C). Biomagnification describes the increasing concentration of toxic molecules at successively higher trophic levels of the food chain; since the portion of the graph labeled *A* is only showing a single trophic level, as opposed to showing the accumulation of toxins as we move up the food chain, (D) can be eliminated. Thus, the correct answer is (B).

13. **D**   Like the previous question, this question requires you to analyze the given graph and identify the process occurring in the section of the graph labeled *B*. Biological oxygen demand is a measure of the rate at which bacteria absorb oxygen from the water; this is not demonstrated in the section of the graph labeled *B*, so eliminate (A). Bioaccumulation is the absorption of pollutants from the environment that have entered the food chain, a process that is demonstrated in the portion of the graph labeled *A*; eliminate (B). Biomass augmentation is not a scientific concept, so eliminate (C). Biomagnification describes the increasing concentration of toxic molecules at successively higher trophic levels of the food chain; thus, the correct answer is (D).

14. **B**   Here, you need to know the characteristics of the soil horizons. The A horizon is known as the topsoil, or the zone of leaching, which is dark in color due to the accumulation of organic matter and the most intensively weathered soil layer; eliminate (A). The B horizon, also known as the zone of illuviation, is where organic matter, soil, clay, and minerals that are washed out of the upper horizons accumulate; keep (B). The C horizon consists of unconsolidated material that is loose enough to be dug up with a shovel and has much less biological material than any of the horizons above it; eliminate (C). The O horizon is composed of organic matter at various stages of decomposition, including humus (not to be confused with tasty hummus!), which is stable residue left after the majority of the organic matter has decomposed; eliminate (D). Accordingly, the correct answer is (B).

15. **D**   This question requires you to use your knowledge about the different types of wind that exist. Areas of relatively still air because the air is constantly rising rather than blowing are known as doldrums; eliminate (A). Trade winds are caused by the surface currents of the Hadley cells and were named for their ability to quickly propel trading ships across the ocean; eliminate (B). Operating on the same thermodynamic principles as trade winds, westerlies are caused by the surface currents of Ferrel cells and are a

result of the Coriolis effect; eliminate (C). Jet streams are high-speed currents of wind that occur in the tropopause and influence local weather patterns. Thus, the correct answer is (D).

16. **A**   In order to answer this question, you need to know the formula for population change over time. Make sure that you read the answer choices carefully—they are all very similar, so make sure that you've committed the formulas used in the AP Environmental Science course to memory! The rate of population change over time can be found by subtracting the sum of the death rate and emigration rate from the sum of the birth rate and immigrant rate and dividing by ten. The correct answer is (A).

17. **C**   The AP Environmental Science Exam requires you to know a variety of different types of figures—and this is an example of one type! A food chain is usually depicted as a series of steps, in which the bottom step represents the producer and the top step represents either a secondary or tertiary consumer, and arrows indicate the transfer of energy through the different levels; eliminate (A). An energy pyramid depicts the amount of energy available at each trophic level organized from least to greatest; eliminate (B). A phylogenetic tree, which can be broad and encompass many types of species (like the figure provided) or specific and depict the evolutionary relationships that exist between two species, is a figure used to model evolution; keep (C). A food web represents the feeding relationships in ecosystems more realistically than a food chain, depicting the multiple, complex relationships that exist in an ecosystem; eliminate (D). Accordingly, the correct answer is (C).

18. **D**   This question requires that you know about the potential safety issues associated with nuclear energy. Thermal pollution occurs when the water used to cool turbines is returned to local bodies of water at a much higher temperature than when it was removed; this is an issue with nuclear energy, so eliminate (A). Highly radioactive waste is an issue because unusable cores, piping, and spent fuel rods need to be stored for centuries, as they contain radioactive elements like plutonium-239; eliminate (B). Since some of the by-products of the nuclear fission reaction can be remade into fission bombs that spread damaging radioactive isotopes, the development of nuclear weapons is a potential safety issue with nuclear energy; eliminate (C). Since nuclear power plants produce *no* greenhouse gas emissions during their operations, the correct answer is (D).

19. **B**   Genetically modified organisms (GMOs) are organisms that have been altered via genetic engineering by adding genes from one species to another in order to encourage desirable characteristics. While GMOs can have desirable characteristics, they also have their problems. Among other things, GMOs discourage biodiversity, can pose new allergen risks, can encourage the rise of pesticide-resistant pests, and can increase antibiotic resistance. Accordingly, since GMOs can *increase* antibiotic resistance, the correct answer is (B).

20. **A**   The Environmental Protection Agency defines point source pollution as any single identifiable source of pollution from which pollutants are discharged, such as a pipe, ditch, ship, or factory smokestack, noting that factories and sewage treatment plants are two of the most common types of point source pollutants. Accordingly, an oil spill from a tanker accident, industrial discharge from a factory pipe, and effluent from a wastewater treatment plant are examples of point source pollutants, as they are all single, identifiable sources of pollution. Conversely, runoff from agricultural fields has no single, identifiable source of pollution and would be considered nonpoint source pollution. Thus, the correct answer is (A).

21. **C** This question requires knowledge regarding the different weather phenomena that occur. The image provided shows that when the land mass is warmer than the surrounding ocean, moist air blows from above the ocean to the land, causing moisture to condense and become rain. The rain shadow effect occurs when a moisture-rich air mass moves in from a large body of water, climbs in altitude (cooling as it rises) when encountering obstructions like mountains, and then precipitates out on the ocean side of the mountain; this is not the process depicted in the provided image, so eliminate (A). The lake effect occurs when land masses warm more quickly than surrounding lakes or bays, causing air masses over the land to rise, and resulting in a breeze; this is not the process depicted in the provided image, so eliminate (B). Monsoons occur when land heats up and cools down more rapidly than the surrounding water does, creating a low-pressure system. As the moist air rises over the land, the air cools, and the moisture is released. This process mirrors the process depicted in the provided image, so keep (C). El Niño is a climate phenomenon occurring in the Pacific Ocean, during which trade winds move the warm surface waters away from the west coasts of both Central and South America, resulting in the weakening or reversal of normal trade winds because of the reversal of high- and low-pressure regions on either side of the tropical Pacific; this is not the process depicted in the provided image, so eliminate (D). Thus, the correct answer is (C).

22. **D** Here, you need to identify advantages of wind power. Wind power offers numerous advantages. Wind turbines require less land and water than traditional power plants that utilize fossil fuels; eliminate (A). Wind, unlike fossil fuels, is a renewable resource, and is both abundant and non-exhaustible; accordingly, eliminate (B). Wind turbines do not produce greenhouse gas emissions, unlike fossil fuels, which intensify climate change through the emission of greenhouse gasses during the combustion process; eliminate (C). Wind power is dependent on both the availability and strength of the wind, which makes it an intermittent and less predictable energy source than some other energy sources; thus, the correct answer is (D).

23. **A** This question requires you to be familiar with concepts related to agriculture; specifically, salinization. Salinization refers to a significant accumulation of salts on the soil's surface that results in land unusable for agricultural purposes; keep (A). The cutting of furrows between crop rows and filling them with water is a form of irrigation known as furrow irrigation; eliminate (B). Slash-and-burn agriculture is the practice of cutting down and burning an area of vegetation prior to sowing crop; eliminate (C). Integrated pest management is the combination of environmentally sensitive methods used to keep the pest population down to an economically viable level; eliminate (D). Thus, the correct answer is (A).

24. **B** In order to answer this question correctly, you need to be familiar with economics and resource utilization. Marginal costs refer to the change in total cost that comes about when the quantity of making a product increases; eliminate (A). Externalities refer to unwanted or unanticipated consequences of using resources; keep (B). Tangible properties refer to the physical value, such as food or shelter, of a given resource; eliminate (C). Intangible properties refer to the nonphysical value, such as recreational opportunity or spiritual value, of a given resource; eliminate (D). Accordingly, the correct answer is (B).

25. **D** This question requires you to implement your knowledge about the purity of different types of coal. Bituminous coal contains 45%–86% carbon and is the most abundant coal produced in the United States. Subbituminous coal contains 35%–45% carbon and has a lower heating value than bituminous coal.

Lignite coal contains 25%–35% carbon, having the lowest energy content of all coal ranks. Anthracite is the purest type of coal, containing between 86%–97% carbon, but accounted for less than 1% of coal mined in the United States. Accordingly, the purest type of coal is anthracite, and the correct answer is (D).

26. **A** In order to tackle this question, you need to know about the different types of forestry management. The management of forest plantations for the purpose of harvesting timber is referred to as silviculture; keep (A). The removal of select trees in an area is known as selective cutting; eliminate (B). The creation of a mutualistic symbiotic relationship by planting trees and crops together is known as agroforestry; eliminate (C). The removal of all trees in an area is known as clear-cutting; eliminate (D). Thus, the correct answer is (A).

27. **B** This question, and the two that follow, require you to implement your knowledge regarding the world's ecosystems. Due to the differences in geographic areas on Earth, ecosystems are broken into two broad categories: biomes, which are ecosystems based on land, and aquatic life zones, which are ecosystems based in aqueous environments; accordingly, eliminate (A) and keep (B). Ecotones refer to the transitional area where two ecosystems meet; eliminate (C). Ecozones, which are also referred to as ecoregions, are smaller areas within ecosystems that share similar features; eliminate (D). Thus, the correct answer is (B).

28. **A** Like the previous question, this one requires you to implement your knowledge about different biomes. The question asks you to identify the ecosystem that one would most likely find in the polar region, which based on the chart is an extremely cold area of the Earth. Tundra is an ecosystem located in the northern latitudes of North America, Europe, and Russia, receives less than 25cm of rain annually, has little vegetation beyond herbaceous plants, and has soil that is permafrost; this is consistent with that of a polar region, so keep answer choice (A). Taiga, also known as coniferous forests, is located in the northern parts of North America and Northern Eurasia, receives between 20–60cm of rain annually, has vegetation primarily consisting of coniferous trees, and has acidic soil; while these biomes are colder than many others, they are not found in the polar region, so answer choice (B) can be eliminated. Temperate rainforests are located in North America, South America, South Africa, Europe, Russia, Northeast Asia, and Oceania, receive over 140cm of rain annually, have vegetation consisting of coniferous and broadleaf trees, mosses, ferns, and shrubs, and have soil richer than that in tropical rainforests; these are not located in the polar region, so eliminate (C). Tropical rainforests are located in South America, West Africa, and Southeast Asia, receive between 200–400cm of rain annually, have vegetation consisting of tall trees, vines, epiphytes, and plants that have adapted to low sunlight, and have poor soil quality; these biomes are not located in the polar region, so eliminate (D). Accordingly, the best answer here is (A).

29. **C** Here, you need to identify the biome located in both temperate and tropical regions whose major vegetation is hardwood trees. Based on the chart provided, you can see that temperate and tropical regions share similar types of trees. Tropical rainforests are located in tropical, not temperate, regions and have vegetation consisting of tall trees, vines, epiphytes, and plants that have adapted to low sunlight; since these biomes are not located in the temperate region and don't have hardwood trees as major vegetation, (A) can be eliminated. Chapparal, or scrub forests and shrubland, are located in temperate, not tropical, regions, and have vegetation that consists of small trees with large, hard evergreen leaves and spiny shrubs;

since these biomes are not located in the tropical region and don't have hardwood trees as major vegetation, (B) can be eliminated. Deciduous forests are located in both temperate and tropical regions, and they have hardwood trees as their major vegetation; since this fits the characteristics we are looking for, keep (C). Coniferous forests, or taiga, are located in subpolar regions, and they have coniferous trees as their primary vegetation; since these biomes are not located in the temperate or tropical regions and don't have hardwood trees as major vegetation, (D) can be eliminated. Accordingly, the correct answer is (C).

30. **C** This question requires you to interpret the information in the figure provided. The legend tells you that an open circle on a line indicates when a region's population will start to shrink. Looking at the figure, we see that three lines have open circles; comparing those lines to the legend we see that Asia and Oceania, Latin America and Caribbean, and Europe are the countries that will have declining populations in 2100. Thus, the correct answer is (C).

31. **C** In order to tackle this question, check each statement for validity against the figure provided. According to the figure, the population of Asia and Oceania is expected to reach 5,272 million people before beginning to shrink; eliminate (A). Based on the figure, North America is expected to experience a *lower* percent increase in population size between 1950–2100 than Africa; eliminate (B). The figure shows that the lines for Europe and Africa intersect around 1995; thus, the population sizes of Africa and Europe were roughly equal in 1995, and you want to keep (C). According to the figure, Europe will experience population decline sooner than any other geographic area shown; eliminate (D). Therefore, the correct answer is (C).

32. **B** The Dust Bowl refers to the period of time in the 1930s when droughts in the Great Plains transformed the region into a giant dust bowl. While drought was a primary cause of the Dust Bowl, unsustainable farming and agricultural practices also contributed to the land's destruction. In response, the Soil Conservation Act of 1935 was passed, which led to the creation of the Soil Conservation Service, now known as the Natural Resources Conservation Service. The Food Security Act of 1985 discouraged the conversion of wetlands to non-wetlands; eliminate (A). The Water Salinization Act of 1938 does not exist; eliminate (C). The Soil and Water Conservation Act of 1977 established soil and water conservation programs to aid landowners, while also setting standards to evaluate U.S. soil and water resources; eliminate (D). Thus, the correct answer is (B).

33. **C** This question requires you to use your knowledge regarding the use of taxes as both policy tools and methods to protect the environment. Green taxes have three primary goals: to generate revenue to correct past pollution damage and reduce future pollution, to change consumption behaviors, and to use the funds received from pollution taxes for restoration. While green taxes may increase the market share of green technologies, as companies may be incentivized to develop these technologies in order to reduce their rates of taxation, this is an indirect, not primary, goal of green taxes. Therefore, the correct answer is (C).

34. **C** Here, you need to know the different ways in which species are labeled when referring to the loss of biodiversity. A species is designated as endangered if it is one that is likely to become extinct; keep (A) for now. An extinct species is one that no longer exists; eliminate (B). A species that is designated as critically endangered is one that is under a very high risk of extinction unless conservation actions are

taken; keep (C). A species is designated as vulnerable if the species is likely to become endangered if no actions are taken; eliminate (D). Since the question asks about a species that is under a very high risk of extinction, as opposed to one that is likely to become extinct, the best answer here is (C).

35. **D** This question requires you to know the primary cause of coral bleaching, which occurs when coral polyps expel the symbiotic algae that live in their shells and provide approximately 90% of the corals' energy. While all of the answer choices provided contribute to coral bleaching, the primary cause of coral bleaching is rising water temperatures; thus, the correct answer is (D).

36. **B** Here, you need to identify whether the type of radioactive waste placed in the EPA categories, which are based on place of origin, produces low- or high-level ionizing radiation. Nuclear reactor waste and waste from the mining and processing of uranium ore produce high levels of ionizing radiation; since these are categorized correctly, you can eliminate (A) and (C). Radioactive waste from industrial or research industries, which includes clothes, gloves, tubes, needles, and animal carcasses, is considered to be low-level radioactive waste; accordingly, you can eliminate (D). Waste from the reprocessing of spent nuclear fuel gives off high levels of ionizing radiation; since this is classified as low-level radiation in the question, the correct answer is (B).

37. **D** Make sure that you familiarize yourself with the types of figures that the College Board will ask about on this exam! A food chain is usually represented as a series of steps, in which the bottom step is the producer and the top step is a secondary or tertiary consumer, and arrows depict the transfer of energy through different trophic levels; this is not what this graph is depicting, so eliminate (A). An energy pyramid depicts the amount of energy (in kilocalories) available at each trophic level from greatest to least; eliminate (B). A phylogenic tree is a figure used to model evolution; since this figure does not depict the evolutionary process, eliminate (C). A food web shows the feeding relationships in ecosystems in a realistic manner, where the arrows point from the species consumed to the consumer; the correct answer is (D).

38. **B** A food web shows the feeding relationships in ecosystems in a realistic manner, where the arrows point from the species consumed to the consumer. In this food web, we see an arrow pointing from a snake to a kite, indicating that snakes are food for kites. Accordingly, the correct answer is (B).

39. **A** For the exam, make sure that you know about the different types of consumers, or organisms that obtain food energy from secondary sources via plant or animal matter! A primary consumer includes herbivores, which only consume producers like plants and algae; since rabbits only consume plants, keep (A). A secondary consumer is an organism that consumes primary consumers; since rabbits do not consume other consumers, eliminate (B). Tertiary consumers are organisms that consume secondary consumers; again, since rabbits do not consume other consumers, eliminate (C). Detritivores are organisms that derive energy from consuming nonliving organic matter like dead animals or fallen leaves; since rabbits are consuming living plant matter, they are not detritivores, and (D) can be eliminated. Accordingly, the correct answer is (A).

40. **C** It's imperative that you know about the different policies, legislation, and agreements that have been made regarding the environment for the AP Exam! The Kyoto Protocol required the 38 participating

developed countries to cut their greenhouse gas emissions back to 5% below 1990 levels. While the U.S. signed the agreement, it did not ratify the agreement and, in turn, the U.S. is not bound to abide by the protocol; eliminate (A). The National Environmental Policy Act created the Council on Environmental Quality that resulted in the creation of the Environmental Protection Agency from the consolidation of various environmental agencies. The National Environmental Policy Act also mandates that federal agencies prepare environmental impact statements; eliminate (B). The Pollution Prevention Act was designed to promote source reduction and stop pollution from being produced; eliminate (D). The Montreal Protocol cut the emissions of CFCs that damage the ozone layer and was later amended to include other key ozone-depleting chemicals; the correct answer is (C).

41. **B** The provided graph shows the severity levels of algal blooms in Western Lake Erie from 2002–2023. As shown on the graph, the objective is to have a severity level of less than or equal to 3, with any values greater than 3 not meeting the objective. Accordingly, (B) is correct.

42. **B** This question requires you to interpret the data on the graph and make a calculation to find the percent of the reported years that the severity of algal blooms in Western Lake Erie met the objective severity level. As the graph shows, the objective is to have a severity of algal blooms that is less than or equal to 3. First, you need to find the number of years that Western Lake Erie met or exceeded the objective level. This objective was met in seven of the 22 reported years: 2002, 2004, 2005, 2006, 2007, 2012, and 2020. Then, you need to translate the number of years into a percentage; in this case $7/22 \times 100 = 31.8\%$. Accordingly, the percent of the reported years that the severity of algal blooms in Western Lake Erie met the objective severity level is approximately 32%, or (B).

43. **D** In order to answer this question, tackle each statement to determine whether it is supported by the data in the provided graph and find the one that is false. According to the graph, the severity of algal blooms in Western Lake Erie was higher in 2015 than in any other recorded year; since this is true, eliminate (A). Based on the graph provided, the severity level of algal blooms in 2010 was approximately 6.5 and the objective level is 3. Therefore, the severity of algal blooms in Western Lake Erie was approximately two times greater than the objective level in 2010; since this is true, eliminate (B). The graph shows that the severity of algal blooms in 2008 was approximately 6, while the severity of algal blooms in 2023 was approximately 5; since the severity level of algal blooms in Western Lake Erie was lower in 2023 than it was in 2008, you can eliminate (C). Finally, the graph shows that the severity level of algal blooms in Western Lake Erie was lowest in 2002 and 2005, *not* 2002 and 2006; accordingly, this answer is false and the correct answer is (D).

44. **A** Here you need to use your knowledge regarding the different types of species and their roles in a given ecosystem. Indicator species are used as a standard to evaluate the health of an ecosystem; keep (A). Indigenous species originate and live or occur naturally in an area or environment; eliminate (B). Keystone species are those whose presence contributes to an ecosystem's diversity and whose extinction would lead to the extinction of other forms of life; eliminate (C). Invasive species are non-native to a given area and have been introduced to an ecosystem; eliminate (D). Thus, the correct answer is (A).

45. **A** This question requires you to identify the amount of energy transferred from one trophic level to the next in a food chain. The 10% rule states that roughly 10% of energy is transferred from one trophic level to the next, while the remaining 90% is used for activities like respiration, digestion, escaping predators, and powering the organism's survival. Accordingly, the correct answer is (A).

46. **D** This question is focused on types of air pollutants and requires you to identify the name of the six pollutants that do the most harm to human health and welfare, as identified by the Environmental Protection Agency. Primary pollutants are those that are released directly into the lower atmosphere and are toxic; while some, but not all, of the six pollutants identified by the EPA are primary pollutants, this is not the name given to the six pollutants and (A) can be eliminated. Secondary pollutants are those that are formed through the combination of primary pollutants in the atmosphere, such as acid rain; again, while some, but not all, of the six pollutants the EPA has identified are secondary pollutants, this is not the name of those six and (B) can be eliminated. Volatile organic compounds are released as a result of various industrial processes, such as dry cleaning, using industrial solvents, or using propane. While these are of growing concern to environmentalists, they do not encompass the six pollutants identified by the EPA; eliminate (C). The EPA refers to the six pollutants that do the most harm to human health and welfare as criteria pollutants (formerly referred to as the dirty half dozen); the correct answer is (D).

47. **C** Pathogens are bacteria, viruses, or other microorganisms that can cause disease. There are five main categories of pathogens: viruses (and other subcellular infectious particles, such as prions), bacteria, fungi, protozoa, and parasitic worms. Accordingly, you can eliminate (A), (B), and (D). Eosinophils are a type of white blood cell that supports your immune system and helps protect your body from parasitic infections. Therefore, eosinophils are not a category of pathogens, and the correct answer is (C).

48. **C** This question requires you to differentiate among the different types of infectious diseases. The plague is a disease transferred to humans via the bite of an infected organism or through contact with contaminated fluids or tissues, has three distinct forms, and causes fever, chills, headaches, and nausea; eliminate (A). Dysentery is a type of gastroenteritis caused by food and water contamination that results in diarrhea with blood, fever, and abdominal pain; eliminate (B). Zika is a virus caused by bites from infected mosquitoes, can be transmitted through sexual contact or from mother to fetus, and causes fever, red eyes, joint pain, headache, and rash; keep (C). Tuberculosis is an airborne bacterial infection that attacks the lungs and causes a chronic cough with bloody mucus, fever, night sweats, and weight loss; eliminate (D). Accordingly, the correct answer is (C).

49. **D** The College Board loves questions about age-structure diagrams, so make sure you familiarize yourself with the ways in which to interpret them! This question requires you to analyze the provided age-structure diagrams and find the true statement. Based on the provided diagrams, we can see that China's proportion of individuals in the 19 and younger age range is *lower* than that of the United States; after all, the United States has been stable for much longer than China, whose population has seen lower birth rates in recent years. Since this is false, you can eliminate (A). Based on the diagram, Afghanistan's population of individuals in the 30–34 age range was lower than China's population. Make sure that you pay attention to the labels on the axes here as the scales on the graphs are different; eliminate (B).

The diagram of the United States shows a country that is experiencing zero growth, while the diagram of Afghanistan is showing that of a rapidly growing population. Accordingly, in 2020, the United States had a *slower*-growing population than Afghanistan did; eliminate (C). Age-structure diagrams classify individuals aged 45 and older to be in the post-reproductive stage. According to the diagram, China's post-reproductive population in 2020 was greater than that of Afghanistan; since this is correct, the correct answer is (D).

50. **A** In order to answer this question, you need to be able to interpret the age-structure diagram *and* associate it to the appropriate phase of demographic transition model (DMT). The diagram of Afghanistan is showing that of a rapidly growing population and a demographic that experiences high birth rates and high death rates. According to the DMT, Phase I is the preindustrial state, Phase II is the transitional state, Phase III is the industrial state, and Phase IV is the postindustrial state. In the preindustrial state, the population exhibits a slow rate of growth and has both high birth and death rates due to poor living conditions. During the transitional state, the birth rates are high, but death rates are lower due to improved living conditions. During the industrial state, population growth is still relatively high, but the birth rate lowers to become similar to the death rate. Finally, during the postindustrial state, the population approaches and reaches a zero-growth rate. Based on the description of the phases of the DMT, Afghanistan best aligns with that of a country in the preindustrial state, or Phase I of the DMT. Thus, the correct answer is (A).

51. **B** This question requires you to assess the age-structure diagram of the United States and then assess the countries provided in the answer choice to find the most similar. The age-structure diagram of the United States shows that of zero growth—the population has remained stable without any major, or anomalous, changes for some time. The United States represents a country that is in the postindustrial state. Japan, while in the postindustrial state of the DMT, is experiencing negative population growth; accordingly, the age-structure diagram would not match that of the United States and you can eliminate (A). The United Kingdom is both in the postindustrial state of the DMT and is experiencing zero growth; keep (B). Mali is currently experiencing rapid population growth, high birth rates, and high death rates; since these are characteristics of a preindustrial state, you can eliminate (C). Egypt exemplifies a country in the industrial state, experiencing population growth, but struggling with high infant mortality rates; eliminate (D). Accordingly, the correct answer is (B).

52. **B** This question requires you to use your knowledge of evolutionary vocabulary. Natural selection occurs when a habitat selects certain organisms to live and reproduce, while others die; eliminate (A). Genetic drift is the accumulation of changes in the frequency of alleles over time due to changes that occur as a result of random chance; keep (B). Selective pressure refers to any cause that reduces the reproductive success in a portion of the population; eliminate (C). Microevolution occurs when a population displays small-scale changes over a relatively short period of time; eliminate (D). Thus, the correct answer is (B).

53. **A** This question requires that you use the knowledge you've learned about water quality in both the text and in your water quality lab. The most important factors in judging water quality are pH, which measures the acidity or alkalinity; the turbidity, which measures the density of suspended particles in the water;

and biological oxygen demand, which is a measure of the rate at which bacteria absorb oxygen from the water. Accordingly, you can eliminate (B), (C), and (D). Temperature is not one of the primary factors when judging water quality, so the correct answer is (A).

54. **D** Here you need to recall the HIPPCO acronym used to represent the major human factors that can cause extinction and decrease biodiversity. HIPPCO stands for **h**abitat destruction or fragmentation, the introduction of **i**nvasive species, **p**opulation growth, **p**ollution, **c**limate change, and issues associated with **o**verharvesting and overexploitation. Accordingly, you can eliminate (A), (B), and (C); the correct answer choice is (D).

55. **C** You will definitely be asked about various pieces of legislation on the AP Exam, so make sure that you familiarize yourself with the relevant legislation that has been passed over the course of America's existence. The 1987 Montreal Protocol cut the emissions of chlorofluorocarbons that damage the ozone layer; eliminate (A). The 1973 Convention on International Trade in Endangered Species of Wild Flora and Fauna bans the capture, exportation, or sale of endangered and threatened species; eliminate (B). The 1990 Pollution Prevention Act (PPA) was designed to promote source reduction; keep (C). The 1940 National Environmental Policy Act mandates that federal agencies prepare environmental impact statements; eliminate (D). Thus, the correct answer is (C).

56. **A** Make sure that you familiarize yourself with the ways in which the global population can impact environmental sustainability! One of the best ways that citizens can exert political influence and guide their leaders toward policies for sustainability is by voting. While posting on social media and spending less on unsustainable products may have an influence on sustainability by spreading the message of sustainable living, these activities will not influence policy-making. Furthermore, donating to politicians will not influence their actions, but simply fund their campaigns. Accordingly, the best answer is (A).

57. **B** Population dispersion refers to how individuals of a population are spaced within a region. Uniform dispersion refers to a condition in which members of a population are uniformly spaced throughout a geographic region; here, we see a cluster of elephants, so eliminate (A). Random dispersion refers to a condition in which the position of each individual is neither determined nor influenced by the positions of other members of the population; eliminate (C). Linear dispersion is not one of the three types of dispersions, so eliminate (D). Clumped dispersion refers to the most common dispersion pattern for populations, in which individuals stick together; since the elephants are grouped together in the image provided, the correct answer is (B).

58. **C** Make sure that you're familiar with all of the different rights and uses of the world's ocean resources for the exam! Riparian rights refer to the private water-use rights tied to the ownership of land bordering a natural river; keep (C). Prior appropriation are water rights given to those who have historically used the natural water resources in a certain area; eliminate (A). The International Whaling Commission instituted recent policies that allow Norway and Japan to capture a certain number of whales annually; eliminate (B). The U.N. changed the previous 12-mile limit rule and granted an extension of each country's jurisdiction to 200 miles from shore for the purpose of offshore fisheries in the 1960s; eliminate (D). The correct answer is (C).

59. **B** Here you need to know about the consequences related to mining ore and minerals—and there are many! First, mining can disrupt the ecosystem and harm the surrounding land; eliminate (A). Second, mineral extraction can leave pollutants resulting from the surface exposure of underground minerals; eliminate (C). Finally, mining can result in acid mine drainage that severely harms local water systems; eliminate (D). *Nuclear energy*, not coal mining, can result in radioactive waste that is highly dangerous to an ecosystem and human health. Thus, the correct answer is (B).

60. **A** This question requires you to read and interpret the data in the provided graph, which shows the global monthly mean of CH_4 between 1994–2023. Here, you need to find the time span that experienced the greatest percent increase. In order to find this, look at the steepness of the line. A line that is moving up and to the right indicates that the amount of CH_4 is increasing, with a steeper slope indicating a greater percent increase. A line that is moving down and to the right indicates that the amount of CH_4 is decreasing, with a steeper slope indicating a greater percent decrease. A line that is moving up horizontally indicates that the amount of CH_4 has remained the same. Between 2020–2023, the amount of CH_4 increased; keep (A). Between 2010–2014, the amount of CH_4 also increased; however, since the slope of the line segment in (A) is greater, you can eliminate (B). Between 2000–2004, the amount of CH_4 remained approximately the same; eliminate (C). Between 1992–1996, the amount of CH_4 also increased; however, the slope of the line segment in (A) is greater than that of the segment from (D), so you can eliminate (D). The correct answer is (A).

61. **B** Once again, you need to check the veracity of each statement in the answer choices against the data in the graph. According to the graph, in 1984, the global monthly mean of CH_4 was approximately 1,635 ppb; this is true, so eliminate (A). Between 2004–2006, the global monthly mean of CH_4 stayed roughly the same and did *not* experience an increase during this period; this is false, so keep (B). In 1994, the first year for which there is data on the global monthly mean of CH_4, the global monthly mean of CH_4 was approximately 1,635 ppm and in 2003, the last year for which there is data on the global monthly mean of CH_4, the global monthly mean of CH_4 was approximately 1,920 ppm. Accordingly, the global monthly mean of CH_4 was approximately 1,920 − 1,635 = 285 ppm greater than it was when measurements were first recorded. Since the global monthly mean of CH_4 is over 250 ppb greater than it was when measurements were first recorded, this is a true statement, and (C) can be eliminated. In 1994, the global monthly mean of CH_4 was approximately 1,635 ppm and in 2012, the global monthly mean of CH_4 was approximately 1,810 ppb. Accordingly, the global monthly mean of CH_4 was approximately 1,810 − 1,635 = 175 ppm greater than it was when measurements were first recorded. Since the global monthly mean of CH_4 in 2012 is approximately 175 ppb greater than it was in 1994, this is a true statement, and (D) can be eliminated. Thus, the correct answer is (B).

62. **D** Undoubtedly, you will be asked about various pieces of legislation on the AP Exam, so make sure that you familiarize yourself with the relevant legislation that has been passed over the course of America's existence. The Prospectors Regulation Act of 1870 does not exist—it is a made-up act; eliminate (A). The Mineral Lands Act of 1866 encouraged the use of resources and is still in effect for many mining regulations; eliminate (B). The Homestead Act of 1862 encouraged the settlement and exploitation of western lands; eliminate (C). Finally, the General Mining Act of 1872 was created to oversee and control prospecting and mining for economic minerals; (D) is the correct answer.

63. **C**   This question requires you to identify the different factors that have a direct impact on a population's birth rate. A population's religious beliefs, culture, and traditions, the demand for children in the labor force, and a base level of education for women are all factors that have a direct impact on a population's birth rate; eliminate (A), (B), and (D). While the availability of nutritious food can have an indirect impact on a population's birth rate, it doesn't directly impact the birth rate. Accordingly, the correct answer is (C).

64. **A**   The four spheres that make up our planet and regulate life on Earth are the solid earth, the pedosphere, the atmosphere, and the hydrosphere. The solid earth consists of the Earth's solid, rocky outer shell, the pedosphere is more commonly known as soil, the atmosphere is the envelope of gases that surrounds the Earth, and the hydrosphere is the Earth's oceans and freshwater bodies. Accordingly, eliminate (B), (C), and (D). The troposphere is the layer of the Earth's atmosphere that is closest to the Earth's surface and is where all the weather we experience takes place. Thus, the correct answer is (A).

65. **C**   Species of fish, mammals, and birds that are caught during fishing operations, but are not the target fish, are known as by-catch. By-catch is caught with drift nets, which float through the water and indiscriminately catch everything in their path; long lining, which uses long lines with baited hooks; and bottom trawling, during which the ocean floor is scraped by heavy nets; eliminate (A), (B), and (D). Capture fisheries consist of fish that are caught in the wild, rather than raised in captivity, for consumption. Thus, the correct answer is (C).

66. **C**   In order to answer this question, you need to analyze the provided graphs, making sure you look at the appropriate graph for each question. Looking at this first graph, we see that the number of species of organisms that were going extinct in 2016 was not higher than in any other year; eliminate answer choice (A). Again, looking at the top graph, we see that the number of species of organisms going extinct has increased over the recorded period; eliminate answer choice (B). The number of organisms that were going extinct was approximately 200 in 2015 and 800 in 2019. The percent change between the two years can be calculated as follows:

$$\frac{Difference}{Original} \times 100 = \frac{800\text{-}200}{200} \times 100 = \frac{600}{200} \times 100 = 3 \times 100 = 300\%$$

Accordingly, this statement is true; keep (C). Looking at the second graph, we see that the number of threatened plant species in 2012 was *not* lower than in any other recorded year; eliminate (D). Thus, the correct answer is (C).

67. **D**   Here you need to identify which type of organisms were facing the greatest threat in 2015. Since the question asks for the type of organism, you will need to look at the second graph. Based on the information in the second graph, we see that plants, invertebrates, and vertebrates were all facing the same threat level in 2015. Accordingly, the correct answer is (D).

68. **B**   This question requires you to use your knowledge about the different forms of energy that exist. Kinetic energy is referred to as energy in motion, and is the motion of waves, electrons, atoms, molecules, substances, and objects; eliminate (A). Radiant energy, also known as electromagnetic energy, is the only form of energy that can travel through empty space, and includes sunlight, x-rays, and radio waves; keep (B). Potential energy is energy held by an object and often referred to as stored energy; eliminate

(C). Elastic energy is energy that is stored in an object due to a force that temporarily changes its shape; eliminate (D). Accordingly, the best answer here is (B).

69. **D** Make sure that you know your units of energy and power for the exam! Here you are asked to identify the given unit that is *not* a unit of energy. Joules (J), calories (cal), British thermal units (BTUs), and kilowatt hours (kWh) are units of energy, while watts (W) and horsepower (hp) are units of power. Accordingly, since watts are a unit of power, the correct answer is (D).

70. **C** In order to answer this question, you need to be familiar with both renewable and nonrenewable energy sources and the role they play in the world. Look at each statement and use POE to identify the correct one! Nuclear energy provides *approximately 10%* of the world's electricity; eliminate (A). Wind power is the most abundant nonrenewable energy source in the United States; since wind power is a *renewable* energy source, eliminate (B). Fracking, or hydraulic fracturing, is used to extract natural gas and oil that lies deep underground; since this is a true statement, keep (C). The United States consumes more oil, in millions of barrels per day, than any other developed country; eliminate (D). Thus, the correct answer is (C).

71. **A** Make sure that you know all of the different types of mathematical and graphical models for the AP Exam! A phylogenetic tree is a graphical model that is used to model evolution; eliminate answer choice (B). The Rule of 70 is a mathematical model that says that the time it takes for a population to double can be approximated by dividing 70 by the current growth rate of the population; eliminate (C). The demographic transition model is a graphical model used to predict population trends based on the birth and death rates of a population; eliminate (D). The IPAT model is a mathematical model to describe the impact that humans have on the environment; the correct answer is (A).

72. **C** Urban areas that contain abandoned factories or former residential sites and are unlikely to experience redevelopment due to possible soil and water contamination are known as brownfields; keep (C). Brown zones is not a term used on the AP Environmental Science Exam; eliminate (A). Tailings refer to piles of waste material, or gangues, that are accrued during the mining process; eliminate (B). Greenbelts are open or forested areas built at the outer edges of cities, in which development is not allowed; eliminate (D). Thus, the correct answer is (C).

73. **D** For this question, make sure that you check each statement against the graph provided and use POE. The U.S. consumption of nuclear energy has *increased* since 1970; eliminate (A). Renewable energy comprises approximately 12% of U.S. primary energy consumption; eliminate (B). The U.S. consumption of petroleum has *increased* since 1950; eliminate (C). Fossil fuels comprised approximately 80% of U.S. primary energy consumption in 2020; the correct answer is (D).

74. **B** Here, look at the portion of the graph on the right side to determine when the U.S. consumption of coal was highest. According to the graph, consumption of coal was highest in 2005, and the correct answer is (B).

75. **D** For this question, you need to determine which energy source had approximately the same rate of consumption in both 1950 and 2020. Nuclear energy experienced a significant increase in consumption during this time period, so (A) can be eliminated. Likewise, natural gas consumption experienced an increase in consumption from 1950 to 2020, so (B) can be eliminated. Petroleum also experienced an

increase in consumption during this period, so eliminate (C). While coal consumption did experience a significant increase in use over the period, the subsequent decline in consumption led to approximately the same consumption rate in both 1950 and 2020. Accordingly, the correct answer is (D).

76. **A** For the AP Exam, you will need to know a few of the reactions that occur in the atmosphere and contribute to climate change. The reaction that occurs in the upper atmosphere when intense UV radiation breaks apart CFC molecules and releases chlorine atoms that then form chlorine monoxide is $Cl + O_3 \rightarrow ClO + O_2$; eliminate (B). The reaction that occurs when ozone is naturally created through the interaction of sunlight and atmospheric oxygen is $O_2 + UV \rightarrow O + O + O + O_2 \rightarrow O_3$; eliminate (C). The process of photosynthesis is modeled as $6CO_2 + 6H_2O \rightarrow C_6H_{12}O_6 + 6O_2$; eliminate (D). The reaction that frees chlorine from chlorine monoxide, which destroys the ozone, is $Cl + O_3 \rightarrow ClO + O_2$; so, the correct answer is (A).

77. **C** Make sure that you know all about the different causes and effects of climate change for the AP exam! Human activities have led to increased emissions of greenhouse gases; eliminate (A). Ocean warming will likely cause changes in coastlines, ocean currents, and sea surface temperatures; eliminate (B). There will be increased crop yields in *cold* environments, which will likely be offset by the loss of croplands in other areas; keep (C). Increased ocean acidification will damage corals by decreasing their ability to calcify and form shells; eliminate (D). Accordingly, the correct answer is (C).

78. **C** A biodiversity hot spot is a highly diverse region that faces severe threats and has already lost 70% of its original natural habitat by area. Accordingly, the correct answer is (C).

79. **B** Over the past couple of centuries, there have been a number of environmental activists who have produced works of literature that have influenced the practice of environmentalism. *The Population Bomb* was penned by Paul Ehrlich in 1968, and warned of the many problems that would arise along with the rapidly increasing human population; eliminate (A). One of the earliest environmental activists was Henry David Thoreau, who described his retreat from society and the quiet years he spent living on Walden Pond studying nature in his book *Walden*; eliminate (C). George Perkins published his book *Man and Nature*, which helped the American public understand that there are limits to natural resources; eliminate (D). Rachel Carson penned *Silent Spring* in 1962, which raised the environmental consciousness of Americans, showing them that the air was dirty, the water was polluted, and hazardous wastes were collecting in landfills nationwide. Accordingly, the correct answer is (B).

80. **A** The 1976 Resource Conservation and Recovery Act covered a myriad of items, including the following: (1) the creation of the underground storage tank program to regulate underground storage tanks containing hazardous substances and petroleum products; (2) the establishment of a system for controlling hazardous waste from the time it is generated until its ultimate disposal; (3) the development of comprehensive plans to manage nonhazardous industrial solid waste and municipal solid waste, the setting of criteria for municipal solid waste landfills, and the prohibition of open dumping of solid waste. The Comprehensive Environmental Response, Compensation, and Liability Act (CERCLA), commonly known as Superfund, established a trust fund to provide for cleanup when no responsible party could be identified; thus, the correct answer is (A).

Section II—Free-Response Questions

Remember, you must write your responses in paragraph form!

Question 1: Human Activity and Climate Change   10 points

(a)   i. **Identify** the highest carbon dioxide concentration.

Accept one of the following:

- Current
- Present
- 2024

(1 point)

ii. **Identify** the dependent variables in each graph.

- CO_2 concentration (PPM)
- Temperature change (°C)

(1 point)

iii. **Describe** the relationship between atmospheric CO_2 concentration and Antarctic temperature.

Accept one of the following:

- As atmospheric CO_2 concentration increases, so too does Antarctic temperature. Likewise, as atmospheric CO_2 concentration decreases, so too does Antarctic temperature.
- Atmospheric CO_2 concentration and Antarctic temperature increase and decrease at the same time.

(1 point)

iv. **Explain** one negative environmental effect of increased atmospheric CO_2 concentration and increases in Antarctic temperature.

Accept one of the following:

- Increased atmospheric CO_2 concentration and increases in Antarctic temperature could contribute to global warming and, in turn, the melting of the ice sheets and glaciers that will lead to rising sea levels. Rising sea levels threaten coastal areas, resulting in increased flooding, erosion, and the potential displacement of communities.
- Elevated carbon dioxide levels are absorbed by the oceans, which can lead to ocean acidification. Ocean acidification negatively impacts marine life, including corals and plankton, leading to disrupted marine ecosystems and negative effects on biodiversity.
- The warming of Antarctica can lead to ecosystem shifts and disrupt local biodiversity. While native species may adapt and survive, invasive species may thrive in these conditions, altering food webs and the overall structure and function of ecosystems.
- Changes in temperature and ocean conditions may disrupt marine food chains, affecting the quantity and dispersion of key species. Such changes can have cascading effects on predators and prey, leading to imbalances in marine ecosystems and potential negative declines in commercial fisheries.

- Warmer temperatures in Antarctica can lead to changes in weather patterns, including the intensity and frequency of wet weather events. Specifically, this can lead to more severe weather events, such as hurricanes and monsoons, with negative implications for coastal regions and ecosystems.
- Antarctica plays a significant role in regulating global climate patterns. Changes in temperature can disrupt the Southern Ocean's circulation and impact atmospheric circulation patterns. In turn, this could lead to shifts in weather systems, increased extreme climate events, and changes to natural ecosystems.

(1 point)

(b) i. **Describe** one human activity that has led to the global warming trend described by the scientists.

Accept one of the following:
- The combustion of fossil fuels for energy production is a major contributor to global warming. As this process releases large quantities of carbon dioxide and other greenhouse gases into the atmosphere, this exacerbates the natural greenhouse effect and traps more heat.
- Deforestation, driven by agriculture, logging, and urbanization, reduces the number of trees available to absorb carbon dioxide through photosynthesis. Since trees act as carbon sinks, their removal leads to increased carbon dioxide concentrations and exacerbates the greenhouse effect.
- Industrial processes release significant amounts of greenhouse gases. For example, the production of many chemicals involves the release of carbon dioxide during the transformation of raw materials, contributing to increased greenhouse gas emissions.
- Agricultural practices, such as the use of synthetic fertilizer or livestock farming, also contribute to global warming. The use of synthetic fertilizers releases nitrous oxide and livestock farming produces methane, both greenhouse gases that contribute to global warming.
- Improper waste management practices, including the decomposition of organic waste in landfills, contribute to the release of methane. Landfills that are not properly designed to capture and utilize methane emissions allow such gas to escape into the atmosphere, contributing to global warming.
- The use of chlorofluorocarbons (CFCs) and hydrochlorofluorocarbons (HCFCs) in both industrial and consumer products adds ozone-depleting greenhouse gases that contribute to global warming.

(1 point)

ii. **Identify** one of the six criteria pollutants that contribute to global warming.

Accept one of the following:
- carbon monoxide
- lead
- ozone
- nitrogen dioxide
- sulfur dioxide
- particulates

(0.5 point)

iii. **Identify** one way that increased levels of atmospheric CO_2 and increased global temperatures can negatively affect human health.

Accept one of the following:

- Increased air pollution and respiratory issues
- Exposure to airborne pollutants that contain harmful chemicals
- Heat-related illnesses associated with increasing temperatures
- Vector-borne diseases resulting from changing weather patterns

(0.5 point)

(c)   i. **Explain** TWO ways that countries can reduce their greenhouse gas emissions.

Accept two of the following:

- Shifting from fossil fuel-based energy sources to renewable energy sources, such as solar, wind, hydroelectric, and geothermal power, can significantly reduce greenhouse gas emissions. Countries should invest in and promote the use of clean energy technologies.
- Implementing energy efficiency measures across various sectors, including industry, transportation, and development, can reduce overall energy consumption and, in turn, lower greenhouse gases. Countries should adopt energy-efficient technologies, improve industrial processes, and promote energy conservation practices.
- Planting trees and reforestation, or establishing new forests, can act as carbon sinks, absorbing and storing carbon dioxide from the atmosphere. Countries should engage in such activities to help sequester carbon, enhance biodiversity, and improve ecosystem services.
- Adopting sustainable agricultural practices can help reduce emissions from the agricultural sector. Countries should promote practices such as agroforestry, precision farming, and organic farming, which can enhance soil health, reduce methane emissions from livestock, and minimize the use of synthetic fertilizers that contribute to nitrous oxide emissions.
- Encouraging the use of public transportation, cycling, and walking, as well as promoting the adoption of low-emission vehicles can help reduce emissions from the transportation sector. Countries should adopt policies supporting the development of public transportation infrastructure and provide incentives for the use of electric vehicles to aid this effort.
- Carbon capture and storage involves capturing carbon dioxide emissions from industrial processes and energy production before they are released into the atmosphere and storing them underground. Countries should promote technologies that reduce emissions from sectors that are challenging to decarbonize, such as heavy industry and energy production.

(2 points)

ii. **Explain** TWO ways that individuals can reduce greenhouse gas emissions.

Accept two of the following:

- Individuals can reduce their carbon footprint by being mindful of energy use. This includes turning off lights and electronic devices when not in use, using energy-efficient appliances, and adjusting thermostats to conserve energy. Behaviors like using energy-efficient light bulbs and unplugging chargers when not in use contribute to overall energy conservation.

- Opting for sustainable transportation options, such as walking, cycling, carpooling, or using public transportation, can significantly reduce individual carbon emissions associated with personal travel. Additionally, choosing fuel-efficient or electric vehicles helps decrease the environmental impact of transportation.

- Water heating and treatment contribute to energy consumption and emissions. Individuals can reduce their water footprint by fixing leaks, using water-saving appliances, and adopting water conservation habits. Conserving water reduces the energy required for water treatment and heating, indirectly lowering greenhouse gas emissions.

- Individuals can minimize waste and promote recycling and composting to reduce the environmental impact of waste disposal. Landfills produce methane, a potent greenhouse gas, as organic waste decomposes. By reducing, reusing, and recycling, individuals contribute to lower emissions associated with waste management.

- Making informed choices about the products we buy can have a positive impact on emissions. Individuals can support companies and products that prioritize sustainability, use eco-friendly materials, and have a lower environmental impact. This includes choosing products with minimal packaging and considering the lifecycle impact of items.

(2 points)

Question 2: Genetic Engineering, Crop Management, and Agricultural Practices
10 points

In agriculture, genetic engineering is the process of improving plants by adding genes from one species to another to encourage desirable characteristics.

(a) i. **Identify** one benefit of genetically engineered plants.

Accept one of the following:
- Disease resistant
- Drought resistant
- Pest resistant
- Faster growth
- Higher crop yields
- Longer shelf life

(1 point)

ii. **Explain** one unintended consequence of genetically engineered plants.

Accept one of the following:

- Genes introduced through genetic engineering may *crossbreed with wild plant species or closely related crops.* If the modified genes confer certain advantages, such as resistance to pests or herbicides, they may persist in the wild population or in non-genetically modified crops, potentially altering their characteristics.
- Gene flow can lead to the creation of hybrid plants that carry a combination of traits from both genetically engineered and non-engineered plants. This may result in *the expression of unintended or unpredictable traits,* and the ecological consequences of these hybrids are not always fully understood.
- Genetic contamination has the potential to *affect biodiversity by introducing modified genes into natural ecosystems.* If the modified genes confer a selective advantage, they may influence the competitiveness of wild plants or non-genetically modified crops, potentially leading to changes in the composition of plant populations.
- Genes for herbicide resistance introduced into genetically modified crops may transfer to weed species in the vicinity, *creating herbicide-resistant weed populations.* This can pose challenges for weed control and may lead to the increased use of herbicides, potentially contributing to the development of more resistant weed species.
- The introduction of genetically engineered plants into ecosystems may lead to *unforeseen ecological interactions.* The complexity of ecosystems makes it challenging to predict all possible outcomes of introducing genetically modified organisms. Unexpected interactions may have cascading effects on other species and ecosystem dynamics.
- Genetically engineered crops may *unintentionally contaminate seed banks of non-genetically modified varieties.* This can occur through the mixing of seeds during harvesting, transportation, or storage. Contamination of seed banks poses challenges for preserving and maintaining the genetic diversity of traditional and non-genetically modified crop varieties.

(1 point)

(b) i. **Identify** the crop that had the fastest increase in percent of planted crop acres between 1996 and 2023.

Accept the following:
- HT Soybeans

(1 point)

ii. **Describe** the relationship in the percentages of planted crop acres for herbicide-tolerant corn and insect-resistant corn between 1996 and 2023.

Accept one of the following:

- From 1996–2007, Bt corn comprised a greater percent of planted crop acres, but was surpassed by HT corn in 2007.
- Both HT and Bt corn crops were initially planted at the same percentage, but HT corn now comprises a greater percentage of planted crop acres than does Bt corn.

(1 point)

iii. **Explain** how improvements in the development of herbicide-tolerant and insect-resistant varieties influenced the adoption of genetically engineered crops in the United States, based on the data in the graph.

Accept one of the following:

- As genetic engineering technology improved, so too did the percent of planted crop acres for genetically modified crops in the U.S.
- The percent of planted crop acres for genetically modified crops in the U.S. has increased over time, mirroring the growth of technology.

(1 point)

iv. **Describe** TWO disadvantages of genetically modified crops, such as those in the graph provided.

Accept two of the following:

- Genetically modified (GM) crops may have unintended *environmental impacts*. For example, the cultivation of insect-resistant crops can lead to the development of resistance in target pests, potentially resulting in the need for increased pesticide use or the emergence of secondary pest problems. Additionally, the introduction of GM crops may have unintended effects on non-target organisms and biodiversity.
- The widespread adoption of a small number of genetically modified crops may contribute to a *reduction in genetic diversity* within cultivated plant species. This loss of diversity can make crops more susceptible to diseases and environmental changes, potentially compromising long-term food security.
- There is a risk of gene flow from genetically modified crops to related wild plants or non-GM crops. This can result in the unintentional spread of transgenes, leading to the development of hybrid plants with altered characteristics. Gene flow can *raise ecological concerns and have implications for biodiversity and ecosystem dynamics.*
- Concerns exist about the potential long-term health effects of consuming genetically modified crops. While many studies suggest that GM crops approved for human consumption are safe, there is ongoing debate and research regarding the potential risks, allergenicity, and unintended health effects associated with the consumption of genetically modified organisms.
- Pests and weeds targeted by genetically modified crops can *develop resistance over time*. For example, insects may evolve mechanisms to tolerate or overcome the effects of insect-resistant crops, and weeds can become resistant to herbicides used with herbicide-resistant crops. This resistance poses challenges for sustainable pest and weed management.
- The widespread adoption of genetically modified crops can lead to *economic dependence on a small number of seed varieties and the companies that produce them.* This concentration of power in the hands of a few large biotechnology companies may raise concerns about control over agricultural practices, seed prices, and intellectual property rights.

(2 points)

(c)   i. **Define** Integrated Pest Management.

> • Integrated Pest Management (IPM) is a sustainable and comprehensive approach to managing pests in agriculture and other settings. IPM focuses on minimizing the impact of pests while promoting environmental sustainability and economic viability.

(1 point)

ii. **Identify** TWO methods used in IPM.

Accept two of the following:
• Introducing natural insect predators to the area
• Diversifying crops
• Crop rotation
• Intercropping
• Using mulch to control weeds
• Using traps
• Constructing barriers
• Releasing pheromone or hormone interrupters

(1 point)

iii. **Propose** one reasonable method, other than IPM and genetic engineering, to reduce crop pest populations and maintain a high crop yield.

Accept one of the following:
• Use a method of pest control that employs a variety of biological, physical, and chemical methods to control the crop pest population.
• Reduce stubble or crop residues in fallow fields that can harbor the crop pest population.
• Apply the pesticide when the crop pest population is most susceptible.
• Use intercropping rather than a monoculture to reduce the amount of habitat for the pests.

(1 point)

Question 3: Analyze a Problem, Propose a Solution, and Do Calculations   10 points

(a)   **Describe** why petroleum is considered a nonrenewable energy source.

Accept one of the following:
• Nonrenewable energy sources like oil exist in a fixed/finite/limited amount.
• The formation rate of oil is less than the consumption/combustion rate of oil.

(1 point)

(b) **Describe** one potential environmental advantage of replacing a car with an oil-based internal combustion engine with an electric vehicle.

Accept one of the following:

- One of the most significant environmental advantages of replacing a traditional oil-based internal combustion engine with an electric vehicle is the potential for a *substantial reduction in greenhouse gas emissions*. Electric vehicles (EVs) produce zero tailpipe emissions, reducing the overall carbon footprint associated with transportation when the electricity they use comes from low-carbon or renewable sources.
- Electric vehicles contribute to *improved air quality* in urban areas because they produce no tailpipe emissions of pollutants such as nitrogen oxides (NOx) and particulate matter. This can lead to decreased respiratory and cardiovascular health issues in populations living in or around urban centers.
- Electric vehicles *can contribute to a reduction in the demand for traditional fossil fuels*, such as gasoline and diesel. This helps *decrease the environmental impact associated with extracting, refining, and transporting these fossil fuels, thereby lessening the negative effects on ecosystems and habitats.*
- By reducing greenhouse gas emissions associated with transportation, the widespread adoption of electric vehicles can contribute to mitigating climate change. *Climate change poses various health risks, including increased heat-related illnesses, changing patterns of infectious diseases, and disruptions to food and water supplies. Electric vehicles play a role in addressing these health risks by helping to reduce the overall carbon footprint.*

(1 point)

(c) **Describe** one potential advantage to human health of replacing a car with an oil-based internal combustion engine with an electric vehicle.

Accept one of the following:

- Electric vehicles produce zero tailpipe emissions, which significantly reduces the concentration of air pollutants such as nitrogen oxides (NOx) and particulate matter in urban environments. This *improvement in air quality can lead to better respiratory health, reducing the incidence of respiratory diseases such as asthma and other respiratory conditions.*
- Traditional internal combustion engine vehicles emit pollutants such as carbon monoxide (CO), volatile organic compounds (VOCs), and fine particulate matter. By transitioning to electric vehicles, *individuals experience lower exposure to these harmful emissions, leading to a potential decrease in respiratory and cardiovascular health issues.*
- Electric vehicles are generally quieter than traditional vehicles with internal combustion engines. *The reduction in noise levels can contribute to lower stress levels, improved sleep, and better overall mental health, particularly in urban and suburban environments where traffic noise is a common source of stress.*
- Electric vehicles, especially those used for short trips or commuting, can encourage more physical activity such as walking or cycling. As the adoption of electric vehicles has increased, cities have started to alter their infrastructure to be less vehicle-centric and more pedestrian friendly. Furthermore, as individuals use alternative transportation modes for short distances, they engage in more physical exercise, leading to positive impacts on cardiovascular health and overall well-being.

- Electric vehicles are typically quieter than traditional internal combustion engine vehicles. *The reduced noise levels contribute to lower noise pollution in urban and suburban environments.* This has positive implications for both human well-being and wildlife, as excessive noise can disturb ecosystems and negatively impact various species.
- Electric vehicles generally have simpler drivetrains and require less maintenance than internal combustion engine vehicles. This can result in *decreased exposure of automotive technicians and workers to hazardous substances found in traditional vehicle components, such as engine fluids and exhaust emissions. The transition to electric vehicles can improve occupational health in the automotive industry.*

(1 point)

(d) **Describe** TWO economic advantages of using solar power, rather than oil, to produce electricity.

Accept two of the following:
- *One of the primary economic advantages of using solar power for electricity generation is the absence of ongoing fuel costs.* Unlike oil-based power generation, which requires the continuous purchase and transportation of fossil fuels, solar power relies on sunlight, which is freely available. Once a solar power system is installed, the operating costs are minimal, leading to long-term cost savings.
- *Solar power provides a more stable and predictable source of energy compared to oil.* The price of oil can be highly volatile due to geopolitical factors, supply and demand fluctuations, and other market dynamics. By harnessing solar energy, businesses and households can reduce their exposure to the uncertainties and price fluctuations associated with the oil market.
- *While the initial upfront costs of installing solar power systems can be significant, the lifecycle costs are generally lower compared to oil-based electricity generation.* Solar panels have minimal ongoing maintenance requirements and can have a lifespan of 25 years or more. In contrast, oil-based power plants require ongoing maintenance, fuel procurement, and are subject to wear and tear, contributing to higher overall lifecycle costs.
- *The solar industry has the potential to stimulate economic growth and create jobs.* As the demand for solar technologies increases, there is a need for manufacturing, installation, maintenance, and research and development, leading to job opportunities in various sectors. This can contribute positively to local economies and enhance workforce development.
- *Investing in solar power provides an opportunity to diversify the energy mix, reducing dependence on a single energy source like oil.* Diversification can enhance energy security, especially in regions heavily reliant on oil imports. It also helps mitigate the economic risks associated with oil price shocks and supply disruptions.
- *Many governments offer financial incentives, tax credits, and subsidies to encourage the adoption of solar power.* These incentives can significantly reduce the initial costs of installing solar systems for businesses and homeowners, making solar power more economically attractive and accelerating its widespread adoption.

(2 points)

(e)   Some oil refineries are also major contributors to ground water and surface water contamination, as some refineries use deep-injection wells to dispose of wastewater generated inside the plants. These wastes end up in aquifers and groundwater. **Propose** a solution to reduce the contamination levels of ground water and surface water caused by these refineries.

Accept one of the following:

- *Invest in advanced wastewater treatment technologies within the refinery to effectively treat and purify wastewater before disposal.* This includes implementing biological treatment systems, advanced oxidation processes, and other state-of-the-art technologies that can remove contaminants and minimize the environmental impact.

- *Implement closed-loop water recycling systems within the refinery to minimize the need for freshwater intake and reduce the volume of wastewater requiring disposal.* This approach involves treating and reusing water within the refinery processes, thereby decreasing the amount of wastewater generated and the potential for contamination.

- *Implement Zero Liquid Discharge systems, in which the refinery aims to eliminate the discharge of liquid waste. ZLD systems focus on recovering and reusing water from the wastewater, leaving behind solid residues that can be safely disposed of.* This approach significantly reduces the environmental impact of refinery wastewater discharges.

- *Explore and adopt alternative processes and materials that are less harmful to the environment.* This may involve using less toxic chemicals in refining processes or employing greener technologies that generate fewer harmful by-products. Reducing the toxicity of the wastewater at its source can have a positive impact on groundwater and surface water quality.

- *Implement rigorous monitoring programs to regularly assess the quality of wastewater leaving the refinery.* This includes regular testing for a wide range of contaminants. Strict compliance with environmental regulations and standards is essential to ensure that the refinery is meeting its environmental responsibilities and minimizing the risk of water contamination.

- *Allocate resources to research and development initiatives focused on improving wastewater treatment technologies and finding innovative solutions to minimize environmental impact.* Investing in sustainable practices and technologies can lead to long-term benefits for both the refinery and the surrounding environment.

(1 point)

Atmospheric Concentration of Methane Produced From Oil in 2000 and 2016

Year	Average Annual Global Methane Emissions (Mt)
2000	46.6
2022	45.6

(f) **Calculate** the percent change in the average annual global methane emissions from 2000 to 2022. **Show** your work.

- Setup: $\dfrac{(46.6 \text{ Mt} - 45.6 \text{ Mt})}{46.6 \text{ Mt}}$

- Solution: 2.145%, 2.15%, or 2.1%

(1 point for correct setup and 1 point for solution)

(g) Zeolite, an abundant and inexpensive type of clay, treated with copper has been found to be extremely effective at absorbing methane from the air. If 1,000 kg of zeolites decreases the average annual global methane emissions by 0.8 Mt, **calculate** the average annual global methane emissions if 10,000 kg of zeolites had been introduced in 2022. **Show** your work.

- Setup: 45.6 Mt – (10 × 0.8 Mt) =

  45.6 Mt – (8 Mt) =

  37.6 Mt

- Solution: 37.6 Mt or 38 Mt

(1 point for correct setup and 1 point for solution)

Part III
About the AP Environmental Science Exam

- The Structure of the AP Environmental Science Exam
- How the AP Environmental Science Exam Is Scored
- Overview of Course Units, Practices, and Big Ideas
- How AP Exams Are Used
- Other Resources
- Designing Your Study Plan

THE STRUCTURE OF THE AP ENVIRONMENTAL SCIENCE EXAM

The AP Environmental Science Exam will be 2 hours and 40 minutes long. You will have 1 hour and 30 minutes to answer 80 multiple-choice questions and 1 hour and 10 minutes to answer three free-response questions:

Current Exam			
Question Type	**Number of Questions**	**Exam Weight**	**Timing**
1. Multiple-Choice Questions	80	60% of score	1 hour, 30 mins
2. Free-Response Questions	3	40% of score	1 hour, 10 mins

The free-response questions will require you to:

- **Q1: Design and analyze an investigation**

- **Q2: Analyze an environmental issue and propose a solution** from a model/visual representation or quantitative data

- **Q3: Analyze an environmental issue and propose a solution using calculations** from an environmental scenario

Nine AP Exams Go Digital in 2025
College Board announced that in May 2025, nine AP exams will go fully digital! For the latest information regarding the specific AP subjects going digital and testing accommodations, visit the College Board website at apcentral. collegeboard.org/exam-administration-ordering-scores/digital-ap-exams.

Section I: The Multiple-Choice Section

Multiple-choice questions will come in two types: individual questions and set-based questions referring to the same diagram or data presentation. This first type of question that we are going to discuss is the set-based question, which you may already be familiar with.

Questions 1–3 refer to the diagram of the atmosphere below:

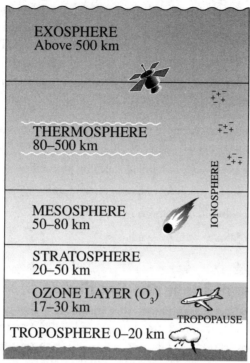

Diagram not to scale

1. Which layer contains the Earth's daily weather?

   (A) Troposphere

   (B) Stratosphere

   (C) Thermosphere

   (D) Mesosphere

2. The highest layer of the atmosphere heated by the IR radiation from the Earth is the

   (A) Stratopause

   (B) Mesosphere

   (C) Thermosphere

   (D) Troposphere

3. The approximate distance of the Mesosphere is which of the following?

   (A) 17 to 30 km

   (B) 50 to 80 km

   (C) 30 to 70 km

   (D) 20 to 50 km

All of the questions in the set above are related to the one diagram presented, so it is important that you reference the image to answer these particular questions. Within these types of set-based questions on the exam, you can expect to see stimulus material including diagrams, models, maps, data tables, charts, and graphs. We'll discuss the science underlying the questions later on, but in case you're interested—the answers are (A), (C), and (B).

The second type of multiple-choice question is more like the traditional multiple-choice questions you are used to seeing:

7. Salt intrusion into freshwater aquifers, beach erosion, and disruption of coastal fisheries all might occur as a result of

    (A) rising ocean levels as global warming proceeds

    (B) more solar ultraviolet radiation on the Earth

    (C) more chlorofluorocarbons in the atmosphere

    (D) reduced rates of photosynthesis

Section II: The Free-Response Section

Environmental Science is interdisciplinary. This means that it draws from several sciences (biology, chemistry, and physics) and the humanities (government, economics, and social studies). The free-response section of the exam will include three questions that will test your ability to do the following: design an investigation, analyze an environmental problem, propose a solution, and work through calculations. A free-response question may look like this:

3. According to the United States Energy Information Administration, the consumption of natural gas in the United States increases 8 percent per year. The United States receives its supplies of gas from a variety of international and domestic locations. Natural gas is used in the home, for industry, and for power generation.

    (a) **Calculate** the approximate number of years it will take to double the consumption of natural gas. Show all work.

    (b) **Describe** one method for the recovery and transportation of natural gas.

    (c) **Describe** two benefits to the environment that would occur by switching from coal to natural gas-fired electric power generation.

    (d) Some people advocate increasing the use of coal instead of natural gas for the production of electricity. **Explain** one argument that the proponents of coal might use to justify their position.

As you can see, for this multipart free-response question, your answers should not be one-dimensional; they will need to encompass many different subjects and areas of thought.

We'll talk more about how to go about writing your responses to these questions in Chapter 2, but for now, remember that you have only 70 minutes to answer all three questions. This translates to about 23 minutes per question, so any practice you can get before test day will be invaluable! Fortunately for you, we have put sample free-response questions and answers in each chapter of this book to give you that practice.

HOW THE AP ENVIRONMENTAL SCIENCE EXAM IS SCORED

What Will My Score Look Like, and What Will It Mean?

After taking the test in May, you will be able to access your AP score in July via your College Board online account. Your score will be a single number from 1 to 5 based on a scale that helps determine how you will qualify to receive college credit and placement. According to the College Board, here's what those numbers mean.

Score	Percentage 2023	Credit Recommendation	College Grade Equivalent
5	8.3%	Extremely Well Qualified	A
4	28.4%	Very well qualified	A–, B+, B
3	17%	Qualified	B–, C+, C
2	26.4%	Possibly Qualified	–
1	19.9%	No Recommendation	–

*Data from The College Board–May 2023 test administration.

Those percentages are pretty intimidating, huh? Quite a few students get either a 1 or 2 on this exam. Why is this? Well, it's probably because these students were not willing to put the time and energy into studying and reviewing the necessary topics.

However, by purchasing this book, you've already proven that you aren't one of those students. No one said this test would be easy, but it is definitely manageable.

How Much Will Each Section Count Toward My Final Score?

As we mentioned, a computer will grade the multiple-choice section of your exam. Your final grade will be made up of the two sections, which will be given different weights: 60 percent of the grade will come from the 80 multiple-choice questions, and the remaining 40 percent will come from the three free-response questions, with about 13 percent coming from each of your responses.

Your Multiple-Choice Score

For this section of the test, you will be scored only on the number of questions you answer correctly. That's right, only the ones you get right are counted! Since there is no penalty for wrong answers, it is in your best interest to answer as many multiple-choice questions as possible. Although random guessing is better than nothing, you should use Process of Elimination (POE) to narrow down the choices first. We'll get into the details of POE in Chapter 1. If you are running out of time,

> **What is LOTD?**
> LOTD stands for letter of the day, which is the letter answer choice you choose when you are stuck on a multiple-choice question. For AP Environmental Science, your letter of the day can be A, B, C, or D.

remember, you need to fill in all the bubbles before time is up. If you don't have time for POE, just choose your letter of the day and move on.

Your Free-Response Score

Each AP free-response question is scored on a scale from 0 to 10, with 10 being the best score. The scores of all three free-response questions are added together to obtain your free-response score.

At the beginning of the free-response question grading, rubrics are formulated by the chief reader, a university professor who also attends development committee meetings, and other AP Environmental Science leaders. These rubrics are refined again and again to provide the best possible scoring rubric for each question. AP Readers (appointed college professors and experienced AP teachers) will grade each test accurately according to the rubric.

As you can imagine, reading thousands of test papers in eight days is an awesome task; therefore, anything you can do to make your free-response answers easier to read is appreciated! A well-organized and readable exam may receive a higher score because the reader can actually find the information they need in order to award the earned points.

There are several different ways to earn those 10 points. However, you can never earn all 10 points unless you answer all parts of the free-response question. So, if the free-response question has 4 parts, you cannot score 10 points unless you answer all 4 parts of that question correctly!

We've said that the highest possible score on each question is a 10. Answers that receive a 10 demonstrate a clear and thorough knowledge of the material and include a superb response to all parts of the question. If the answer requires a calculation, the student performs the calculation showing all work and labeling all units. If it includes placing points on a graph or graphing an answer, the graph is drawn correctly, and the points are placed in the correct place, with all elements (including the x- and y-axes) labeled correctly.

Answers that receive a 9 also demonstrate a thorough knowledge of the material, but are docked a point due to sloppy work, such as not labeling graphs or not showing work when doing calculations. Answers that score a 7 or 8 do not give a satisfactory response to one part of the question, while answers that score a 3, 4, or 5 give an unsatisfactory response for more than one part.

In all cases, even if you don't think you know the answer to one or more questions, you should thoroughly read the question, complete the rest of the free-response questions in the section, and then come back to it.

To look at real questions from past exams and get late-breaking information directly from the College Board, visit apcentral.collegeboard.org/courses/ap-environmental-science/exam.

Your Final Score

Your final 1-to-5 score is a combination of your section scores. Remember that the multiple-choice section counts for 60 percent of the total and that the free-response questions count for 40 percent. However, the bottom line is that both sections are very important, and you must concentrate on doing your best on both parts.

Keep in mind that even if you do not get college credit for this course, you will not have wasted your time. Research shows that just taking an AP course helps your college performance. Small consolation, perhaps, but as we mentioned earlier, because you're making this effort to prepare properly, you'll most likely get due credit.

But How Do I Get Credit If (I Mean, When) I Score a 4 or 5?

Remember, you can choose to send your score online for free to one school via your College Board AP account. After that, you must pay a fee to send your score online to additional schools. We recommend you keep a copy of your score and take it when you register for classes at the college or university you will attend.

Also, you should probably keep the syllabus and your laboratory notebook from your AP Environmental Science course. You may need to show it to college or university counselors in order to get college credit for a science laboratory course.

OVERVIEW OF COURSE UNITS, PRACTICES, AND BIG IDEAS

Below outlines the nine units and topics the College Board suggests be covered in your AP course. We've made sure this book follows the exact outline of the course units, covering all necessary subjects to prepare for the exam.

Unit 1: The Living World: Ecosystems (6–8%)
Introduction to Ecosystems
Terrestrial Biomes
Aquatic Biomes
The Carbon Cycle
The Nitrogen Cycle
The Phosphorus Cycle
The Hydrologic (Water) Cycle
Primary Productivity
Trophic Levels
Energy Flow and the 10% Rule
Food Chains and Food Webs

Unit 2: The Living World: Biodiversity (6–8%)
Introduction to Biodiversity
Ecosystem Services
Island Biogeography
Ecological Tolerance
Natural Disruptions to Ecosystems
Adaptations
Ecological Succession

Unit 3: Populations (10–15%)
Generalist and Specialist Species
K-Selected, *r*-selected Species
Survivorship Curves
Carrying Capacity
Population Growth and Resource
  Availability
Age Structure Diagrams
Total Fertility Rate
Human Population Dynamics
Demographic Transition

Unit 4: Earth Systems and Resources (10–15%)
Plate Tectonics
Soil Formation and Erosion
Soil Composition and Properties
The Earth's Atmosphere
Global Wind Patterns
Watersheds
Solar Radiation and the Earth's Seasons
The Earth's Geography and Climate
El Niño and La Niña

Unit 5: Land and Water Use (10–15%)
The Tragedy of the Commons
Clearcutting
The Green Revolution
Impacts of Agriculture Practices
Irrigation Methods
Pest Control Methods
Meat Production Methods
Impacts of Overfishing
Impacts of Mining
Impacts of Urbanization
Ecological Footprints
Introduction to Sustainability
Methods to Reducing Urban Runoff
Integrated Pest Management
Sustainable Agriculture
Aquaculture
Sustainable Forestry

Unit 6: Energy Resources and Consumption (10–15%)
Renewable and Nonrenewable Resources
Global Energy Consumption
Fuel Types and Uses
Distribution of Natural Energy
Resources
Fossil Fuels
Nuclear Power
Energy from Biomass
Solar Energy
Hydroelectric Power
Geothermal Energy
Hydrogen Fuel Cell
Wind Energy
Energy Conservation

Unit 7: Atmospheric Pollution (7–10%)
  Introduction to Air Pollution
  Photochemical Smog
  Thermal Inversion
  Atmospheric CO_2 and Particulates
  Indoor Air Pollutants
  Reduction of Air Pollutants
  Acid Rain
  Noise Pollution

Unit 8: Aquatic and Terrestrial Pollution (7–10%)
  Sources of Pollution
  Human Impacts on Ecosystems
  Endocrine Disruptors
  Human Impacts on Wetlands
    and Mangroves
  Eutrophication
  Thermal Pollution
  Persistent Organic Pollutants (POPs)
  Bioaccumulation and Biomagnification
  Solid Waste Disposal
  Waste Reduction Methods

  Sewage Treatment
  Lethal Dose 50%
  Dose Response Curve
  Pollution and Human Health
  Pathogens and Infectious Diseases

Unit 9: Global Change (15–20%)
  Stratospheric Ozone Depletion
  Reducing Ozone Depletion
  The Greenhouse Effect
  Increases in the Greenhouse Gases
  Global Climate Change
  Ocean Warming
  Ocean Acidification
  Invasive Species
  Endangered Species
  Human Impacts on Biodiversity

In addition to the course units, the College Board also emphasizes four Big Ideas and seven Science Practices connecting the many topics you will learn throughout each unit. These are:

Big Ideas	Science Practices
BIG IDEA 1: Energy Transfer	Practice 1: *Concept Explanation*
	Practice 2: *Visual Representations*
BIG IDEA 2: Interactions Between Earth Systems	Practice 3: *Text Analysis*
	Practice 4: *Scientific Experiments*
BIG IDEA 3: Interactions Between Different Species and the Environment	Practice 5: *Data Analysis*
	Practice 6: *Mathematical Routines*
BIG IDEA 4: Sustainability	Practice 7: *Environmental Solutions*

As you go through your AP course during the school year, make sure you are paying very close attention to how your AP teacher ties in each topic with the Big Ideas and Science Practices.

HOW AP EXAMS ARE USED

Different colleges use AP Exam scores in different ways, so it is important that you go to a particular college's website to determine how it uses AP Exam scores. The three items below represent the main ways in which AP Exam scores can be used:

Are You Preparing for College?

Check out all of the useful books from The Princeton Review, including *SAT Prep, ACT Prep, The Best 390 Colleges,* and more! Visit Penguin Random House online book store penguinrandomhouse. com/series/CTP/college-test-preparation/

- **College Credit**. Some colleges will give you college credit if you score well on an AP Exam. These credits count toward your graduation requirements, meaning that you can take fewer courses while in college. Given the cost of college, this could be quite a benefit, indeed.

- **Satisfy Requirements**. Some colleges will allow you to "place out" of certain requirements if you do well on an AP Exam, even if they do not give you actual college credits. For example, you might not need to take an introductory-level course, or perhaps you might not need to take a class in a certain discipline at all.

- **Admissions Plus**. Even if your AP Exam will not result in college credit or allow you to place out of certain courses, most colleges will respect your decision to push yourself by taking an AP course or even an AP Exam outside of a course. A high score on an AP Exam shows mastery of more difficult content than is taught in many high-school courses, and colleges may take that into account during the admissions process.

OTHER RESOURCES

There are many resources available to help you improve your score on the AP Environmental Science Exam, not the least of which are your teachers. If you are taking an AP class, you may be able to get extra attention from your teacher, such as obtaining feedback on your free-response essays. If you are not in an AP course, reach out to a teacher who teaches science and ask if they will review your essays or otherwise help you with content.

Another wonderful resource is **AP Students**, the official student site of the AP Exams. The scope of the information at this site is quite broad and includes:

- the course exam description, which provides details on what content is covered and sample questions

- the latest AP Environmental Science Exam free-response and scoring guidelines

- access to AP Classroom if you are enrolled in a course (teacher assistance required)

- free-response prompts from previous years and exam tips

The AP Students web page is apstudent.collegeboard.org.

Finally, **The Princeton Review** offers Homework Help for the AP Environmental Science Exam. Our expert instructors can help you refine your strategic approach and add to your content knowledge. For more information, call 1-800-2REVIEW.

DESIGNING YOUR STUDY PLAN

In Part I, you identified some areas of potential improvement. Let's now delve further into your performance on Practice Test 1, with the goal of developing a study plan appropriate to your needs and time commitment.

Read the answers and explanations associated with the multiple-choice questions (starting on page 33). After you have done so, respond to the following questions:

- Review the Overview of Course Units on pages 68–69, and, next to each one, indicate your rank of the topic as follows: "1" means "I need a lot of work on this," "2" means "I need to beef up my knowledge," and "3" means "I know this topic well."

- How many days/weeks/months away is your AP Environmental Science Exam?

- What time of day is your best, most focused study time?

- How much time per day/week/month will you devote to preparing for your AP Environmental Science Exam?

- When will you do this preparation? (Be as specific as possible: Mondays and Wednesdays from 3 p.m. to 4 p.m., for example.)

- Based on the answers above, will you focus on strategy (Part IV) or content (Part V) or both?

- What are your overall goals in using this book?

Study Breaks Are Important
Don't burn yourself out before test day. Remember to take breaks every so often— go for a walk, listen to a favorite playlist, or get some fresh air.

Part IV
Test-Taking Strategies for the AP Environmental Science Exam

Chapter 1
How to
Approach
Multiple-Choice
Questions

THE PRINCETON REVIEW APPROACH

There are basically two ways to prepare for the AP Environmental Science Exam.

- Know absolutely everything about everything. Bad idea.

- Review only what you need to know and tackle the test strategically. Good idea.

The second is The Princeton Review's way—and the best way—to improve your score.

Rather than trying to teach you everything there is to know about environmental science, we at The Princeton Review focus on test-taking strategies. Naturally, we'll review some hard science as well. But rather than cluttering your brain, we'll look only at the environmental science you need to know for the test, explaining and highlighting key concepts along the way.

First, we'll give you some simple, straightforward strategies for tackling multiple-choice questions and for writing free-response answers. Let's now take a closer look at how to approach the multiple-choice section.

The Two-Pass System

Proven Techniques

The two-pass system allows you to pick up easy points from the start!

The AP Environmental Science Exam covers a broad range of topics. There's no way, even with our extensive review, that you will know everything about every topic in environmental science. So, what should you do?

Adopt a two-pass system. The two-pass system entails going through the test and answering the easy questions first. Save the more time-consuming questions for later. (Don't worry—you'll have time to do them later!) First, read the question and decide if it is a "now" or "later" question. If you decide this is a "now" question, answer it in the test booklet. If it is a "later" question, come back to it. Once you have finished all the "now" questions on a double page, transfer the answers to your bubble sheet. Flip the page and repeat the process.

Once you've finished all the "now" questions, move on to the "later" questions. Start with the easier questions first. These are the ones that require calculations or that require you to eliminate the answer choices (in essence, the correct answer does not jump out at you immediately). Transfer your answers to your bubble sheet as soon as you answer these "later" questions.

Watch Out for Those Bubbles!

Because you're skipping problems, you need to keep careful track of the bubbles on your answer sheet. One way to accomplish this is by answering all the questions on a page and then transferring your choices to the answer sheet. If you prefer to enter them one by one, make sure you double-check the number beside the ovals before filling them in. We'd hate to see you lose points because you forgot to skip a bubble!

Process of Elimination (POE)

It makes sense to assume that you need to know your material backward and forward in order to get the right answer. In other words, if you don't know the answer beforehand, you probably won't answer the question correctly. This is particularly true of fill-in-the-blank and essay questions. We're taught to think that the only way to get a question right is by knowing the answer. However, that's not the case on Section I of the AP Environmental Science Exam. You can get a perfect score on this portion of the test without knowing a single right answer—provided you know all the wrong answers!

What are we talking about? This is perhaps the most important technique to use on the multiple-choice section of the exam. Let's take a look at an example.

41. The long-term storage of phosphorus and sulfur occurs in which of the following?

   (A) Bacteria

   (B) Rocks

   (C) Water

   (D) Plants

Applied Strategies
It is often easier to identify a wrong answer than a right answer. Use POE to get rid of bad answers!

Now, if this were a fill-in-the-blank-style question, you might be in a heap of trouble. But let's take a look at what we've got. You see the elements phosphorus and sulfur in the question, which leads you to conclude that we're talking about elements. Right away, you can probably remember that these aren't normally components of water, so you can eliminate (C). Also, plants don't live a long time, so sulfur and phosphorus can't be stored for the long-term in plants, right? Get rid of (D). The same goes for bacteria, so lose (A). You're left with (B), the correct answer.

We think we've illustrated our point: Process of Elimination is the best way to approach the multiple-choice questions. Even when you don't know the answer right off the bat, you'll surely know that two or three of the answer choices are not correct. What then?

Active Guessing

As mentioned earlier, you are scored only on the number of questions you get right, so we know guessing can't hurt you. But can it help you? It sure can. Let's say you guess on five questions; odds are you'll get one right. So, you've already increased your score by one point. Now, let's add POE into the equation. If you can eliminate as many as two answer choices from each question, your chances of getting them right increase and so does your overall score. Remember, don't leave any bubbles blank on test day!

Word Associations

Another way to rack up the points on the AP Environmental Science Exam is by using word associations in tandem with your POE skills. Make sure that you memorize all of the words in the Glossary, which is Chapter 12 of this book. Know them backward and forward. As you learn them, make sure you group them by "association," as you're bound to be tested on them on the AP Environmental Science Exam. What do we mean by "word associations"?

Let's take the example of air pollution. You'll soon see from our review, and possibly your course study, that there are several compounds associated with various types of air pollution. For example, ozone, VOC, and nitrogen oxides are all terms associated with air pollution. Now, take a look below at a typical question about pollution.

2. All of the following are important in smog production EXCEPT

    (A) photochemical reactions

    (B) stratospheric ozone

    (C) tropospheric ozone

    (D) volatile organic compounds

This might seem like a difficult question, but let's think about the associations we just discussed. The question asks us about smog. Choices (C) and (D) are terms that we've associated with air pollution. Therefore, we can eliminate them. Maybe you're unsure about whether or not photochemical reactions are part of air pollution, but since you know for sure that stratospheric ozone has nothing to do with smog production (or for that matter, air pollution), you might guess that (A) is the correct answer (and you'd be right!).

We'll explain what these words mean in Chapter 9, in which we discuss pollution, but the point is that without even racking your brain, you've managed to get this down to two answer choices—not bad! You would have a fifty-fifty chance of guessing correctly on this question.

By combining the associations we'll offer throughout this book with strategic POE techniques, you'll be able to rack up points on problems that might have seemed particularly difficult at first.

Mnemonics—or the Environmental Science Name Game

One of the big keys to simplifying biology is to organize terms into a handful of easily remembered packages. The best way to accomplish this is by using mnemonics. A mnemonic, as you may already know, is a convenient device, such as a rhyme or phrase, for remembering something. Environmental science is all about names: the names of chemicals, processes, theories, and more. How are you going to keep them all straight without a little help?

For example, the major components of air pollution are:

- **S**ulfur dioxide—SO_2

- **P**articulates

- **L**ead—Pb

- **O**zone—O_3

- **N**itrogen dioxide—NO_2

- **C**arbon monoxide—CO

The first letter of each component spells SPLONC, which is otherwise known as Some Pollution Lands On Nature Constantly. Learn the mnemonic and you'll never forget the science!

Mnemonics can be as goofy as you like, so long as they help you remember. Be creative! Remember, the important thing is that you remember the information, not how you remember it.

Identify Question Types

Many of the traps on the AP Environmental Science Exam deal with the way in which the question is asked. Here's information about a few types of multiple-choice questions you may see on the updated exam.

EXCEPT/NOT/LEAST Questions

Some of the multiple-choice questions in Section I may be EXCEPT/NOT/LEAST questions. With this type of question, you must remember that you're looking for the wrong (or the least correct) answer. The best way to approach these is by using POE.

More often than not, the correct answer is a true statement, but is wrong in the context of the question. Cross off the three that apply, and you're left with the one that does not. Here's an example of this type of question.

27. All of the following are components of integrated waste management EXCEPT

    (A) using canvas bags that can be reused rather than disposable bags

    (B) using old appliances for construction of artificial reefs

    (C) using disposable diapers instead of cloth diapers

    (D) using reused glass bottles

If you don't remember anything about integrated waste management, you should at least understand that the question is asking about waste. So, which of the choices does *not* deal with a way to reduce or reuse waste? Well, (C) would result in more, and not less, waste; and it is the correct answer. Remember, the best way to answer these types of questions is to spot all the right statements and cross them off. You'll wind up with the wrong statement, which happens to be the correct answer.

Unspecified One-or-More

Another type of multiple-choice question that might appear on the updated exam is called the Unspecified One-or-More question. These questions are designed to have you select all of the correct answers, though they do not prompt you on how many might be correct. In this case, you need to carefully analyze each answer, independent of the other answers. Be sure to consider each choice carefully, determine which ones are correct, and then look at the answer options to see which one corresponds with the selection of answers you have determined is correct. Here's an example of this type of question.

1. The rain shadow effect may cause which of the following?

 I. Drier conditions on the leeward side of mountain ranges
 II. Warmer conditions on the windward side of mountain ranges
 III. More light on the leeward side of mountain ranges

 (A) I only

 (B) II only

 (C) III only

 (D) I and III only

The correct answer is (A). Mountains cause warm, moist air to rise and compress, ultimately creating rain on the windward side of mountains. As a result, the leeward sides of mountains are much dryer. Therefore, the correct answer must have something to do with precipitation or moisture. In considering each answer independently, statement (I) is clearly correct, but statements (II) and (III) are not.

Chapter 1 Drill

Let's practice some of the different types of multiple-choice questions that you just learned about. For answers and explanations, see Chapter 13.

1. All of the following are ways in which populations of species can be dispersed EXCEPT

    (A) uniform

    (B) random

    (C) symmetrical

    (D) clumped

2. Weathering that takes place because of the activities of living organisms is referred to as

    (A) physical weathering

    (B) ecological weathering

    (C) chemical weathering

    (D) biological weathering

3. Over the past two decades, population growth in the city of Houston has led to the establishment of several new communities in the surrounding regions. Accordingly, it is difficult to determine where the Houston metro ends and communities like Katy and Galveston, Texas, begin. This is an example of

    (A) urban sprawl

    (B) suburban revolution

    (C) gentrification

    (D) brownfields

4. Acid precipitation leads to which of the following?

    I. The leaching of minerals from the soil, detrimentally altering the soil chemistry

    II. An increased rate of Severe Acute Respiratory Syndrome (SARS) in urban areas

    III. Damage to statues, monuments, and buildings

    (A) I only

    (B) III only

    (C) I and III only

    (D) I, II, and III

5. Which of the following is NOT a human factor that can cause extinction and decrease biodiversity?

    (A) Habitat destruction

    (B) Infectious disease

    (C) Climate change

    (D) Pollution

6. Located in regions that are just south of the Arctic ice caps, and extending across North America, Europe, and Siberia in Asia, this type of ecosystem has herbaceous plants as its major vegetation.

    (A) Chapparal

    (B) Coniferous forest

    (C) Deciduous forest

    (D) Tundra

7. The Safe Drinking Water Act

    (A) set up conditions to evaluate the conditions of American soil, water, and related resources, and established soil and water conservation programs to aid landowners and users

    (B) established a federal program to monitor and increase the safety of the drinking water supply

    (C) used regulatory and nonregulatory tools to protect all surface waters in the United States

    (D) created a tax on the chemical and petroleum industries and provided broad federal authority to respond directly to releases of hazardous substances that may endanger public health or the environment

8. In the year 2022, a country had a population of 20 million people with a birth rate of 5 percent and a death rate of 1.5 percent. Assuming that these rates remain stable, and the country experiences no migration, in what year will the population of that country be approximately 40 million?

    (A) 2032

    (B) 2036

    (C) 2042

    (D) 2046

9. Birth rates exceed death rates in all states of the demographic transition model EXCEPT the

   (A) preindustrial state

   (B) transitional state

   (C) industrial state

   (D) postindustrial state

10. Crown fires

   (A) are smoldering fires that occur in bogs or swamps, originate from surface fires, and are difficult to detect and extinguish

   (B) may start on either the ground or forest canopies that have not experienced recent surface fires, spread quickly, and are characterized by high temperatures

   (C) typically only burn the underbrush of a forest, and serve to protect the forest from more harmful fires by removing underbrush and debris

   (D) are controlled fires that are intentionally started in order to reduce the fuel buildup and, in turn, decrease the likelihood of more severe fires

11. Which of the following statements regarding renewable and nonrenewable resources is NOT true?

   (A) Renewable resources can be regenerated quickly.

   (B) 100 years is widely considered the crossover point from renewable to nonrenewable resources.

   (C) Nonrenewable resources include minerals and fossil fuels.

   (D) Wind is considered a renewable resource.

Chapter 2
How to Approach Free-Response Questions

THE ART OF THE FREE-RESPONSE ESSAY

You're given three free-response questions to answer in 70 minutes. That's only about 23 minutes per question. Each of these three questions will present a scenario and ask you to answer several smaller questions (generally 3 to 4). You can get a maximum of 10 points per free-response question, and you need to answer every part of the question to get all 10 points. Each question has a certain grading rubric assigned to it, which is what the free-response readers use to give you points for your responses. The best way to rack up points on this section is to give the graders what they're looking for. Fortunately, we know precisely how to do this.

Now or Later?

For the exam, one of the free-response questions will ask you to design an investigation, and one will ask you to propose a solution to an environmental problem using models. The third question will ask you the same as the second, but this time using calculations, so a calculator can be very useful for this question.

While you do have to answer each of these questions, you do *not* need to answer them in order. The best strategy is to read the scenario (not the sub-questions) and decide if this is a question you want to attempt now or later. Do this before reading the next scenario. If you decide to do it later, move on to the next question.

If you decide to do it now, look at the questions and start answering them. Remember the grading rubric, and make it easy for the grader to give you points. If the question asks for two solutions, label the solutions (for example, "a" and "b") so the grader can easily find them. If the question asks for a calculation, show all your work, including any formulas you are using. If the question asks you to plot something, clearly label the *x*- and *y*-axes, and any relevant point or area on the chart.

> A four-function (with square root), scientific, or graphing calculator may be used on all sections of the AP Environmental Science Exam.

Calculate the Math

Yes, there is math in environmental science and you will be allowed to use a calculator on the exam. So even if you're able to do the math in your head, it helps to know you can now double-check your work on a calculator when solving challenging questions!

Hot-Button Terms

The AP essay graders have a checklist of key terms and concepts that they use to assign points. We like to call these "hot-button" terms. Simply put, for each hot button that you include in your essay, you will receive a predetermined number of points. For example, if the essay question deals with photochemical smog, the AP graders are instructed to give students two points for writing: "In the presence of sunlight and heat, VOCs (volatile organic compounds), NO_x, and ozone combine to form smog"—or something very similar to that. So where do you find these key terms? Funny you should ask—they are at the end of each chapter in this book and in the Glossary in Chapter 12. Make sure you have a grasp of all of the words in those lists, and use them as hot buttons in your essays.

Make an Outline

As you read the question, brainstorm a list of terms and concepts you want to cover. Use the sub-questions to help you with your list of terms and concepts. Next, draft an outline that will help you organize them into some logical order. While you do not get points for organization, a well-organized essay is easier to write (and more importantly, easier to grade). The best way to organize your response is to write a clear, simple outline (just a couple of bullets per section). Outlining should take no more than about two to three minutes.

Of course, if you just composed a list of key scientific terms, you wouldn't be writing an essay. It is important to remember that the three free-response questions are essay questions, and they need to be written in paragraph style. An answer that's written as a list or an outline is not acceptable and will not be scored. On average, you will need to write no more than one or two paragraphs for each question.

If the question asks for two examples, give just that—two examples. If you present more than two examples, the grader may not even count them toward your score. Make sure you read carefully and provide just what the question asks for.

Outline Before You Write
You have about 23 minutes to answer each free-response question. Use the first 2 to 3 minutes to make an outline of your answer. This process will help you organize your thoughts and direct your writing.

Label All Diagrams and Figures

Sometimes it's easier to present a diagram or figure as part of your essay. You may illustrate your answer, but all illustrations should be labeled and discussed in the verbiage of your answer. Remember to label your diagram or figure properly; otherwise, the AP graders will give you no more than partial credit for your work.

Know the Labs Covered in Your AP Course

At least one of the three essay questions will be experiment-based. Sometimes the questions will refer back to a laboratory experiment conducted in your AP class. Consequently, the laboratory component of your course is an integral part of this exam. In Chapter 11, we'll review some of the laboratory experiments you may have performed in your AP Environmental Science class.

Review Your Answers

After answering all parts of a question, give yourself a couple of minutes to review your answers before moving to the next question. Remember, once you are done with a question, you are DONE. Do not go back to a question you have completed—even if you have time at the end of the test.

Practice, Practice, and More Practice!

The only way to get good at writing an essay in 23 minutes is to keep at it. Try out the strategies you've learned on the practice questions found on the next page and in the free-response questions found at the end of each content chapter in Part V.

Chapter 2 Drill

Keeping in mind the things that we just covered when tackling free-response questions, try out some of these questions for practice. For answers and explanations, see Chapter 13.

1. Before 1880, the North American black-tailed prairie dog population thrived in the hundreds of millions and shaped the Great Plains temperate grassland ecosystem for over 200 other species of plants and animals observed living on or near prairie dog burrow colonies. But, by 1972, the population was estimated to be approximately 3,100 with the projection of becoming extinct by 2000. Since 1972, the population has increased from 3,100 individuals to approximately 12,400 in 2012.

   (a) **Calculate** the population growth rate from 1972 to 2012.

   (b) **Identify** and **describe** TWO major causes for the original decline of these species.

   (c) **Identify** and **describe** one likely ecological impact of the loss of black-tailed prairie dogs in the grassland biome.

   (d) Make one economic or one ecological **argument for** protecting the black-tailed prairie dog or another endangered species that you identify and one economic or one ecological **argument against** it.

   (e) **Identify** and **describe** one piece of United States or international legislation that is in place to prevent the decline of species or encourage the regrowth of species population.

2. The diagram below is of diversity downstream from the point of discharge from a sewage treatment plant.

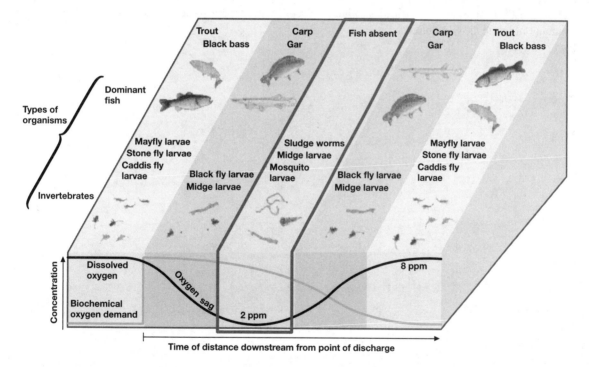

   (a) **Explain** the limiting factor that is responsible for these different zones, creating different biodiversity levels in each area.

   (b) **Explain** the health of the stream as you move downstream from the point of discharge based on what you notice about changes in biodiversity.

   (c) Sewage wastewater treatment plants mimic some of the same elements of wetlands in improving water quality before it reaches streams. **Explain** how wastewater treatment plants perform these equivalent responsibilities in the primary and secondary treatments.

   (d) **Describe** one possible pollutant to target during disinfection of wastewater before its final release back into nature and describe a method of sewage wastewater treatment that is meant to disinfect water.

3. Your growing city of 500,000 people has realized they need to produce another waste stream destination. You are appointed the chairperson of the committee to make the decision of which is the best facility for your community. The two options open for discussion are either the development of a new sanitary landfill or an incineration facility. You must select only one of these options.

(a) **Provide** TWO environmental reasons why your recommendation is better than the alternative.

(b) **Identify** and **describe** one economic reason why your recommendation is better than the alternative.

(c) A local recycling company has concerns about how your recommendation may affect their business. Use the assumptions below to answer the questions that follow. For each calculation, **show** all work.

Recycling company payout	5 cents
Recycling income from wholesaler	7 cents
Mileage of recycling truck	10 miles per gallon
Fuel cost	$2 per gallon
Recycling truck daily distance travel	100 miles
Truck driver pay	$100 per day

   i. **Calculate** the gross income the company would need to make weekly in order to earn a $350 profit.
   ii. **Calculate** how many cans must be recycled to make a profit of $50 per day.

(d) **Describe** TWO conservation measures (other than recycling) that the city could take to reduce the total amount of waste the city outputs.

Chapter 3
Using Time Effectively to Maximize Points

BECOMING A BETTER TEST-TAKER

Very few students stop to think about how to improve their test-taking skills. Most assume that if they study hard, they will test well, and if they do not study, they will do poorly. Most students continue to believe this even after experience teaches them otherwise. Have you ever studied really hard for an exam, and then blew it on test day? Have you ever aced an exam for which you thought you weren't well prepared? Most students have had one, if not both, of these experiences. The lesson should be clear: factors other than your level of preparation influence your final test score. This chapter will provide you with some insights that will help you perform better on the AP Environmental Science Exam and on other exams, as well.

PACING AND TIMING

A big part of scoring well on an exam is working at a consistent pace. The worst mistake made by inexperienced or less savvy test-takers is that they come to a question that stumps them, and, rather than just skip it, they panic and stall. Time stands still when you're working on a question you cannot answer, and it is not unusual for students to waste five minutes on a single question (especially a question involving a graph or the word EXCEPT) because they are too stubborn to cut their losses. It is important to be aware of how much time you have spent on a given question and on the section you are working on. There are several ways to improve your pacing and timing for the test.

- **Know your average pace.** While you prepare for your test, try to gauge how long you take on 5, 10, or 20 questions. Knowing how long you spend on average per question will help you identify how many questions you can answer effectively and how best to pace yourself for the test.

- **Have a watch or clock nearby.** You are permitted to have a watch or clock nearby to help you keep track of time. It is important to remember, however, that constantly checking the clock is in itself a waste of time and can be distracting. Devise a plan. Try checking the clock after every 15 or 30 questions to see if you are keeping the correct pace or need to speed up. This will ensure that you are cognizant of the time but will not permit you to fall into the trap of dwelling on it.

- **Know when to move on.** Since all questions are scored equally, investing appreciable amounts of time on a single question is inefficient and can potentially deprive you of the chance to answer easier questions later on. If you are able to eliminate answer choices, do so, but don't worry about picking a random answer and moving on if you cannot find the correct answer. Remember, tests are like marathons; you do best when you work through them at a steady pace. You can always come back to a question you don't know. When you do, very often you will find that your previous mental block is gone, and you will wonder why the question perplexed you the first time around (as you gleefully move on to the next question). Even if you still don't know the answer, you will not have wasted valuable time you could have spent on easier questions.

- **Be selective.** You don't have to do any of the questions in a given section in order. If you are stumped by an essay or multiple-choice question, skip it or choose a different one. Select the questions or essays that you can answer and work on them first. This will make you more efficient and give you the greatest chance of getting the most questions correct.

- **Use Process of Elimination on multiple-choice questions.** Many times, one or more answer choices can be eliminated. Every answer choice that can be eliminated increases the odds that you will answer the question correctly. Review the section on this strategy in Chapter 1 to find these incorrect answer choices and increase your odds of getting the question correct.

Remember, when all the questions on a test are of equal value, no one question is that important. Your overall goal for pacing is to get the most questions correct. Finally, you should set a realistic goal for your final score. In the next section, we will break down how to achieve your desired score and ways of pacing yourself to do so.

GETTING THE SCORE YOU WANT

Depending on the score you need, it may be in your best interest *not* to try to work through every question. Check with the schools to which you are applying.

Keep in mind, students are assessed only on the total number of correct answers. It is really important to remember that if you are running out of time, you should fill in all the bubbles before the time for the multiple-choice section is up. Even if you don't plan to spend a lot of time on every question and even if you have no idea what the correct answer is, it's to your advantage to fill something in.

> **Answer Every Question**
> Here's something worth repeating: there is no penalty for guessing. Don't leave a question blank; there's at least a 25% chance you might have gotten it right!

TEST ANXIETY

Everybody experiences anxiety before and during an exam. To a certain extent, test anxiety can be helpful. Some people find that they perform more quickly and efficiently under stress. If you have ever pulled an all-nighter to write a paper and ended up doing good work, you know the feeling.

However, too much stress is definitely a bad thing. Hyperventilating during the test, for example, almost always leads to a lower score. If you find that you stress out during exams, here are a few preemptive actions you can take.

- **Take a reality check.** Evaluate your situation before the test begins. If you have studied hard, remind yourself that you are well prepared. Remember that many others taking the test are not as well prepared, and (in your classes, at least) you are being graded against them, so you have an advantage. If you didn't study, accept the fact that you will probably not ace the test. Make sure you get to every question you know something about. Don't stress out or fixate on how much you don't know. Your job is to score as high as you can by maximizing the benefits of what you do know. In either scenario, it is best to think of a test as if it were a game. How can you get the most points in the time allotted to you? Always answer questions you can answer easily and quickly before you answer those that will take more time.

- **Try to relax.** Slow, deep breathing works for almost everyone. Close your eyes, take a few slow, deep breaths, and concentrate on nothing but your inhalation and exhalation for a few seconds. This is a basic form of meditation, and it should help you to clear your mind of stress and, as a result, concentrate better on the test. If you have ever taken yoga classes, you probably know some other good relaxation techniques. Use them when you can.

- **Eliminate as many surprises as you can.** Make sure you know where the test will be given, when it starts, what type of questions are going to be asked, and how long the test will take. You don't want to be worrying about any of these things on test day or, even worse, after the test has already begun.

The best way to avoid stress is to study both the test material and the test itself. Congratulations! By buying or reading this book, you are taking a major step toward a stress-free AP Environmental Science Exam.

Work Hard, Play Hard
Remember to give yourself small rewards as you prepare for the AP Environmental Science Exam.

Part V
Content Review for the AP Environmental Science Exam

HOW TO USE THE CHAPTERS IN THIS PART

You may need to come back to the following chapters more than once. Your goal is to obtain mastery of the content you are missing, and a single read of a chapter may not be sufficient. At the end of each chapter, you will have an opportunity to reflect on whether you truly have mastered the content of that chapter.

Guess What?

You're about to embark on a comprehensive content review of AP Environmental Science. It's a lot to remember, and we can help! Check out the study guides in your Student Tools, which you can access by scanning the code below and registering your book.

Chapter 4
Units 1 and 2: The Living World: Ecosystems and Biodiversity

In this first chapter, we'll review the first *two* units covered on the AP Environmental Science Exam, which AP calls *The Living World: Ecosystems* and *The Living World: Biodiversity*, respectively. Obviously, these two topics are extremely interrelated. According to the College Board, about 6–8% of the test is based directly on each of these topics, so that's a total of about 12–16% of test material devoted to the ideas covered in this chapter. If you are unfamiliar with a topic presented here, consult your textbook for more in-depth information.

The chapter starts with a discussion of what an ecosystem is, a review of the concepts of evolution, and a classification of biomes. Next, we'll discuss the abiotic elements that are essential to life and their natural cycles. Then we'll examine the biotic components of ecosystems—living systems—and how energy is used among them. To review biodiversity, we'll start with a discussion of what biodiversity is, what can affect it, and how it relates to ecosystems. Next, we'll review how ecosystems and biodiversity provide humans with essential services. Finally, we'll review how ecosystems change as a result of disruptions, including the process of ecological succession.

One of the fundamental concepts of environmental science is that of interconnected systems. The Earth itself is a grand interconnected system, encompassing the components, processes, and relationships that result in all the smaller living and nonliving systems we see every day. To begin to understand this complexity, an easy place to start is with the idea of an ecosystem.

ECOSYSTEMS

An **ecosystem** is a system of interconnected elements: a community of living organisms and its environment. It includes both **biotic** (living) and **abiotic** (nonliving) components. Ecosystems are the result of the biotic and abiotic components interacting. Biotic components—life—require resources to flourish, and the availability of those resources influences the interactions between species and their interactions with their environments. Over time, the most fundamental interaction between and among ecosystems is that of evolution—which produces life as we know it.

EVOLUTION

Biodiversity in all forms is the result of **evolution.** Evolution is the change in a population's genetic composition over time.

We use a figure called a **phylogenetic tree** to model evolution. Phylogenetic trees can be very broad, like the one on the next page, which encompasses many types of species, or they can be very specific and describe the evolutionary relationships that exist between two species (or even the genomes of one species!).

While you won't need to know much about evolution for this exam, you will need to have a rough idea of how and why it takes place, so we'll run through that now. Without trying to recreate the evolution of all living organisms, we will limit our discussion to a description of how new species are formed. This process is called **speciation**.

Strictly speaking, a **species** is defined as a group of organisms that are capable of breeding with one another—and incapable of breeding with other species. As you may recall, individual organisms that are better adapted for their environment will live and reproduce, ensuring that their genes are part of their population's next generation. This is what Charles Darwin meant by **evolutionary fitness.**

Phylogenetic Tree

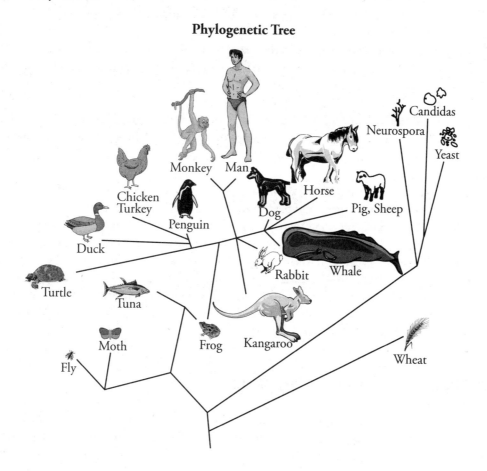

How Evolution Works

When a habitat (an organism's physical surroundings) selects certain organisms to live and reproduce and others to die, that population is said to be undergoing **natural selection**. In natural selection, beneficial characteristics that can be inherited are passed down to the next generation, and unfavorable characteristics that can be inherited become less common in the population. Any cause that reduces reproductive success (fitness) in a portion of the population is a **selective pressure,** and these are what drive natural selection. It is important to remember that natural selection acts upon a whole population, not on an individual organism during its lifetime. What changes during evolution is the total genetic makeup of the population, or **gene pool,** and natural selection is one of the mechanisms by which evolution operates.

The other way evolution operates is **genetic drift**. Genetic drift is the accumulation of changes in the frequency of alleles (versions of a gene) over time due to sampling errors—changes that occur as a result of random chance. For example, in a population of owls there may be an equal

> **Fast Fact**
> The term *survival of the fittest* does not necessarily refer to the fastest or the strongest. It refers to those organisms that produce offspring that will go on to also produce offspring.

chance of a newly born owlet having long talons or short talons, but due to random breeding variances, a slightly larger number of long-taloned owlets are born. Over many generations, this slight variance can develop into a larger trend, until the majority of owls in that population have long talons. These breeding variances could be a result of a chance event—such as an earthquake that drastically reduces the size of the nesting population one year. Small populations are more sensitive to the effects of genetic drift than large, diverse populations.

When a population displays small-scale changes over a relatively short period of time, **microevolution** has occurred. **Macroevolution** refers to large-scale patterns of evolution within biological organisms over a long period of time.

Just as new species are formed by natural selection and genetic drift, other species may become extinct. **Extinction** occurs when a species cannot adapt quickly enough to environmental change and all members of the species die.

Biological extinction is the true extermination of a species. There are no individuals of a biologically extinct species left on the planet (for example, the dodo bird or passenger pigeon). **Ecological extinction** occurs when there are so few individuals of a species that this species can no longer perform its ecological function (for example, alligators in the Everglades in the 1960s or wolves in Yellowstone before re-introduction in the last decade). **Commercial or economic extinction** is when a few individuals exist, but the effort needed to locate and harvest them is not worth the expense (for example, the groundfish population of the Grand Banks off the Maritimes of Canada).

Relationships Between Species

Let's talk more about how species get along together in ecosystems. You probably recall from your biology class that a group of organisms of the same species is called a **population,** and when populations of different species occupy the same geographic area, they form a **community.** Every species within a community has an ecological niche. A species' **niche** is described as the total sum of a species' use of the biotic and abiotic resources in its environment. The niche describes where the species lives, what it eats, and all of the other resources the species utilizes in an ecosystem. Another term you should know for the exam is **habitat**—a habitat is the area or environment where an organism or ecological community normally lives or occurs.

Species can be generalist or specialist. A **specialist** species is one that has a narrow niche and can only live in a certain habitat. A **generalist** species is one that has a broad niche, is highly adaptable, and can live in varied habitats. Specialist species tend to have an advantage when their environments are relatively unchanging, while generalist species have the advantage in habitats that undergo frequent change.

Some species interact quite a bit with other members of their population; for example, some animals form herds, while other species are loners—like bears. The reasons for these different levels of sociability are largely competition, predation, and a general need to exploit the resources in the environment.

Competition arises when two individuals—of the same species or of different species—are competing for resources in the environment. When the two individuals that are competing are of the same species, this is called **intraspecific competition,** and when they are of different species, it's called **interspecific competition.** The resources that are competed for can be food, air, shelter, sunlight, and various other factors necessary for life; individuals may be competing to live in a fallen tree, to catch a running rabbit, or to mate with the most desirable female in the population. The competitor who is "most fit" eventually wins and obtains the resource. That's right—the others are eliminated by competition.

One more thing about competition: when two different species in a region compete and the better adapted species wins, this phenomenon is called **competitive exclusion. Gause's principle** states that no two species can occupy the same niche at the same time and that the species that is less fit to live in the environment will relocate, die out, or occupy a smaller niche. When a species occupies a smaller niche than it would in the absence of competition, this compromised niche is called its **realized niche.** (The niche it would have if there were no competition is known as its **fundamental niche.**) Direct competition can also be avoided in the case of **resource partitioning.** This occurs when different species use slightly different parts of the habitat, but rely on the same resource. For example, there are five species of warblers that can all live in the same pine tree. They can coexist because each species feeds in a different part of the tree: the trunk, at the ends of the branches, and at other sites. Keep in mind that many types of species can engage in both short- and long-term migration, for reasons including food and water availability, temperature changes, mating opportunities, and safety from predation. This means that a given species might be part of several different communities at different times, and might fill a given niche in each of those communities only some of the time. All right, moving on!

Know the three types of relationships between species: competition, predation, and symbiotic relationships.

Although it's relatively easy to observe competition between animals, competition between plants is much more subtle and occurs much more slowly. However, if you have a few years to kill, spend some time in your backyard watching the trees and other plants grow. You'll see that they compete for sunlight and for ground space; they even produce chemicals that inhibit other plants' growth!

The second important type of interspecies interaction is predation. **Predation** occurs when one species (a **predator**) feeds on another (**prey**), and it drives changes in population size. For example, in a year in which rainfall is relatively high in some regions, rabbits have plenty of food; this enables them to reproduce very successfully, and the number of rabbits in a population will increase dramatically. In turn, if the coyote is a predator of the rabbit, coyotes will have plenty of food, and their population will also boom. However, if the following year the rainfall is below average, there will be less grass. Then the population of rabbits will decline, and this will result in a decline in the population of coyotes. As a final note about predation: while it's tempting to think of predation existing only between animals, remember that herbivores prey on plants and zooplankton on phytoplankton!

A third type of relationship that exists between organisms is the symbiotic relationship. **Symbiotic relationships** are close, prolonged associations between two or more different organisms of different species that may, but do not necessarily, benefit each member. There are three types of symbiotic relationships, and you should be familiar with all of these for exam day. In mutualistic symbiotic relationships **(mutualism),** both species benefit; for example, this type of relationship exists between sea anemone and clown fish. The clown fish protects the sea anemone from some of its predators, while the stinging cells of the anemone protect the clown fish; the fish also eats some of the detritus left behind when the anemone feeds. In commensalistic symbiotic relationships **(commensalism)**, one organism benefits while the other is neither helped nor hurt. One example of this type of relationship exists between trees and epiphytes (bromeliads and some orchids). The trees are not affected by the epiphytes growing in them, and the epiphytes benefit by collecting water running down the bark and get better access to light than they would on the ground. Finally, **parasitism** is a relationship in which one species is harmed and the other benefits; for example, the relationship that exists between fleas and dogs.

Now that you've reviewed how the biotic components of ecosystems change, survive, and thrive, let's look at what ecosystems can be found on the planet we call home.

THE WORLD'S ECOSYSTEMS

Because different geographic areas on Earth differ so much in their abiotic and biotic components, we can easily place them in broad categories. The two largest categories are broken down in this way: ecosystems that are based on land are called **biomes,** while those in aqueous environments are known as **aquatic life zones.**

Biomes

Land environments are separated into biomes based on factors such as climate, geology, soils, topography, hydrology, and vegetation. Although it might seem that each biome listed in the table on the following page is very distinct, in reality, biomes blend into each other; they do not have distinct boundaries. The transitional area where two ecosystems meet actually has a name—these areas are called **ecotones.** Another important term that you should be familiar with for the exam is **ecozones** (also called ecoregions), which are smaller regions within ecosystems that share similar physical features.

Types of Ecosystems			
Biome	**Annual Rainfall, Soil Type**	**Major Vegetation**	**World Location**
Deciduous forest (temperate and tropical)	75–250 cm, rich soil with high organic content	Hardwood trees	North America, Europe, Australia, and Eastern Asia
Tropical rainforest	200–400 cm, poor quality soil	Tall trees with few lower limbs, vines, epiphytes, plants adapted to low light intensity	South America, West Africa, and Southeast Asia
Grasslands	10–60 cm, rich soil	Sod-forming grasses	North American plains, prairie, and savanna; Russian steppes; South African velds; Argentinean pampas
Coniferous forest (Taiga)	20–60 cm—mostly in summer, soil is acidic due to vegetation	Coniferous trees	Northern North America, Northern Eurasia
Tundra	Less than 25 cm, soil is permafrost	Herbaceous plants	The northern latitudes of North America, Europe, and Russia
Chaparral (scrub forest or shrubland)	50–75 cm—mostly in winter, soil is shallow and infertile	Small trees with large, hard evergreen leaves, spiny shrubs	Western North America, the Mediterranean region
Deserts (cold and hot)	Less than 25 cm, soil has a coarse texture (i.e., sandy)	Cactus, other low-water adapted plants	30 degrees north and south of the equator
Temperate rainforest	over 140 cm, soil richer than that in tropical rainforests	Coniferous and broadleaf trees, epiphytes, mosses, ferns, and shrubs	North America, South America, South Africa, Europe, Russia, Northeast Asia, Australia, New Zealand
Savanna	10–30 cm almost all in rainy season, soil is porous and has only a thin layer of humus	Grasses with more widely spaced trees	Australia, South America, India, and half of Africa

Not surprisingly, each biome has specific characteristics that determine the types of organisms that are capable of living in it. Some of these characteristics are the type and availability of nutrients, the ecosystems' temperature, the availability of water, and how much sunlight the region receives.

Aquatic Life Zones

Recall that aquatic life zones are the equivalents of biomes in aquatic environments. Aquatic ecosystems are categorized primarily by the salinity of their water—freshwater and saltwater ecosystems fall into separate categories.

Freshwater Biomes

In all natural bodies of water, there exist layers of water that vary significantly in their temperature, oxygen content, and nutrient levels. These layers are affected differently by seasonal changes and other disturbances, and this also contributes to how they are categorized.

In freshwater, the layers are the **epilimnion**, which is the uppermost and thus the most oxygenated, layer; and the **hypolimnion,** which is the lower, colder, and denser layer. The demarcation line between these two layers, at which the temperature shifts dramatically, is the **thermocline.**

The layers of freshwater bodies may also be categorized differently, according to the types of organisms that can live in them. You should definitely be familiar with the following terms for the AP Environmental Science Exam, so take note!

- **Littoral zone:** Begins with the very shallow water at the shoreline. Plants and animals that reside in the littoral zone receive abundant sunlight. These also include turtles, frogs, and other species that travel back and forth from water to land. The end of this zone is defined as the depth at which rooted plants stop growing.

- **Limnetic zone:** Surface of open water; the region that extends to the depth that sunlight can penetrate. Organisms that are residents in this zone tend to be short-lived and rely on sunlight: photosynthesizing phytoplankton use it directly, and they provide energy to zooplankton, insects, and fish.

- **Profundal zone:** The depths: water that is too deep for sunlight to penetrate. Because the profundal zone is aphotic (a place where light cannot reach), photosynthesizing plants and animals cannot live here; instead, organisms adapted to little light, colder temperatures, and less oxygen reside in this less populated zone.

- **Benthic zone:** The surface and sub-surface layers of the river-, lake-, pond-, or streambed, characterized by very low temperatures and low oxygen levels and inhabited by organisms that live on, in, or below the sediment surface, including bottom-feeders, scavengers, and decomposers (including microorganisms such as bacteria and fungi).

Another important type of freshwater body that you should know about is the estuary. An **estuary** is a site where the "arm" of the sea extends inland to meet the mouth of a river. Estuaries are often rich with many different types of plant and animal species, because the freshwater in these areas usually has a high concentration of nutrients and sediments. The waters in estuaries are usually quite shallow, which means that the water is fairly warm and that plants and animals in these locations can receive significant amounts of sunlight. Subcategories of estuarine environments that you should know for the exam include saltwater marshes, mangrove forests, inlets, bays, and river mouths.

Some of the Earth's most important ecologically diverse ecosystems are **wetlands**—areas along the shores of fresh bodies of water, wet inland habitats fed only by rainwater, and ephemeral (seasonally temporary) water bodies. Types of wetlands include marshes, swamps, bogs, prairie potholes (which exist seasonally), and floodplains (which occur when excess water flows out of the banks of a river and into a flat valley). So, those are the main types of freshwater bodies you'll need to know.

Let's look more specifically at the mangrove swamp. **Mangrove swamps** are coastal wetlands (areas of land covered in freshwater, saltwater, or a combination of both) found in tropical and subtropical regions, and they are threatened by activities such as shrimp aquaculture and the degradation of the Western coastlines. Mangroves are characterized by trees, shrubs, and other plants that can grow in brackish tidal waters and are often located in estuaries, which, as you learned earlier, are areas where freshwater meets saltwater. In North America, mangrove swamps are found from the southern tip of Florida along the entire Gulf Coast to Texas; Florida's southwest coast supports one of the largest mangrove swamps in the world.

A huge diversity of animals is found in mangrove swamps. Because these estuarine swamps are constantly replenished with nutrients transported by freshwater runoff from the land, they support a bursting population of bacteria, other decomposers, and filter feeders. These ecosystems also sustain billions of worms, protozoa, barnacles, oysters, and other invertebrates, which in turn feed fish and shrimp, which support wading birds, pelicans, and, in the United States, the endangered crocodile.

The importance of mangrove swamps has been well established. They function as nurseries for shrimp and recreational fisheries, exporters of organic matter to adjacent coastal food chains, and enormous sources of nutrients valuable to plants, wildlife, and ecosystem function. Their physical stability also helps to prevent shoreline erosion, shielding inland areas from severe damage during hurricanes and tidal waves.

The World's Oceans

Before we get into our review of the world's oceans, let's consider another aquatic ecosystem (besides wetlands and estuaries) that's an important source of biodiversity. This one is a saltwater ecosystem. Certain landforms that lie off coastal shores are known as **barrier islands.** Because barrier islands are created by the buildup of deposited sediments, their boundaries are constantly shifting as water moves around them. These spits of land are generally the first hit by offshore storms, and they are important buffers for the shoreline behind them.

In tropical waters, a very particular type of barrier island called a **coral reef** is quite common. These barrier islands are formed not from the deposition of sediments, but from a community of living things. The organisms responsible for the creation of coral reefs are cnidarians, which secrete a hard, calciferous shell; these shells provide homes and shelter for an incredible diversity of species, but they are also extremely delicate and thus very vulnerable to physical stresses as well as changes in light intensity, water temperature, ocean depth, and pH. The increase in ocean temperatures and dissolved CO_2 due to climate change is resulting in more acidic waters resulting in coral bleaching. Coral bleaching occurs when acidic conditions cause the coral to expel the colorful algae which provided them with food.

Like freshwater bodies, oceans are divided into zones based on changes in light and temperature. Study the following terms, and know them cold for the test!

- **Coastal zone**: This zone consists of the ocean water closest to land. Usually it is defined as being between the shore and the end of the continental shelf (the edge of the tectonic plate). Life thrives here due to abundant sunlight and oxygen and the proximity of the sediment surface, allowing for varied niches. In addition, coastal zones border and extend into estuaries, beaches, and marshes, which have their own varied populations of organisms adapted to their conditions.

- **Euphotic zone:** The photic, upper layers of water. The euphotic zone is the warmest region of ocean water; this zone also has the highest levels of dissolved oxygen. Much like the limnetic zone in freshwater biomes, it supports algae as well as fish. Algae in marine biomes supply a large portion of the Earth's oxygen, and also take in carbon dioxide from the atmosphere. Since most red light is absorbed in the top 1 meter of water, and blue light doesn't usually penetrate deeper than 100 meters, photosynthesizers have adapted mechanisms to address the lack of visible light. How deep the euphotic zone extends depends on the turbidity of the water in a given area.

- **Bathyal zone:** The middle region; it is colder and darker and does not receive enough light to support photosynthesis, so the density of organisms that live there is less. It's difficult for many fish to live there because of the lack of nutrients, and those that do often lack eyes since there's so little sunlight. It is populated by organisms such as sponges and sea stars, as well as larger predators such as squid, octopus, sharks, and large whales.

- **Abyssal zone:** This is the deepest region of the ocean. This zone is marked by extremely cold temperatures and low levels of dissolved oxygen, but high levels of nutrients because of the decaying plant and animal matter that sinks down from the zones above. Without plants, the base level of the food chain in this aquatic zone is decomposers. Many of the creatures adapted to live here produce bioluminescence in order to attract prey or mates, and most are adapted to the cold, low oxygen, and intense pressure, using slower metabolisms and the ability to eat more when food is available to help them survive.

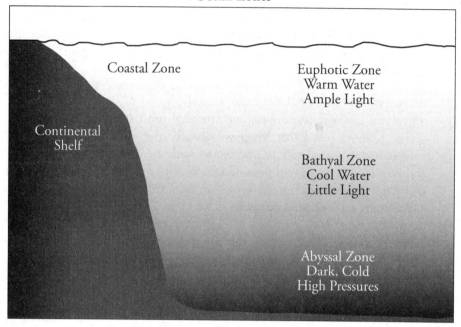

Ocean Zones

Coastal Zone

Euphotic Zone
Warm Water
Ample Light

Continental Shelf

Bathyal Zone
Cool Water
Little Light

Abyssal Zone
Dark, Cold
High Pressures

Both freshwater and saltwater bodies experience a seasonal movement of water from the cold and nutrient-rich bottom to the surface. These **upwellings** provide a new nutrient supply for the growth of living organisms in the photic regions. Therefore, they are followed by an almost immediate exponential growth in the population of organisms in these zones, especially the single-cell algae, which may form blooms of color called algal blooms. These algae can also produce toxins that may kill fish and poison the beds of filter feeders such as oysters and mussels. One notorious recurring toxic algal bloom is referred to as **red tide**; this is caused by a proliferation of dinoflagellates.

Water is densest at 3.98°C, or 39°F. In non-tropical regions of the Earth, after ice melts in spring, the water-surface temperature of lakes and ponds will rise from 0°C to 4°C, whereupon this dense surface water will sink to the bottom of the lake or pond. This will displace water at the bottom of the lake or pond to the surface. This overturn brings oxygen to the bottom and nutrients to the top of the lake or pond and occurs during spring and fall as the temperature of the ecosystem changes from cold to warm or the reverse.

Keep in mind that the worldwide distribution of biomes is dynamic; the distribution of both biomes and aquatic life zones across specific places on Earth has changed in the past and may again shift as a result of global climate changes.

Now that we've outlined what the major biomes and aquatic life zones look like, let's take some time to discuss the abiotic elements of ecosystems: specifically, the elements that bridge the gap between the nonliving and the living—water, nitrogen, carbon, and phosphorus—and how they cycle through the environment, bringing the key components that ecosystems need to function and burgeon with life.

CYCLES IN NATURE

As you may have learned in your biology class, nutrients such as carbon, oxygen, nitrogen, phosphorus, sulfur, and water all move through the environment in complex cycles known as **biogeochemical cycles**. Well, you'll need to know a bit about these cycles for the exam, so we'll go through each of them here.

For the AP Environmental Science Exam, it won't be enough for you to know that water moves from the atmosphere to the soil. You'll need to know the different ways it has of getting there.

As you can probably tell from the collective name of these natural cycles, living organisms, geologic formations, and chemical substances are all involved in these cycles. Keep in mind that when we describe the movement of these inorganic compounds, it's important to understand both the destinations of the compounds and how they move toward their destinations. In other words, you'll need to know that water moves from the atmosphere to the Earth's surface through precipitation, either in the form of snow or rainfall.

But let's talk about a few things that all of these cycles have in common before we go into each one in detail. First of all, the term **reservoir** is used to describe a place where a large quantity of a nutrient sits for a long period of time (in the water cycle, the ocean is an example of a reservoir). The opposite of a reservoir is an **exchange pool,** which is a site where a nutrient sits for only a short period of time (in the water cycle, a cloud is an example of an exchange pool). The amount of time a nutrient spends in a reservoir or an exchange pool is called its **residency time.** In the water cycle, water might exist in the form of a cloud for a few days, but it might exist as part of the ocean for a thousand years! Perhaps surprisingly, living organisms can also serve as exchange pools and reservoirs for certain nutrients; we'll delve into more about this later.

The energy that drives these biogeochemical cycles in the biosphere comes primarily from two sources: the Sun, and the heat energy from the mantle and core of the Earth. The movements of nutrients in all of these cycles may occur via abiotic mechanisms, such as wind, or may occur through biotic mechanisms, such as through living organisms (as we mentioned earlier).

Another important fact to note is that while the **Law of Conservation of Matter** states that matter can neither be created nor destroyed, nutrients can be rendered unavailable for cycling through certain processes—for example, in some cycles, nutrients may be transported to deep ocean sediments where they are locked away interminably.

Though we won't get into a discussion of trace elements here, you should also know that certain trace elements such as zinc, copper, and iron are necessary in small amounts for living organisms. Trace elements can cycle in conjunction with the major nutrients, but there's still much to be discovered about these elements and their biogeochemical cycles. For this exam, just know that there are certain trace elements required by living things that cycle, along with the major elements, through the biosphere.

Let's start with perhaps the best-known biogeochemical cycle: the water cycle.

The Water Cycle

As you might imagine, the water that exists in the atmosphere is in a gaseous state, and when it condenses from the gaseous state to form a liquid or solid, it becomes dense enough to fall to the Earth because of the pull of gravity. This process is formally known as **precipitation.** When precipitation falls onto the Earth, it may infiltrate the surface and percolate through soil and rock until it reaches the water table to become **groundwater,** or it may travel across the land's surface as **runoff** and enter a drainage system, such as a stream or river, which will eventually deposit it into a body of water such as a lake or an ocean. Lakes and oceans are reservoirs for water. In certain cold regions of the Earth, water may also be trapped on the Earth's surface as snow or ice; in these areas, the blocks of snow or ice are reservoirs.

Water is also cycled through living systems. For example, plants consume water (and carbon dioxide) in the process of photosynthesis, in which they produce carbohydrates and oxygen. Because all living organisms are primarily made up of water, they act as exchange pools for water.

Water is returned to the atmosphere from both the Earth's surface and from living organisms in a process called **evaporation.** Specifically, animals respire and release water vapor and additional gases to the atmosphere. In plants, the process of **transpiration** releases large amounts of water into the air. Finally, other major contributors to atmospheric water are the vast number of lakes and oceans on the Earth's surface. Incredibly large amounts of water continually evaporate from their surfaces.

Take a look at the following graphic, which shows all of the forms that water takes in the biosphere and atmosphere.

The Water Cycle

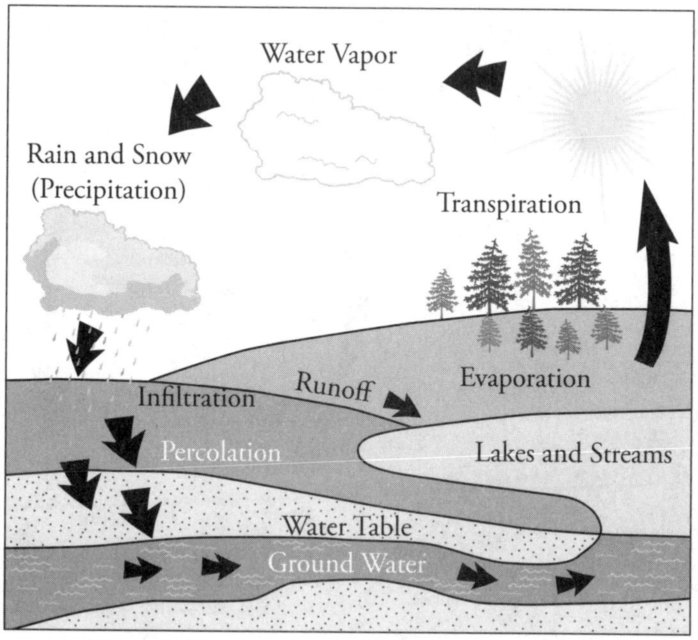

Fast Fact
Cloud formation results from condensation when water vapor becomes liquid or solid.

The Carbon Cycle

Now, let's talk about carbon. The key events in the carbon cycle are **respiration,** in which animals (and plants!) breathe in oxygen and give off carbon dioxide, and **photosynthesis,** in which plants take in carbon dioxide, water, and energy from the sun to produce carbohydrates. In other words, living things act as exchange pools for carbon.

When plants are eaten by animal consumers, the carbon locked in the plant carbohydrates passes to other organisms and continues through the food chain (more on this later in the chapter). In turn, when organisms—both plants and animals—die, their bodies are decomposed through the actions of bacteria and fungi in the soil; this releases CO_2 back into the atmosphere.

One aspect of the carbon cycle that you should definitely be familiar with for the exam is this: when the bodies of once-living organisms are buried deep and subjected to conditions of extreme heat and extreme pressure, this organic matter eventually becomes oil, coal, and gas. Oil, coal, and natural gas are collectively known as fossil fuels, and when fossil fuels are burned, or **combusted,** carbon is released into the atmosphere. Finally, carbon is also released into the atmosphere through volcanic action.

There are three major reservoirs of carbon: the first is the world's oceans, because CO_2 is very soluble in water. The second large reservoir of CO_2 is the Earth's rocks. Many types of rocks—called carbonate rocks—contain carbon in the form of calcium carbonate. Finally, fossil fuels are a huge reservoir of carbon.

The Carbon Cycle

The Nitrogen Cycle

The Earth's atmosphere is made up of approximately 78 percent nitrogen and 21 percent oxygen. (The other components of the atmosphere are trace elements and the greenhouse gases.) Nitrogen is the most abundant element in the atmosphere. For this reason, it might not seem like living organisms would find it difficult to get the nitrogen they need in order to live. But it is! This is because atmospheric N_2 is not in a form that can be used directly by most organisms. Thus, while the atmosphere is the major reservoir in this cycle, a step is needed to convert that nitrogen into usable forms. Most of the other reservoirs in the nitrogen cycle hold nitrogen compounds for relatively short periods of time. In order to keep this rather complicated cycle straight, let's look at it in steps.

Step 1: Nitrogen fixation—In order to be used by most living organisms, nitrogen must be present in the form of ammonia (NH_3) or nitrates (NO_3^-). Atmospheric nitrogen can be converted into these forms, or "fixed," by atmospheric effects such as lightning storms, but most nitrogen fixation is the result of the actions of certain soil bacteria. Fixing is the process that allows nitrogen to be made biologically available, much as photosynthesis makes carbon biologically available. One important soil bacteria that participates in nitrogen fixation is *Rhizobium*. These nitrogen-fixing bacteria are often associated with the roots of legumes such as beans or clover. In the future, we may be able to insert the genes for nitrogen fixation into crop plants, such as corn, and reduce the amount of fertilizer that is used.

Step 2: Nitrification—In this process, soil bacteria converts ammonia (NH_3) or ammonium (NH_4^+) into nitrites (NO_2) and then to one of the forms that can be used by plants—nitrate (NO_3^-).

Step 3: Assimilation—In assimilation, plants absorb ammonium (NH_3), ammonia ions (NH_4^+), and nitrate ions (NO_3^-) through their roots. Heterotrophs, or organisms that receive energy by consuming other organisms, then obtain nitrogen when they consume plants' proteins and nucleic acids.

Step 4: Ammonification—In this process, decomposing bacteria convert dead organisms and other waste to ammonia (NH_3) or ammonium ions (NH_4^+), which can be reused by plants or volatilized (released into the atmosphere).

Step 5: Denitrification—In denitrification, specialized bacteria (mostly anaerobic bacteria) convert ammonia back into nitrites and nitrates, and then into nitrogen gas (N_2) and nitrous oxide gas (N_2O). These gases then rise to the atmosphere.

The Nitrogen Cycle

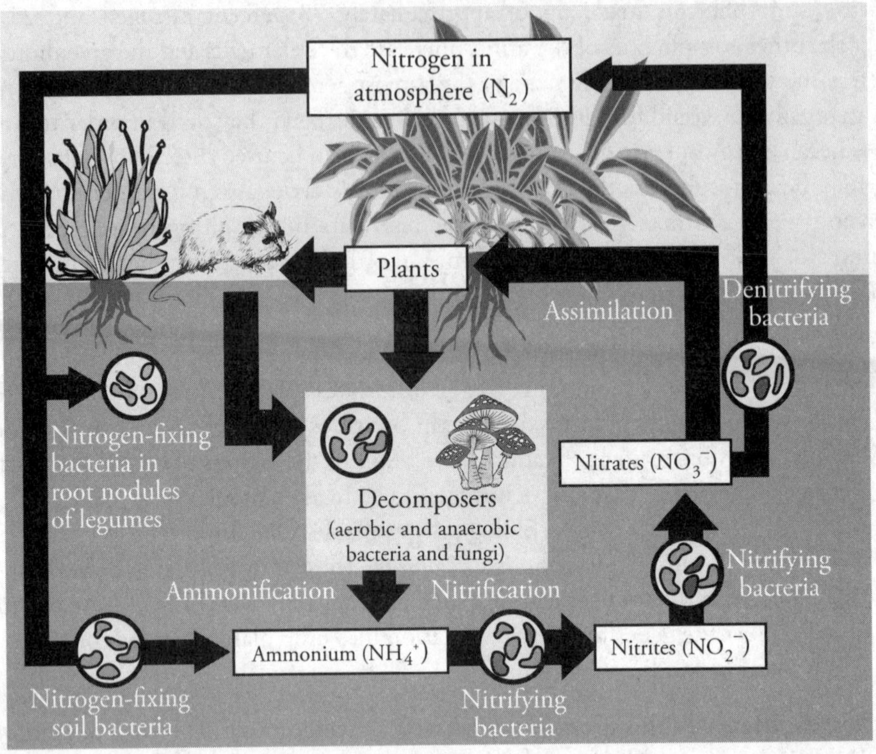

The Phosphorus Cycle

The **phosphorus cycle** is perhaps the simplest biogeochemical cycle, mostly because phosphorus does not exist in the atmosphere outside of dust particles. Phosphorus is necessary for living organisms because it's a major component of nucleic acids, ATP (cellular energy), cell membranes, and other important biological molecules. One important idea for you to remember about the phosphorus cycle is that phosphorus cycles are more local than those of the other important biological compounds.

For the most part, phosphorus is found in soil, rock, and sediments; it's released from these rock forms through the process of chemical weathering. Phosphorus is usually released in the form of phosphate (PO_4^{3-}), which is soluble and can be absorbed from the soil by plants. Symbiotic relationships that form between fungi and plants are known as *mycorrhizae*. In these relationships, *mycorrhizal* fungi colonize the root system of a host plant, which increases the water and nutrient absorption capabilities of the plant, while the plant provides the fungi with carbohydrates formed from photosynthesis. You should know that phosphorus is often a **limiting factor** (any factor that controls a population's growth—food, space, water) for plant growth, so plants that have little phosphorus are stunted.

Phosphates that enter the water table and travel to the oceans can eventually be incorporated into rocks in the ocean floor. Through geologic processes, ocean mixing, and upwelling, these rocks from the seafloor may rise up so that their components once again enter the **terrestrial cycle.** Take a look at the phosphorus cycle shown in the diagram.

Humans have affected the phosphorus cycle by mining phosphorus-rich rocks in order to produce fertilizers. The fertilizers placed on fields can easily leach into the groundwater and find their way into aquatic ecosystems where they can cause eutrophication. **Eutrophication** occurs when a body of water receives excess nutrients. The abundance of nutrients can cause an overgrowth of algae and deplete the water of oxygen.

The Phosphorous Cycle

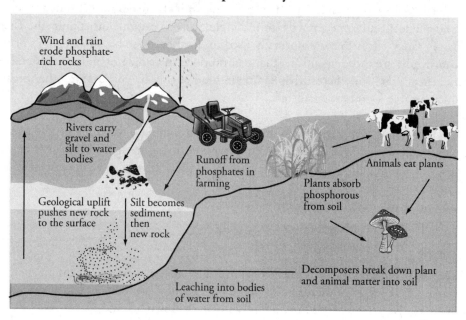

We are almost done with the chemistry. But we need to discuss one more element before we move on to discuss the biosphere, and that's sulfur.

Sulfur

The last biogeochemical cycle we'll talk about is the **sulfur cycle.** Sulfur is one of the components that make up proteins and vitamins, so plants and animals both need sulfur in their diets. Plants absorb sulfur when it is dissolved in water, so they can take it up through their roots when it's dissolved in groundwater. Animals obtain sulfur by consuming plants.

Most of the Earth's sulfur is tied up in rocks and salts or buried deep in the ocean in oceanic sediments, but some sulfur can be found in the atmosphere. The natural ways that sulfur enters the atmosphere are through volcanic eruptions, certain bacterial functions, decomposition in estuaries, and the decay of once-living organisms. When sulfur enters the atmosphere through human activity, it's mainly via industrial processes that produce sulfur dioxide (SO_2) and hydrogen sulfide (H_2S) gases. We'll talk more about sulfur and how it contributes to air pollution in Chapter 9.

FOOD CHAINS AND FOOD WEBS

Now that we've reviewed the abiotic elements essential to ecosystems, it's time to begin our study of the living, biotic components of the Earth. Together, all of the living things on Earth constitute the biosphere.

All living things can be classified by how they obtain food. You might recall that plants and some cyanobacteria are capable of making their own food through photosynthesis, and that some animals (for example, mice) eat plants. Some animals (for example, humans) eat both plants and animals, and some animals (for example, wolves) eat only other animals. There are actually two fancy terms that are normally used to describe these broad categories of organisms: **autotrophs** are those organisms that can produce their own organic compounds from inorganic chemicals, while **heterotrophs** obtain food energy by consuming other organisms or products created by other organisms.

Finally, as unpleasant as it might be to think about, some animals feed only on the remains of other plants and animals! All of these different types of living things fall into specific categories—and you will definitely need to memorize all of these terms before the test, if you don't already know them!

Producers

Producers are organisms that are capable of converting radiant energy, or chemical energy, into carbohydrates. The group of producers includes plants and algae, both of which can carry out photosynthesis. The overall reaction of photosynthesis is shown below.

$$12H_2O + 6CO_2 + \text{solar energy} \longrightarrow C_6H_{12}O_6 + 6O_2 + 6H_2O$$

While most producers make food through photosynthesis, a few autotrophs make food from inorganic chemicals in **anaerobic** (without oxygen) environments, through the process of chemosynthesis. Chemosynthesis is only carried out by a few specialized bacteria, called **chemotrophs,** some of which are found in hydrothermal vents deep in the ocean. This unbalanced reaction is shown below.

$$CO_2 + 4H_2S + O_2 \longrightarrow CH_2O + 4S + 3H_2O$$

At this point, let's discuss a few other environmental science terms that you'll be required to know for the exam. The **Net Primary Productivity** (**NPP**) is the amount of energy that plants pass on to the community of herbivores in an ecosystem. It is calculated by taking the **Gross Primary Productivity,** which is the amount of sugar that the plants produce in photosynthesis, and subtracting from it the amount of energy the plants need for growth, maintenance, repair, and reproduction. NPP is measured in kilocalories per square meter per year ($kcal/m^2/y$). In other words, the Gross Primary Productivity of an ecosystem is the rate at which the producers are converting solar energy to chemical energy (or, in a hydrothermal ecosystem, the rate of productivity of the chemotrophs). Perhaps not surprisingly, the net productivity of an ecosystem is a limiting factor for its number of consumers. A limiting factor is a factor that controls a population's growth. It can be many things: space, available food, water, nutrients, and as we just mentioned, the net productivity of an ecosystem.

Consumers

Consumers are organisms that must obtain food energy from secondary sources, for example, by eating plant or animal matter. There are a number of different types of consumers, all of which you should commit to memory!

- **Primary consumers:** This category includes the herbivores, which consume only producers (plants and algae).

- **Secondary consumers:** Organisms that consume primary consumers are secondary consumers.

- **Tertiary consumers:** Organisms that consume secondary consumers are tertiary consumers.

- **Detritivores:** The organisms in this group derive energy from consuming nonliving organic matter such as dead animals or fallen leaves. They include termites, earthworms, and crabs.

- **Decomposers:** These are organisms that consume dead plant and animal material. The process of decomposition returns nutrients to the environment.

- **Saprotrophs:** These are decomposers that use enzymes to break down dead organisms and absorb the nutrients; they include bacteria and fungi.

Note that one organism may occupy multiple levels of a food chain. When eating a hamburger with toppings, you are a primary consumer because you are eating tomatoes and lettuce, and a secondary consumer by eating the beef.

Let's move on and talk about how energy flows through all of these different types of organisms in ecosystems.

Food Chains

As you probably recall, energy flows in one direction through ecosystems: from the Sun to producers, to primary consumers, to secondary consumers, to tertiary consumers. In an ecosystem, each of these feeding levels is referred to as a **trophic level**. With each successive trophic level, the amount of energy that's available to the next level decreases. In fact, the laws of thermodynamics dictate that only about 10 percent of the energy from one trophic level is passed to the next; most is lost as heat, and some is used for metabolism and anabolism. Interestingly enough, this is why food chains rarely have more than four trophic levels.

Food chains are usually represented as a series of steps, in which the bottom step is the producer and the top step is a secondary or tertiary consumer. In food chains, the arrows depict the transfer of energy through the levels, and in fancier food chains, the relative biomass (the dry weight of the group of organisms) of each trophic level will often be represented.

Here's a simple food chain.

Food Chain

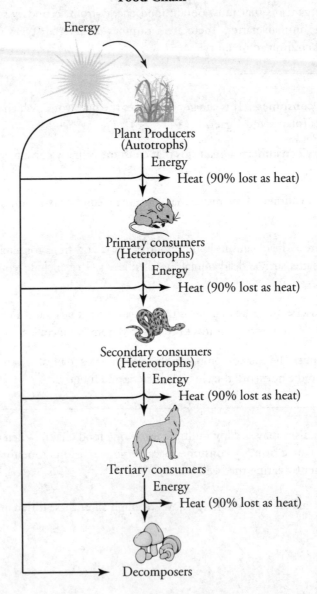

What we're showing here is a typical terrestrial food chain, but keep in mind that there are aquatic food chains as well, with algae and different types of fish.

One final note about food chains: in a food chain, only about 10% of the energy is transferred from one level to the next. The other 90% is used for things like respiration, digestion, running away from predators—that is, it's used to power the organism doing the eating! This is known as the **10% Rule**. In other words, the producers have the most energy in an ecosystem; the primary consumers have less energy than producers; the secondary consumers have less energy than the primary consumers; and the tertiary consumers will have the least energy of all. The amount of energy (in kilocalories) available at each trophic level organized from greatest to least is an **energy pyramid**.

Energy Pyramid

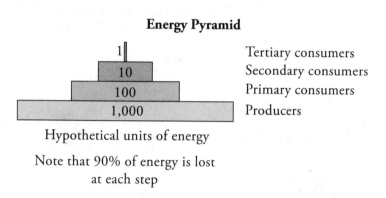

1	Tertiary consumers
10	Secondary consumers
100	Primary consumers
1,000	Producers

Hypothetical units of energy

Note that 90% of energy is lost at each step

Consider the following example that shows an energy pyramid for an aquatic food chain. In this example, the producers, or phytoplankton, start with 100,000 grams of energy, but only 10,000 grams of energy are transferred to the primary consumers, or zooplankton. With each step up the food chain, 90% of energy is lost and only 10% is transferred.

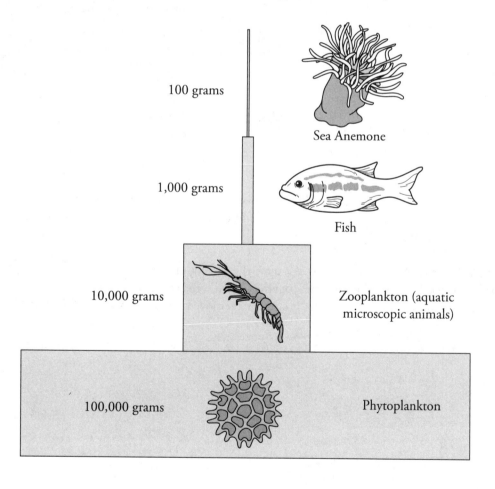

100 grams — Sea Anemone

1,000 grams — Fish

10,000 grams — Zooplankton (aquatic microscopic animals)

100,000 grams — Phytoplankton

Food Webs—Tangled Food Chains

As you're probably already aware, food chains are an oversimplified way of demonstrating the myriad feeding relationships that exist in ecosystems. Because there are so many different types of species of plants and animals in ecosystems, their relationships in real-world ecosystems are much more complicated than can be depicted in a single food chain. Therefore, we use a **food web** in order to represent feeding relationships in ecosystems more realistically.

Food Web

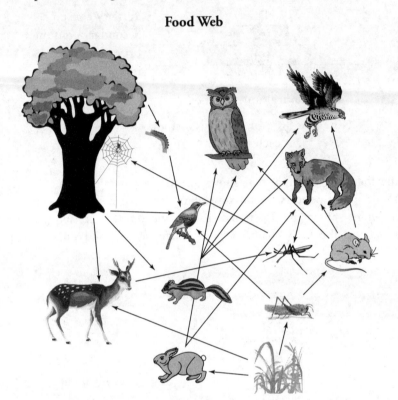

Again, this is a typical terrestrial food web, but keep in mind that very complicated aquatic food webs exist as well! Let's take a step back for a minute and discuss the setting for food chains and food webs—ecosystems.

When something changes in an ecosystem, the effects of that change can quickly spread, partly because food webs link species together. Food webs contain positive and negative feedback loops, so that when one species is added or removed, the rest of the food web is affected, sometimes drastically.

BIODIVERSITY

The term **biodiversity** is used to describe the number and variety of organisms found within a specified geographic region, or ecosystem. It also refers to the variability among living organisms, including the variability within and between species and within and between ecosystems. Therefore, when we talk about the biodiversity of an area, we must specify the aspect of biodiversity that we're describing, or else the term is too vague to be comprehensible. **Species richness** refers to the number of different species found in an ecosystem. In general, however, biodiversity in an ecosystem is a good thing. The more biodiversity in a certain species within an ecosystem, the larger and more diverse the species' gene pool, and the greater its chance of adaptation and thus survival.

What happens when an ecosystem is threatened or habitat is lost? This habitat loss tends to lead to a loss of specialist species, which then can lead to a loss of generalist species. Additionally, species that require large territories tend to suffer, and their numbers are reduced. When a large proportion of a population is lost, this leads to a bottleneck, which can reduce genetic diversity within the species. A disturbance in an ecosystem will affect the total biomass, species richness, and net productivity over time. Ecosystems with larger numbers of species (species richness) tend to recover more easily from disruptions.

This is a reflection of the fact that the more genetically diverse a given population is, the better it can cope with an environmental disturbance. So you can see that ecosystems and biodiversity are inextricably linked.

One important law to be familiar with for this test is the **Law of Tolerance.** The Law of Tolerance describes the degree to which living organisms are capable of tolerating changes in their environment. Living organisms exhibit a range of tolerance, and even individuals within a population tolerate changes to their environment differently. This concept is the basis for natural selection, which drives evolution.

Another important law for you to know is the **Law of the Minimum,** which states that living organisms will continue to live, consuming available materials until the supply of these materials is exhausted.

One special case of the interrelatedness of habitat, biodiversity, and adaptation is the **theory of island biogeography**, which is a field that studies species richness and diversification in isolated communities: oceanic islands, and also other isolated ecosystems, such as mountain peaks, oases, seamounts, and fragments of habitat separated by human development. The number of species found on an island or in an isolated area is determined by two factors: immigration and extinction. Since islands are often colonized by new species arriving from elsewhere, immigration is a main factor. Once established on an island or in an isolated ecosystem, many species evolve to become specialists as an adaptation to the limited resources available. Immigration becomes a factor again when invasive species arrive, since the typically-generalist invasives may outcompete the native specialists and threaten their long-term survival.

In addition to understanding how ecosystems and biodiversity function together and how disturbances affect them, you should also have an understanding of how the complex systems at play in the natural world benefit humans.

ECOSYSTEM SERVICES

Ecosystem services are benefits that humans receive from the ecosystems in nature when they function properly. There are four categories: **provisioning services:** providing humans with water, food, medicinal resources, raw materials, energy, and ornaments; **regulating services:** waste decomposition and detoxification, purification of water and air, pest and disease control and regulation of prey populations through predation, and carbon sequestration; **cultural services:** use of nature for science and education, therapeutic and recreational uses, and spiritual and cultural uses; and **supporting services** (the ones that make other services possible): primary production, nutrient recycling, soil formation, and pollination.

Human disruptions to ecosystem services can detrimentally affect our ability to benefit from them, resulting in ecological and economic consequences for us.

HOW ECOSYSTEMS CHANGE

Believe it or not, oftentimes the biotic balance in a community is maintained by a single species, known as the **keystone species.** The name *keystone* comes from the last stone placed in an arch bridge, which is the key stone. A keystone species is a species whose very presence contributes to an ecosystem's diversity and whose extinction would consequently lead to the extinction of other forms of life. For example, fig trees are the keystone species in a tropical forest; likewise, wolves were introduced back into Yellowstone Park because without wolves to control the number of herbivores, the ecosystem had drastically changed. As a general rule, if the keystone species is removed from an ecosystem, then the ecosystem completely changes.

Indicator species are species that are used as a standard to evaluate the health of an ecosystem. They are more sensitive to biological changes within their ecosystems than are other species, so they can be used as an early warning system to detect dangerous changes to a community. Trout are a common indicator species, because they are particularly sensitive to pollutants in water. The disappearance of trout from a particular habitat is a warning that that habitat is becoming polluted.

Indigenous species are those that originate and live or occur naturally in an area or environment. With increasing frequency, however, new species are being introduced into ecosystems by chance, by accident, or with intention. While some introduced species cannot find a niche and die out, many others are quite happy in their new environment, and compete successfully with the indigenous species. One example of this is gray squirrels, which were introduced to England in 1876. The gray squirrel competed with England's native species of squirrel, the red squirrel, and today there are fewer than 30,000 red squirrels alive in England. Another example of the harm that introduced species can do was seen when, in 1904, a fungus was introduced accidentally into the deciduous forests of the eastern United States. This fungus caused a blight that killed nearly all of the chestnut trees by the early 1950s.

Although some people don't like to use the term **invasive species** because they feel that it's derogatory, it is often used to describe introduced species. Two other examples of invasive species are zebra mussels, which were introduced into the Great Lakes when ships dumped ballast water into the lakes, and the quickly growing vine kudzu, which was originally introduced in the southeastern United States in order to control the problem of erosion.

Ecological Succession

Communities are not static; they are constantly changing. Species of plants and animals are continually coming and going, evolving and dying out. Some of the changes that take place in a geographic area are predictable ones that can be described as **ecological succession.**

If ecological succession begins in a virtually lifeless area, such as the area below a retreating glacier, it is called **primary succession. Secondary succession** is ecological succession that takes place where an existing community has been cleared (by disturbance events such as fire, tornado, or human impact), but the soil has been left intact. Succession in a disturbed ecosystem will affect the total biomass, species richness, and net productivity over time. The organisms in the first stages of either type of succession are referred to as **pioneer species**, and typically have wide ranges of environmental tolerance. These pioneers, over time, usually adapt to the particular conditions of the habitat. This may result in the origin of new species. The communities in each stage of succession facilitate the environmental changes that will allow the next stage to take over. The final stage of succession, in which there is a dynamic balance between the abiotic and biotic components of the community, is referred to as the **climax community.**

How does a new habitat full of bare rocks eventually turn into a forest? The first stage of the job usually falls to a community of lichens. Lichens are hardy organisms. They can invade an area, land on bare rocks and erode the rock surface, and over time turn them into soil. Lichens are pioneer organisms. Once lichens have made an area more habitable, other organisms can settle in. Lichens are replaced (out-competed!) by mosses and ferns, which in turn are replaced by tough grasses, then low shrubs, then conifers, then short-lived hardwood trees such as dogwood and red maple trees, and finally long-lived hardwood trees. Refer to the flowchart of ecological succession for a deciduous forest. Note that the stages are classified by the major new plant group, but remember that with the introduction of each new plant species comes an array of different animal species that exploit it.

Ecological Succession

Bare rock
↓
Lichen, Algae, Mosses, Bacteria
(Break down rock and leave organic debris which together form soil)
↓
Grasses
(Add organic matter to soil and anchor it in place)
↓
Small herbaceous plants
(Continue to add organic matter to soil)
↓
Small bushes
(Add shelter and shade for other plants)
↓
Conifers
(Create additional habitats)
↓
Short-lived hardwoods such as dogwood and red maple
(Can tolerate shade of conifers but are short-lived and vulnerable to damage)
↓
Long-lived hardwoods
(More specialized, hardier hardwoods such as oak and hickory)

When the size of an organism's natural habitat is reduced, or when, for example, development occurs that isolates the habitat, this process is called **habitat fragmentation.** Habitat fragmentation can be quite damaging. As you know, ecosystems are not isolated; they abut each other and meet at wide and overlapping boundaries, called ecotones. At these boundaries, there is greater species diversity and biological density than there is in the heart of ecological communities, and this is called the **edge effect.** Some species can only live on the edge of certain habitats, and if the boundaries of a habitat are changed, a new edge is created, damaging both the edge and interior habitats.

When the changes taking place in a geographic area result from less-predictable events and have more drastic consequences for an ecosystem, they may be considered disruptions. Human-made disruptions such as pollution, habitat destruction, and depletion of natural resources will be a major focus in the coming chapters. But what about natural disruptions? Some natural disruptions to ecosystems have environmental consequences that can exceed those caused by humans. The Earth's climate has changed, over the course of geological time, many times and for varied reasons, including internal causes (changes in the type and distribution of species and the effects they produce on climate and changes in ocean-atmosphere circulations) and external ones (changes in the Earth's orbit, solar output, volcanism, plate tectonics and the resulting size and configuration of continents, and asteroid impacts). Some of these factors cause periodic or episodic change, while others cause change at random times, and the timescales involved vary greatly.

When global climate change occurs, the effects are far-reaching. For example, changes in the amount of glacial ice on Earth have caused sea levels to vary quite significantly over geological history, which in turn changes the size and shapes of landmasses. When big changes occur in the environment, habitats can change on enormous scales, which in turn can cause extinctions, bottlenecks, and short- and long-term migrations among the species inhabiting the affected ecosystems.

Whew. You're done with this chapter! Before moving on to the next chapter (Populations), answer all of the questions in the drill following the key terms list—and don't forget to use the techniques you learned in Part III.

CHAPTER 4 KEY TERMS

Make sure you know these words and how to use them in your essays.

Ecosystems
ecosystem
biotic
abiotic

Evolution
evolution
phylogenetic tree
speciation
species
evolutionary fitness
natural selection
selective pressure
gene pool
genetic drift
microevolution
macroevolution
extinction
biological extinction
ecological extinction
commercial/economic extinction
population
community
niche
habitat
specialist
generalist
competition
intraspecific competition
interspecific competition
competitive exclusion
Gause's principle
realized niche
fundamental niche
resource partitioning
predation
predator
prey
symbiotic relationships
mutualism
commensalism
parasitism

The World's Ecosystems
biomes
aquatic life zones
ecotones
ecozones/ecoregions
deciduous forest
tropical rainforest
grasslands
coniferous forest/taiga
tundra
chaparral
deserts
temperate rainforest
savanna
epilimnion
hypolimnion
thermocline
littoral zone
limnetic zone
profundal zone
benthic zone
estuary
wetlands
mangrove swamps
barrier islands
coral reef
coastal zone
euphotic zone
bathyal zone
abyssal zone
upwellings
red tide

Cycles in Nature
biogeochemical cycles
reservoir
exchange pool
residency time
Law of Conservation of Matter
precipitation
groundwater
runoff
evaporation
transpiration
respiration
photosynthesis

Want Printable Lists?
Head over to your Student Tools (your Princeton Review online companion for this book), where you can find printable versions of all Key Terms lists.

combusted
nitrogen fixation
nitrification
assimilation
ammonification
denitrification
phosphorus cycle
limiting factor
terrestrial cycle
eutrophication
sulfur cycle

Food Chains and Food Webs
autotrophs
heterotrophs
producers
anaerobic
chemotrophs
Net Primary Productivity (NPP)
Gross Primary Productivity
consumers
primary consumers
secondary consumers
tertiary consumers
detritivores
decomposers
saprotrophs
trophic level
food chains
10% Rule
energy pyramid
food web

Biodiversity
biodiversity
species richness
Law of Tolerance
Law of the Minimum
theory of island biogeography

Ecosystem Services
ecosystem services
provisioning services
regulating services
cultural services
supporting services

How Ecosystems Change
keystone species
indicator species
indigenous species
invasive species
ecological succession
primary succession
secondary succession
pioneer species
climax community
habitat fragmentation
edge effect

Chapter 4 Drill

Directions: Each of the questions or incomplete statements below is followed by four suggested answers or completions. Select the one that is best in each case. For answers and explanations, see Chapter 13.

1. Natural selection

   I. occurs when a habitat selects certain organisms to live and reproduce, while allowing others to die

   II. is driven by selective pressures that reduce reproductive success in a portion of the population

   III. acts upon a whole population, rather than an individual organism

   (A) I only

   (B) III only

   (C) I and III only

   (D) I, II, and III

2. Which ecosystem receives between 10–30 cm of rain, has porous soil that contains only a thin layer of humus and vegetation consisting of grasses with widely spaced trees, and is located in Australia, South America, India, and parts of Africa?

   (A) Savanna

   (B) Grassland

   (C) Chapparal

   (D) Desert

3. The relationship between a sea anemone and a clown fish is an example of

   (A) predation

   (B) mutualism

   (C) commensalism

   (D) parasitism

4. Which of the following zones is characterized by shallow water at the shoreline and is home to turtles, frogs, and other species that travel back and forth between water and land?

   (A) Littoral zone

   (B) Limnetic zone

   (C) Profundal zone

   (D) Benthic zone

5. The bathyal zone is characterized by

   (A) abundant sunlight and high levels of dissolved oxygen

   (B) warm waters and the highest levels of dissolved oxygen

   (C) cool waters and little sunlight

   (D) cold waters and no sunlight

6. Plants release large amounts of water into the air through the process of

   (A) precipitation

   (B) transpiration

   (C) evaporation

   (D) infiltration

7. A plant that experiences stunted growth is likely lacking sufficient

   (A) nitrogen

   (B) sulfur

   (C) carbon

   (D) phosphorous

8. An earthworm that consumes nonliving organic matter is known as a

   (A) autotroph

   (B) decomposer

   (C) detritivore

   (D) saprotroph

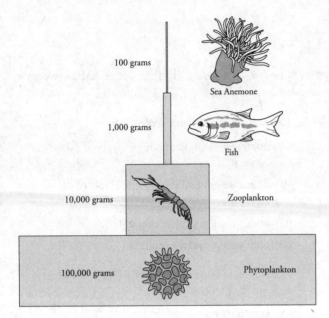

100 grams   Sea Anemone

1,000 grams   Fish

10,000 grams   Zooplankton

100,000 grams   Phytoplankton

10. Which of the following would NOT be considered a primary consumer?

(A) Frog

(B) Caterpillar

(C) Bee

(D) Rabbit

11. All of the following are factors that affect a population's birth rate EXCEPT the population's

(A) base level education for women

(B) religious beliefs and culture

(C) standard hygiene practices

(D) demand for children in the workforce

9. The image above depicts

(A) a food chain

(B) an energy pyramid

(C) a food web

(D) ecological succession

Free-Response Question

1. Scientists designed an experiment to learn about the functioning of the hydrologic cycle and the phosphorus cycle in a forest. Using two areas of the same size and geologic features, they cut all the trees down from one plot and did not disturb the other plot. They were able to accurately measure the amount of water that flowed out of the two plots as well as measure the amounts of phosphorus found in the runoff.

   (a) **Describe** what the differences would be in the volume of water running off the two plots and **explain** ONE reason why. Assume that the two areas received the same amounts of precipitation.

   (b) **Describe** the differences in the levels of phosphate found in the runoff of the two plots. Assume that both plots started off with the same amount of phosphorus in the soil.

   (c) **Identify** one negative effect that might occur in a stream that receives the runoff water and sediment.

   (d) When a tropical rain forest is cut down and used as farmland, the fertility of the soil only lasts a few years. **Explain** why there is little organic matter in rain forest soil and what would happen to that material after deforestation.

Summary

o An ecosystem is a system of interconnected elements, which includes both biotic (living) and abiotic (nonliving) components.

o The biodiversity on Earth is the result of speciation and natural selection of traits. Extinction allows for niches to become available and new traits or species to evolve to use those niches.

o Organization of living things can be thought of as follows:

Biosphere $\longrightarrow$ Ecosystems $\longrightarrow$ Communities $\longrightarrow$ Populations (of species with unique niches)

o Species relationships within a community include the following:
  • competition
  • predation
  • parasitism
  • resource partitioning
  • commensalism
  • mutualism

o There are several major biomes on Earth (deciduous forest, tropical rainforest, grasslands, coniferous forest/taiga, tundra, chaparral, deserts, temperate rainforest, and savanna) and aquatic life zones that make up the biosphere. Terrestrial biomes are defined by their average rainfall and annual temperature. The species that survive in each zone have a specific range of tolerance. Ecotones are the regions where different biomes overlap.

o Freshwater biomes and ocean zones are stratified by temperature, light, oxygen content, and nutrient levels. Salinity and seasonal changes in temperature influence water density and thus vertical movement between the layers.

o The five major biogeochemical cycles you should know are:
  • water
  • carbon
  • nitrogen
  • phosphorus
  • sulfur

o Each cycle has both a natural cycle and anthropogenic influence.

o Food chains and webs are made up of the following different categories of species:
   • autotrophs
   • producers
   • detritivores
   • heterotrophs
   • consumers
   • decomposers

o All food chains move energy through the various trophic levels that make up the system.
   • All energy originates from the Sun.
   • The 10% Rule describes the movement of energy between the levels.

o Communities can have species with very specific roles, such as:
   • keystone species
   • indicator species
   • indigenous species

o Ecological succession describes how ecosystems recover after a disturbance in terms of a disturbance event (primary or secondary succession) and the stage of succession (pioneer species, mid-succession species, climax communities).

Chapter 5
Unit 3:
Populations

In this chapter, we'll review Unit 3 of the AP Environmental Science course, *Populations*. According to the College Board, about 10–15% of the test is based directly on the ideas covered in this chapter. If you are unfamiliar with a topic presented here, consult your textbook for more in-depth information.

We'll start by discussing some important characteristics of populations and then lead you through a section on how and why populations grow. Next, we'll get to the heart of the topic in a section specifically devoted to human population growth. We'll review statistics used to study human populations, age-structure diagrams, and the demographic transition model.

Remember to use the techniques you learned in Part III as you complete the drills—the more practice you have using those techniques, the better prepared you'll be on test day.

TERMS USED TO DESCRIBE POPULATIONS

A **population** is defined as a group of organisms of the same species that inhabits a defined geographic area at the same time. Individuals in a population generally breed with one another, rely on the same resources to live, and are influenced by the same factors in their environment.

Two important characteristics of populations are the density of the population and how the population is dispersed. **Population density** refers to the number of individuals of a population that inhabit a certain unit of land or water area. An example of population density would be the number of squirrels that inhabit a particular forest. **Population dispersion** is a little more complicated; this term refers to how individuals of a population are spaced within a region. There are three main ways in which populations of species can be dispersed, and you should know all of them for the test.

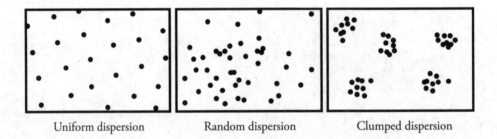

| Uniform dispersion | Random dispersion | Clumped dispersion |

- **Uniform:** The members of the population are uniformly spaced throughout their geographic region. This is seen in forests, in which trees are uniformly distributed so that each receives adequate light and water. Uniform dispersion is often the result of competition for resources in an ecosystem.

- **Random:** The position of each individual is not determined or influenced by the positions of the other members of the population. An example is seen in species of plants that are interspersed in fields or forests—the location of their growth is random and relative to other species, not their population. This type of dispersion is relatively uncommon.

- **Clumped:** The most common dispersion pattern for populations. In this type of dispersion, individuals "flock together." This makes sense for many species—many species of plants tend to grow together in a location or habitat that is near their parents and suits their requirements for life; fish swim in schools to avoid predation; and birds and many other animals migrate in groups.

POPULATION GROWTH

So, we know what populations are and how they're dispersed, but how do populations grow? What determines whether they will or will not grow? When populations do grow, what are the trends? These are all questions that you'll need to be able to answer on test day. Let's review the basics of population size and growth before we get into a more specific discussion of how human population growth occurs.

The **biotic potential** of a population is the amount that the population would grow if there were unlimited resources in its environment. This is not a practical model for population growth simply because in reality the amount of resources in the environments of populations is limited.

As we reviewed in the last chapter, in every ecosystem, members of a population compete for space, light, air, water, and food. The **carrying capacity** (K) of a particular species in a particular environment is defined as the maximum population size for the species that can sustainably be supported by the available resources in that environment. As you might expect, a given geographic region will have different carrying capacities for populations of different species—because different species have different requirements for life. For example, within a certain area, you would expect a population of bacteria to be quite a bit larger—in terms of the number of individuals—than a population of zebras. This is because individual bacteria are much smaller than individual zebras; thus, each bacterium requires fewer resources to live than each zebra. These differences in population size may be driven not just by the different sizes of individual organisms of each species, but by each species' resource requirements and the particular array of resources available in the area.

Population Growth Graphs

If we looked at the growth of a population of bacteria in a petri dish with plenty of food, the curve produced by plotting the increase in their number over time would be in the shape of a J, because the bacteria would grow exponentially. The exponential growth curve is shown below.

Exponential (Unrestricted) Growth

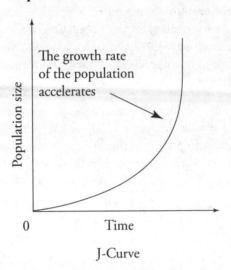

J-Curve

Now, we said that this **exponential population growth** rate is seen where resources are unlimited, but in nature, such ideal conditions are rare and fleeting. In reality, resource availability and the total resource base are limited and finite on any timescale. In a more realistic model for population growth, after the initial burst in population, the growth rate generally drops, and the curve ultimately resembles a flattened S.

This type of growth, which is a much better model for what exists in natural settings, is called **logistic population growth**. The logistic growth model basically says that when populations are well below the size dictated by the carrying capacity of the region they live in, they will grow exponentially, but as they approach the carrying capacity, the resource base of the population shrinks relative to the population itself. This leads to increased potential for unequal distribution of resources, which will ultimately result in increased mortality, decreased fecundity, or both. The result is that population growth declines to, or below, carrying capacity, and the size of the population will eventually become stable. This logistic growth is shown on the next page.

Logistic (Restricted) Growth

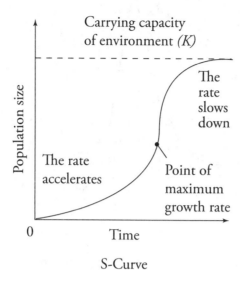

S-Curve

If the slowdown of population growth is the result of increased mortality rather than decreased fecundity, the reality tends to be a little messier than the word *slowdown* implies: it will probably involve **overshoot,** which occurs when a population exceeds its carrying capacity. There are environmental impacts of population overshoot, including resource depletion. If resource depletion is severe enough, the carrying capacity of the environment may be lowered. The severity of these effects varies, but resource depletion generally leads to dieback of the population, which can be severe to catastrophic, because the lack of available resources leads to famine, disease, and/or conflict. Once the dieback occurs, the population once again falls below carrying capacity; if the events were not too catastrophic, the environment can recover and the reduction in carrying capacity may not be permanent.

The graph on the following page shows some possible scenarios involving population overshoot. *K* (original) shows the initial carrying capacity of the environment, while *K* (reduced) indicates the new carrying capacity after some resource depletion. Three different results are shown: for population *c,* the resource depletion is not too intense, a small dieback occurs, and the population recovers. After some oscillation (small amounts of overshoot and dieback), the population becomes stable around the environment's carrying capacity.

For population *b,* the resource depletion is more severe; the environment's carrying capacity is permanently lowered. The dieback is correspondingly more severe as well, but once the environment has time to recover and the new carrying capacity is established, the population oscillates and becomes stable around the reduced carrying capacity. Population *a* represents the most catastrophic scenario: the population is unable to recover from its dieback and goes extinct.

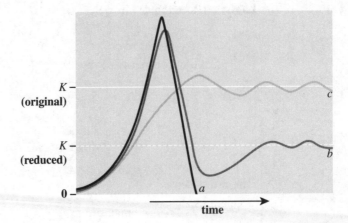

We can predict long-term population growth rates using a model called the Rule of 70. The **Rule of 70** says that the time it takes (in years) for a population to double can be approximated by dividing 70 by the current growth rate of the population. For example, if the growth rate of a population is 5 percent, then the population will double in 14 years ($\frac{70}{5} = 14$). The Rule of 70 can be used to estimate the number of years for any variable to double—the **doubling time.**

Not surprisingly, the rate of growth of a population depends on the species. Recall from Chapter 4 that a species can be either generalist (having a broad niche, highly adaptable, and able to live in varied habitats) or specialist (having a narrow niche and only able to live in a certain habitat). Specialist species tend to have an advantage when their environments are relatively unchanging, while generalist species have the advantage in habitats that undergo frequent change.

Likewise, species can be divided into two groups based on their reproductive strategies: the *r*-selected pattern or the *K*-selected pattern. Here's the difference: ***r*-selected organisms** have populations below the carrying capacity of their environment, which means that population growth is constrained only by the species' own biological limits. Competition for resources in *r*-selected species' habitats is usually relatively low. These organisms tend to be small and have short life spans; they mature and reproduce early in life and have many offspring at once—they may have so many that they reproduce only once in a lifetime. Thus they have a high capacity for reproductive growth. Some examples of *r*-selected species are bacteria, algae, and protozoa. In these species, little or no care is given to the offspring, but due to the sheer numbers of offspring in the population, enough of the offspring will survive to enable the population to continue. On the other hand, ***K*-selected organisms** have populations whose growth is limited by the carrying capacity of the environment; they live in stable environments where competition for resources is relatively high. These organisms tend to be large and have longer life spans. They mature and reproduce later in life after years of parental care, produce fewer offspring per reproduction event (though they tend to reproduce more than once in their lifetimes), and devote significant time and energy to the nurture of offspring. For these species, it is important to preserve as many members of the offspring as possible because they produce so few: parents have a tremendous investment in each individual offspring. Some examples of *K*-selected species are humans, lions, and cows. Invasive species tend to be *r*-selected, while the species most adversely affected by invasives tend to be *K*-selected. Many species lie on the continuum between these two strategies, and some can change strategies in different conditions or at different times, but the groups are useful for broad comparisons.

Survivorship curves represent the number of individuals in a population born at a given time (called a **cohort**) that remain alive as time goes on. Different species have different survivorship curves depending on their life cycles and life strategies, including *r*- and *K*-selection. Survivorship curves differ for *K*-selected and *r*-selected species: *K*-selected species typically follow a Type I or Type II curve, while *r*-selected species tend to follow a Type III curve.

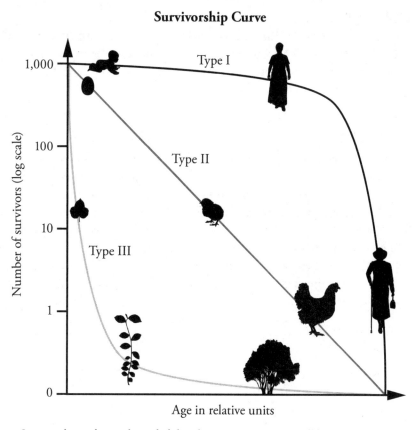

Survivorship Curve

Survivorship indicates the probability that a given organism will live to a certain age.

Type I (*K*-selected)—The convex shape of this curve indicates that most individuals in the population survive into adulthood, with a sharp increase in mortality as the population approaches the species' maximum age.

Type II—Mortality and survival rates are fairly constant throughout life. Many bird species, mice, and some species of lizards exhibit this straight-line pattern.

Type III (*r*-selected)—The convex shape indicates that most offspring die young, but if they live to a certain age, they will live a longer life. Species with this curve produce high numbers of offspring that encounter bottlenecks to survival that wipe out most young, and parents provide little or no nurture to their young. Examples include plants that produce millions of seeds throughout their lifetimes and most marine invertebrates. A clam, for example, produces millions of eggs, but the larvae are highly vulnerable to dying off from ocean currents and predators. The individuals that live long enough to develop their shell, however, will live to advanced age.

Most actual populations exhibit some combination of these patterns. For example, at different points in human history or in different societies, infant mortality has been unusually high, resulting in a sharp dip in the survivorship curve before it flattens out to the typical convex shape. Crustaceans like crabs and lobsters are most vulnerable while molting (replacing the hard shell). Since these species molt regularly throughout their lives, their survivorship curves show a stair-step pattern.

Population Cycles

When we observe populations in their natural habitats, there are two patterns that are more specific and involve more factors than just overshoot and dieback: the boom-and-bust cycle and the predator-prey cycle. These two patterns aren't explicitly tested on the AP exam, but relate to patterns that are. Let's look into these a little deeper.

Boom-and-Bust Cycle

The **boom-and-bust cycle** is very common among *r*-strategists. In this type of cycle, there is a rapid increase in the population and then an equally rapid drop-off. These rapid changes may be linked to predictable cycles in the environment (temperature or nutrient availability, for example). These cycles may reflect regular changes in rainfall, temperature, or nutrient availability over the course of the year. Or they may reflect longer and less regular cycles. When the conditions are good for growth, the population increases rapidly. When the conditions for that population worsen, its numbers rapidly decline. You might say that their strategy is "get it while the getting's good." Study the graph below so you can see this type of cycle in action.

Boom-and-Bust Cycle

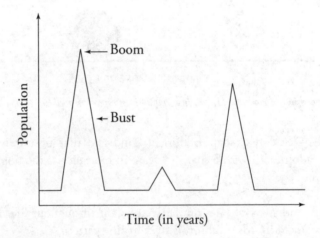

Predator-Prey Cycle

Remember the rabbit and coyote populations from the last chapter? We discussed how in a year of relatively high rainfall, rabbits have plenty of food, which enables them to reproduce very successfully. In turn, because the coyote is a predator of the rabbit, coyotes would also have plenty of food, and their populations would also rise rapidly. However, if the rainfall is below average a few years later, then there would be less grass, the population of rabbits would decline, and the coyote population would decline in turn. The graph of the predator-prey relationship looks like the following.

Predator-Prey Cycle

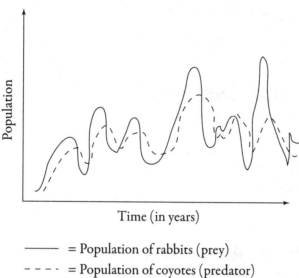

Time (in years)

——— = Population of rabbits (prey)

- - - - = Population of coyotes (predator)

Something important to notice in this graph is that the coyote population does not change at exactly the same time as the rabbit population. The coyote population actually rises *after* the rabbit population does. That is because the rabbit population has to have time to build up to fairly high levels before the coyotes can find enough to eat. When there is enough food, the coyote mothers have enough energy to give birth to and feed their pups. Only then can the coyote population increase.

The predator-prey cycle also plays a role in understanding why many endangered species are large carnivores. Large predator populations can suffer directly if humans alter their natural habitats, but they can also suffer indirectly if humans kill off their prey. If the prey population falls so low that the predator cannot find food, then the predator population will decline, sometimes to the point of extinction.

> **Remember Your Bio Definitions!**
> A carnivore is an animal that consumes only other animals, an herbivore consumes only plants, and an omnivore consumes both plants and animals.

Factors Influencing Population Growth

There are population-limiting factors that are purely the result of the size of the population itself. For example, in many populations of species in nature, birth and death rates are influenced by the density of the population. Other **density-dependent factors** that influence population size are increased predation (which occurs because there are more members of the population to attract predators); competition for food or living space; disease (which can spread more rapidly in overcrowded populations); and the buildup of toxic materials.

Some population-limiting factors operate independently of the population size. These **density-independent factors** will change the population's size regardless of whether the population is large or small. Independent factors include fire, storms, earthquakes, and other catastrophic events.

Now that you have a basic understanding of how and why populations change in size, let's move on to discuss human populations more specifically.

HUMAN POPULATIONS

You might have heard something about human population growth as you read the news or studied biology and earth science in school. But do you know how many humans are on the planet now? Do you know how fast the human population is growing? This information isn't explicitly tested on the AP exam anymore, but having a general sense of it will help you think about human populations.

How Many People Are There in the World?

According to the United Nations, the world's population passed 8 billion on November 15, 2022. The birth rate has actually fallen in the United States and most developed (industrialized) countries worldwide. But that only means that, in prosperous countries, the population is increasing more slowly, and overall the world's population is still increasing. Take a look at the following table, which lists the world's most populated countries as estimated for January 2023.

Top 10 Most Populous Countries	
Country	**Estimated Population (January 2024 estimated)**
India	1.43 billion
China	1.42 billion
United States	340 million
Indonesia	278 million
Pakistan	242 million
Nigeria	226 million
Brazil	217 million
Bangladesh	173 million
Russia	144 million
Mexico	128 million

Source: The World Population Review.

The following graph shows how the world's human population has increased since the 1950s, and how it is predicted to increase, though less rapidly, into the 2060s.

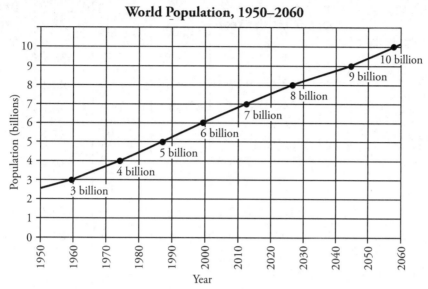

Projected global population growth 1950–2060. Source: U.S. Census Bureau, International Data Base

The population of many of the countries shown in the table on the previous page is currently increasing in size. We can determine the rate of population change per year of a country by using the following formula.

$$\text{Population Change Over Time} = \frac{(\text{birth rate} + \text{immigration rate}) - (\text{death rate} + \text{emigration rate})}{10}$$

All rates in the formula are raw numbers per 1,000 people per year. For example, the birth rate (the **crude birth rate**) is equal to the number of live births per 1,000 members of the population in a year, and the death rate (or **crude death rate**) is equal to the number of deaths per 1,000 members of the population in a year. A simpler version of the above formula, called the **Actual Growth Rate**, considers only the birth and death rates, excluding immigration and emigration: (birth rate – death rate) / 10.

The table that follows gives the 2023 estimated growth rates for a selection of 11 countries, to show the range in this figure. Also shown are the crude birth and death rates as well as key demographic information on median age and population density per unit area. Note that some countries are experiencing negative growth. The statistics on median age are one indication of quality of life, while the information on population density suggests that not all densely populated countries are poor and not all sparsely populated countries are prosperous.

2023 Growth Rates Chart

Country	Rank in World Population	Growth Rate % Per Year	Birth Rate Per 1,000	Death Rate Per 1,000	Median Age	Population Density (per km²)
India	1	0.86%	8.52	7.07	38.4	151
China	2	0.12%	16.57	7.35	28.7	481
United States	3	0.47%	10.9	10.3	38.5	37
Indonesia	4	0.77%	16.65	8.96	31.1	148
Nigeria	6	2.44%	37.47	13	18.6	246
Russia	9	−0.17%	9.8	14.6	40.3	9
Mexico	10	0.71%	15.56	9.33	29.3	66
Japan	12	−0.47%	6.8	11.1	48.6	338
United Kingdom	21	0.39%	10.2	10.4	40.6	280
Canada	38	0.93%	9.4	8.1	41.8	4
Latvia	151	−1.11%	9.2	15.2	44.4	29

Source: World Population Review. All data are estimates.

How Do Human Populations Change?

Populations can also change in number as a result of migration into and out of the population. Two important vocabulary words to describe human migration are **emigration,** which is the movement of people out of a population, and **immigration,** which is the movement of people into a population. In the Annual Growth Rate formula, immigration and emigration would also need to be expressed as rates per thousand in the population. Keep in mind that, in general, emigration and immigration are only small factors in the changes in size of human populations; however, the United States, unlike many other highly developed countries, has the third-largest population due to immigration.

The most significant additions to human populations are due to births, plain and simple. The term **total fertility rate (TFR)** is used to describe the number of children a woman will bear during her lifetime, and this information is based on an analysis of data from preceding years for the population in question.

Total fertility rates are predictions that provide a rough estimate, but they can't be depended on because they assume that the conditions of the past will be the conditions of the future.

Not surprisingly, a number of factors affect the total fertility rates in a population, and as a result, the population's birth rate. Among these are:

- the availability of birth control

- the demand for children in the labor force

- the base level of education for women

- the existence of public and/or private retirement systems

- the population's religious beliefs, culture, and traditions

The **replacement birth rate** of a human population refers to the number of children a couple must have in order to replace themselves in a population. While you might automatically think that the answer is always 2, in reality it is slightly higher to compensate for the deaths of children, the existence of non-child-bearing females in the population, and other factors. In developing countries, the replacement birth rate can be as high as 3.4 because of higher mortality rates! If the fertility rate is at replacement levels, a population is considered relatively stable.

The **infant mortality rate** is the number of deaths of children under 1 year old per 1,000 live births. Obviously, whether mothers have access to good healthcare and nutrition has the greatest effect on the infant mortality rate. Other factors are sanitation, clean drinking water, environmental conditions, and political infrastructure. Changes in these factors can lead to changes in the infant mortality rate over time. As we mentioned earlier, despite the relatively recent drop in total fertility rates worldwide, the world's population is still increasing. This is because many members of the human population are future parents, so even if they only reproduce at a replacement rate, there will be an overall increase in the total population. Now, let's look at a graph of how the overall world population growth rate has changed since 1750 and is projected to change through 2100.

World Projected Population Growth Rate, 1750–2100

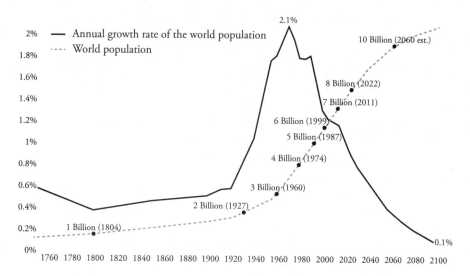

Source: OurWorldinData.org and the United Nations

Though not shown on this graph, the human population has actually been growing exponentially for more than three centuries. But what did we learn earlier in this chapter? That no population can grow exponentially indefinitely…we'll talk more about that in a bit. For now, let's discuss some factors that affect the growth rates of human populations.

Perhaps not surprisingly, there is a strong empirical correlation between the education level of women and the growth rate of populations. Additionally, the reason that religion and culture are predictors of birth rates is that in some countries, certain groups have a proclivity toward reproduction for religious reasons. The reason that the world's population has grown so considerably, especially in the past 100 years, is not because of an increased number of births, but because of the significant drop in the world death rate. People are living longer lives, and there are far fewer infant deaths today than there were 100 years ago. This is due, in large part, to the

Industrial Revolution, which improved the standard of living for millions living in industrialized nations. Other causes of the extension of the human life span are the development of clean water sources and better sanitation, the creation of dependable food supplies, and better health care. In general, the overall health of a population can be estimated by examining the expected life span of individuals and the mortality rate of infants.

On the other hand, there are obviously factors that limit human population growth. We've already seen that the availability of birth control and education for women limit the total fertility rate; negative density-independent effects, such as major storms, fires, heat waves, or droughts can also be limiting factors. Additionally, density-dependent effects of a population nearing its carrying capacity, such as decreased access to clean water and air, lack of resources including food, disease outbreaks, and territory size can also limit growth, showing that the carrying capacity of a given area, and of the Earth as a whole, are ultimate factors that will have to be reckoned with.

Age-Structure Pyramids

Age-structure pyramids (also called **age-structure diagrams**) are useful for graphically representing populations. Some age-structure diagrams group humans into three categories by age: those who are **pre-reproductive** (0–14), those who are **reproductive** (15–44), and those who are **post-reproductive** (45 and older). Each of these groups at the same stage of life is also called a cohort. Age pyramids, such as the one shown on the next page, group members of the population strictly by age, with each decade representing a different group. The x-axis contains the information relating to the percent or number of individuals in each of the age groups.

We can use age-structure pyramids to predict population trends; for example, when the majority of a population is in the post-reproductive category, the population size will decrease in the future because most of its members are incapable of reproducing. The opposite is true if the majority of a population is in the pre-reproductive category; these populations will increase in size as time goes on. Take a look at the shapes of the example age-structure diagrams representing Niger and the United States. As you can see, Niger has a large number of pre-reproductive and reproductive members in its population, while the United States has a fairly even distribution. From this, we can see that the population of Niger should increase significantly over time—it has what is referred to as **population momentum**—while the population of the United States should grow more slowly.

You should be able to identify the growth rate of a country based upon its age-structure pyramid. Here are the four main types:

Age-Structure Pyramid

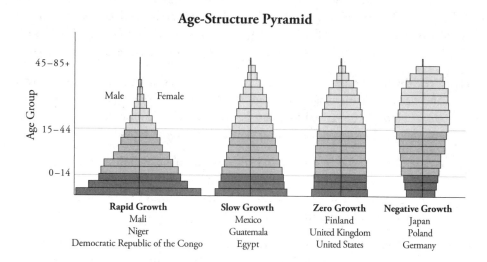

The *x*-axis contains the information relating to the percent or number of individuals in each of the age groups. (On the AP Exam, when a question asks for the percent or number of individuals in a particular age group, be sure to count both males and females.)

The Demographic Transition Model

In 1794, Thomas Malthus wrote *An Essay on the Principle of Population.* In it, he argued that populations have a propensity to increase, and that that tendency would only be curbed when the "means to subsistence"—the necessary resources to maintain a human population, such as food—grew to be in short supply. He argued that this was inevitable, since population multiplies geometrically while food production does so arithmetically. He predicted that when the lower classes of a society started to suffer hardship, famine, and disease, the increase of a population might be curbed. When good conditions returned, populations would naturally increase again.

When the Green Revolution took place in the 1950s and 1960s, the rapid increase in global food supply was again followed by rapid population growth. At this point, many theorists began to worry about human overpopulation—the idea that humans might overshoot the carrying capacity of the Earth as a whole and suffer some sort of catastrophe (usually termed a **Malthusian catastrophe,** after Thomas Malthus). When they began to look at the environmental impact of humans, not just on food supply, but on the whole spectrum of ecosystem services, it became clear that the rate of population growth was not sustainable. Though some predictions of certain disaster did not come to pass since the Green Revolution and genetic modifications helped the food supply surpass predicted levels, the basic idea that continued population growth at the exponential rates that were being seen could not continue was sound. Population trends needed to change.

However, a change began to take place as nations industrialized more and more. As nations began to become "developed"—a term to describe countries with high economic indicators and standards of living—these countries saw a concurrent change in their population characteristics. Since countries seemed to follow a pattern of industrialization and development, and generally their population characteristics tended to follow a specific pattern as this happened, the idea was codified into the demographic transition model.

Fast Fact

The most significant impact on the movement from a preindustrial stage to postindustrial is the education of girls. Education empowers girls and women to make choices and become independent. This leads to a lower TFR and better economic growth.

The **demographic transition model** is used to predict population trends based on the birth and death rates of a population. In this model, a population can experience zero population growth via two different means: as a result of high birth rates and high death rates; or as a result of low birth rates and low death rates. When a population moves from the first state to the second state, the process is called **demographic transition.** The four states that exist during this transition are the following:

1. **Preindustrial state:** In this state, the population exhibits a slow rate of growth and has a high birth rate and high death rate because of harsh living conditions. Harsh living conditions can be considered environmental resistance, an umbrella term for conditions that slow a population's growth.

2. **Transitional state:** In this second state, birth rates are high, but due to better food, water, and health care, death rates are lower. This allows for rapid population growth. Birth rates remain high due to cultural or religious traditions and a lack of education for women.

3. **Industrial state:** In the third state, population growth is still fairly high, but the birth rate drops, becoming similar to the death rate. Many developing countries are currently in the industrial state—these countries tend to have higher infant mortality rates and more children in the workforce than developed countries.

4. **Postindustrial state:** In the final state, the population approaches and reaches a zero growth rate. Populations may also drop below the zero growth rate (as we saw for Japan, Poland, and Germany in the table of growth rates).

Check out the following graph to better understand the demographic transition model.

Demographic Transition Model

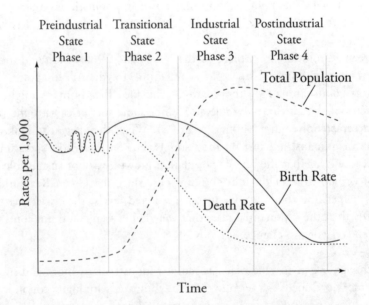

You will most likely see questions that will ask you specifically to describe the demographic transition model or apply it to hypothetical or real population states, so make sure you understand this material. If you don't, go back to your textbook for further explanation!

THE IMPACT OF HUMANS ON HUMAN POPULATIONS

As you're probably well aware, humans have the greatest impact on the environment of any living species on Earth, and the increase in our population over the last few centuries has seriously and dramatically changed the face of the Earth. We'll study further how humans have changed the Earth in Chapter 10; meanwhile, the following information explores how humans are impacted by our growing population and by our impact on the planet. This information is not explicitly tested but is useful for understanding the interplay between humans and the biosphere.

Four of the most significant factors that have contributed to the increase in the world human population are improved nutrition, the availability of clean water, newly implemented systems for sanitary waste disposal, and better medical care.

Another significant factor is the increase in food production. About 38% of the Earth's land surface is currently devoted to agricultural production; approximately one-third is used for crop production and the remaining consists of meadows and pastures used for grazing livestock. As we mentioned earlier, this enormous amount of food production takes its toll on the land, and we now face excessive and harmful erosion, in addition to a variety of environmental problems that have resulted from the wide-scale use of irrigation. Finally, the widespread use of pesticides and fertilizers for increased crop yields leaves large amounts of harmful chemical residues in the soil and water.

In response to these problems, the agricultural industry is continuing to invent and promote soil conservation techniques, organic farming, more efficient irrigation methods, and genetically modified crops. However, these new techniques will need to be implemented in all countries in order to be effective globally. In many cases, new techniques introduce new problems—for example, crop sizes may increase, but pesticides must then be used in order to protect the larger crops.

The use of **genetically modified organisms (GMOs)** is also controversial. Inserting strands of DNA that code for resistance to pests or for larger crop size may lead to less genetic diversity. This in turn can lead to the likelihood of crops being susceptible to future pests and diseases. Also, there is no clear data regarding the effect these GMOs may have on human health in the future.

What Happens When There Aren't Enough Resources?

Our bodies need certain nutrients to keep them healthy and to help resist disease. Some nutrients, or **macronutrients,** are needed in large amounts. These include proteins, carbohydrates, and fats. Other nutrients are needed in smaller amounts; these are called **micronutrients**. Micronutrients include vitamins, iron, and minerals such as calcium. When people are deprived of food, one result is the onset of hunger. Technically speaking, **hunger** occurs when insufficient calories are taken in to replace those being expended. **Malnutrition** is poor nutrition that results from an insufficient or poorly balanced diet; those whose diets lack essential vitamins and other components often suffer from it. A third term used to describe those who aren't receiving sufficient resources is **undernourished.** Undernourished people have not been provided with sufficient quantity or quality of nourishment to sustain proper health and growth.

According to the Food and Agriculture Organization, or FAO, between 691 and 783 million people were affected by hunger in 2022. A comparison of food insecurity among rural, suburban, and urban populations reveals that global food insecurity, at both levels of severity, is lower in urban areas. Moderate or severe food insecurity affected 33.3 percent of adults living in rural areas in 2022 compared with 28.8 percent in suburban areas and 26.0 percent in urban areas. On the other hand, 39.6 percent of adults in the United States are considered obese and, globally, 1.9 billion people are overweight. Why does this dichotomy exist?

While the reasons for hunger are many and complex, the simplest explanation for the problem is poverty. Our planet produces sufficient food to feed today's world population, but many people lack the money to buy food or the resources to produce it.

All over the world, human communities are trapped in a cycle of poverty, resource degradation, and high fertility. For example, in the first third of the 20th century, Asia, Africa, and Latin America produced enough grain that it was not necessary for them to import it from other countries. However, because of their constantly increasing populations, all of these countries are now importing grain; this is an ominous sign of impending problems with hunger in these countries.

Encouragingly, China, Thailand, and Indonesia are working hard to implement government reforms that will increase the quality of life for their citizens. In China, for example, as a result of reform and development in rural areas, the number of people in the country without enough food and clothing has decreased from 250 million in 1978 to 23.65 million in 2006. Furthermore, new initiatives, such as the Zero-Hunger Initiative for West Africa, the Asia-Pacific Zero Hunger Challenge, and the Hunger-Free Latin America and the Caribbean Initiative aim to eradicate hunger and achieve food security by 2030.

Hunger in America

Despite its high obesity rate, there are hungry people even in the United States, one of the richest countries in the world—lots of hungry people. Thankfully, the number of hungry people in the United States is less now than it was when international leaders set hunger-cutting goals at the 1996 World Food Summit. At this summit, government leaders pledged to cut the number of Americans living in hunger from 30.4 million to 15.2 million by 2010, but this goal has not been met. According to the United States Department of Agriculture, 12.8% of American households were considered "food insecure" in 2022. Additionally, approximately 12.5% of the U.S. population relies on food stamps. That's 41.9 million Americans!

Poverty or Government Warfare?

According to Alex de Waal, author of *Mass Starvation: The History and Future of Famine,* the majority of famines in modern times are not caused by insufficient resources to produce or buy food, but are the result of deliberate warfare by certain governments against their own people.

U.S. Households with Children by Food Security Status of Adults and Children 2022

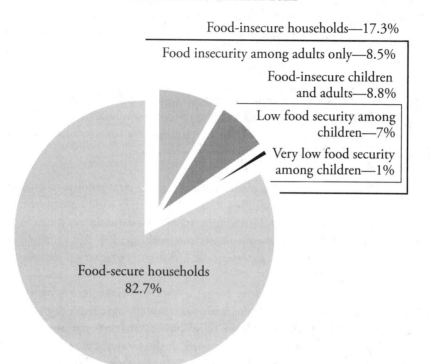

Food-insecure households—17.3%

Food insecurity among adults only—8.5%

Food-insecure children and adults—8.8%

Low food security among children—7%

Very low food security among children—1%

Food-secure households 82.7%

Source: USDA, Economic Research Service using data from U.S. Department of Commerce, U.S. Census Bureau, 2022 Current Population Survey Food Security Supplement.

But why are American people hungry when they live in a nation that's known as the world's breadbasket? Again, the main reason is poverty. Many neighborhoods in which the majority of the citizens who reside there are low-income are often called **food deserts**. This is because access to fresh, healthy food is difficult. The residents rely on low-quality processed foods for subsistence. Lack of education on healthy food choices also contributes to poor nutrition.

In the mid-1990s, a call for welfare reform resulted in the passing of the Personal Responsibility and Work Opportunity Reconciliation Act (PRWORA). The premise of this welfare reform, according to its proponents, was that people who are able to work should be encouraged to find employment, so that they will not remain dependent on government assistance. The act limited the number of people who qualified for food stamps and limited the duration that people could receive food stamps and public support.

While at the time it was generally agreed that welfare reform was necessary, many families are now reaching their deadline for public assistance. Since the implementation of this act, and in the future, it will be of crucial importance for state and local groups to find ways to support the truly needy.

In the United States, there are a number of charitable agencies that provide food at no or low cost to those in need. One example is Feeding America, which makes use of food that would otherwise go to waste. Feeding America receives food from food processors and distributors and redistributes it via food banks. The organization helps to feed more than 45 million Americans each year.

Many grocery stores, restaurants, and individuals waste a lot of food. Food is often discarded because of the "sell by" date, when actually it is still fine to eat. Fruits and vegetables are discarded if they do not look perfect. The dumpsters behind delis and bagel stores are filled with bread at the end of the day, which could have gone to a shelter or food pantry.

Global Hunger Policies—For Good or Ill

While social reform is a viable solution to the problem of hunger in the United States, often the only solution in developing countries is to enable communities to become self-sufficient in food procurement. These destitute communities either need monetary resources that will allow them to purchase necessary food supplies or the resources that would enable them to produce their own food.

Another issue that must be dealt with in the struggle against global hunger is that many countries just can't produce the food they need in order to feed their citizens. In these cases, the first viable solution mentioned above—that of providing monetary resources to people so they can purchase food—is rendered irrelevant.

The World Trade Organization (WTO) controls the policies of international trade. Unfortunately for smaller, poorer nations, the economically strong nations of the world have more influence over the creation of policies by the WTO. Oftentimes, trade policies undercut prices for developing nations, which makes it difficult for them to enter the world market. Also, many of the poorest countries have corrupt governments that prevent the distribution of food aid to their own citizens.

Another problem for developing countries is that a trade imbalance exists between them and developed nations. In many poorer nations, national resources have been degraded in an effort to reduce the national debt. The people in these poverty-stricken nations have nothing to export—except labor. Relatively recently, companies in the United States and other developed nations have begun outsourcing jobs to developing countries. Many Americans are appalled at the terrible conditions under which these overseas laborers work—their hours are abnormally long and they are paid almost nothing. Despite these conditions, the competition between poor communities to secure contracts with companies overseas is fierce because often the alternative is continued poverty.

Now you're ready for the key terms review and the following drills. Remember to use our techniques as you go through them.

CHAPTER 5 KEY TERMS

Know these terms backward and forward so that you can spout them in your sleep.

Terms Used to Describe Populations

population
population density
population dispersion
uniform
random
clumped

Population Growth

biotic potential
carrying capacity *(K)*
exponential population growth
logistic population growth
overshoot
Rule of 70
doubling time
r-selected organisms
K-selected organisms
survivorship curves
cohort
type I
type II
type III
boom-and-bust cycle
density-dependent factors
density-independent factors

Human Populations

crude birth rate
crude death rate
Actual Growth Rate
emigration
immigration
total fertility rate (TFR)
replacement birth rate
infant mortality rate
age-structure pyramids (age-structure
  diagrams)
pre-reproductive
reproductive
post-reproductive
population momentum
Malthusian catastrophe
demographic transition model
demographic transition
preindustrial state
transitional state
industrial state
postindustrial state

The Impact of Humans on Human Populations

genetically modified organisms (GMOs)
macronutrients
micronutrients
hunger
malnutrition
undernourished
food deserts

Chapter 5 Drill

Directions: Each of the questions or incomplete statements below is followed by four suggested answers or completions. Select the one that is best in each case. For answers and explanations, see Chapter 13.

1. Which type of population dispersion is exemplified by schools of fish that swim together in the ocean to avoid predation?

   (A) Random

   (B) Uniform

   (C) Linear

   (D) Clumped

2. The maximum population size of a certain species in a certain environment that can be sustainably supported by the environment's available resources is referred to as

   (A) biotic potential

   (B) carrying capacity (K)

   (C) logistic population density

   (D) population overshoot

3. All of the following are K-selected species EXCEPT

   (A) humans

   (B) lions

   (C) frogs

   (D) horses

4. r-selected organisms

   I. tend to follow a Type III survivorship curve

   II. produce low numbers of offspring that encounter few bottlenecks to survival

   III. provide little to no care to their offspring

   (A) I only

   (B) II only

   (C) II and III only

   (D) I and III only

5. What is the actual growth rate of country Z, given that country Z has a birth rate of 32.4 and a death rate of 9.8?

   (A) 2.26%

   (B) 4.22%

   (C) 5.78%

   (D) 7.74%

6. The idea that humans might overshoot the carrying capacity of the Earth and suffer some sort of devastation is referred to as

   (A) the Tragedy of the Commons

   (B) the Malthusian catastrophe

   (C) the Tragedy of Free Access

   (D) the Hadley effect

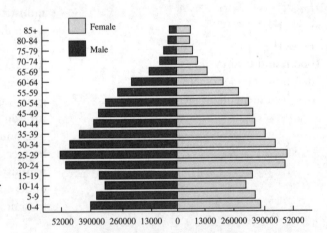

7. All of the following statements are true of the age-structure pyramid shown above EXCEPT

   (A) the population is experiencing greater birth rates than death rates

   (B) the population has fewer individuals in the reproductive stage than in the pre-reproductive stage

   (C) the population is experiencing negative growth

   (D) the population has more individuals in the reproductive stage than in the post-reproductive stage

8. Countries like Germany and Russia, which are experiencing zero or negative birth rates, are in which state of the demographic transition model?

(A) Preindustrial state

(B) Transitional state

(C) Industrial state

(D) Postindustrial state

9. In which state of the demographic transition model does a country's birth rate drop, becoming similar to its death rate?

(A) Preindustrial state

(B) Transitional state

(C) Industrial state

(D) Postindustrial state

10. Which of the following is NOT a macronutrient?

(A) Iron

(B) Carbohydrates

(C) Proteins

(D) Fats

11. Poor nutrition that results from an insufficient or poorly balanced diet is known as

(A) hunger

(B) undernourishment

(C) malnutrition

(D) starvation

Free-Response Question

1. A habitat's carrying capacity imposes limits on the growth of populations and their consumption of resources.

    (a) **Describe** carrying capacity by defining the term. **Explain** how carrying capacity can impose limits on a population with TWO examples.

    (b) **Explain** how a population's consumption of natural resources might be controlled. **Identify** TWO examples of how nature slows down the consumption of natural resources by a population.

    (c) **Describe** TWO ways human activity can raise a habitat's carrying capacity for humans.

Summary

○ Population patterns: The distribution of a population is influenced by the density a particular species can handle based on its resource use and interspecific competition.

○ Population growth: The biotic potential of a population is the amount that the population would grow if there were unlimited resources.

○ Limits in the ecosystem define the maximum population size, known as the carrying capacity *(K)*. If the population overshoots its carrying capacity, resource depletion can lead to dieback.

○ The rate of growth of the population depends on the species.
  • *K*-selected species: populations are limited by carrying capacity; stable environments; high competition; large in size; long lifespans; intensive parental care; later maturity; fewer offspring per reproductive event
  • *r*-selected species: populations are below carrying capacity; low competition; small in size; short lifespans; mature early; many offspring from fewer reproductive events; little to no care involved in raising offspring

○ Survivorship curves show the pattern of mortality along the lifespans of a population's members.
  • Type I: most individuals in the population survive into adulthood, with a sharp increase in mortality as the population approaches the species' maximum age.
  • Type II: mortality and survival rates are fairly constant throughout life.
  • Type III: most offspring die young, but if they live to a certain age, they will live a longer life.

○ Cyclic patterns in population size can be influenced either by the abiotic environmental changes or the population changes of their predators or prey.

○ Human populations: There are 8 billion humans on Earth. Although the growth rate has decreased, the population is still increasing.

○ Actual growth rate is determined by the difference in birth rate and death rate. Population change over time is also influenced by immigration and emigration rates.

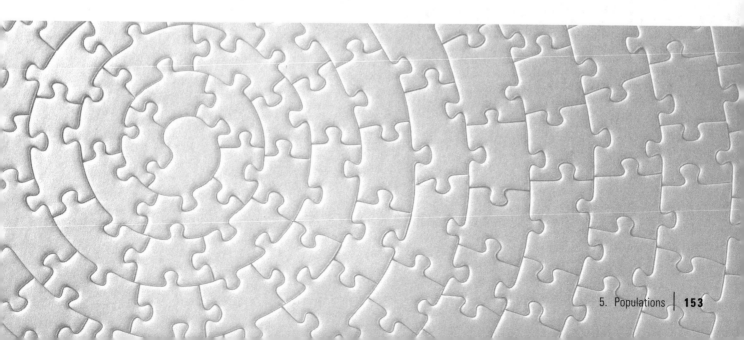

o Total fertility rate (TFR) is greatly influenced by:
  • education and employment opportunities for women (most important)
  • child labor
  • existence of a retirement system
  • cultures, traditions, and religious beliefs

o Replacement birth rates must remain just above 2 due to deaths before reaching reproductive age and females in the population who do not bear children; rates are almost always higher in developing countries due to higher infant and child mortality rates.

o Age-structure pyramids are essential diagrams to be able to read.
  • Large pre-reproductive cohorts imply population growth.
  • Large post-reproductive cohorts imply population decrease.
  • Pyramids which depict little difference between cohorts suggest a stable population with little to no population change.

o Demographic transition
  • Populations typically achieve zero net population growth toward the end of the industrial state, moving from a period of high birth/high death rates to low birth/low death rates.
  • During the industrial state, however, there is often a large growth in population because the decreasing birth rate lags behind the decreasing death rate.

Chapter 6
Unit 4: Earth Systems and Resources

In this chapter, we review Unit 4 of the AP Environmental Science course, *Earth Systems and Resources*. Specifically, we'll cover planet Earth's structure and the materials from which it is made, especially those used by humans as resources for our needs and economic gain. According to the College Board, about 10 to 15% of the test is based directly on the content covered in this chapter. If you are unfamiliar with a topic presented here, consult your textbook for more in-depth information.

Let's take a step back and review the fundamental themes of this chapter. First of all, a **resource** is strictly defined as any substance, capability (such as work performed by humans or animals), or other asset that is available in a supply that can be accessed and drawn on as needed. Resources are utilized for the effective functioning of an organism or community, often for economic gain. Within the scope of Environmental Science, we typically understand resources to mean natural resources—resources that occur in nature. That is to say, they exist apart from humans, without human effort or intervention prior to our exploitation of them. What natural resources are we talking about? Well, all of the Earth's substances and materials that humans rely on to live—namely the land and water and the things growing there or found there. Natural resources can also refer to properties of the natural world such as gravity, magnetic power, and other forces of potential use to humans.

The chapter starts with a brief Welcome to Planet Earth providing basic information on the planet's age and location in the solar system. The remaining topics correspond to the four physical spheres that make up our planet and regulate life on Earth—the materials and structure of these spheres as well as the mechanics of how each one works. We also discuss the importance of each sphere to humans. The four spheres are:

- The **Solid Earth**—the Earth's solid, rocky outer shell
  - Topics: the movement of tectonic plates, volcanoes and earthquakes, and the different types of rock
  - The upper shell of the solid Earth is called the **Lithosphere** and is the part that interacts most with the other spheres

- The **Pedosphere**—more commonly known as soil
  - Topics: soil's characteristics, formation, layers, and development

- The **Atmosphere**—the envelope of gases that surrounds the Earth
  - Topics: the greenhouse effect, climate, and weather events

- The **Hydrosphere**—the Earth's oceans and freshwater bodies
  - Topics: fresh- and saltwater bodies and ocean currents

We've broken down discussion of the four spheres into neat sections for review, and we'll go through everything you need to know about these systems for test day.

The four physical spheres provide resources that support the **Biosphere**, the fifth of the Earth's spheres, which comprises all the living organisms that inhabit the planet and draw on the physical resources of the other four spheres. You already know a lot about that sphere—it was the main topic of Chapter 4!

As illustrated here, all five spheres interact to shape the variety of landforms, biomes, and phenomena that make environmental science such an endlessly fascinating area of study!

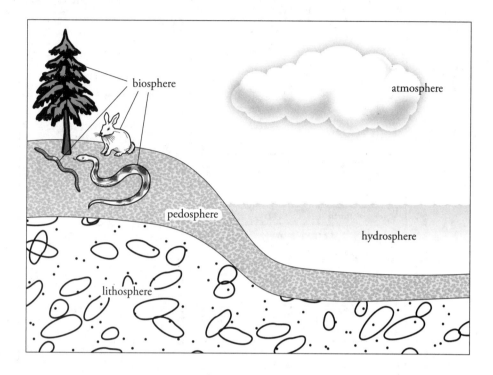

WELCOME TO PLANET EARTH

The first thing you should know about the Earth is its history. The Earth is thought to be between 4.5 and 4.8 billion years old. That amount of time is pretty inconceivable to humans, but the following geologic time scale will help you get a sense of the vast amount of time that has gone by since the Earth was formed. You will not be responsible for memorizing all of the eons, eras, periods, and epochs for this exam, but you should be familiar with the major ones— they will come in handy.

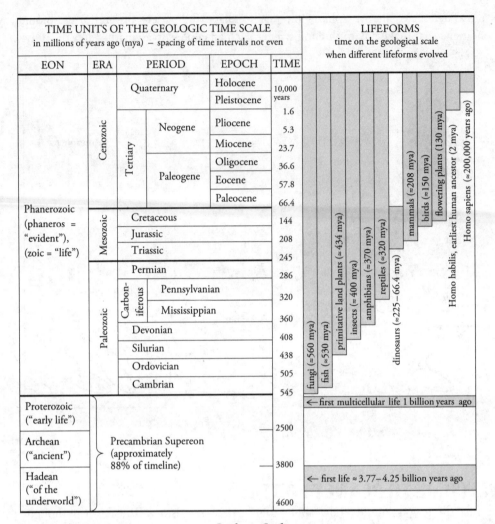

TIME UNITS OF THE GEOLOGIC TIME SCALE in millions of years ago (mya) – spacing of time intervals not even					LIFEFORMS time on the geological scale when different lifeforms evolved
EON	ERA	PERIOD	EPOCH	TIME	
Phanerozoic (phaneros = "evident"), (zoic = "life")	Cenozoic	Quaternary	Holocene	10,000 years	
			Pleistocene		
		Tertiary — Neogene	Pliocene	1.6	
				5.3	
			Miocene	23.7	
		Tertiary — Paleogene	Oligocene	36.6	
			Eocene	57.8	
			Paleocene	66.4	
	Mesozoic	Cretaceous		144	
		Jurassic		208	
		Triassic		245	
	Paleozoic	Permian		286	
		Carboniferous — Pennsylvanian		320	
		Carboniferous — Mississippian		360	
		Devonian		408	
		Silurian		438	
		Ordovician		505	
		Cambrian		545	
Proterozoic ("early life")		Precambrian Supereon (approximately 88% of timeline)		2500	
Archean ("ancient")				3800	
Hadean ("of the underworld")				4600	

Lifeforms labels: fungi (≈560 mya); fish (≈530 mya); primitive land plants (≈434 mya); insects (≈400 mya); amphibians (≈370 mya); reptiles (≈320 mya); dinosaurs (≈225–66.4 mya); mammals (≈208 mya); birds (≈150 mya); flowering plants (130 mya); Homo habilis, earliest human ancestor (2 mya); Homo sapiens (≈200,000 years ago)

← first multicellular life 1 billion years ago

← first life ≈ 3.77–4.25 billion years ago

Geologic Scale

Here are some important takeaways from this table.

- We are currently in the Holocene Epoch.

- The Quaternary and Tertiary are the two most recent geologic periods.

- Non-avian dinosaurs lived during the Mesozoic Era. (Note that evolutionary biologists consider birds to be avian, or flying, dinosaurs. Pterosaurs such as pterodactyls are a group of reptiles less closely related to dinosaurs and birds.)

- The Precambrian eons represent the vast majority of the geologic time scale.

- The next epoch will be called the Anthropocene, to recognize humankind's accelerating effects on the planet's physical resources, climate, and life forms. Many geologists propose that we have already entered this new, post-Holocene epoch.

Where Is the Earth in the Solar System?

The Earth is the third planet from the Sun in our solar system, which contains a total of eight currently known and recognized planets. From the Sun outward, the planets are Mercury, Venus, Earth, Mars, Jupiter, Saturn, Uranus, and Neptune.

Each planet has its own orbit around the Sun in the shape of an ellipse (a "stretched" circle). And you probably already know that it takes the Earth about 365.25 days, or 1 year, to complete its orbit of the Sun.

The Solid Earth

Planet Earth is made up of three concentric zones of rocks that are either solid or liquid (molten). The innermost zone is the core. The core has two parts: a solid inner core and a molten outer core. The inner core is composed mostly of nickel and iron and is solid due to the tremendous pressure from overlying matter. The outer core is composed mostly of iron, also mixed with nickel as well as some lighter elements, and is semi-solid due to lower pressure. Surrounding the outer core is the **mantle**, which is made mostly of solid rock. Near the top of the mantle lies a layer of slowly flowing rock called the **asthenosphere**. The **lithosphere,** a thin, rigid layer of rock, is the Earth's outer shell. The lithosphere includes the rigid upper mantle above the asthenosphere and the **crust**, the solid surface of the Earth. Think of it as floating atop the asthenosphere like a cracker atop a thick layer of hot pudding.

The following diagrams show the chemical and physical properties of the Earth's layers and a detail of the lithosphere.

The Earth's Layers

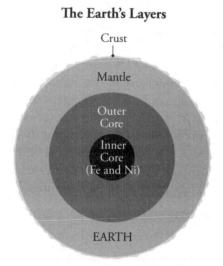

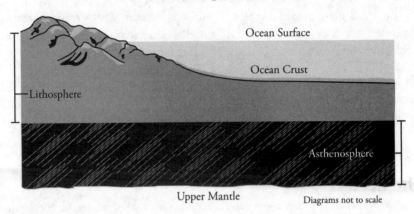

Lithosphere

Ocean Surface

Ocean Crust

Lithosphere

Asthenosphere

Upper Mantle

Diagrams not to scale

Tectonic Plates

While the information from the previous few pages (the geologic time scale, the Earth's position in the solar system, and the Earth's layers) is not explicitly tested on the AP Environmental Science Exam, an understanding of how the Earth's crust, specifically, is structured—that is, an understanding of tectonic plates, is required. Scientists theorize that during the Paleozoic and Mesozoic Eras, the continents were joined together, forming a supercontinent known as **Pangaea**. Roughly 200 million years ago, Pangaea began to break apart.

> **Fast Fact**
>
> In ancient Greek, the word *pan* means whole, while the word *gaia* means Earth. Thus, Pangaea means the whole Earth!

Today, it is believed that the Earth's crust is composed of several large pieces of lithosphere—called **tectonic plates**—that move slowly over the mantle of the Earth. There are a total of a dozen or so tectonic plates that move independently of one another. The majority of the land on Earth sits above six giant plates; the remainder of the plates lie under the ocean as well as the continents.

Some plates consist only of ocean floor, such as the Nazca plate, which lies off the west coast of South America, while others contain both continental and oceanic material. One example of the latter is the North American plate, where the United States is located; this plate extends out to the mid-Atlantic ridge. There is even a plate that is located exclusively within the Asian continent; its boundaries nearly coincide with those of Turkey. The largest plate is the Pacific plate—it primarily consists of ocean floor, but also includes Mexico's Baja Peninsula and southwestern California. The major plates of the Earth are shown on the map on the following page.

The Earth's Plates

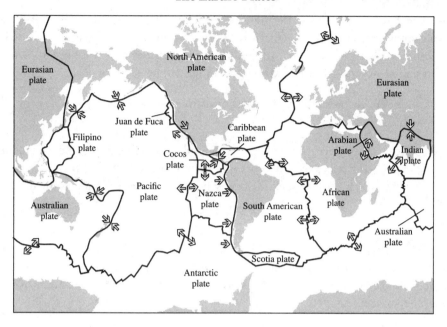

The edges of the plates are called **plate boundaries**, and the places where two plates abut each other are where events like sea floor spreading and most volcanoes and earthquakes occur. There are three types of plate boundary interactions:

- **Convergent boundary:** Two plates are pushed toward and into each other. One of the plates slides beneath the other, pushed deep into the mantle.

- **Divergent boundary:** Two plates move away from each other. This creates a gap between plates that may be filled with rising magma (molten rock). When this magma cools, it forms new crust.

- **Transform fault boundary:** Two plates slide against each other in opposite directions—as when you rub your hands back and forth to warm them up. These are also called simply transform boundaries.

So, what happens when plates collide? It depends on whether the collision happens between two oceanic boundaries, between two continental boundaries, or at an oceanic-continental boundary. Converging ocean-ocean and converging ocean-continent boundaries often result in **subduction,** in which a heavy ocean plate is pushed below the other plate and melts as it encounters the hot mantle. Converging continent-continent boundaries result in orogeny, the uplifting of plates that form large mountain chains as they crunch into each other. Examples include the Himalayas (which were created by a collision between the plate carrying India and the Asian plate), the Urals of western Russia, the Alps of southern Europe, and the Appalachian Mountains of the Eastern U.S.

One important result of plate movement is the creation of volcanoes and earthquakes. Let's examine those next.

Volcanoes and Earthquakes

Volcanoes are mountains formed by pressure from magma rising from the Earth's interior. **Active volcanoes** are those that are currently erupting or have erupted within recorded history (that is, within the last 10,000 years), while **dormant volcanoes** have not been known to erupt during this period. It's thought that **extinct volcanoes** will never erupt again.

Active volcanoes are categorized by the kind of tectonic events that produce them. They are associated with the following:

- **Subduction zones** occur at convergent boundaries between oceanic and continental plates, or sometimes between two oceanic plates. The subducting plate is recycled into new magma, which rises through the overlying plate to create volcanoes inland.

- **Rift valleys** occur at divergent boundaries, usually between two oceanic plates. New ocean floor is formed as magma fills in the gap between separating plates. Thick magma rising from rift valleys is made of basaltic minerals and forms pillow lava upon contact with the cold ocean water. Rift valleys may also occur between continental plates; a prominent example is the Great Rift Valley of eastern Africa, which gave rise to Mount Kilimanjaro and other volcanoes.

- **Hot spots** do not form at plate boundaries. Instead, they are found in the middle of tectonic plates, in locations where columns of unusually hot magma melt through the mantle and weaken the Earth's crust. The Hawaiian islands continue to form over a hot spot beneath the Pacific plate. Volcanoes over oceanic hot spots are basaltic, resulting in milder eruptions; while volcanoes over continental hot spots are characterized by rhyolitic rocks, which produce more violent eruptions.

There are four types of volcanoes:

- **Shield volcanoes** have broad bases and are tall with gentle slopes. They generally form over oceanic hot spots and usually have mild eruptions with slow lava flow. Sometimes, however, when water enters the vent, they can be very explosive, forming pyroclastic flows, a fluidized mixture of hot ash and rock.

- **Composite volcanoes** have broad bases and are also tall but with steeper slopes. They are formed at subduction zones and are associated with violent eruptions that eject lava, water, and gases as superheated ash and stones.

- **Cinder volcanoes** are small, short, and steeply sloped cones. They form when molten lava erupts and cools quickly in the air, hardening into porous rocks (called cinders or scoria) that fracture as they hit the Earth's surface. Cinder volcanoes generally form near other types of volcanoes.

- **Lava domes** are small and short with steep slopes and rounded tops. They are formed from lava that is too viscous to travel far but instead hardens into a dome shape. This type of volcano occurs near or even inside other types of volcanoes.

Earthquakes are the result of vibrations (often due to sudden plate movements, such as stress overcoming a locked fault) deep in the Earth that release stored energy. The Earth's tectonic plates move slowly all the time, at about the same pace as fingernail growth, but earthquakes are very sudden movements. They often occur as two plates slide past one another at a transform boundary. The **focus** of the earthquake is the location at which it begins within the Earth, and the initial surface location of the event is the **epicenter.** The size, or magnitude, of earthquakes is measured by using an instrument known as a **seismograph,** which was devised by Charles Richter in 1935. The **Richter scale** measures the amplitude of the highest S-wave of an earthquake. Observed values range from 0 to 9.5, although theoretically there is no maximum value. Each increase in Richter number corresponds to an increase of approximately 33 times the energy of the previous number. As technology has improved over time, and more seismograph stations were installed worldwide, it became clear that the Richter scale was only accurate for some distances and frequency ranges. Accordingly, scientists expanded upon Richter's original idea and developed the Moment Magnitude Scale, which measures the total energy released during an earthquake, including body wave magnitude (Mb) and surface wave magnitude (Ms). In turn, the Moment Magnitude scale can provide a valid estimate of the size of an earthquake over the complete range of magnitudes—a characteristic absent from the original Richter scale.

> **Fast Fact**
> An **S-wave**, also known as a **shear wave**, is a seismic body wave that shakes the ground up and down or side to side, perpendicular to the direction the wave is moving.

In January 2010, an earthquake of magnitude 7.0 struck the nation of Haiti. The quake's epicenter occurred in the boundary region separating the Caribbean plate and the North American plate. Official estimates from the U.S. Geological Service indicated 222,570 people killed, 300,000 injured, 1.3 million displaced, 97,294 houses destroyed, and 188,383 houses damaged across the Port-au-Prince area and much of southern Haiti. This includes at least four people killed by a local tsunami in the Petit Paradis area near Léogâne. **Tsunamis** are very large ocean waves, or chains of waves, caused by the movement of the Earth during an earthquake or volcanic eruption and can be extremely destructive.

In March 2011, Japan suffered an earthquake of 9.0 magnitude off the eastern coast, near Sendai. This quake caused a massive tsunami wave that was 33 feet high. Both the earthquake and tsunami caused extensive damage: many buildings, roads, and railways were destroyed; major fires occurred; villages, thousands of homes, and people were washed away; and at least three nuclear power plants experienced dangerous explosions.

The Rock Cycle

Rocks are all around us, in the soil, our buildings, and the ore used in industry. So, where do all those rocks come from? The answer is: other rocks. The oldest rocks on Earth are 3.8 billion years old, while others are only a few million years old. This means that rocks are recycled. These transformations are described as the **rock cycle**. In the rock cycle, time, pressure, and the Earth's heat interact to create three basic types of rocks.

- **Igneous** rock results when rock is melted (by heat and pressure below the crust) into a liquid and then resolidifies when cooled. The molten rock (**magma**) comes to the surface of the Earth, and when it emerges it is called lava; cooled lava becomes solid igneous rock. An example of an igneous rock is basalt.

- **Sedimentary** rock is formed as sediment (eroded rocks and the remains of plants and animals) builds up and is compressed. Sedimentary rock forms under water as sediments or dissolved minerals deposit on a stream bed or ocean floor. They are compressed as more material is deposited and then cemented together. An example of a sedimentary rock is limestone.

- **Metamorphic** rock is formed as a great deal of pressure and heat produces physical and/or chemical changes in existing rock. This can happen as sedimentary rocks sink deeper into the Earth and are heated by the high temperatures found in the Earth's mantle. An example of a metamorphic rock is slate, which results from the metamorphosis of shale.

The diagram below illustrates the rock cycle.

The Rock Cycle

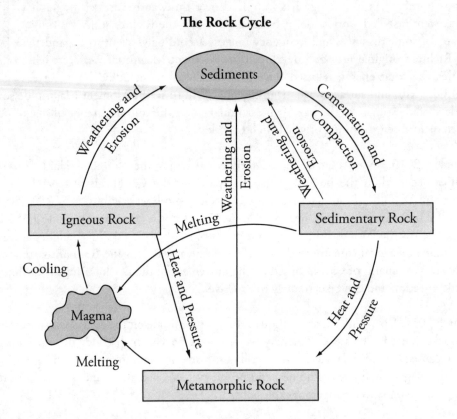

SOIL

One very important but often underappreciated player in the Earth's interdependent systems is soil. Soil plays a crucial role in the lives of the plants, animals, and other organisms and acts as an essential link between the abiotic (nonliving) components of the world and its biotic (living) components. Soil is sometimes called the pedosphere and exists at the interface of the other four systems or spheres: the lithosphere, atmosphere, hydrosphere, and biosphere. As we've already seen in Chapter 4, soil plays an active role in the cycling of nutrients. Additionally, soil performs the important task of protecting water quality—soil effectively filters and cleans water that moves through it. Let's take a moment to review the major characteristics of soil that you'll be expected to know for the test.

Soil Is More Than Just Dirt

Although we may be tempted to think of soil as simply "dirt," soil is actually a complex, ancient material teeming with living organisms. Many soils are tens of thousands of years old! A typical soil has the following composition: about 45% is mineral in the form of rocks broken down into tiny particles. About 25% is air and 25% water, which fill the pores between soil particles. About 5% is organic matter, both living and dead. In just one gram of soil, there may be as many as 50,000 protozoa, as well as bacteria, algae, fungi, and larger organisms such as earthworms and nematodes.

Soils can be categorized according to a number of physical and chemical features, including color and texture. There are many ways to test soil (by looking at its chemical, physical, and biological properties) that can help people make decisions about its use (such as for irrigation and fertilization). The United States Department of Agriculture (USDA) divides soil particles into three size classes. The class with the smallest particles is **clay,** which has particles less than 0.002 mm in diameter. The next largest size class is **silt,** with particles 0.002–0.05 mm in diameter. **Sand** is the coarsest soil, with particles 0.05–2.0 mm in diameter. Sand particles are usually too spherical in shape and too large to easily stick together; sand particles have larger pore spaces, with more room for water but less ability to hold it. Clay particles more easily adhere to each other because they are small and flat in shape; there is less room for water but the water is held more tightly. Soils with a high proportion of clay particles are extremely compact.

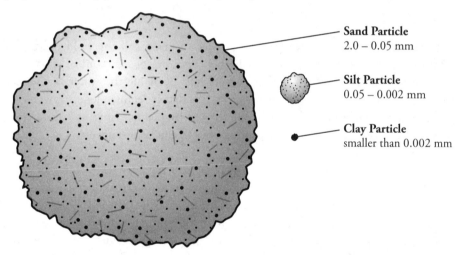

Sand Particle
2.0 – 0.05 mm

Silt Particle
0.05 – 0.002 mm

Clay Particle
smaller than 0.002 mm

A soil's texture is defined by its proportion of these three particle-size classes. Together, the mixture of particle sizes in a given soil (its **composition**) affects its **porosity** (the number of "holes" it has), its **permeability** (how easily fluids such as water and air move through it), and its **fertility** (how well plants can grow in it). A soil texture triangle is a diagram that allows for the identification and comparison of soil types based on their percentages of clay, silt, and sand. A soil's texture is called **loam** if it has a proportion of the three size classes considered optimal for plant growth: 7–27% clay, 28–50% silt, and less than 52% sand. **Water holding capacity** is the total amount of water a given soil can hold.

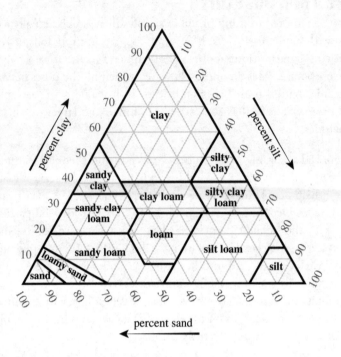

Water holding capacity is the total amount of water a given soil can hold. Water holding capacity varies among soil types: soils with higher amounts of clay and/or organic matter can hold the most water. However, it also matters how available the water in the soil is to plants. One reason loam is optimal for plant growth is that it retains plenty of water and has a texture that allows that water to be available for use by plants. Thus, soil water retention contributes to the fertility of soil and the productivity of the land it covers.

pH Pop Quiz!
Q: Which of the following tastes is associated with acidic food?
(A) Sour
(B) Salty
(C) Sweet
(D) Bitter
Turn the page for the answer.

Another very important characteristic of soil types is soil acidity or alkalinity. Recall that the pH of a substance ranges from 0–14 and is a measure of the concentration of hydronium (H_3O^+) ions in aqueous solution. Therefore, **soil pH** is actually the pH of the soil's water component and, in the laboratory, is commonly measured by suspending soil in water or a dilute neutral salt solution. Most soils fall into a pH range of about 4–8, meaning that most soils range in pH from being neutral to slightly acidic. Soil pH is important because it affects the solubility of nutrients, and this in turn determines the extent to which these nutrients are available for absorption by plant roots. If the soil in a region is too acidic or basic, certain soil nutrients will not be able to be used by the regional plants. One last thing about soils that the College Board wants you to know: when the soil solution becomes more acidic, it more readily dissolves heavy metals such as mercury (Hg) or aluminum (Al) from soil minerals. Leached into the ground water, these ions can travel to streams and rivers and harm both plants and aquatic animal life. For example, aluminum ions can damage the gills of fish and cause them to suffocate.

Where Does Soil Come From?

Basically, soil is a combination of organic material and rock that has been broken down by chemical and biological **weathering**—which is, unsurprisingly, the process by which rock and other material decomposes. Therefore, it should also not be surprising to learn that the types of minerals found in soil in a particular region will depend on the identity of the base rock of that region. Note that **erosion** is a distinct process, by which broken-down material is removed from one place and transported to another, across the Earth's surface, usually by wind or water. Sometimes the same process can be responsible for both erosion and weathering, as when a glacier grinds material from surface rock and drags it along with its own movement, depositing it far away.

Water, wind, temperature, and living organisms are all prominent agents of weathering, and all weathering processes are placed into the following three rather broad categories.

- **Physical weathering** (also known as **mechanical weathering**): Any process that breaks rock down into smaller pieces without changing the chemistry of the rock. The forces responsible for physical weathering are typically wind and water.

- **Chemical weathering:** Occurs as a result of chemical reactions of rock with water, air, or dissolved minerals. Chemical processes result in minerals that are broken down or restructured into different minerals. This type of weathering tends to dominate in warm or moist environments. One example of chemical weathering is rust, which forms when iron and other metallic elements come in contact with water.

- **Biological weathering:** Weathering that takes place as the result of the activities of living organisms, which may act through physical or chemical means. When tree roots enlarge the cracks in rocks as they grow, that is a physical process. When plant roots or lichens growing on rocks release organic acids that dissolve minerals, that is a chemical process.

Soil is made up of distinct layers with very different characteristics. Let's discuss those next.

Soil Layers

Soil comprises distinct layers known as **horizons,** each of which has distinct physical, chemical, and biological properties. Study the following diagram which illustrates and describes the different horizons. Not every soil contains all of these horizons.

- **O horizon:** This layer is made up of organic matter at various stages of decomposition. It includes animal waste, leaves and other plant tissues (such as dead roots), and the decomposing bodies of organisms. The stable residue left after most organic matter has decomposed is a dark, crumbly material called **humus.**

- **A horizon:** This is the topsoil—the topmost mineral horizon and the most intensively weathered soil layer. Its dark color is due to accumulation of organic matter from the O horizon. In soils lacking an E horizon, this may also be called the zone of **leaching**.

- **E horizon:** This is the eluviated horizon. It is light in color and coarse in texture; no organic matter has traveled down from the A horizon, while clays and minerals like iron and aluminum oxides have been washed out by leaching and **eluviation**. The E horizon isn't found in all soils: it's mostly found in soils developed under forest.

Eluviation is the movement of water-borne minerals, humus, and other materials from higher soil layers to lower soil layers. This is due to the downward movement of water via gravity. **Illuviation** is the deposition of these materials in a lower soil horizon. **Leaching** is similar to eluviation but refers specifically to dissolved (not suspended) organic and chemical compounds, and implies loss of these substances from the soil profile by draining into the groundwater.

- **B horizon:** Sometimes called the subsoil, this is where organic matter, clay, and minerals washed out of the upper horizons accumulate. Thus, it is called the zone of accumulation or the zone of **illuviation**.

- **C horizon:** This layer is the parent material—unconsolidated material, loose enough to be dug up with a shovel. Weathering at this depth is minimal so the soil retains identifiable features of the parent material (rock from which the A and B horizons formed). This horizon has much less biological activity than the horizons above it.

- **R horizon:** Beneath the soil lies a layer of consolidated (cemented), unweathered rock. Because it hasn't been weathered, this isn't part of the soil, strictly speaking. If the material from which horizons A through C formed was not transported from elsewhere, the R horizon has the same source minerals as the parent material above. The R stands for regolith, a word meaning "blanket or surface rock" in ancient Greek.

Soil Horizons

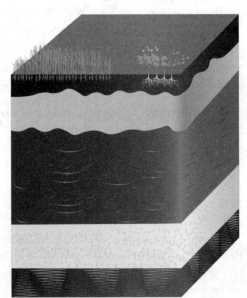

O organic matter
A topsoil

E eluviated horizon

B subsoil

C parent material

R bedrock

Soil Development

pH Pop Quiz!
A: Acidic foods, such as vinegar and lemon juice, taste sour because of their acidic nature. The answer is (A).

- **Four Processes of Soil Development.** Soil development is an intricate business! The development of distinct soil horizons is accomplished by four basic processes that work on soil minerals and particles, organic matter, and soil chemistry. These four processes include **additions** of materials from off-site; **losses** through erosion or leaching or biological activity (plant uptake); vertical **translocations** or movement from one soil horizon to another; and **transformations** of materials in place by weathering or chemical activity.

- **Six Soil-Forming Factors.** The processes of soil development are influenced by six soil-forming factors. It takes hundreds to thousands to millions of years for a soil to develop its characteristic layers or profile. To wrap it all up in one big sentence: any soil you see is a dynamic formation produced by the effects of climate and biological activity (organisms), as modified by topography (relief) and human influences, acting on parent materials over time. The six soil-forming factors can be remembered by the awkward acronym **Cl–O–R–P–T–H.** Here is a summary of how each factor works.

  o **Climate**: Climate involves differences in temperature and precipitation across the globe, and both heat and water facilitate chemical and biochemical reactions. Seasonal fluctuation of heat and moisture affects processes such as freeze-thaw cycles that weather rock. Climate also helps to determine what organisms grow in a particular location.

  o **Organisms** (Biological Activity): Different local conditions support different organisms, which influence the soil in a multitude of ways. Microorganisms perform biochemical functions such as decomposition of organic matter and transformation of minerals into different forms. Animals move soils, consume vegetation, and add nutrients through waste and decomposing bodies. Plants perform physical weathering through root growth, take up soil nutrients and water, alter soil chemistry in various ways, and add nutrients when they die and decompose.

  o **Relief** (Topography): Topographical relief affects where water moves on the landscape and also the depth of the water table in a given location. Relief similarly affects erosion—which locations are likely to lose surface material through the action of wind and rain, and which locations are likely to accumulate eroded material. Topographical relief also leads to differences in how much sun different locations receive. Through these characteristics, relief influences which organisms grow in a particular location.

  o **Parent Material**: This is the starting point for soil development. Its mineral properties, hardness, and topographical form affect how it is weathered into soil. As parent material varies from location to location, so will the soil that develops at each location. As an example, parent material rich in quartz, such as granite and sandstone, weathers into sandy soil. Shale weathers into soil richer in silt and clay.

  o **Time**: More time equals more change! Hard parent material weathers more slowly and softer material more quickly. A flat, stable topographic position develops horizons more quickly than do slopes and depressions where material is lost and gained.

  o **Human Influence**: The effects of human activity must increasingly be acknowledged as a factor in soil development. Use of fertilizer, pollution, and acid rain alter soil chemistry on a broad scale. Construction activities such as digging and plowing tend to mix soils and blur the distinctions between horizons. Human activities also lead to compaction (through the traffic of vehicles and machinery), erosion (through removal of stabilizing vegetation), and salinization (increase in salt content, through irrigation and depletion of groundwater).

THE ATMOSPHERE

In the broadest definition, the atmosphere is a layer of gases that's held close to the Earth by the force of gravity. The inner four layers of the atmosphere reach an altitude that's just about 12.5% of the Earth's radius. The layer of gases that lies closest to the Earth is the **troposphere;** it extends from the Earth's surface to about, on average, 12 km (7.5 miles) at the poles and 20 km (12.4 miles) at the equator. The troposphere is where all the weather that we experience takes place. The layer also contains 99% of the atmosphere's water vapor and clouds. Generally, the troposphere is well-mixed from bottom to top—with the exception of periodic temperature inversions. The troposphere gets colder with altitude, decreasing 6.5°C for every kilometer of altitude (or 3.5°F for every thousand feet).

Make sure you have a solid grasp of the order and characteristics of the layers of the Earth's atmosphere!

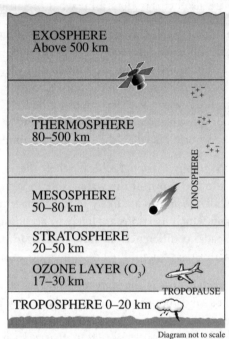

Diagram not to scale

Because of its density, the troposphere contains about 75–80% of the Earth's atmosphere by mass. You've probably heard about the troposphere before in the news because of the **greenhouse effect.** The troposphere contains the air we breathe, which is made up of 78% nitrogen and 21% oxygen. The remaining 1% includes the so-called "greenhouse" gases (GHGs). The proportion of these gases in the troposphere is minuscule, but their effects on conditions on Earth are disproportionately significant. The most important of them are water vapor (H_2O), carbon dioxide (CO_2), and methane (CH_4). As the sun's rays strike the Earth, some of the solar radiation is reflected back into space; however, greenhouse gases in the troposphere intercept and absorb a lot of this radiation. This warming effect of greenhouse gases was a good thing, until their concentration in the atmosphere shot up after the Industrial Revolution. We'll further investigate the greenhouse effect in Chapter 10, but for now take a look at the following figure.

The Greenhouse Effect

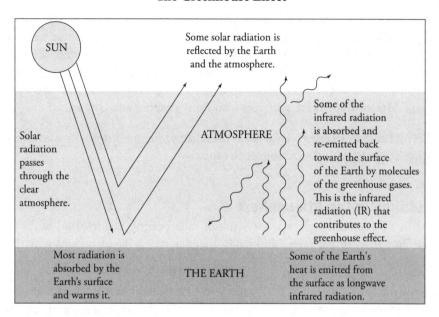

Crowning the troposphere is the **tropopause,** which is a layer that acts as a buffer between the troposphere and the next layer up, the stratosphere. This buffer zone is where the jet streams, air currents that are important drivers of weather patterns and important factors in planning airline routes, travel.

The **stratosphere** sits on top of the tropopause and extends about 20–50 km (7.5–31 miles) above the Earth's surface. As opposed to those in the troposphere, gases in the stratosphere are not well mixed and temperatures increase with distance from the Earth. This warming effect is due to the **ozone layer**, a thin band of ozone (O_3) that exists in the lower half of this layer. The ozone traps the high-energy radiation of the sun, holding some of the heat and protecting the troposphere and the Earth's surface from this radiation. The stratosphere is similar to the troposphere in gas composition, only less dense and drier, with a thousand times less water vapor. Commercial jets may also fly in the lower part of this layer.

Above the stratosphere are two layers called the mesosphere and the thermosphere. The **mesosphere** extends to about 80 km (50 miles) above the Earth's surface and is the area where meteors usually burn up. Temperatures again decrease here, to the coldest point in the atmosphere at the top of this layer, around –90°C (–130°F).

The **thermosphere** extends from 80 to around 500 km above the Earth (50–435 miles). Gases are very thin (rare) and it's in this layer that the spectacular and colorful auroras (northern lights and southern lights) take place. The furthest layer is the **exosphere**, extending to 10,000 km (6,200 miles) or more above the Earth, although the upper limit of this layer is not definitively settled. The concentration of gases is thinnest here. Human-made satellites orbit in the exosphere and in the upper thermosphere. The **ionosphere** is not a distinct layer but dispersed throughout the upper mesosphere, the thermosphere, and the lower exosphere. The ionosphere comprises regions of ionized gases that absorb most of the energetic charged particles from the sun—the protons and electrons of the solar wind. Interestingly, the ionosphere also reflects radio waves, making long-distance radio communication possible. You'll need to know how the climates that we experience on Earth are created by the atmosphere, so let's go into this next.

Climate

The Earth's atmosphere has physical features that change from day to day as well as patterns that are consistent over a space of many years. The day-to-day properties such as wind speed and direction, temperature, amount of sunlight, pressure, and humidity are referred to as **weather**. The patterns that are constant over many years (30 years or more) are referred to as **climate**. The two most important factors in describing climate are average temperature and average precipitation amounts. **Meteorologists** are scientists who study weather and climate.

The weather and climate of any given area are the result of the sun unequally warming the Earth (and the gases above it) as well as of the Earth's rotation.

Air Circulation in the Atmosphere

The motion of air around the globe is the result of solar heating, the rotation of the Earth, and the physical properties of air, water, and land. There are three major reasons why the Earth is unevenly heated.

- More of the sun's rays strike the Earth at the equator in each unit of surface area than strike the poles in the same unit area. This is because the angle of the sun's rays strikes the Earth more directly at the equator.

- The tilt of the Earth's axis points regions toward or away from the sun. When pointed toward the sun, those areas receive more direct or intense light than when pointed away. This causes the seasons.

- The Earth's surface at the equator is moving faster than at the poles, because the circumference is larger but the rotation time is the same. Because an object or air mass nearer to the equator is moving more rapidly (from east to west), it will maintain this eastward momentum as it moves away from the equator to where the surface is moving more slowly, winding up further east. Therefore, winds moving north from the equator near the surface are deflected to the right (east), and winds moving south from the equator are deflected to the left (east). Conversely, winds blowing toward the equator will be deflected to the west because they are moving eastward more slowly than is the surface in the lower latitudes where they are moving to. So, in the Northern Hemisphere, this westward deflection will be to the right, and in the Southern Hemisphere, it will be to the left. This deflection pattern is known as the **Coriolis effect**. The resulting wind patterns are known as the **prevailing winds:** belts of air that distribute heat and moisture unevenly around the globe.

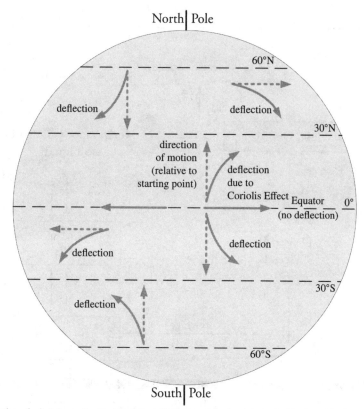

Deflection of wind due to the Coriolis Effect. Deflection is to the right in the Northern Hemisphere and to the left in the Southern Hemisphere. Winds heading due east and due west are still deflected, except at the equator. This is due to the interaction of rotational momentum in a circle, centrifugal force (directed outward from the Earth's axis), and gravity (directed inward toward the Earth's center).

Solar energy warms the Earth's surface. The heat is transferred to the atmosphere by radiation heating. The warmed gases expand, become less dense, and rise, creating vertical air flow called **convection currents**. The warm currents can also hold a lot of moisture compared to the surrounding air. As these large masses of warm, moist air rise, cool air flows along the Earth's surface to occupy the area vacated by the warm air. This flowing air or **horizontal airflow** is one way that surface winds are created.

As warm, moist air rises into the cooler atmosphere, it cools to the **dew point**, the temperature at which water vapor condenses into liquid water. This condensation creates clouds. If condensation continues and the water drops get bigger, they can no longer be held up by the convection in the Earth's atmosphere and they fall as **precipitation** (which can be frozen or liquid). This cold, dry air is now denser than the surrounding air. This air mass then sinks to the Earth's surface, where it is warmed and can gather more moisture, thus starting the **convection cell** rotation again.

Convection Cell

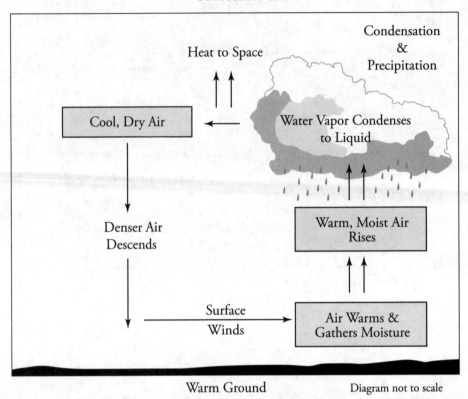

Heat to Space

Condensation
&
Precipitation

Cool, Dry Air

Water Vapor Condenses
to Liquid

Denser Air
Descends

Warm, Moist Air
Rises

Surface
Winds

Air Warms &
Gathers Moisture

Warm Ground

Diagram not to scale

On a local level, this phenomenon accounts for land and sea breezes. On a global scale, these cells are called **Hadley cells**. A large Hadley cell starts its cycle over the equator, where the warm, moist air evaporates and rises into the atmosphere. The precipitation in that region is one cause of the abundant equatorial rainforests. The cool, dry air then descends about 30 degrees north and south of the equator, forming the belts of deserts occurring around the Earth at those latitudes.

Hadley Cell

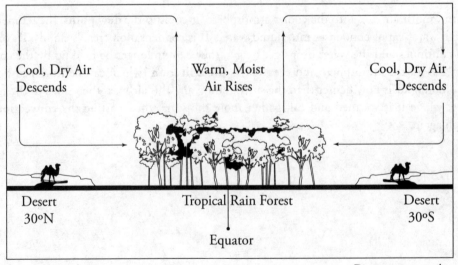

Cool, Dry Air
Descends

Warm, Moist
Air Rises

Cool, Dry Air
Descends

Desert
30°N

Tropical Rain Forest

Desert
30°S

Equator

Diagram not to scale

Seasons

The motion of the Earth around the sun and the Earth's axial tilt of 23.5 degrees together create the seasons that we experience on Earth. The Earth's tilt means that sunlight hits most directly and for the longest number of hours per day on the parts of the Earth that face the sun most directly, and those parts change across the Earth's orbit. In other words, the main source of energy for a given place on Earth (the sun's rays) varies depending on latitude and season: season because of the tilt, and latitude because the highest solar radiation per unit area is received at the equator and decreases toward the poles. Any place will receive the most solar radiation on its longest summer day and the least on its coldest winter day. The Earth's **revolution**, or trip around the sun, takes 365.25 days. The calendar that we use is based on the Earth's revolution around the sun: every four years, we have a leap year with one extra day to account for the accumulation of four quarter-days. As the Earth revolves around the sun, it is spinning on its **axis**; this spinning action is called **rotation**. On Earth, it takes 24 hours, or one day, to make a complete rotation.

Interestingly, because of the Earth's tilt, the sun rises and sets just once a year at the North and South Poles. Approximately six months of the year at the poles are daytime, while the other six months are nighttime.

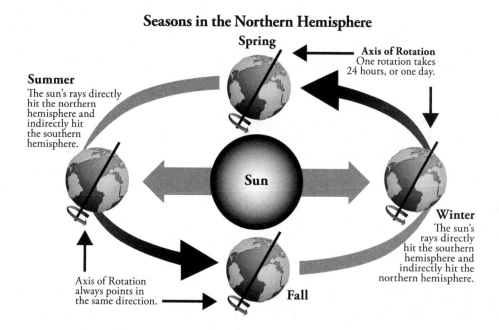

Seasons in the Northern Hemisphere

Spring

Axis of Rotation
One rotation takes 24 hours, or one day.

Summer
The sun's rays directly hit the northern hemisphere and indirectly hit the southern hemisphere.

Sun

Winter
The sun's rays directly hit the southern hemisphere and indirectly hit the northern hemisphere.

Axis of Rotation always points in the same direction.

Fall

Albedo (Reflectance)

Another important property that affects the climates of different regions on Earth is **albedo,** the percentage of **insolation** (incoming solar radiation) reflected by a surface. The lower the surface albedo, the more solar radiation is absorbed. An albedo value of 0 corresponds to zero reflectance and absorption of all radiation, whereas an albedo value of 1 corresponds to reflection of all incoming radiation. Snow and ice have high albedo values, while land and trees have lower albedo values. Changes in albedo can lead to alterations in temperature. For example, snowfall may raise the albedo of an area, leading to an increase in the reflection of solar radiation and a decrease in temperature.

Let's move on to discuss wind. You might be thinking, "I know what wind is, so I can skip this section!" Don't skip it! The AP Environmental Science Exam may ask you about the specific types of winds and air movements coming up next—so it's better to be safe than sorry.

Types of Winds

So, what is "wind"? Why does everyone refer to wind when they're discussing weather? Well, the term wind is widely used to refer to air currents, and we already know that air currents tend to flow from regions of high pressure to regions of low pressure. But let's review some important details you'll need to know about wind before we move on to our review of the hydrosphere. Formally speaking, **wind** is air that's moving as a result of the unequal heating of the Earth's atmosphere. It is part of the Earth's circulatory system and moves heat, moisture, soil, and even pollution around the planet.

One crucial wind-related phenomenon that you'll need to know about for the AP Environmental Science Exam is trade winds. **Trade winds** were named for their ability to quickly propel trading ships across the ocean. The trade winds that blow between about 30 degrees latitude and the equator are steady and strong, and travel at a speed of about 11 to 13 mph. They are caused by the surface currents of the Hadley cells, described above, along with the Earth's direction of rotation (counterclockwise if viewed toward the North Pole). In the Northern Hemisphere, the trade winds blow from the northeast and are known as the Northeast Trade Winds; in the Southern Hemisphere, the winds blow from the southeast and are called the Southeast Trade Winds.

Another important type of moving air mass, called a **westerly** (named for the direction from which it originates), travels north and east in the Northern Hemisphere and south and east in the Southern Hemisphere in the latitudes between 30 degrees and 60 degrees north and south of the equator. The movement of air that accounts for the westerlies, called the Ferrel cell, is the reverse of the Hadley cell but operates on the same thermodynamic principles. The eastward movement of westerlies is a result of the Coriolis effect. **Polar easterlies** are formed by similar forces: in polar easterlies, winds between latitudes of 60 degrees and the North Pole blow from the north and east, and winds between 60 degrees and the South Pole blow from the south and east.

It's important that you can distinguish among the different types of wind and blow through the wind questions on test day!

Between the wind belts mentioned above, air movement is less predictable, and often no wind blows at all for days. For example, between about 30 degrees to 35 degrees north and 30 degrees to 35 degrees south of the equator lie the regions known as the **horse latitudes** (or the subtropical high). Subsiding dry air and high pressure result in very weak winds in these regions. Some people say that sailors gave the regions of the subtropical highs the name "horse latitudes" because ships relying on wind were unable to sail in these areas—so, afraid of running out of food and water, sailors would throw their horses (and other live cargo) overboard to save on food and water and to make the ship lighter and easier to move.

Similarly, the air near the equator is relatively still because air there is constantly rising rather than blowing. For this reason, early sailors called this region the **doldrums.** The region of the doldrums, occurring between 5 degrees north and 5 degrees south of the equator, is also known as the Intertropical Convergence Zone, or ITCZ for short. The trade winds converge there, producing convectional storms that give the ITCZ some of the world's heaviest precipitation.

The last type of moving air system that you'll need to be familiar with for the exam is the **jet stream.** Jet streams are high-speed currents of wind that occur in the tropopause; these fast-moving air currents have a large influence on local weather patterns.

Winds Around the World

Now, let's move on to review the types of weather that result from all of these moving air masses.

Weather Events

Weather and climate are affected not only by the insolation in a given area but also by geologic and geographic factors, such as terrain (mountains, plains, distance from the ocean) and ocean temperatures. **Monsoons**, or seasonal winds that are usually accompanied by very heavy rainfall, occur when land heats up and cools down more quickly than water does. In a monsoon, hot air rises from the heated land and a low-pressure system is created. The rising air is quickly replaced by cooler moist air that blows in from over the ocean's surface. As this air rises over land, it cools, and the moisture it carries is released in a steady seasonal rainfall. This process happens in reverse during the dry season, when masses of air that have cooled over the land blow out over the ocean. Monsoons primarily occur in coastal areas. Check out the helpful illustration on the next page.

How a Monsoon Forms

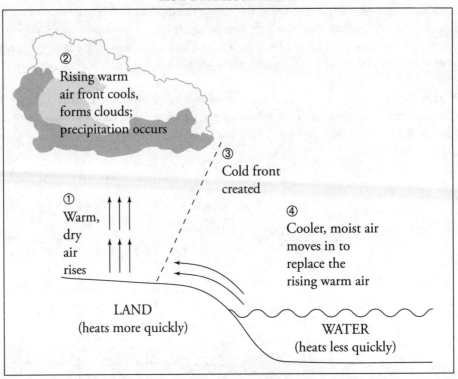

② Rising warm air front cools, forms clouds; precipitation occurs

③ Cold front created

① Warm, dry air rises

④ Cooler, moist air moves in to replace the rising warm air

LAND (heats more quickly)

WATER (heats less quickly)

On a smaller—local or regional—scale, this effect can be seen on the shores of large lakes or bays. In these areas, again the land warms faster than does the water during the day, so the air mass over the land rises. Air from over the lake moves in to replace it, and this creates a breeze. At night, the reverse happens: the land cools more quickly than the water, and the air over the lake rises. The air mass from the land moves out over the lake to replace the rising air, and this creates a breeze as well. This small-scale monsoon effect is called the **lake effect** (something of a misnomer since the phenomenon isn't limited to inland lakes). If you live near one of the Great Lakes or in the Bay Area of San Francisco, you may have experienced the lake effect yourself!

As we mentioned above, the air that moves in from over the ocean or a large body of water contains large amounts of water. If an air mass is forced to climb in altitude—if, for instance, it encounters an obstruction such as a mountain—the air will be forced to rise. When the air mass rises, it will cool, and water will precipitate out on the ocean side of the mountain. By the time the air mass reaches the opposite side of the mountain, it will be virtually devoid of moisture. This phenomenon is known as the **rain shadow effect** and is responsible for the impressive growth of the Olympic rainforest (within Olympic National Park) on the coast of Washington State. Interestingly, Olympic National Park has a temperate rainforest on its west side where annual precipitation is about 150 inches (making it perhaps the wettest area in the continental United States) and forests on its much drier east side, where the annual precipitation is around 15 inches.

How the Rain Shadow Effect Works

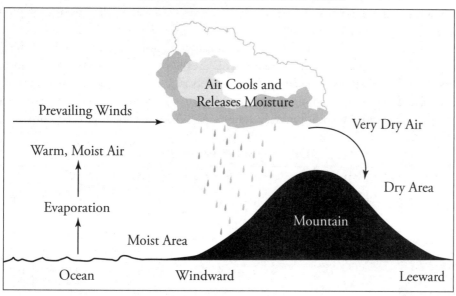

Remember trade winds from the last section? Well, they occur in steady and somewhat predictable wind patterns, but they may cause local disturbances when they blow over very warm ocean water. When this occurs, the air warms and forms an intense, isolated, low-pressure system while also picking up more water vapor from the ocean surface. The wind will circle around this isolated low-pressure air area (counterclockwise in the Northern Hemisphere and the opposite in the Southern Hemisphere—once again due to the Coriolis effect!). The low-pressure system will continue to move over warm water, increasing in strength and wind speed; this will eventually result in a tropical storm.

Certain tropical storms are of sufficient intensity to be classified as **hurricanes.** Hurricanes must have winds with speeds greater than 120 km/hr. The rotating winds of a hurricane remove water vapor from the ocean's surface, and heat is released as the water vapor condenses. This addition of heat energy continues to contribute to the increase in wind speed, and some hurricanes have winds traveling at speeds of nearly 400 km per hour! A major hurricane contains more energy than that released during a nuclear explosion, but since the force is released more slowly, the damage is generally less concentrated. Another important note about this type of storm is that they are referred to as hurricanes in the Atlantic Ocean, but they are called **typhoons** or **cyclones** when they occur in the Pacific Ocean. Go figure!

El Niño is a climate variation that takes place in the tropical Pacific about once every three to seven years, and it lasts for about one year. Under normal weather conditions, trade winds move the warm surface waters of the Pacific away from the west coast of Central and South America. As a result, the cold ocean water that lies under the displaced water moves to the surface (causing the thermocline, or line of demarcation between two layers of water with different temperatures, to rise), bringing nutrients with it and keeping the temperature of the coastal water relatively cool. This phenomenon is called upwelling.

During El Niño, the normal trade winds are weakened or reversed because of a reversal of the high and low pressure regions on either side of the tropical Pacific. This reversal of pressure systems is known as the **Southern Oscillation.** Without these regular trade winds off the Central and South American coast, the process of upwelling slows or stops, and the water off the coast becomes warmer and contains fewer nutrients. This means that during El Niño, the northern United States and Canada experience warmer winters and a less intense hurricane season; the eastern United States and regions of Peru and Ecuador that are typically dry have higher-than-average rainfall; and the Philippines, Indonesia, and Australia are drier than normal.

Differing Wind Patterns

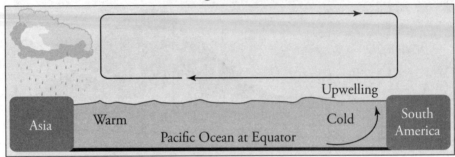

Normal Conditions

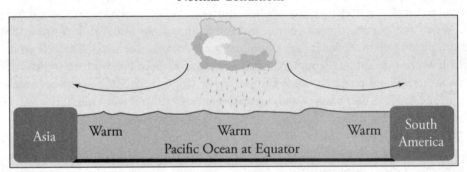

El Niño Conditions

One environmentally important effect that El Niño has on humans is that, because of the suppression of upwelling, the offshore fish populations of certain coastal areas decline. In countries like Peru, which relies heavily on fishing, El Niño has devastating economic effects.

The reverse of El Niño is known as **La Niña.** The Coriolis effect contributes to La Niña conditions. As air moves toward the equator to replace rising hot air, the moving air deflects to the west and helps move the surface water, allowing the upwelling. During La Niña, the surface waters of the ocean surrounding Central and South America are colder than normal. The term *El Niño* comes from the fact that traditionally these conditions were observed to begin around Christmas time (*"el niño"* means "the boy" in Spanish). The alternations of atmospheric conditions that lead to El Niños and La Niñas are referred to as **ENSO** (El Niño/Southern Oscillation) events. Keep in mind that El Niño and La Niña are large-scale climate patterns, but they are influenced by geological and geographic factors and affect different locations in different ways.

THE HYDROSPHERE

Water covers about 71 percent of planet Earth. Most of the water on the Earth's surface is saltwater. On average, the saltwater in the world's oceans has a salinity of about 3.5 percent. This means that for every 1 liter (1,000 ml) of sea water, there are 35 grams of salts (mostly, but not entirely, sodium chloride) dissolved in it. (1 ml of water weighs approximately 1 gram.) In fact, 1 cubic foot of seawater would evaporate to leave about 2 pounds of sea salt! However, **sea water** is not uniformly saline throughout the world. The planet's freshest sea water is in the Gulf of Finland, part of the Baltic Sea. The most saline open sea is the Red Sea, where high temperatures and confined circulation result in high rates of surface evaporation.

Freshwater is water that contains only minimal quantities of dissolved salts, especially sodium chloride. All freshwater ultimately comes from precipitation of atmospheric water vapor, which reaches inland lakes, rivers, and groundwater bodies directly, or after melting of snow or ice. Let's start with a discussion of freshwater before discussing the world's oceans.

Freshwater

Freshwater is deposited on the surface of the Earth through precipitation. Water that falls on the Earth and doesn't move through the soil to become groundwater moves along the Earth's surface via gravity, forms small streams, and then eventually forms larger ones. The size of the stream will continue to increase as water is added to it, until the stream becomes a river, and the river will flow until it reaches the ocean. The land area that drains into a particular stream is known as a **watershed,** or drainage basin. A particular watershed has some characteristics that define it: its area, length, slope, soil and vegetation types, and how it's separated from adjoining watersheds.

As water moves into streams, it carries with it sediment and other dissolved substances, including small amounts of oxygen. Turbulent waters are especially laden with dissolved oxygen and carbon dioxide, such as those found at the source, or headwaters, of a stream. As a general rule, the more turbulent the water, the more dissolved gases it will contain.

Freshwater Bodies

Freshwater bodies include streams, rivers, ponds, and lakes. As you probably know, freshwater that travels on land is largely responsible for shaping the Earth's surface. Erosion occurs when the movement of water etches channels into rocks and soil. The moving water then carries eroded material farther downstream.

Because of obstructions on land, moving water does not move in a straight line. Instead, it follows the lowest topographical path, and as it flows continuously along the same path, it cuts farther into its banks to eventually form a curving channel. As the water travels around these bends, its velocity decreases, and the stream drops some of its sedimentary load. Water always follows the path of least resistance as it travels from the highlands to the sea.

Rivers drop most of their sedimentary load as they meet the ocean because their velocity decreases significantly at this juncture. At these locations, landforms called **deltas**—which are made of deposited sediments—are created.

Ocean Currents

Ocean currents play a major role in modifying conditions around the Earth that can affect where certain climates are located. As the sun warms water in the equatorial regions of the globe, prevailing winds, differences in salinity (saltiness), and the Earth's rotation set masses of ocean water in motion. For example, in the Northern Hemisphere, the **Gulf Stream** carries sun-warmed water northward along the east coast of the United States and across the Atlantic Ocean as far as Great Britain. This warm water displaces the colder, denser water in the polar regions, which can move south to be re-warmed by the equatorial sun. Northern Europe is kept 5° to 10°C warmer than it would be were the current not present.

Oceanographers also study a major current, the **ocean conveyor belt**, that moves cold water in the depths of the Pacific Ocean while creating major upwellings in other areas of the Pacific.

Ocean Circulation

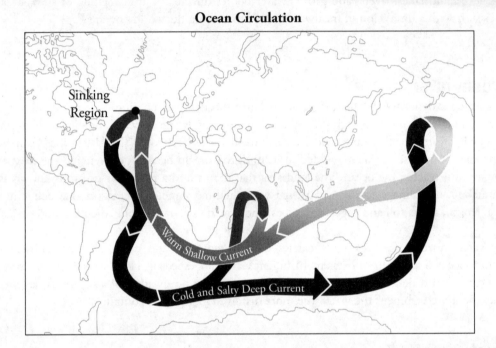

Whew. You're done with this chapter! Before moving on to the next chapter (Land and Water Use), answer all of the questions in the following drill—and don't forget to use the techniques you learned in Part III.

CHAPTER 6 KEY TERMS

Use this list to review the key terms in this chapter. Keep these terms in mind when you are brainstorming hot buttons for your essays.

Fundamental Themes

resource
solid Earth
lithosphere
pedosphere
atmosphere
hydrosphere
biosphere

Welcome to Planet Earth

mantle
asthenosphere
crust
Pangaea
tectonic plates
plate boundaries
convergent boundary
divergent boundary
transform fault boundary
subduction
active volcanoes
dormant volcanoes
extinct volcanoes
subduction zones
rift valleys
hot spots
shield volcanoes
composite volcanoes
cinder volcanoes
lava domes
earthquakes
focus
epicenter
seismograph
Richter scale
S-wave (shear wave)
tsunamis
rock cycle
igneous
magma
sedimentary
metamorphic

Soil

clay
silt
sand
composition
porosity
permeability
fertility
loam
water holding capacity
soil pH
weathering
erosion
physical (mechanical) weathering
chemical weathering
biological weathering
horizons
O horizon
humus
A horizon
leaching
E horizon
eluviation
B horizon
illuviation
C horizon
R horizon
additions
losses
translocations
transformations
Cl-O-R-P-T-H
climate
organisms
relief
parent material
time
human influence

The Atmosphere

troposphere
greenhouse effect
tropopause
stratosphere
ozone layer
mesosphere
thermosphere
exosphere
ionosphere
weather
climate
meteorologists
Coriolis effect
prevailing winds
convection currents
horizontal airflow
dew point
precipitation
convection cell
Hadley cells
revolution
axis
rotation
albedo

insolation
wind
trade winds
westerly
polar easterlies
horse latitudes
doldrums
jet stream
monsoons
lake effect
rain shadow effect
hurricanes
typhoons (cyclones)
El Niño
Southern Oscillation
La Niña
ENSO

The Hydrosphere

sea water
freshwater
watershed
deltas
Gulf Stream
ocean conveyor belt

Chapter 6 Drill

Directions: Each of the questions or incomplete statements below is followed by four suggested answers or completions. Select the one that is best in each case. For answers and explanations, see Chapter 13.

1. The soil mantle of the Earth is known as the

   (A) atmosphere

   (B) hydrosphere

   (C) lithosphere

   (D) pedosphere

2. The Earth is approximately

   (A) 6,000 years old

   (B) 45 million years old

   (C) 4.5 billion years old

   (D) 45 billion years old

3. Two tectonic plates move away from each other at a

   (A) transform fault boundary

   (B) friction boundary

   (C) divergent boundary

   (D) convergent boundary

4. Volcanoes that form at subduction zones and are associated with violent eruptions of lava, water, and gases are known as

   (A) shield volcanoes

   (B) cinder volcanoes

   (C) composite volcanoes

   (D) lava domes

5. All of the following statements about the rock cycle are true EXCEPT

   (A) in the rock cycle, time, pressure, and water interact to create igneous, sedimentary, and metamorphic rock

   (B) limestone is a type of sedimentary rock

   (C) igneous rock is created when magma solidifies when cooled

   (D) metamorphic rock is formed when high pressure and heat produce physical and/or chemical changes in existing rock

6. Soil

   I. consists of clay, silt, and sand particles
   II. that has a high proportion of sand particles is very compact
   III. composition impacts its fertility, permeability, and porosity

   (A) I only

   (B) II only

   (C) II and III only

   (D) I and III only

7. The soil horizon that is known as the zone of leaching is the

   (A) A horizon

   (B) B horizon

   (C) E horizon

   (D) O horizon

8. All of the following are types of weathering processes that create soil EXCEPT

   (A) biological weathering

   (B) chemical weathering

   (C) physical weathering

   (D) temporal weathering

9. The layer of the Earth's atmosphere that exists between 50–80 km above Earth's surface, and is where meteors usually burn up, is known as the

   (A) thermosphere

   (B) stratosphere

   (C) mesosphere

   (D) troposphere

10. The alternations of atmospheric conditions that lead to El Niño and La Niña are referred to as

(A) ELLA events

(B) ENSO events

(C) cyclones

(D) typhoons

11. The albedo value of snow is

(A) zero

(B) low

(C) medium

(D) high

Free-Response Question

1. Global climate change due to anthropogenic greenhouse gas emissions is having a huge effect on local climate and weather all over the Earth.

   (a) **Explain** how the greenhouse effect works and why increases in greenhouse gases intensify it.

   (b) The changes in weather and climate involved in global climate change will interact with the pattern of the seasons to change local weather and shift biomes worldwide. **Describe** THREE reasons for the pattern of the seasons and **explain** how increased temperatures might affect these patterns in a given area.

   (c) The intensification of the greenhouse effect is the strongest in the polar regions. **Identify** TWO factors that make this so.

   (d) One example of a local climate change effect is the intensification of weather patterns such as hurricanes. **Describe** what a hurricane is and how it develops.

Summary

o The lithosphere is constantly changing due to plate tectonics, volcanoes, earthquakes, and the rock cycle.

o The soil that makes up part of the lithosphere is the combination of rock and organic matter formed by:
   • physical weathering
   • chemical weathering
   • biological weathering
   • decomposition of organic matter

o As rock weathers into soil, the soil develops horizons or stacked layers, each with its own properties of texture, color, and mineral/chemical properties. These horizons take hundreds to thousands of years or longer to develop. Here are the soil-forming factors that interact to make this happen:
   • climate
   • organisms (biological activity)
   • relief (topographical relief, that is)
   • parent material
   • time
   • human influence

o There are two zones of the atmosphere that are essential for life on Earth:
   • The troposphere contains greenhouse gases (GHGs) that regulate the temperature of the Earth and is where all weather takes place.
   • The stratosphere contains the ozone layer, which protects the Earth from harmful ultraviolet radiation.

o Climate is dictated by a region's average temperature and precipitation, which is influenced by air circulation from convection cells and the Earth's tilt, daily rotation around its axis, and annual revolution around the sun.

o Unique local weather events include:
   • monsoons
   • lake effect
   • rain shadow effect
   • El Niño

o The movement of ocean water around the globe is due to differences in density between layers, prevailing winds, regional air temperatures, and the Earth's rotation. The resulting currents in turn influence weather patterns and thus terrestrial environments.

Chapter 7
Unit 5: Land and Water Use

In this chapter, we'll review Unit 5 of the AP Environmental Science course, *Land and Water Use*. According to the College Board, about 10–15% of the test is based directly on the ideas covered in this chapter. If you are unfamiliar with a topic presented here, consult your textbook for more in-depth information.

As we'll see in this chapter, humans use the land and water for countless reasons. We will begin our discussion with a description of the resources of the world—including what happens if people don't get enough resources and who has too few. We'll then go through the ways land and water are used for agriculture, forest resources, water use, mining, and urban development. We'll end with a (short!) discussion of the economics behind our resource use and an introduction to the idea of sustainability. Let's begin!

SHARE AND SHARE ALIKE?

When people talk about managing common property resources such as air, water, and land, the **Tragedy of the Commons** often comes to mind. This is an important concept introduced by the English economist William Forster Lloyd in 1833 and later applied to the field of natural resource management by Garrett Hardin in a 1968 paper published in *Science* magazine. In his paper, Hardin referenced the example used by Lloyd, in which a piece of open land, a commons, was to be used collectively by the townspeople for grazing their cattle. Each townsperson who used the land continued to add one cow or ox at a time until the commons was overgrazed. Hardin eloquently says, "Each [person] is locked into a system that compels him to increase his herd without limit—in a world that is limited. Ruin is the destination toward which all [people] rush, each pursuing his own best interest in a society that believes in the freedom of the commons. Freedom in a commons brings ruin to all."

The Tragedy of the Commons serves as a foundation for modern conservation. **Conservation** is the management or regulation of a resource so that its use does not exceed the capacity of the resource to regenerate itself. This is different from **preservation,** which is the maintenance of a species or ecosystem in order to ensure their perpetuation, with no concern as to their potential monetary value.

In this chapter, we'll continue to show how human economics influence how we interact with the Earth's resources. Bear in mind that natural resources are drawn from the biotic and abiotic components of functioning ecosystems, and so our exploitation of those resources necessarily affects the functioning of those ecosystems. Human impact in turn affects the ability of those ecosystems to continue providing the resources. When we humans exploit a resource for the functioning of society or for economic gain, we are placing an economic value on it; therefore, natural resources are described in terms of their value as **ecosystem capital** or natural capital.

Some Terms Used to Describe Resources

Let's start by discussing the two main types of resources.

- **Renewable resources** are resources that can be regenerated quickly, such as plants and animals. Water is an abiotic substance that's renewable because it can be used over and over again and because sources of water are replenished naturally through the water cycle. Certain natural sources of energy—such as the sun, the wind, and the tides—are also considered renewable because their occurrence in nature is perpetual and not depleted with use. The time necessary for hardwood trees to mature (about 50 years) is widely considered the cross-over point from renewable resources to nonrenewable resources. But, in purely practical terms, a resource is renewable if it can be replenished within the time it takes to draw down its supply. Bear in mind that even renewable resources must be carefully managed in order to conserve their sources and insure an ongoing supply.

- **Nonrenewable resources** are resources that do not regenerate quickly, such as minerals and fossil fuels. Nonrenewable resources are typically formed by very slow geologic processes, so we consider them incapable of being regenerated within the realm of human existence.

There are a couple more terms you should know before we dive into our review of the major resources available to humans on Earth; these are consumption and production. The **consumption** of natural resources refers to the day-to-day use of environmental resources such as food, clothing, and housing. On the other hand, **production** refers to the use of environmental resources for profit. An example of this might be a fisherman who sells his fish in a market. Got those terms? Let's move on.

AGRICULTURE

How do resources relate to your dinner? Well, 77 percent of the world's food comes from croplands, 16 percent comes from grazing lands, and 7 percent comes from ocean resources. Despite the importance of our ever-increasing population, fewer people than ever in the history of the United States now farm the land. Why is this? The short answer is that it has a lot to do with increasing urbanization and industrialization. Now that machines are readily available to work the land and harvest crops, farms have become more like factories—currently only 2 percent of the United States population is directly employed in agriculture. Farms in the United States today are quite a bit larger than farms of the past. According to the USDA, in 2021, the average farm size was 445 acres, while in the early 20th century the average farm size was about 100 acres.

The use of machinery in farming has allowed farmers to work more land more efficiently; however, one of the drawbacks of the machinery is the amount of fossil fuel needed to power it. As the cost of fuel rises, the cost of food will also rise.

This rise in agricultural productivity can be tied to new pesticides and fertilizers, expanded irrigation, and the development of new high-yield seed types. However, it has also resulted in a significant decrease in the genetic variability of crop plants and led to huge problems in erosion.

Traditional Agriculture and the Green Revolution

Throughout most of history, agriculture all over the world was such that each family grew crops for itself, and families relied primarily on animal and human labor to plant and harvest crops. This process is called **traditional subsistence agriculture,** and it provides enough food for one family's survival. Traditional subsistence agriculture is currently practiced by about 42 percent of the world's population, predominantly in developing nations. Such intensive mixed farming allows people to settle permanently and subsist without having to migrate seasonally. Extensive subsistence agriculture results in low amounts of labor inputs per unit of land.

Do not confuse the Green Revolution, which is about farming, with the Green Movement, which is about conservation.

One form of traditional agriculture that's still practiced in many developing countries today is a method called **slash-and-burn,** a practice that dates back to early humankind and is especially common in the tropics. In slash-and-burn, an area of vegetation is cut down and burned before being planted with crops. Tropical soils are typically thin and poor, and whatever fertility they hold is rapidly depleted by the deforestation and subsequent farming. Therefore, the farmer must leave the area after a relatively short time and find another location to clear. Practiced indiscriminately on a broad scale, slash-and-burn agriculture has led to rapid deforestation of the tropical rainforest.

The **Green Revolution**, which occurred in the 1950s and 1960s, is generally thought of as the time after the Industrial Revolution when farming became mechanized and crop yields in industrialized nations boomed. Such innovations also allowed farmers in developing countries to increase crop production on small plots of land. Later, there was a second green revolution, which promoted integrated pest management and organic methods, such as fertilizers that are not synthetic.

Fertilizers and Pesticides

One factor that contributed to the Green Revolution was an increase in the use of fertilizers and pesticides. Interestingly, when the non-native settlers (the first white settlers) planted their first corn crops, certain tribes of Native Americans taught them to plant fish leftovers (the inedible parts of the fish) along with the corn seed. The fish acted as a natural fertilizer for the crops. As you can see, manures and other organic materials have been used as fertilizers by farmers for many years. However, the development of inorganic (chemical) fertilizers brought about the huge increases in farm production seen during the Green Revolution. It's estimated that if chemical fertilizers were suddenly no longer used, then the total output of food in the world would drop about 40 percent!

Of course, there are downsides to the widespread use of chemical fertilizers, including the reduction of organic matter and oxygen in soil; the large amounts of energy needed to produce, transport, and supply the fertilizers; and the fact that once the fertilizers are washed into watersheds, they are dangerous pollutants.

Likewise, the increased use of pesticides in the Green Revolution has significantly reduced the number of crops lost to insects, fungi, and other pests, but these chemicals have also had an effect on ecosystems in and surrounding farms. It's estimated that the average insect pesticide will only be useful for 5–10 years before its target pest evolves to become immune to its effects through natural selection; therefore, new pesticides must constantly be developed. However, even with this constant development, crop loss due to pests has not decreased since 1970, although the use of pesticides has tripled!

Because the use of pesticides is so prevalent in the United States, Congress passed the Federal Insecticide, Fungicide, and Rodenticide Act (FIFRA) in 1947 and amended it in 1972. This law requires the EPA to approve the use of all pesticides in the United States.

Integrated Pest Management

When dealing with pests, **integrated pest management (IPM)** uses a combination of several methods and is a more environmentally sensitive approach than chemical pesticides. Rather than try to get rid of every single pest on the farm, IPM tries to keep the pest population down to an economically viable level. Some of the methods include introducing natural insect predators to the area, intercropping, using mulch to control weeds, diversifying crops, crop rotation, releasing pheromone or hormone interrupters, using traps, and constructing barriers. People using IPM consider using chemical pesticides only in the worst-case scenario. While IPM has great advantages in terms of reducing the risk that traditional pesticides pose to wildlife, water supplies, and human health, it can be complex, requiring a lot of work to implement, and it can also be quite expensive.

Irrigation

Another major contributor to the increased crop yields seen in the Green Revolution was advanced irrigation techniques, which allowed crops to be planted in areas that normally would not have enough precipitation to sustain them. However, repeated irrigation can cause serious problems, including a significant buildup of salts on the soil's surface, which makes the land unusable for crops. To combat this **salinization** of the land, farmers have begun flooding fields with massive amounts of water in order to move the salt deeper into the soil. The drawback to this, however, is that the large amounts of water can waterlog plant roots, which will kill the crops, and this process also causes the water table of the region to rise. Furthermore, the water for these irrigation farms comes from underground water tables called aquifers. These aquifers are being depleted at a rapid rate and large-scale, grain-producing countries such as India, China, and the United States are examples of those caught in this predicament.

Another drawback to the Green Revolution resulted from the dramatic increase in irrigation worldwide; the largest human use of freshwater, 70%, is for irrigation! **Furrow irrigation,** which involves cutting furrows between crop rows and filling them with water, is inexpensive but loses about 1/3 of the water used to evaporation and runoff. **Flood irrigation** (mentioned above), which involves flooding a field with water, can lead to waterlogging and loses about 20% of the water to evaporation and runoff. **Spray irrigation** involves pumping water into spray nozzles and spraying fields; it loses only about 1/4 of the water but requires energy to run and can be expensive. One problem is waterlogging: if too much water is left to sit in soil, it can raise the water table of the groundwater, causing plants to have trouble absorbing oxygen through their roots. Additionally, over-irrigated soils undergo salinization. In salinization, the soil becomes waterlogged; when it dries out, salt forms a layer on its surface. This eventually leads to **land degradation.** In order to combat this problem, researchers have developed **drip irrigation,** which allots an area only as much water as is necessary and delivers the water directly to the roots using perforated hoses that release small amounts of water: this is far more efficient, with only about 5% of water lost to evaporation and runoff, but is more expensive.

Genetically Engineered Plants

The third and last significant contributor to the Green Revolution was the introduction of genetically engineered plants. In genetic engineering, scientists try to improve plants by adding genes from one species to another to encourage desirable characteristics, such as longer shelf life, disease/drought/pest resistance, faster growth, and higher crop yields. One beneficial example of this method was the development of golden rice, which contains vitamin A and iron. The introduction of this rice addresses two of the serious health problems that are seen in developing nations: vitamin A deficiency, which can result in blindness and other serious health problems; and iron deficiency, which leads to anemia.

Main GMO Crops Grown in the U.S.

Corn   Soybeans   Cotton   Potatoes   Canola

Alfalfa   Apples   Sugar Beets   Papaya

However, there are many problems that arise from **genetically modified organisms (GMOs)**, as well. Therefore, GMOs have become a very controversial topic. Because this is a relatively new technology, scientists don't know exactly how GMOs will affect the planet ecologically. Genetically modified plants discourage biodiversity, which may harm beneficial insects and organisms, could pose new allergen risks, may increase antibiotic resistance, and could encourage the rise of new pesticide-resistant pests. Many farmers and consumers are also concerned that cross-pollination can contaminate other crops, including organic farms that choose not to use GMOs, or cause unwanted mutations with unknown results.

Monotonous Monoculture

Believe it or not, three grains provide more than half of the total calories that are consumed worldwide! These three crops are rice, wheat, and corn, and the phenomenal increase in the yield of these crops was a result of genetic engineering. Genetic engineers discovered a way to cause plants to divert more of their photosynthetic products (called **photosynthate**) to grain biomass rather than plant body biomass.

It's estimated that of the roughly 50,000 plant species that could possibly be used for food, only 10,000 have been used historically with any regularity. Today, 90 percent of the caloric intake worldwide is supplied by just fifteen plant species and eight terrestrial animal species! In other words, today's agriculture represents a major reduction in agricultural biodiversity.

Much of the farming that occurs today is characterized by **monoculture.** In a monoculture, just one type of plant is planted in a large area. Monocultures became common in the era of early political civilizations, when farms produced a staple crop in order to feed whole societies and armies. As we discussed earlier, this has proved to be an unwise practice for numerous reasons. **Plantation farming,** which is practiced mainly in tropical developing nations, is a type of industrialized agriculture in which a monoculture cash crop such as bananas, coffee, or vegetables, is grown and then exported to developed nations.

Soil Problems for (and Caused by) Humans

In order to be able to grow all of the foods that humans consume, we must have enough **arable**—suitable for plant growth—soil to meet our agricultural needs. Soil fertility refers to soil's ability to provide essential nutrients, like nitrogen (N), potassium (K), and phosphorus (P), to plants. Humus (remember, it's in the O layer!) is also an extremely important component of soil because it is rich in organic matter.

Remember that soils composed of a balanced mixture of the three particle sizes (clay, silt, and sand) are described as **loamy,** and these types of soil are considered the best for plant growth. Another important characteristic for agricultural purposes is **soil structure**, or the extent to which it aggregates or clumps. Soil **aggregates** are formed and held together by such substances as clay particles and organic matter—plants and roots, the root-like filaments of fungi, and sticky substances released by bacteria and fungi. The most fertile soils have good structure.

Soil is considered a nonrenewable resource due to the great length of time required to form arable soil. It takes 500 to 1,000 years to form a single inch of soil, and at least 3,000 years to form enough fertile soil to support crop growth.

Unfortunately, certain agricultural activities can change the texture and structure of soil; for example, repeated plowing tends to break down soil aggregates, leaving "plow pan" or "hard pan," which is hard, unfertile soil.

Whereas communities traditionally planted many different types of crops in a field, in modern agriculture monoculture, or the planting of just one type of crop over a large area, predominates. Over the history of agriculture, a significant decrease in the biodiversity of crop species has taken place—both in the number of crop species and in the genetic makeup of individual species. This creates numerous problems. First of all, a lack of genetic variation makes crops more susceptible to pests and diseases. Secondly, the consistent planting of one crop in an area eventually leaches the soil in that area of the specific nutrients that the plant needs in order to grow. One way to prevent this phenomenon is to practice **crop rotation**, in which different crops are planted in the area in each growing season. Another practice that farmers will use to increase sustainability is **polyculture**, planting several crops on the same plot of land simultaneously. This increases biodiversity.

Other problems with modern agriculture include its reliance on large machinery (which can damage soil through compaction), and the fact that as an industry, agriculture is a huge consumer of energy. Energy is consumed both in the production of pesticides and fertilizers and in the use of fossil fuels to run farm machinery.

Soil Erosion

The small rock fragments that result from weathering may be moved to new locations in the process of erosion, and bare soil (upon which no plants are growing) is more susceptible to erosion than soil covered by organic materials.

Because of the constant movement of water and wind on the Earth's surface, the erosion of soil is a continual and normal process. However, when erosion removes valuable topsoil or deposits soil in undesirable places, it can become a problem for humans. Eroded topsoil usually ends up in bodies of water, posing a problem for both farmers, who need healthy soil for planting, and people in general, who rely on bodies of water to be uncontaminated with soil runoff.

The most significant portion of erosion caused by humans results from logging and from agriculture—especially slash-and-burn agriculture, which we've already discussed. The removal of plants in an area makes the soil much more susceptible to the agents of erosion.

Unfortunately, human activities—unsustainable agricultural practices, overgrazing, urbanization, and development deforestation—have significantly increased the levels of erosion in the upper layers of soil. These processes will continue to create problems for farmers searching for arable land until new techniques that preserve the integrity of soil are introduced and utilized.

Soil Degradation

Have you ever read *The Grapes of Wrath* by John Steinbeck? Well, the story in this book took place in the 1930s, when droughts in the Great Plains reduced the area to a giant **Dust Bowl**. Although the drought was the major cause of the Dust Bowl, farming practices used at that time also contributed to the destruction of the land.

Much of what we know about soil conservation was established relatively recently. The Soil Conservation Act was passed in 1935 and led to the creation of the Soil Conservation Service. These developments came in response to the Dust Bowl of the 1930s, which was a period of unprecedented dust storms caused by severe drought and ill-advised farming practices. The Soil Conservation Service was a federal agency founded by Hugh Hammond Bennett. Its mission was to promote sustainable soil conservation practices among farmers and other landowners and to help restore ecological balance across the nation's landscape. The agency is now called the Natural Resources Conservation Service.

Soil Conservation

In order to conserve soil resources, several best management practices have been developed. These practices return organic matter to the soil, slow down the effects of wind, and reduce the damage to the soil from tillage (plowing). Here are some of the more common methods.

- Use of animal waste (manure), compost, and the residue of plants to increase the amount of organic matter in the soil.

- The practice of organic agriculture, a method of farming that utilizes compost, manure, crop rotation, and non-chemical methods to enhance soil fertility and control pests. Organic producers avoid or strictly limit the use of chemical fertilizers and pesticides as well as genetically modified organisms.

- Modification of tillage practices to reduce the breakup of soil and to reduce the amount of erosion. These include no-till farming, contour plowing, and strip planting.

- Use of trees and other wind barriers to reduce erosion from wind.

The practice of **contour plowing,** in which rows of crops are plowed across a hillside, prevents the erosion that can occur when rows are cut up and down on a slope. **Terracing** also aids in preventing soil erosion on steep slopes. Terraces are flat platforms that are cut into the hillside to provide a level planting surface; this reduces the soil runoff from the slope. Additionally, **no-till methods** are quite beneficial; in no-till agriculture, farmers plant seeds without using a plow to turn the soil. Soil loses most of its carbon content during plowing. Plowing accelerates the decomposition of organic matter in the soil, decreasing soil fertility and releasing carbon dioxide gas into the atmosphere. (And as you know, increased levels of CO_2 in the atmosphere have been associated with global climate change!) **Perennial crops**—crops that grow back without replanting each year—are another way to reduce the need to till (by eliminating replanting) and keep erosion at bay. A **windbreak** is made up of one or more rows of trees or shrubs planted near crops in such a way as to provide shelter from eroding winds.

Finally, crop rotation can provide soils with nutrients when legumes are part of the cycle of crops in an area. An alternate to crop rotation is **intercropping** (also called **strip cropping**), which is the practice of planting bands of different crops in a field. This type of planting can also prevent some erosion by creating an extensive network of roots. As you might be aware, plant roots hold the soil in place and reduce or prevent soil erosion. Another way to prevent soil degradation is to add nutrients to the soil using green manure or limestone. **Green manure** is made by leaving plants (uprooted or simply sown) to wither and then serve as mulch: they are plowed under and incorporated into the soil before they can rot, providing valuable nutrients. Specific cover crops can be grown for this purpose. Alternatively, pulverized limestone can be used as a soil conditioner to neutralize soils with too much acidity.

Soil Laws

While the following soil laws are not required for the exam, they are relevant to the discussion of soil!

Date	Name of Legislation	What It Did
1977	Soil and Water Conservation Act	This act established soil and water conservation programs to aid landowners and users; it also set up conditions to continue evaluating the condition of U.S. soil, water, and related resources.
1985	Food Security Act	Nicknamed the Swampbuster, this act discouraged the conversion of wetlands to nonwetlands. In 1990, federal legislation denied federal farm supplements to those who converted wetlands to agriculture and provided a restoration of benefits to those who converted lands to wetlands.

The Livestock Business

Perhaps not surprisingly, the introduction of all these new agricultural techniques has significantly affected the livestock business. **Free-range grazing,** which simply implies that animals are able to move about outdoors and eat the foods they are adapted to eat, is a new term for the traditional way livestock animals were fed: by grazing on the land. In contrast, new meat production industry techniques include **feedlots,** or **concentrated animal feeding operations (CAFOs),** in which animals are confined and concentrated into smaller spaces in order to keep costs down and quickly get livestock ready for slaughter. They tend to be crowded and are often fed grains or feed rather than grass. Feedlots often require the use of antibiotics to prevent the spread of disease among animals densely packed together, and create problems in disposing of animal waste, which can contaminate ground and surface water. Manure is not used as fertilizer due to difficulty with transport. It has instead become the most widespread source of water pollution in the United States. Free-range animals tend to be free from antibiotics and their waste can be used as fertilizer, but since this method requires large areas of land, the meat produced is more expensive for consumers. As long as the grazing area is sufficient for the number of animals, livestock grazing is a sustainable practice. If, however, grass is consumed by animals at a faster rate than it can regrow, land is considered **overgrazed.** Overgrazing is harmful to the soil because it leads to erosion and soil compaction. Overgrazing can cause desertification: the degradation of low-precipitation regions toward being increasingly arid until they become deserts. One solution to the problem of overgrazing is similar to crop rotation. **Rotational grazing** is the regular rotation of livestock between different pastures in order to avoid overgrazing in a particular area. Another solution involves the overall control of herd numbers.

Various tracts of public lands are available for use as rangeland, and cooperation between government agents, environmentalists, and ranchers can help avoid problems of overgrazing on these lands. The Bureau of Land Management is responsible for managing federal rangelands.

Grazing animals also consume 70 percent of the total grain crop consumed in the United States, making them expensive food stuff. Meat production is less efficient than agriculture: it takes approximately 20 times more land to produce the same number of calories from meat as from plants. One possible solution is for people to consume less meat overall: this could reduce emissions such as carbon dioxide, methane, and nitrogen oxides; conserve water and reduce water pollution; reduce the use of antibiotics and growth hormones; and improve topsoil.

FOREST RESOURCES

Many environmentalists are concerned about the deforestation that is taking place in North America. It is interesting to note that the number of trees growing in North America is approximately the same as 100 years ago, but only 5 percent of the original forests are left. The numbers are approximately the same because of the number of trees growing in national parks and tree plantations. What does this mean? It means that most of the trees in North America are young, and that most forests have been harvested and replanted, and have undergone significant succession.

Deforestation

Deforestation, or the removal of trees for agricultural purposes or purposes of exportation, is a major issue for conservationists and environmentalists. Worldwide, industrialized countries have a higher demand for wood and less deforestation, while developing countries exhibit a smaller demand for wood, but more deforestation. This can be partly explained by the fact that the deforestation that occurs in developing countries primarily takes place because land is being cleared for pastures and farms. Industrialized countries also import lumber from developing countries.

Nearly all of the deforestation that takes place in North America is done in order to create space for homes and agricultural plots. In sites where deforestation is occurring, the impact on resident ecosystems is significant. Take, for instance, Canada's Vancouver Island. On this island, whole mountainsides have been stripped bare of the centuries-old forests that once existed. While the lumber industry tries to offset this destruction by planting new trees, the saplings, which won't be harvestable for another 50 years, are no substitute for forests of 300-foot giant redwoods. Remember how we talked about ecological succession in Chapter 4? Where do you think all of the plants and animals that relied upon this old growth forest ecosystem (which was a climax community) went to live?

Despite the moral questionability of this habitat destruction, the lumber industry will not be asked to leave Vancouver Island. This is because it's the island's most important source of income. Fifty cents of every dollar the island earns comes from lumbering—this number easily beats the island's income from tourism, which is the runner-up. This type of deforestation, also called clear-cutting, has other consequences as well. The areas affected experience a great deal of runoff due to the loss of root structure, which leads to more erosion. The soil is washed into freshwater streams and rivers and makes the environment less suited for salmon. The loss of shade also leads to higher stream temperatures, which also affect aquatic organisms.

Another environmentally negative by-product of deforestation is seen in countries with tropical forests. In these forests, when trees are removed and farms are placed in the cleared land, the already-poor soil is further degraded, and the area can only support crops for a short time. Usually, once the soil will no longer support a crop, the land will be used for grazing, but the soil becomes more and more depleted over time until humans have no use for it.

Additionally, any forest is made up of trees, which absorb pollutants and store carbon dioxide. Cutting and burning trees releases carbon dioxide (along with preventing those trees from absorbing it in the future), so deforestation contributes to climate change. The negative repercussions of clearing tropical rainforests—the losses in biodiversity, and the erosion and depletion of nutrients in the soil, and the release of carbon dioxide—seem to outweigh the economic gains in many people's opinions. However, for those who would like to take a stand by refusing to purchase wood from tropical rainforests, it is often difficult to determine which wood products come from tropical rainforests cleared for slash-and-burn agriculture and which come from sustainable forests. Various organizations, such as the nonprofit group the Forest Stewardship Council, have developed certifying procedures based on standards that will encourage only the use of wood from sustainable forests.

How Can We Use Forests Sustainably?

There are three major types of forests, which are categorized based on the age and structure of their trees. An **old growth forest** is one that has never been cut; these forests have not been seriously disturbed for several hundred years. Not surprisingly, the controversies that revolve around the issue of deforestation are primarily centered on instances in which deforestation is occurring in old growth forests. As we mentioned in the last section, old growth forests contain incredible biodiversity, with myriad habitats and highly evolved, intricate niches for a multitude of organisms. **Second growth forests** are areas where cutting has occurred and a new, younger forest has arisen naturally. About 95 percent of the world's forests are naturally occurring, and the remaining forests are known as **plantations** or **tree farms.** Plantations are planted and managed tracts of trees of the same age (because they were planted by humans at the same time) that are harvested for commercial use.

It makes sense that those in the forestry business would be concerned about finding a way to promote sustainable forestry, because without forests they have no way of perpetuating their income. From an economic viewpoint, the forest must be managed to continually supply humans' need for wood. The management of forest plantations for the purpose of harvesting timber is called **silviculture.** This relatively modern field has a basic tenet to create a sustainable yield; to do this, humans must harvest only as many trees as they can replace through planting. There are two basic management plans that attempt to uphold this tenet.

- **Clear-cutting** is the removal of all of the trees in an area. This is typically done in areas that support fast-growing trees, such as pines. Obviously, this is the most efficient way for humans to harvest the trees, but it has major impacts on the habitat, as in our previous example of Vancouver Island.

- **Selective cutting** is the removal of select trees in an area. This leaves the majority of the habitat in place and has less of an impact on the ecosystem. When selective cutting is used, it's quite difficult to remove these trees from the forest, though. This type of **uneven-aged management** is more common in areas with trees that take longer to grow or if the forester is only interested in one or more specific types of trees that grow in the area. Another type of uneven-aged management occurs in **shelter-wood cutting**. For shelter-wood cutting, mature trees are cut over a period of time (usually 10–20 years); this leaves some mature trees in place to reseed the forest.

In the case of **agroforestry,** trees and crops are planted together. This creates a mutualistic symbiotic relationship between the trees and crops—the trees create habitats for animals that prey upon the pests that harm crops, and their roots also stabilize and enrich the soil. Of course, two other options that can also have a great impact in mitigating deforestation are reforestation (planting new forests) and reuse of existing wood.

National Forest Policy

The federal government owns about 28 percent of all land in the United States. The need to preserve some of the land was recognized by President Lincoln when he set aside a park in Yosemite, California as a land grant (the precursor to the National Park System). In 1916, the National Park System was created in part to manage and preserve forests and grasslands. Today, in addition to the National Park System, there are several ways the federal government controls forested land: Wilderness Preservation Areas are open only for recreational activities with no logging permitted. The National Forest System, Natural Resource Lands, and National Wildlife Refuges are the other groups of federally controlled lands that allow logging with a permit.

One more point about managing treed areas: recent times have seen an increase in the number of greenbelts, nationally. **Greenbelts** are open or forested areas built at the outer edges of cities. Since no one is permitted to build in them, they can increase the quality of life for people living nearby. They also border cities, putting limits on their growth. Sometimes, satellite towns are built outside the greenbelts and interconnected with the cities by highways and mass transportation methods; in this way, we can add green spaces in urban areas.

Natural Events (That Create Problems for Humans) in Forests

Tree diseases (usually caused by fungal pathogens) and insect pests of trees are two natural problems in forested areas. These can create problems for humans (in addition to the trees) because, oftentimes, they affect the quality of the food and the number of trees that are available for use. Some of the most devastating pathogens and diseases are non-native invasive species, introduced—intentionally or not—through human travel and commerce. Humans manage these natural events in many different ways: by removing infected trees, by removing select trees or planting them sparsely to provide adequate spacing between them, by using chemical and natural pest controls, such as integrated pest management, by carefully inspecting imported trees and tree products, and by developing pest- and disease-resistant species of trees through genetic engineering.

Forest fires are another natural occurrence. There are three major types of fires that occur in forests, and you should be familiar with them for the test.

- **Surface fires** typically burn only the forests' underbrush and do little damage to mature trees. These fires actually serve to protect the forest from more harmful fires by removing underbrush and dead materials that would burn quickly and at high temperatures if they accumulate, escalating more severe fires.

- **Crown fires** may start on the ground or in the canopies of forests that have not experienced recent surface fires. They spread quickly and are characterized by high temperatures because they consume underbrush and dead material on the forest floor. These fires are a huge threat to wildlife, human life, and property.

- **Ground fires** are smoldering fires that take place in bogs or swamps and can burn underground for days or weeks. Originating from surface fires, ground fires are difficult to detect and extinguish.

One final note about forest fires: most people believe that forest fires are a bad thing despite the fact that they are part of the natural life of a forest. Some trees and plants even need fire in order for their seeds to germinate. The U.S. Forest Service started an advertising campaign to warn people about the ravages of fires and soon adopted "Smokey Bear" to help get the message out. This policy reduced the number of fires, but it also created conditions for more destructive fires. Under natural conditions, fires burn every few years and consume the fuel (dry leaves, needles, and wood) on the forest floor. However, if there are fewer fires, the amount of fuel can build up to very high levels. When this large amount of fuel ignites, the fires are much hotter and the flames much larger, causing more damage than if the fuel supplies are low. One way to solve the fuel buildup issue is to implement **controlled burns,** also called **prescribed burns.** These are small fires started when the conditions are just right and which lower the amounts of fuel. Obviously, this practice must be implemented with great caution and can be quite controversial.

Have you got all that information about forests? Let's move on to another vast resource that exists on Earth—water.

WATER USE

The following information about human use of water resources is not explicitly tested on the exam, but is helpful to know when you are thinking about water and about resource use. As you know, we all need water in order to live. In particular, communities need water for many different industries, including fisheries, recreation, transportation, and agriculture. Agriculture is one of the biggest water-users of all—about 70 percent of the global demand for water is for crop irrigation. Industry accounts for about 17 percent of all water use, and domestic use accounts for the remainder.

Since the 1950s, global water use has tripled—mostly due to population growth and improvements in the global standard of living. One way that humans have recently dealt with potential water shortages in communities is through **interbasin transfer**. During interbasin transfer, water is transported very long distances from its source, through aqueducts or pipelines. An example of this type of engineering is the pipeline that now exists between the western and eastern slopes of the Rocky Mountains in Colorado. Known as the Big Thompson Project, 213,000 acre-feet of water is delivered annually to the eastern slope of Colorado. However, this method has several negative effects. It can result in different geographic areas arguing over water rights. It can also have serious environmental repercussions; interbasin transfer can increase the salinity of the water body being exploited and even change the local climate of an ecosystem.

In North America especially, humans rely on groundwater as a primary source of water for everyday use. **Groundwater** refers to any water that comes from below the ground—that is, from wells or from **aquifers,** which are underground beds or layers of earth, gravel, or porous stone that hold water. Water found in an **unconfined aquifer** is free to flow both vertically and horizontally. A **confined aquifer**, however, has boundaries that don't readily transport water. Our reliance on and use of groundwater has several detrimental environmental effects; for example, it can result in a depressed water table and the drying up of local groundwater sources. In the late 1990s, a drought in Florida resulted in such a severe reduction in the aquifers that roads collapsed from lack of subterranean structural support. This subsidence (or sinking) of the Earth's surface is another serious consequence of groundwater withdrawal.

Additionally, aquifers can become **compacted**—meaning that the mineral grains making up the aquifer collapse on each other and the material is unable to hold as much water. Furthermore, in most urban areas, humans have rendered the groundwater incapable of being replenished by building structures and roads that are impermeable to precipitation.

Global Water Needs

Scientists differentiate between countries that are water-stressed and those that are water-scarce. Countries that are **water-stressed** have a renewable annual water supply of about 1,000–2,000 m³ per person, but countries that are **water-scarce** have less than 1,000 m³ per person and lack sufficient freshwater resources to meet demand. Currently, approximately 4 billion individuals experience severe water scarcity for at least one month a year, and approximately 700 million people in 43 countries experience severe water scarcity year-round. Many of the countries that are currently considered water-scarce are developing countries that have rapidly increasing populations—which means that their water-scarcity problems will grow over time.

Water scarcity is affected by national and regional politics, civil strife, and other issues affecting access and distribution; and lists of water-scarce countries differ by the source reporting. However, among the countries currently experiencing the most severe water scarcity are Yemen, Libya, Jordan, Western Sahara, and Djibouti. Unfortunately, more and more countries are expected to become water-scarce by the year 2050.

Water Use in the United States

The United States is not considered water-scarce, but certain regions of the United States are considered water-stressed. Additionally, water use in the United States is out of control—we use water more quickly than it can possibly be replenished, so water scarcity is definitely in our future if we continue to use water at our present, furious rate.

The hydrologic cycle supplies the water that we use for all of our activities. Water used in our homes, manufacturing, cooling equipment that generates electricity, and irrigating croplands are a few examples. To give you a sense of water use in the United States, take a look at the chart below that shows trends in water withdrawals and population from 1950–2015.

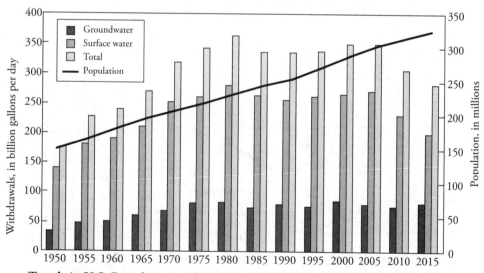

Trends in U.S. Population and Freshwater Withdrawals by Source, 1950–2015

Source: U.S. Geological Survey

As you see, the nation's water use peaked in about 1980 and has been fairly steady since then. Many of the stresses making greater water use likely have risen since 1980, such as population, irrigation of crops to feed this larger population, and more industry, yet total water use has not risen. The fact that water use has leveled off despite the increase in these stresses shows that water conservation efforts and greater efficiencies in using water have had a positive effect in the last 35 years. Nonetheless, we often hear environmental news stories covering droughts, proposed emergency water relief plans, and rules about lawn watering.

What Are We Doing About It?

Water is a tricky business. It's difficult for politicians and lawmakers to put restrictions on water use because many people think that water should be free. After all, it falls from the sky; we can take a bucket from the lake down the street and no one will arrest us for stealing. For the AP Environmental Science Exam, you should know about certain concepts of human water rights. The first is the idea of riparian right. **Riparian** means of, on, or relating to the banks of a natural course of flowing water, and **riparian right** is the right of people who have legal rights to use that area. Alternately, in **prior appropriation,** water rights are given to those who have historically used the water in a certain area. In other words, prior appropriation can be thought of as water squatters' rights!

It has been proposed that, in order to solve current global water crises, we simply take tons of ocean water and desalinate it—this is a fairly simple process physically, but unfortunately it isn't currently economically viable on a large scale, because it takes a great deal of energy to remove the salt through distillation or reverse osmosis. As water becomes scarcer globally, it will be important for countries to think of ways to regulate the use of water, as estimated global water consumption is set to continue rising. As global water crises become more common, research into the economic viability of desalination has increased.

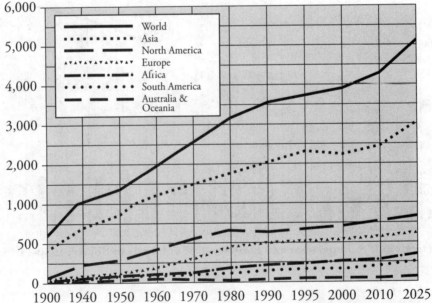

Global Water Consumption 1900–2025

(by region, in billions of m³ per year)

OCEAN RESOURCES

The term **fishery** is used in several ways, but it is primarily defined as the industry or occupation devoted to the catching, processing, or selling of fish, shellfish, or other aquatic animals. In the economic sense, a fishery is the sum of all activities on a given marine resource.

Worldwide, about 1.4 billion people depend on fish as their main source of food, and at least 600 million livelihoods depend at least partially on fisheries and aquaculture. Incredibly, about 200 million metric tons of fish are harvested each year—approximately 75 percent of this total amount is consumed as food by humans, and the other 25 percent is used for other purposes.

For many years, nations were subject to what is known as the 12-mile limit—this limited each nation's territorial waters to just 12 miles from shore. However, in the late 1960s, the depletion of a number of offshore fisheries inspired the United Nations to host a series of international conferences to address the problems of fish scarcity. The result of this conference was that nations were authorized to extend their limits of jurisdiction to 200 miles from shore. The depletion of marine fisheries worldwide came to be seen as a further example of the Tragedy of the Commons on an international scale. A new term was coined to recognize this shift: the **Tragedy of Free Access.**

Today, fishermen must go farther and farther out to sea to catch fish and need to rely on more sophisticated methods for finding them. Sonar mapping, thermal sensing, and satellite navigation are just a few of the advances that have aided fishermen as fish become scarcer and harder to locate.

By-Catch and Overfishing

Most of the fish that are harvested worldwide come from **capture fisheries;** they are caught in the wild and not raised in captivity for consumption. Some of the techniques that have been developed in order to improve fishing yields are creating problems that relate to overfishing. One of these problems is known as by-catch. **By-catch** refers to species of fish, mammals, and birds that are caught during fishing operations but are not the target fish. Some fishing methods that result in by-catch are **drift nets,** which float through the water and indiscriminately catch everything in their path; **long lining,** which is the use of long lines that have baited hooks and will be taken by numerous aquatic organisms; and **bottom trawling,** in which the ocean floor is literally scraped by heavy nets that scrape away or smash everything in their path, including corals and other delicate marine life on underwater mountains known as seamounts. Some advances that have been made in the fishing industry in an attempt to mitigate the problems of by-catch are restrictions on the use of drift nets, the installation of ribbons on bait hooks that scare away birds and prevent them from being caught, and bans on bottom trawling.

The National Oceanic and Atmospheric Administration worked with fisheries that used trawling for shrimp to design an apparatus called a Turtle Excluder Device (TED). It is located at the end of the trawling net and will eject large organisms such as sea turtles and sharks from the net while keeping most of the shrimp.

How Many Fish Are Left?

It has recently been reported that about 90 percent of the major fish stocks of the world are depleted, overexploited, or fully exploited. Close to another 8 percent of the stocks are moderately overexploited, and only about 2 percent are underexploited; and this is mostly due to over-fishing and the lack of regulatory oversight over the worldwide fishing industry.

One partial solution to the problem of overfishing is **aquaculture,** which is the raising of fish and other aquatic species in captivity for harvest. In general, the fish that are raised in captivity are those with the highest economic value—for example, salmon and shrimp. Various different methods are used in aquaculture—some fish are raised totally in captivity and then harvested, while others (like salmon) are initially hatched in captivity, but then released into the wild and captured later. Some saltwater aquaculture is performed in shallow coastal areas, though this is generally for raising seaplants and mollusks.

While aquaculture, also known as **fish farming**, does help to meet worldwide demands for fish, it is not a panacea for all of our fishery problems. One concern about aquaculture is the possibility of the accidental release of farmed fish into the wild, which has the potential to introduce new diseases to ocean fish and contaminate the native gene pool. Carp released from fish farming in Southern states are currently causing issues in the Mississippi River and the Great Lakes. Another problem lies in the fact that many fish that are raised in captivity are carnivorous and are fed captured wild fish, which defeats the purpose of the attempt to kill fewer wild fish!

Most of the public outcry about the endangered animals of the sea has centered on two groups: dolphins and whales. Dolphins are a high-profile by-catch, and as you may have noticed, many cans of tuna now advertise as having been caught using "dolphin safe" nets. The slogan "Save the Dolphins" has been frequently employed by international marine conservation groups. However, that slogan is impossible to obey unless humans first work to save the natural *habitat* of these creatures.

The International Whaling Commission (1974) regulates whaling. Recent policies implemented by the IWC allow the capture of a certain number of whales annually—by Norway for human consumption and by Japan for scientific use, although whales are also eaten in Japan. One rationale proposed for eating whale meat is that whales eat many fish that could instead be caught by humans. Another industry that has recently been criticized for damaging whales' ecosystems is the tourism industry—whale-watching tours are said to disrupt whale migration patterns and cause the whales undue stress.

The importance of mangrove swamps has been well established. They function as nurseries for shrimp and recreational fisheries, exporters of organic matter to adjacent coastal food chains, and enormous sources of nutrients valuable to plants, wildlife, and ecosystem function. Their physical stability also helps to prevent shoreline erosion, shielding inland areas from severe damage during hurricanes and tidal waves.

MINING

Mining is the excavation of earth for the purpose of extracting ore or minerals. We can divide mineral resources into two main groups according to how they're used. **Metallic minerals** are mined for their metals (for example, zinc), which can be extracted through smelting and used for various purposes. **Nonmetallic minerals** are mined to be used in their natural state—nothing is extracted from them. Examples of nonmetallic minerals are salt and precious gems. Here are two more terms you should know for the exam, if you don't already: a **mineral deposit** is an area in which a particular mineral is concentrated. An **ore** is a rock or mineral from which a valuable substance can be extracted at a profit.

The cost of extracting minerals depends on numerous factors, including the location and size of the mineral deposit. Additionally, the impetus for mining certain deposits more than others is often purely based on the value of the mineral resource. Understandably, the higher the value of the resource, the more money and effort will be put into mining it.

Environmental concerns about mining do not center on the depletion of mineral resources from the Earth's surface. Instead, they revolve around the damage that is done during the extraction process. The extraction of a mineral from the Earth generally disrupts the ecosystem and scars the land. Sometimes the extraction leaves pollutants that result from the surface exposure of underground minerals, from transformation of these minerals during mining, from chemicals or other substances introduced during extraction, or even from the machinery used for extraction. One example of this is the deposition of iron pyrite and sulfur in the mining of coal. The acid forms as water seeps through mines and carries off sulfur-containing compounds. The chemical conversion of sulfur-bearing minerals occurs through a combination of biological (bacterial) and inorganic chemical reactions, and the result is the buildup of extremely acidic compounds in the soil surrounding the deposit. These compounds create acid mine drainage that can severely harm local stream ecosystems. In mining processes, waste material is called **gangue,** and piles of gangues are called **tailings.**

As the more accessible ores are mined to depletion, mining operations are forced to access lower-grade ores. Accessing these ores requires increased use of resources that can cause increased waste and pollution.

Surface mining is the removal of large portions of soil and rock (this layer is called **overburden**, and it is whatever material lies above an area of scientific interest) in order to access the ore underneath. An example is **strip mining**, which involves removal of the vegetation from an area, which makes the area more susceptible to erosion. This type of mining is only practical when the ore is relatively close to the surface, which is why it's used mainly for coal mining. This is the least expensive—and least dangerous—method of mining for coal. However, because strip mining requires removing massive amounts of topsoil, it has a much greater impact on the surrounding environment than underground mining. The most extreme form of strip mining, mountaintop removal, transforms the summits of mountains and destroys ecosystems. This method is mostly associated with coal mining in the Appalachian Mountains. As coal reserves get smaller, due to a lack of easily accessible reserves, it becomes necessary to access coal through subsurface mining, which is very expensive. With **shaft mining**, vertical tunnels are built to access and then excavate minerals that are underground and otherwise unreachable.

Another environmental drawback to mining is that the refinement of these minerals often requires extensive energy input. For example, it takes approximately 15.7 kW of electricity to produce one kilogram of pure aluminum from its ore. On the other hand, recycling aluminum requires only 5 percent of the energy that's required to smelt it, and generates only 5 percent of the greenhouse gases. Recycle those soda cans!

After minerals have been extracted from their ore, they may be used in their rough form or further processed. Aluminum, for example, must be further refined after it is mined. Coal is an exception. After mining, it is transported to a power plant and burned in its original state. Sometimes two metals are combined to form a product; this is the case with stainless steel, which is a combination of iron and either nickel or chromium, and regular steel, which is 95.5 percent iron and 0.5 percent carbon. Because of the energy expended in mining and extraction, the steel industry is responsible for much of the air pollution that exists today!

Fortunately, air, land, and water harmed by mining can be reclaimed through **mine restoration** projects. In 1977, Congress passed the Surface Mining Control and Reclamation Act (SMCRA), which created one program to help coal mines manage pollutants and another to guide the reclamation of abandoned mines.

HOUSING AND COMMUNITY

The majority of humans live in some type of community, and the largest percentage of the human population lives in relatively large communities and urban centers.

Since the development of ancient civilizations, humans have lived together in large centralized communities, or cities. A couple of ancient cities that you may be familiar with are Rome (in what is now Italy) and Athens (in Greece). However, never before have the urban centers of the world grown as quickly as they are growing now. If we traced the growth of urban areas in the United States, we would find that before the Civil War (around the 1850s), only about 15 percent of the population lived in cities. Around the time of World War I (1920), that number grew to encompass about 50 percent of the total population of the United States, and today it hovers around 80 percent.

Globally, a little over half of the world's population today lives in an urban area. In the United States, this is partly due to the fact that our aging population has largely moved into the cities to have greater access to health services, employment opportunities, and cultural activities.

When considering those who live in urban areas, we also count those who reside in satellite communities, or **suburbs.** Interestingly, people who live in the suburbs, on average, occupy eleven times more space than do those who live in the city. One of the advantages of living in the suburbs is that people have their own land space—a backyard—which they need not share with others.

The term used to describe the emigration of people out of the city and into the suburbs is **urban sprawl.** In some areas of the United States, urban sprawl takes over vast tracts of land. In Colorado, for example, population growth has resulted in a number of new communities between Denver and Boulder; when traveling between the two cities, it is now difficult to determine where the Denver metro area ends and the city of Boulder begins.

When urban areas grow too large and become too dense, distributing water to all citizens becomes increasingly difficult. Coupled with this is the strain on the water supply—more people means more water use. In many of these newly crowded areas, water shortages have led to the implementation of restrictions on water usage.

Additionally, urbanization greatly impacts the ecosystem in which a city is located. The sheer number of people using resources in such a densely packed cluster puts a strain on those resources far more than in any other location—and not just the water supply. The burning of fossil fuels for industry, transportation, and electric power releases greenhouse gases and affects the carbon cycle. The sheer weight of buildings and roads compacts the soil. The impervious surfaces that make up so much of a city's footprint—concrete, pavement, etc.—do not allow water to reach the soil, disrupting the natural flow of water and causing runoff, which can lead to flooding without more infrastructure to redirect flow. Methods to increase water infiltration (reducing runoff) include replacing traditional pavement with permeable pavement, planting trees (which redistribute water more evenly), increasing the use of public transportation (to reduce road usage), and building up, not out (decreasing a city's overall "footprint").

Runoff matters because without the normal amount of water flowing into it, soil can become salinized, and saltwater can intrude into the water table. Even the ubiquitous front lawn is often at complete odds to what the natural ecosystem of an area would look like, and the pesticides used to keep lawns looking up to social standards add to the number of pollutants present.

Another problem that results from the increase in the populations of cities is what to do with all of the waste that's created. When you think about it, almost all human activities create waste—when you go to your local coffee shop and get a cup of coffee, you probably don't think much of it. However, if you get a cup of coffee every morning in a paper cup, five days a week for the 52 weeks of the year, then at the end of the year you've accumulated more than 250 cups! That's a significant pile of garbage. This all has to go somewhere, and with the large populations in cities, landfills continually fill up and new landfills are required to replace them. (Not to mention that landfills also release greenhouse gases!)

There are also urban areas that contain abandoned factories or former residential sites. These are referred to as **brownfields**. Any type of redevelopment of these areas is hindered by the possibility that the soil and water are contaminated.

Transportation Alternatives

While many people find the suburbs a pleasant place to live, ecologists and city planners have recently come to realize that this urban sprawl may reduce quality of life for all urban dwellers. One major concern of policy-makers and citizens in metro areas is what to do about transportation. Ideally, people would be encouraged to use mass transit or participate in carpools rather than drive separately in personal vehicles. Having fewer cars on the road decreases air pollution from automobile emissions and makes for less congestion on roadways.

Other environmentally conscious or "green" modes of transportation include bicycles, motor scooters, and electric bikes. Larger cities often opt to build subway systems, but they are extremely expensive to develop and are only cost-effective when there are enough people who will pay to use them. However, city buses are an option for both large and small cities. Fleets of buses are less expensive than subways to create and maintain, and although they contribute to road congestion, they decrease congestion by accommodating more people per vehicle. In addition, many cities have buses that use hybrid fuel systems to reduce harmful greenhouse gases.

Rapid rail or light rail systems are more common in Japan and Western Europe than in the United States, but as of recently they're being considered as an option for cities that lack subways. Rapid rail systems work by magnetic levitation; suspended above a track, a train moves along as a result of strong attractive and repulsive magnetic forces.

Building Sustainable Cities

In order for cities to be sustainable, city planners and developers must build and manage cities to work with, and within, their natural settings, instead of merely placing buildings and structures in these settings.

There are certain cities in the United States and elsewhere in the world that are setting examples of progressive thinking in conservation and ecology. For example, the city of Boulder, Colorado, has long been recognized for its forward-thinking, green policies. Bicycle paths cover the city, allowing cyclists to move freely from one area of the community to another. Buses move around the city and in and out of Denver and the surrounding communities, which enables people to commute to work without using their cars and creating more emissions. For those who need to drive to work, the city encourages carpools and provides parking areas for those who carpool. Additionally, the city's strong recycling programs help reduce the amount of material that's added to landfills. The city is ringed by open spaces that can be used by the city population for recreation. These areas are also leased to local ranchers for grazing cattle. Boulder's citizens have voted for tax increases to purchase additional communal green space for the city!

Other cities that have been held up as models for city planning are Curitiba, Brazil, and Portland, Oregon. Curitiba has an excellent mass transit system as well as bicycle paths and pedestrian walkways. The city provides recycling programs, job training, health care, and environmental education for its citizens. Likewise, in the 1970s, the state of Oregon became determined to head off urban sprawl, and the city of Portland established zoning policies and restrictive growth policies for urban areas that were adopted statewide. City developers were encouraged to invest in established neighborhoods rather than develop undisturbed areas. The city of Portland established Metro, a regional body that deals with land use, city planning, and the development of natural areas. Light rail systems were developed, and Metro began to encourage neighborhood self-sufficiency in order to keep the number of people who need to commute for food or other supplies to a minimum.

Now and in the future, it will be important for city planners to deal with new problems created as a result of urban sprawl. City planners and developers must take environmental concerns into consideration: providing green spaces and transportation alternatives and planning for the supply of water are all relatively new challenges for those involved in building cities. According to the Global Health Observatory (GHO), the global population went from 34% urban in 1960 to 56.2% urban in 2022, and the urban population continues to grow.

Big Cities in Less-Developed Countries

So far, we've made it sound like the cities of the world are dealing well with the boom in their population, but this is not true globally. Some cities, called **megacities**, have grown in excess of 10 million people very rapidly. In less-developed countries, this increase in the population size of major cities has many very negative effects. Among the worst effects is a deficiency of housing or habitable areas for the burgeoning population. As a result, people are homeless, become "squatters," or make their homes in areas that are completely undeveloped—areas that have no water, electricity, or stable, durable housing.

Some of the reasons people in less-developed countries are moving to cities are similar to those of people in developed countries: for example, cities have more opportunities for employment. However, these people often have other motivations that drive them out of the country, such as war, religious or cultural persecution, or the degradation of their environment.

Ecological Footprint

One concept that you should definitely be familiar with for this exam is that of the ecological footprint. An **ecological footprint** is used to describe the environmental impact of a population or individual person. It is defined as the amount of the Earth's surface that's necessary to supply the needs of, and dispose of the waste of, a particular population or individual. Americans have one of the largest ecological footprints: we require about 9.7 hectares per capita (per person). One hectare is 10,000 square meters, or about 2.5 acres. America's amount is comparatively enormous—the ecological footprint of Indonesia is only 1.1 hectares per capita. In general, affluent populations have a much higher ecological footprint than non-affluent ones.

We can use a mathematical model to describe the impact that humans have on the environment. Nicknamed the **IPAT model**, it is written as

$$I = P \times A \times T$$

In the model, I = the total impact, P = population size, A = affluence, and T = level of technology.

While you probably won't be asked to use this model to calculate the impact of populations on the exam, it's a good idea to know this formula exists and that the variables of a population's size, affluence, and level of technology all influence its environmental impact.

ECONOMICS AND RESOURCE UTILIZATION

The study of how people use limited resources to satisfy their wants and needs is called economics. While economics is not an explicitly tested subject on the AP Environmental Science Exam, it is closely interrelated with many aspects of environmental science. As you can imagine, some of our wants and needs are tangible, having a physical value (food, shelter, and clean air are examples), while others are intangible (such as recreational opportunity and the spiritual value of a forest's beauty). A resource can have both **tangible** and **intangible** properties.

A forest has value for supplying jobs and wood and for removing CO_2 from the atmosphere (tangible), as well as for its ability to nurture the human spirit and inspire artistic expression (intangible). When private citizens, governments, and corporations make a decision on how to use the forest, they must weigh the benefits (more jobs or lumber) against the costs—tangible and intangible—of cutting down the trees (less recreation space, the loss of biodiversity, decrease in CO_2 removal). This process is called **cost-benefit analysis**. While it may be easy to assign a monetary value to many tangible assets (like the amount of lumber that can be harvested from a forest), others are harder to quantify (such as ecosystem services like clean air, clean water, and biodiversity). It's even harder to assign a monetary value to most intangible assets (like the beauty of a forest), although impact on other intangibles (like recreational and tourism value) can be quantified. While cost-benefit analysis helps make decisions on how to use resources, you can see that the process is very difficult, and it can lead to different estimates by different groups.

Economists also want to figure out the cost of each step in a process. From our forest example, what is the cost to the economy of removing one more acre from the forest; or what is the benefit to us if we add one more acre to the forest? The additional costs are termed **marginal costs**; the added benefits are called **marginal benefits**. It is important to remember that resources are not free and unlimited. Some resources must be expended in order for us to use them. While we may benefit from more acres to hike in, the lumber company will suffer from not having as many trees to cut. In other words, marginal benefits and costs help us understand tradeoffs. By preserving a forest, we trade more hiking space with less profit for local economies.

As we use resources, there are often unwanted or unanticipated consequences of our using those resources, or **externalities**. These can be positive, when the result is good, and negative when the result is bad for the environment. Consider buying an air conditioner, for example. When you buy the A.C., there are costs—you pay for the labor, raw materials, and electricity to run it. After you buy the A.C., the dealer uses some of that money to pay employees to clean up litter on a highway. That cleanup benefits everyone (positive externalities), even those who do not buy the air conditioner. On the other hand, there are also negative externalities. If you run the A.C. a lot (perhaps during the day), you use a lot of electricity (and you can see just how much more when your electric bill spikes in the summertime). That electricity is generated by burning coal, and that causes acid rain. The damage done by the acid rain harms everyone—a negative externality.

The use of economics (cost-benefit analysis, marginal costs and benefits, and externalities) to make choices about dealing with environmental issues is morally neutral. These economic factors do not say anything about the ethics or fairness of those choices. There are situations in which we make decisions not based on the best balance between marginal costs and benefits, but on what is best for everyone. Take water pollution, for example. If a toxic chemical "X" is in a stream, there is a cost to clean it up. We make the decision to clean up most of chemical "X," even if the marginal costs exceed the marginal benefits, because removing chemical "X" will keep all of the people healthy. The perspective needed to think about benefit to humans and to the environment and balance these needs with economics is sustainability.

THE IMPORTANCE OF BEING SUSTAINABLE

To environmentalists, sustaining environmental quality usually means working in the biotic and abiotic environments in a way that ensures they are capable of functioning sustainably. Generally speaking, sustainability refers to humans using resources in such a way as to not deplete those resources for future generations, following environmental indicators such as biological diversity, food production levels, average global surface temperatures and CO_2 concentrations, human population, and, of course, resource depletion. The aim is to cultivate systems that provide sustainable yield—the amount of a renewable resource that can be taken without reducing the available supply. However, along with maintaining a sustainable environment, maintaining the health and happiness of the human species would also be a part of most environmentalists' goals for the Earth. The human species cannot exist in an unsustainable environment; after all, humans are part of a larger ecosystem, just as all other species of living things are. However, our advantage—or rather, our responsibility—lies in the fact that we are the most technologically advanced and capable species on the planet. We are also the ones causing the most damage.

As environmentally literate, reasonable citizens, we know that we're sometimes obliged to make choices that may not make everyone happy, but we strive to make choices that will ultimately benefit the greatest possible number of people.

Highly developed countries comprise only 20% of the world population, but consume more than half of the world's energy resources. If every country consumed global resources to this extent, we would need more resources to live than the Earth can supply. This is because most of the resources that we rely upon are limited—recall the fossil fuels we burn, the way that we use water, and the rate at which we produce and dispose of waste.

We're done discussing resources. As we've alluded to many times, as populations increase, more pressure is placed on the Earth's natural resources, and along with this comes the need for humans to find ways to develop and maintain those natural resources for direct human use. With this in mind, let's move on to the next chapter and review energy resources and consumption.

> What does the term *sustainable* mean to you? How much would you be willing to sacrifice in order to sustain environmental quality? These are questions that all citizens should ask themselves before entering the voting booth.

CHAPTER 7 KEY TERMS

Here are your key terms for Chapter 7. Learn them, love them, and commit them to memory.

Share and Share Alike?
Tragedy of the Commons
conservation
preservation
ecosystem capital
renewable resources
nonrenewable resources
consumption
production

Agriculture
traditional subsistence
   agriculture
slash-and-burn
Green Revolution
integrated pest management
   (IPM)
salinization
furrow irrigation
flood irrigation
spray irrigation
land degradation
drip irrigation
genetically modified
   organisms (GMOs)
photosynthate
monoculture
plantation farming
arable
loamy
soil structure
aggregates
crop rotation
polyculture
Dust Bowl
contour plowing
terracing
no-till methods
perennial crops
windbreak
intercropping (strip
   cropping)
green manure

free-range grazing
concentrated animal feeding
   operations (CAFOs)
   (feedlots)
overgrazed
rotational grazing

Forest Resources
deforestation
old growth forest
second growth forests
plantations (tree farms)
silviculture
clear-cutting
selective cutting
uneven-aged management
shelter-wood cutting
agroforestry
greenbelts
surface fires
crown fires
ground fires
controlled burns (prescribed
   burns)

Water Use
interbasin transfer
groundwater
aquifers
unconfined aquifer
confined aquifer
compacted
water-stressed
water-scarce
riparian
riparian right
prior appropriation

Ocean Resources
fishery
Tragedy of Free Access
capture fisheries
by-catch

drift nets
long lining
bottom trawling
aquaculture (fish farming)

Mining
mining
metallic minerals
nonmetallic minerals
mineral deposit
ore
gangue
tailings
overburden
strip mining
shaft mining
mine restoration

Housing and Community
suburbs
urban sprawl
brownfields
megacities
ecological footprint
IPAT model

Economics and Resource Utilization
tangible
intangible
cost-benefit analysis
marginal costs
marginal benefits
externalities

The Importance of Being Sustainable
sustainability
sustainable yield

Chapter 7 Drill

Directions: Each of the questions or incomplete statements below is followed by four suggested answers or completions. Select the one that is best in each case. For answers and explanations, see Chapter 13.

1. The amount of the Earth's surface that is covered by water is approximately

   (A) 12 percent
   (B) 36 percent
   (C) 50 percent
   (D) 75 percent

2. Which of the following correctly describes the process of clear-cutting?

   (A) Some mature trees are left to provide shade for younger trees.
   (B) Only trees with commercial value are cut down.
   (C) A few mature trees are left to reseed the land after cutting.
   (D) All the commercially usable trees in an area are cut down.

3. Moderate irrigation with groundwater over a long period of time can cause

   (A) salinization
   (B) waterlogging
   (C) desertification
   (D) succession

4. All of the following are problems created by the deforestation of rainforests EXCEPT

   (A) increased erosion
   (B) loss of biodiversity in the area
   (C) changes in local rainfall level
   (D) an increase in the availability of grazing land

5. Greenbelts are useful to

   (A) slow the process of urban growth
   (B) get more crops out of farmland
   (C) maintain borders around a person's home property
   (D) prevent erosion

6. Which of the following government agencies is responsible for the management of federal rangeland?

   (A) The U.S. Park Service
   (B) The U.S. Bureau of Mines
   (C) The Bureau of Land Management
   (D) The Environmental Protection Agency

7. Which of the following is NOT a renewable resource?

   (A) Air
   (B) Soil
   (C) Copper ore
   (D) Water

8. Nations have overfished international waters and have depleted many commercially important fish species. This is a good example of which of the following?

   (A) International agreements
   (B) The Tragedy of the Commons
   (C) The Rule of 70
   (D) Trade barriers

9. Which of the following best describes industrialized agriculture?

   (A) Consumes large amounts of fossil fuels, pesticides, and water
   (B) Uses human labor and draft animals to grow crops
   (C) Intersperses rows of crop plants with rows of trees
   (D) Uses little water or fossil fuels; relies on human labor

10. Which of the following are problems that have emerged with the overuse of pesticides?

  I. Better crop yield
  II. Pesticide-resistant pests
  III. Improved human health

  (A) I only

  (B) II only

  (C) III only

  (D) I and III only

11. The acid most commonly found in mine drainage is

  (A) carbonic acid

  (B) sulfuric acid

  (C) hydrochloric acid

  (D) acetic acid

12. Which of the following best describes the goal of environmentally sustainable economic growth?

  (A) Allowing rapid population growth so there will be more workers

  (B) Exploration to find more natural resources

  (C) Increasing the quality of goods without depleting the natural resources needed to make the goods

  (D) Cutting down forests and replacing them with rangeland

Free-Response Question

1. The irrigation of farmland is vital to the production of the world's food supply. In China, 87 percent of the water withdrawn is used for irrigation. In the United States, this figure approaches 41 percent. Most of the water is applied to the land in a process called gravity irrigation, in which the water is simply allowed to flow, via the force of gravity, into the fields.

   (a) **Identify** one positive and one negative aspect of gravity irrigation.

   (b) **Identify** one alternative to gravity irrigation, and **identify** one positive and one negative effect of that practice.

   (c) Massive irrigation programs can also impact underground water supplies. **Describe** one negative impact that irrigation might have on those supplies.

   (d) Dams are often used to create irrigation water reservoirs. **Describe** TWO positive and TWO negative effects that a large dam would have on the immediate area around it.

Summary

o The Tragedy of the Commons is essential to understanding the management of shared resources. Resources have both an innate value to the function of their ecosystem and an economic value to the humans who use them, known as ecosystem capital.

o Resource replenishment:
- Renewable resources are perpetual or regenerate quickly, in less than the average human lifetime (≈50 years), or before they can be depleted by human consumption.
- Nonrenewable resources renew at an insufficient rate for human use.

Agriculture

o Traditional subsistence agriculture:
- This type of agriculture is most common in developing nations.
- It provides enough food for one's family.
- The slash-and-burn method is often used, especially in undeveloped countries of the tropics.

o Industrial agriculture (Green Revolution):
- Industrial agriculture is most common in developed nations due to initial cost.
- Chemical fertilizers and pesticides may increase food production but also require lots of energy to produce and can contaminate water supplies; additionally, organisms can develop immunity to them.
- Technologies for irrigation have allowed for food to be produced where it otherwise could not, but involves problems such as salinization and water wars.
- Monoculture and plantation farming make crops easier to grow due to mechanized planting, but reduce native biodiversity.
- GMOs can increase economic benefits but discourage biodiversity and may have unknown health results.
- Increased tilling and livestock overgrazing can cause major soil erosion.

Forest Resources

○ Deforestation is the largest concern in forestry management, primarily in developing countries, for agricultural space and lumber exports. Minimal deforestation takes place in industrialized countries, where it is primarily for development.

○ Even if forests are replanted, the age of the forests is impacted and that can change the native biome.

○ Several national forest policies and programs have been established to protect the age and diversity of forests with different policies on use and management.

○ Forest fires can be essential to plant regeneration and forest health but can also burn uncontrolled and cause great economic loss.

Water Use

○ Human water demands are influenced by activities such as:
  • irrigation for agriculture
  • domestic and public use
  • livestock, aquaculture
  • thermoelectric supply
  • industry
  • mining

○ Not all regions of the world have equal water supplies. Therefore, the possible activities and their proportional demand on the water resources will differ by region.

Ocean Resources

○ Improved technologies, fishing strategies, and human population growth have all contributed to harvesting fish faster than the populations replenish themselves.
  • Unintentional catches known as by-catch also increase the number of fish caught, yet they are often wasted instead of used.
  • Aquaculture is a partial solution, though it has potential health impacts to both humans and the ecosystem.

Mining

o   Minerals have great value and multiple uses to humans, but their extraction and use also have a significant environmental impact:
  • extraction pollution
  • soil erosion and/or habitat loss
  • extraction and refinement require extensive energy input

Housing and Community

o   A larger and larger proportion of humans live in urban environments, resulting in urban sprawl and problems with water distribution, accumulation of waste, and the disruption of local ecosystems.

o   The ecological footprint of a person varies greatly from country to country. With the growing world population, rising expectations of prosperity, and increasing reliance on technology, our species' collective footprint is now exceeding the space available on Earth.

Economics and Resource Utilization

o   The cost-benefit analysis of a resource is complicated to assess because of the combination of its tangible and intangible values to humans and its intrinsic value to the ecosystem. Often, the price of a resource is not representative of its true cost because some costs of the product or unanticipated consequences are not included. These are called externalities.

Chapter 8
Unit 6: Energy Resources and Consumption

In this chapter, we'll review Unit 6 of the AP Environmental Science course, *Energy Resources and Consumption.* According to the College Board, about 10–15% of the test is based directly on the ideas covered in this chapter. If you are unfamiliar with a topic presented here, consult your textbook for more in-depth information.

Unlike the essential elements we discussed in earlier chapters, energy flows on a one-way path through the atmosphere, hydrosphere, and biosphere, and is essential for living organisms in many of its forms. At the most fundamental level, **energy** is defined as the capacity to do work. There are three types of energy: **potential energy** is energy at rest—it's stored energy— while **kinetic energy** is energy in motion. You might recall from your physics class that potential energy can be converted to kinetic energy. The third type of energy is **radiant energy**— for example, sunlight—and it is the only form of energy that can travel through empty space. Two terms that describe the movement of energy around the Earth are **convection,** which is the transfer of heat by the movement of the heated matter, and **conduction,** which is the transfer of energy through matter from particle to particle. Keep in mind that different energy sources are capable of storing types of energy that differ in quality. For example, both wood and coal will burn to produce heat, but coal produces more heat because it contains higher **energy quality**. **Net energy yield** refers to the comparison between the energy cost of extraction, processing, and transportation and the amount of useful energy derived from the fuel.

> It is important to note that energy cannot be created or destroyed, but it can be transferred to another form. Some energy is lost as heat, which increases entropy.

As you saw when we reviewed weather, convection and conduction are very important processes that drive the movement of water in the hydrosphere of the Earth. This is just one way in which the flow of energy around the Earth affects every process—geological or biological—that takes place.

In this chapter, we'll begin with a discussion of the basic units of energy and review the two laws of thermodynamics that you'll need to know for the test. After that, we'll begin our discussion of the Earth's energy resources—the Earth provides humans with resources of energy, just as it does the physical, natural resources that we learned about in the last chapter. Let's begin!

UNITS OF ENERGY

For this exam, you'll be expected to recognize the following units of energy and power:

- **Energy Units:** joule (J), calorie (cal), British thermal unit (BTU), and kilowatt hour (kWh), which is a measure of watts × time
- **Power Units:** watt (W) and horsepower (hp)

Remember, watts are equal to volts × amperes. You should also be intimately familiar with the First and Second Laws of Thermodynamics, so let's review those before we move on—and make sure you memorize them before test day!

Laws of Thermodynamics

1. The First Law of Thermodynamics says that energy can neither be created nor destroyed; it can only be transferred and transformed. One example of such a transformation occurs in photosynthesis. In photosynthesis, radiant energy from the Sun is converted to chemical energy in the form of the bonds that hold together atoms in carbohydrates.

2. The Second Law of Thermodynamics says that the entropy (disorder) of the universe is increasing. One corollary of this Second Law of Thermodynamics is that, in most energy transformations, a significant fraction of energy is lost to the universe as heat; for example, as we reviewed in Chapter 4, in food chains only about 10 percent of the energy from one trophic level is available for the next higher energy level upon consumption.

Okay, those are the basics about energy that you'll need to know for the test. Now, let's begin our review of the energy resources that exist on Earth. Recall from the previous chapter that resources can either be renewable or nonrenewable, depending on how long it takes for them to be replenished after use. Energy resources can also be renewable—those that can be replenished naturally, at or near the rate of consumption, and reused or nonrenewable—those that exist in a fixed amount and involve energy transformation that cannot be easily replaced.

NONRENEWABLE ENERGY

Perhaps surprisingly, one of our biggest uses of energy is in the production of electricity. In other words, we use tons of energy each year to produce electricity—another form of energy!

In general, electricity is produced in the following way: an energy source provides the power that heats up water, transforming it into steam, which then turns a turbine. Hence, the turbine converts kinetic energy (from the steam) into mechanical energy (the spinning of the turbine). Now here's where the generator comes in. The generator consists of copper wire coils and magnets, one of which is stationary (stator) and the other of which rotates (rotor). As the turbine spins, it causes the magnets in the generator to pass over the wire coils (or vice versa), generating a flow of electrons through the copper wire and thus producing an alternating current that passes into electrical transmission lines. In lieu of steam, flowing water or wind can also provide the power needed to turn the turbine and produce electricity.

So, where do we get the energy that we use to heat up that water in the first step of the creation of electricity? Well, the three main sources for electricity production are:

- fossil fuels (provide 60 percent of the world's electricity)

- renewable energy sources (provide 30 percent of the world's electricity)

- nuclear energy (provides 10 percent of the world's electricity)

Quick Quiz!
Q: Which of the following energy sources contributes the least to global warming?

(A) Coal
(B) Natural Gas
(C) Oil
(D) Solar

Turn the page for the answer.

Let's go through each of the types of energy above and see where they come from, what effect their use has on the Earth, and how sustainable they are.

Fossil Fuels

During the Industrial Revolution (in the early 18th century), steam was produced almost exclusively through the burning of firewood and coal—and this, in turn, provided the energy for most mechanical processes. In addition, fossil fuels can be made into specific fuel types for specialized uses (for example, gasoline for use in motor vehicles). Today, oil is our primary power source. About 33 percent of total global energy production comes from oil products; the runner-up to oil is coal, and the runner-up to coal is natural gas. Together, the combustion of these three fossil fuels provides 81 percent of the world's energy. That **combustion** (burning) is a chemical reaction between the fuel (oil, coal, or natural gas) and oxygen that yields carbon dioxide and water and releases energy.

As developing countries become more developed, generally their reliance on fossil fuels for energy increases; and as a result, as the world as a whole becomes more industrialized, the demand for energy, and specifically for fossil fuels, increases.

Fossil fuels, as the name indicates, are formed from the fossilized remains of once-living organisms. Over vast amounts of time, this organic matter was exposed to intense heat and pressure. Eventually, these forces broke down the organic molecules into oil, coal, and natural gas.

Oil, or petroleum, is made of long chains of hydrocarbons; coal contains a mixture of carbon, hydrogen, oxygen, and other atoms. Natural gas is made mostly of methane gas (CH_4) with a mixture of other gases.

Generally, oil and natural gas are formed in the same areas. These materials are found deep in the Earth under both land and ocean floor, where they are stored in the pores (spaces) between rocks. Coal is found in long continuous deposits, called **seams,** at various depths underground. The seams represent areas where large amounts of plant remains were buried and eventually transformed into coal. We will cover the process of coal mining on the next page.

Certain types of geologists locate fossil fuel reserves. They plan and supervise the extraction of these fuels from the Earth. Using knowledge of geology and rock formations, these scientists make predictions about which sites are most likely to have fossil fuel deposits. They use **exploratory wells** to drill and sample a particular area. If an exploratory well hits a fossil fuel reserve, it can provide an estimate of the amount of fuel that can be obtained from that area; this is called the **proven reserve**. It is important to know that although exploratory wells can provide a fairly precise estimate of the size of a reserve, these numbers are just educated guesses (not so proven after all!). The amount of a resource that can be extracted from a reserve is dependent on the technologies available and the cost of extraction. If extraction costs are too high, it is not economically feasible to extract the resource. If a coal seam is buried very deeply, for instance, it may cost more money and fuel to extract it than the value of the seam.

Quick Quiz!

Answer: Solar power contributes the least to global warming. Coal, oil, and natural gas release carbon dioxide and toxins into the atmosphere when burned, increasing the rate of global warming.

Fast Fact

Petroleum and coal have different origins. Petroleum used today was formed from ancient marine organisms. Coal was formed from the organisms in ancient swamps. Fossil fuels are still forming today, but they will not be available for use for many lifetimes.

It is also important to keep in mind that the global distribution of natural energy resources (such as ores, coal, crude oil, and gas) is not uniform and depends on regions' geologic history. Thus, some political regions have vastly more access to these resources than do others. Most renewable energy sources (to be discussed later in this chapter) are unevenly distributed in a similar way; however, different regions may be comparatively poor in one resource (such as sunlight) while rich in another (such as wind or geothermal energy).

What About Oil?

When oil is pumped up fresh from a reserve, it is called **crude oil**. Crude oil varies greatly from reserve to reserve. It can range from thin to viscous (thick); from high sulfur to low sulfur; it can even vary in color and odor.

There are three different methods of extracting oil. In primary extraction, the oil can be easily pumped to the surface. When some oil wells are tapped for the first time, there is a large release of oil and gas, a *gusher*, due to the pressure in the reserve. When the oil is harder to extract, people rely on pressure extraction, which uses mud, saltwater, or even CO_2 to push out the oil from the reserve. The final method utilizes steam, hot water, or hot gases to partially melt very thick crude oil and make it easier to extract. Oil reserves can also be found in rock (shale oil) and surface sands (tar sands). Tar sands—which are a combination of clay, sand, water, and bitumen—are the dirtiest source of oil extraction, and the method of extraction and processing needed to get oil from tar sands is detrimental to the environment and uses a great deal of energy. There was much controversy about the Keystone Pipeline, which ran from Alberta, Canada, south to the refineries along the Gulf Coast. Those in favor of the pipeline said it would bring much-needed jobs to the areas around the pipeline. Those who disapproved of it listed the pollution it could bring, including contamination of groundwater.

> **Fast Fact**
> Crude oil is used for the production of not only fuel products but also many items used every day in the form of plastic and petroleum jelly (Vaseline).

In 2010, the United States suffered one of its worst environmental disasters when an explosion on the Deep Water Horizon drilling rig caused oil to spill from the well into the Gulf of Mexico for three months. This was the largest marine oil spill in the history of the oil industry. Eleven men were killed during the explosion, and the spill caused a great deal of damage to both marine and wildlife habitats and local economies along the coast.

Drilling for oil is only moderately damaging to the environment because little land is needed to drill. However, since oil is transported thousands of miles by tankers, pipelines, and trucks, a lot of environmental damage can occur during transportation.

What About Coal?

Let's spend some time reviewing coal. The qualities of different types of coal are ranked by the number of BTUs that they produce upon burning. The purest coal is called **anthracite,** which is almost pure carbon. The second purest coal is **bituminous,** followed by **subbituminous,** and finally **lignite**—the least pure coal. It's heat, pressure, and depth of burial that determine which types of coal are formed by given deposits and the resulting qualities. Coal mining occurs through one of two processes—strip mining or underground mining—both of which can be hazardous and have serious environmental impacts. **Underground mining** involves sinking shafts to reach underground deposits. In this type of mining, networks of tunnels are dug or blasted, and humans enter these tunnels to manually retrieve the coal. After production stops at these mines, cave-ins can occur, causing massive slumping or **subsidence. Strip mining** involves the removal of the Earth's surface, all the way down to the level of the coal seam. The coal is then removed, the **overburden** (the earth that was removed) is replaced and topped with soil, and the area is contoured and re-vegetated. Most states require strip-mine owners and operators to completely reclaim areas that are mined by taking all of the steps outlined above. However, the process of mining and removing the coal from the Earth leaves hazardous slag heaps containing sulfur that can leach out and enter the water table.

Coal Power Process

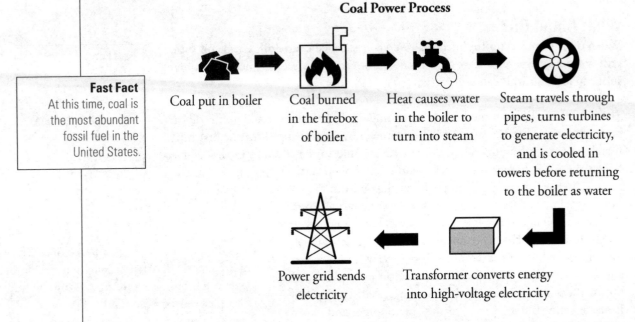

Coal put in boiler

Coal burned in the firebox of boiler

Heat causes water in the boiler to turn into steam

Steam travels through pipes, turns turbines to generate electricity, and is cooled in towers before returning to the boiler as water

Power grid sends electricity

Transformer converts energy into high-voltage electricity

Fast Fact
At this time, coal is the most abundant fossil fuel in the United States.

The use of coal to produce electricity has several disadvantages. For one, when it is burned in the production of electricity, carbon dioxide, nitrogen oxides, mercury, and sulfur dioxide—all of which contribute to air pollution—are released as by-products. However, some of these by-products can be removed through the actions of **scrubbers,** which contain alkaline substances that precipitate out much of the sulfur dioxide. The neutral compound formed in the scrubber (calcium sulfate) is eliminated in waste sludge. Two other waste products produced by the burning of coal are **fly ash** and **boiler residue**—you should be familiar with both of these terms for the exam.

Another problem with coal is that it often contains a significant amount of the element sulfur, both in the form of iron sulfide (pyrite) and as organic sulfur. Sulfur is another contributor to air pollution. While iron sulfide can be removed by grinding the coal into small lumps and washing it, organic sulfur is only released during the combustion (burning) of coal. However, scrubbers can remove organic sulfur from the flue gases after the coal is burned. Another solution to this problem is to burn the coal with limestone—the liberated sulfur then combines with the calcium in limestone to form calcium sulfate, which prevents it from being released through the flue. There are different types of smokestack scrubbing: wet scrubbing and baghouse or cyclo scrubbing. Wet scrubbing uses a fine mist of water to transform sulfur oxides (SO_x) from an air pollution issue to either a water pollution issue or to a commercial product, sulfuric acid. This is similar to how a rain storm will cleanse the air of pollen and dust that cause allergies in some people. Dry scrubbers are very similar to large vacuum cleaners that either filter (baghouse scrubbers) or spin (cyclo scrubbers) particulates out of the effluent gases. Electrostatic precipitators use an electric charge to attract dust particulates to metal surfaces where they can be gathered and disposed of as solid waste. This

Fast Fact
Electrostatic precipitators help to remove 98% of particulate matter from flue emissions.

is similar to your television screen being the dustiest surface in your living room due to electric properties of the device.

What About Natural Gas?

The third fossil fuel you need to know about is natural gas. Natural gas is made mostly of methane (CH_4) as well as pentane, butane, and several other gases in small quantities. As you learned earlier, natural gas is produced by the actions of heat and pressure over long periods of time. It is also produced by living organisms (mostly by anaerobic bacteria). Methane-producing bacteria can be found in landfills, swamps, and the intestines of various animals. Here's an interesting fact: while the largest source of methane is wetlands, the second largest source is our flatulent livestock.

Currently, natural gas is used for heating homes and cooking. It can also be burned to generate electricity. Some power plants are designed to switch between oil and natural gas fuels depending on the cost. The engines of cars and trucks can be modified to burn natural gas instead of gasoline. There is a landfill operator in the state of New Jersey who tested a process of trapping methane from a landfill, liquefying it, and then using the liquid methane to power the trucks that bring garbage to the landfill.

Because of its simple molecular structure, natural gas produces only carbon dioxide and water when it burns. It does not produce the oxides of nitrogen and sulfur associated with burning coal or oil. Before you get really excited about natural gas, you should be aware of its dangers. In an uncontrolled release (like a leak), it can cause violent explosions. It is also more difficult to transport than coal or oil. Because a tank can hold a small amount of gas, producers liquefy it by putting the gas under high pressure (**L**iquefied **N**atural **G**as). This process requires energy. Natural gas can also be transported by pipes. However, pipes carry the risk of leaks and explosions, and some habitats are damaged during the building of the pipe system.

Furthermore, methane is a potent greenhouse gas. Its chemical structure makes it 30 times more powerful than carbon dioxide at trapping heat in the atmosphere. And, as climate change continues to warm the Earth, the biochemical reactions by which certain bacteria produce methane will accelerate. This is a particular danger in habitats such as swamps and other wetlands as well as freshwater sediments. This situation, in which increasing concentrations of a greenhouse gas in the atmosphere further accelerate the production of greenhouse gases, is just one of many feedback loops that makes the climate change unleashed by human activity so scary and difficult to quantify!

How Much Fossil Fuel Is Left?

In order to understand how long our accessible fossil fuel supplies will last, you should know how quickly we are using up those fuels. Let's take oil, the most widely used fuel, as an example. The table on the next page shows the amount of oil (including crude oil, other petroleum liquids, and biofuels) that selected countries use each day. This data is from 2022, the last year for which the U.S. Energy Information Administration has this information.

Current Day Application

Fracking, also called **hydraulic fracturing,** is a process by which natural gas and oil are extracted from rock that lies deep underground. A deep well is drilled and then millions of gallons of toxic fracking fluid—a mix of water, sand, and harsh chemicals—are injected at a high enough pressure to fracture the rock and release the oil or gas. It's a highly controversial practice that has been linked to earthquakes in the states of Arkansas, Ohio, and Pennsylvania. It requires a large amount of water, which has to be safely stored after use due to chemical contamination. Fracking can also cause groundwater contamination and the release of volatile organic compounds.

Fast Fact
The increase in hydraulic fracturing and deepwater drilling has been the result of new technology that allows for horizontal drilling. Several areas currently being surveyed for future drilling are major fisheries, such as the Grand Banks.

Country	2022 Oil Consumption (millions of barrels per day)
United States	20
Saudi Arabia	3.6
Russia	3.6
Canada	2.3
China	14.5

Source: U.S. Energy Information Administration

Fast Fact
In 1956, Hubbert predicted that the lower 48 states would reach peak oil production by 1970 and then decline. In fact, advances in technology have increased the amount of oil available in the United States today.

As you can see, the United States is by far the largest consumer of oil. A quick bit of addition shows that these five countries alone consume approximately 44 million barrels of oil each day! As you can imagine, that leads some scientists to ask questions about how long our supplies of oil (and the other fossil fuels) will last. One well-known authority on the future of oil production, the late M. King Hubbert, stated that the end of oil as a cheap and easily available form of energy is in the near future and that we must begin to develop alternative fuel sources. This theory is commonly known as **Hubbert peak** or "peak oil."

Nuclear Energy

Nuclear energy is the world's primary non-fossil fuel, nonrenewable energy source. In the United States, about 18 percent of electrical energy is provided by nuclear power plants. Worldwide, 436 nuclear power plants produce approximately 10 percent of the world's electrical energy. The United States and France lead the world in the creation of new nuclear facilities, with the U.S. leading the pack and currently operating 93 nuclear power plants.

Let's talk a little bit about what exactly these nuclear power plants do. The key reaction in the production of nuclear energy is **fission,** a nuclear reaction in which the nucleus of an atom is struck by a neutron and then splits into two smaller, lighter nuclei. The current nuclear plant technology involves the use of uranium-238, which is enriched with 3 percent uranium-235. Nuclear fission releases a large amount of heat, which is used to generate steam, which powers a turbine and generates electricity. **Breeder reactors** generate new fissionable material faster than they consume such material. Furthermore, they can use a more abundant form of uranium, uranium-238, or an alternative called thorium-232.

However, the products of nuclear fission are much more **radioactive** than the fuel used; in other words, they have unstable nuclei which emit energy over time as particles or photons. Radioactivity can be measured in curies. Radioactive materials differ in terms of rate of decay—this is generally thought of in terms of **half-life,** which is the time it takes for half of a given radioactive sample to degrade on average. A radioactive element's half-life can be used to calculate a variety of things, including the rate of decay and the radioactivity level at specific points in time. For example, if a given sample of a radioactive substance has an activity level of 10 curies and a half-life of 5 days, then after 5 days it will have an activity level of 5 curies; after 10 days it will have an activity level of 2.5 curies; after 15 days it will have an activity level of 1.25 curies; and so on.

Uranium-235 remains radioactive for a long time—its half-life is 703.8 million years—which leads to problems with the disposal of nuclear waste. So as it stands, nuclear power is considered a cleaner energy source than fossil fuels because it does not produce air pollutants, but it does release thermal pollution and hazardous solid waste.

However, the future of nuclear power will probably involve **nuclear fusion,** which is the process of fusing two nuclei—likely from the hydrogen isotopes tritium and deuterium, which have two and one neutrons, respectively (the common form of hydrogen has no neutrons). Fusion would have several advantages over fission as a source of nuclear power—less radioactivity and nuclear waste, more easy-to-obtain fuel supplies, and increased safety—but while research into this process has been ongoing since the 1940s, it hasn't produced a viable way to harness nuclear fusion into a usable power source yet.

In the United States, there are two types of nuclear reactors. They are known as boiling water reactors and pressurized water reactors.

- **Boiling Water Reactors**—These reactors use the heat of the reactor core to boil water into steam. This steam is piped directly to the turbines. The steam spins the turbines, which generate the electricity. The water is cooled back to a liquid (by a heat exchanger), then pumped back to the core to be turned into steam again. This reactor uses two water circulation systems; one system makes steam and carries it to the turbine and the other cools the water from the core so it can be turned back into steam.

Boiling Water Reactor

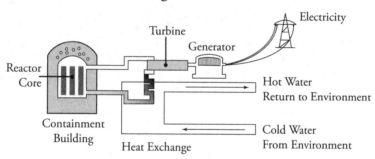

- **Pressurized Water Reactors**—These reactors also produce electricity by generating steam but contain three water circulation systems. Similar to boiling water reactors, they use the heat produced from the reactor core to heat the first water supply. However, the first water supply is kept under high pressure to prevent the water from boiling. It is then passed through the reactor heat exchanger, where heat from the first water supply is transferred to the second water supply. The second water supply is not kept under high pressure and forms steam to spin the turbines. A third water supply cools the steam from the turbines to regenerate the second water supply.

Pressurized Water Reactor

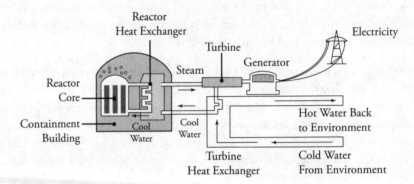

Some arguments in favor of the use of nuclear power include the fact that the production of nuclear energy produces no sulfur dioxide or nitrogen oxide and less carbon dioxide than does the production and combustion of fossil fuels.

Since the first testing of atomic weapons, people have been concerned about safety issues surrounding nuclear energy. The nuclear reactor accident that occurred at the Three Mile Island facility in Pennsylvania in 1979, the devastating explosion that occurred at the Chernobyl facility in Ukraine in 1986, and the catastrophic failure at the Fukushima nuclear power plant in Japan in 2011 brought some major safety concerns to the public's attention.

The chart below shows a number of these issues.

Safety Issue	Description
Meltdown	Reactor loses coolant water, and thus the very hot core melts through the containment building. The radioactive materials could then get into the groundwater.
Explosion	Gases generated by an uncontrolled core burst the containment vessel and spread radioactive materials into the environment.
Nuclear weapons	Some of the by-products of the fission reaction can be remade into fission bombs, or "dirty bombs," that spread damaging radioactive isotopes.
Highly radioactive waste	No longer usable cores, piping, and spent **fuel rods** need to be stored for many centuries. The "spent" fuel can contain radioactive elements like plutonium-239, which has a half-life of 2.13×10^6 years.
Thermal pollution	The water used to cool turbines is returned to local bodies of water at a much higher temperature than when it was removed unless first cooled.
Radioactive elements	Gamma rays produced by radioactive decay can damage cells and DNA, which can cause breast, thyroid, and stomach cancer, and leukemia. Damage to the immune system can also result.
Concern for one's safety	People suffer from mental stress, anxiety, and depression caused by concerns for their safety.

At this point, the cost for building a new nuclear power plant in the United States is prohibitive due to changing regulations; and worldwide, more than 100 nuclear power plants have been decommissioned. The main reason for this is the problem of the nuclear waste that's created.

The United States' spent fuel is currently stored on site at nuclear power plants, but the plants are running out of space. While Yucca Mountain (in Nevada) was previously proposed as a final destination storage site in the past, the plan has stalled due to pushback from Nevada residents. This has prompted the government to consider other storage options, such as Salado Salt in New Mexico, or deep holes in the Earth's crust.

RENEWABLE ENERGY

Obviously, the advantage to discovering or developing renewable energy sources lies in the fact that these types of energy sources won't run out. Global renewable energy production has increased, with the most recent United Nations report (2021) stating that nearly 30 percent of electricity currently comes from renewable sources. The goal is to reach 90 percent by 2050.

Biomass is one of the most consistently and widely used renewable energy sources today, and was even more widely used in the past. Biomass includes wood, charcoal (wood that has been baked to remove water and impurities), and animal waste products. A related fuel, **peat**, is partially decomposed organic material harvested from a type of wetlands called peatlands or bogs. These fuel sources are very accessible and thus are often used in developing countries. Although biomass is renewable, it is only renewable when it is used at a pace that allows time for replacement.

Using biomass as an energy source is beneficial because it is low-cost, readily available, and better for the environment than fossil fuels. Biomass can also help reduce the amount of waste in landfills, as it relies on the burning of organic matter. However, burning most biomass fuels releases carbon dioxide, carbon monoxide, nitrogen oxides, particulates, and volatile organic compounds. In addition, the overharvesting of trees for fuel (if wood is used) can result in deforestation.

Biomass Power Process

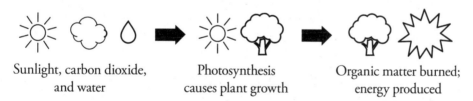

| Sunlight, carbon dioxide, and water | Photosynthesis causes plant growth | Organic matter burned; energy produced |

One interesting fuel that's recently been developed is called gasohol, a type of synfuel. **Gasohol** is a gasoline extender made from a mixture of 90 percent gasoline and 10 percent ethanol, which is often obtained by fermenting agricultural crops or crop wastes. So, as you can see, it is partly derived from organic substrate. Gasohol has higher octane than gasoline and burns more slowly, coolly, and completely, thus resulting in reduced emissions of some pollutants. Despite those advantages, it also vaporizes more readily than gasoline, and has the potential to aggravate ozone pollution in warm weather. Ethanol and, similarly, methanol from organic sources are surface carbons and therefore add no net carbon to the atmosphere. However, because ethanol is carbon-based, carbon dioxide is released during its combustion. Ethanol-based gasohol is also expensive and energy-intensive to produce—one bushel of corn produces only two and a half gallons of ethanol.

This is clearly an inefficient way to use biomass. However, researchers are developing new and better ways to use biomass as a fuel. Biodiesel is a great example. While it is made primarily from virgin oils (such as soybean oil), it can also be made from waste vegetable oil. More and more restaurants are beginning to use their waste oil byproducts to produce biodiesel. In addition, it is possible to use algae to produce biodiesel. The major benefit of algae as a source is that they can be grown in sewage water, allowing the production of biodiesel without using land normally used for food production. It is important to note, though, that the cost of producing any of these forms of biodiesel is currently greater than that of producing traditional diesel fuel from nonrenewable energy sources. Without government subsidies, it is tough to make biodiesel prices competitive with those of fossil fuels.

Hydroelectric Energy

Hydroelectric power works via the systematic placement of river dams that spin turbines to produce electricity. On a smaller scale, turbines can be placed directly in small rivers, where the flowing water spins the turbine and generates power. One advantage of hydroelectric power is that its production releases no pollutants. However, hydroelectric power does produce thermal pollution. In addition, it requires that rivers be dammed, and this can change the rates at which rivers flow and also lead to the destruction of habitats. On the other hand, as water is held behind dams, new habitats, in the form of wetlands, are created.

Fast Facts
The reservoir formed behind the dam leads to the destruction of habitats. Vegetation is drowned and will decompose, therefore still supplying CO_2 as a waste product.

Hydroelectric Power Process

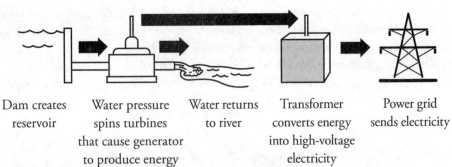

Dam creates reservoir

Water pressure spins turbines that cause generator to produce energy

Water returns to river

Transformer converts energy into high-voltage electricity

Power grid sends electricity

There are other problems associated with dams that you need to know about. One of them is silting. As water sits behind the dam, the normal sediments it carries have time to sink to the bottom. This puts additional weight on the structure and means that dams have to be built strong enough to hold back the many tons of sediment. This also means that the sediment is not passed farther down the river. The sediment that used to fertilize the flood plains of the river is now trapped behind the dam. In addition, the reservoir usually has a greater surface area than the preceding lake or river; this increased surface area can actually result in a higher rate of evaporation and water loss than before.

Generally, a river has many dams along its length. The addition of each new dam leads to less and less water downstream and a change in the natural course of the river. In California, there is continued unrest because of the rationing of water to aquaculture.

Another problem is that fish that spawn in the normally silty river no longer have a place to do so. Salmon and other anadromous fish breed in the streams where they hatched from eggs. Dams prevent the salmon from returning to their hatching streams. **Fish ladders** are structures on or around artificial and natural barriers, such as dams, locks, and waterfalls, that facilitate fishes' natural migration and movements by enabling fish to pass around the barriers by swimming and often leaping up a series of relatively low steps. While fish ladders do let some fish return upriver, the number of fish that get through is so limited that the populations will still decline.

One more important note about hydroelectric power: the development of this as an alternative, renewable energy source is limited—simply because there are a limited number of rivers of sufficient flow and drop in the world that can be used for these purposes.

Solar Energy

While it is important to remember that we already obtain through the Sun the energy we need to live—producers capture the Sun's energy and convert it into chemical energy—**solar energy** also has the potential to supply many of our external energy needs as well.

The use of solar energy actually dates back to Roman times; the Romans developed window glass, which allowed sunlight to come in and trapped solar heat indoors. Another interesting historical note is that the Swiss scientist Horace Bénédict de Saussure built a solar reflector in 1767 that could heat water and cook food.

Passive solar energy collection is the use of building materials, building placement, and design to passively collect solar energy (such as through windows) that can be used to keep a building warm or cool. On the other hand, **active collection** is the use of devices, such as solar panels, that collect, focus, transport, or store solar energy. Solar panels absorb solar energy and pass on the energy to tubes in which water is circulating; this heated water can be stored for later use. Direct collection of solar energy via **photovoltaic cells** (PV cells) produces electricity, which is then stored in batteries. When sunlight hits the PV cells, electrons are energized and can flow freely, producing an electric current. After this, one of two things can happen. If the home or building is connected to a regional electric grid, the energy produced is fed into the grid; this results in the electricity meter on the building actually spinning backward! Homeowners who have installed solar panels receive a credit against further charges from their electricity providers when the energy that they've fed into the grid exceeds the amount of energy the household uses. If the home or building is not connected to the local electric grid, the energy stored can be stored in batteries to be utilized later.

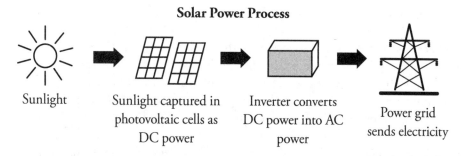

Solar Power Process

| Sunlight | Sunlight captured in photovoltaic cells as DC power | Inverter converts DC power into AC power | Power grid sends electricity |

While the use of solar energy produces no air pollutants, the production of photovoltaic cells does require the use of fossil fuels. The advantages of solar panels are that photovoltaic cells use no moving parts, require little maintenance, and are silent. However, not every location receives enough sunlight to make solar panels worthwhile. Also, the initial financial outlay for solar power is significant, although money is saved when the home is disconnected from the regional grid. Additionally, some states (such as New Jersey) give homeowners financial assistance for the installation of solar systems in their homes. Eventually, new technology should significantly lower the cost of solar systems. While many PV cells are installed on rooftops or in backyards to power individual homes and businesses, larger arrays of PV cells can generate more power for commercial or industrial use. The desert is a particularly profitable place to locate such an array given the abundance of sunshine. However, large solar energy farms do have the potential to negatively impact desert ecosystems.

Wind Energy

People have been using wind to produce energy for centuries. As early as the 17th century, windmills were so abundant in Schermerhorn (which is northwest of Amsterdam) that their turning paddles could be heard as far as 20 miles away! Windmills work in this way: wind turns the blades, or paddles, of the windmill and this drives a shaft that's connected to several cogs. The cogs then turn wheels that can perform mechanical work, such as grinding grain or pumping water. Although the Dutch windmill is a picturesque symbol of **wind power,** the modern wind **turbine** looks more like an airplane propeller. The wind that blows into the wind turbine spins the blades, and this, in turn, causes the machinery inside the base of the windmill to rotate. The base of the windmill is called the **nacelle,** and it houses a gearbox and generator as well as machinery that controls the turbine. Wind turbines can be designed to utilize the energy from wind at all speeds, or to function only when the wind is at a certain velocity.

Wind Power Process

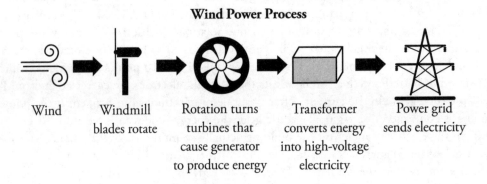

| Wind | Windmill blades rotate | Rotation turns turbines that cause generator to produce energy | Transformer converts energy into high-voltage electricity | Power grid sends electricity |

Wind energy is the fastest-growing alternative energy source, and modern wind turbines are usually placed in groups called **wind farms** or parks. In the United States, the largest of these wind farms is located in Altamont Pass, California; this farm has several thousand wind turbines. Wind-generated power has been increasing at a rate of more than 30 percent per year and is projected to supply a full 20 percent of the world's energy needs by 2030. Although, in the United States, wind farms are predominately in California and Texas at this time, many locations have enough prevailing winds to make production of electricity from wind power feasible. Wind farms can also be located offshore in the ocean, and although they're currently only located near to shore, in the future they may be placed on floating docks in deep water.

At this time, wind power is more costly than using fossil fuels because of the initial outlay of capital that must be invested in order to build the windmills; windmills are also considered by many to be annoyingly loud and unattractive. However, perhaps the biggest problem with this type of renewable energy source is that alternate energy sources must be in place for times when there is no wind. In the 1990s, one other public concern about the use of wind turbines was that birds would be cut up and killed by the blades, but now we know that as long as wind farms are not located in the middle of migration routes, only one or two birds per turbine per year are killed—and this is far fewer than the number of birds killed by other types of towers. Finally, one tremendous advantage of using wind energy is that it produces no harmful emissions. Let's move on and talk about another type of renewable energy source—geothermal energy.

Geothermal Energy

Geothermal energy is a form of energy that's obtained from within the Earth; it's energy that's produced by harnessing the Earth's internal heat. The interior of the Earth is still warm due to radioactive decay. Therefore, geothermal energy indirectly gains its energy from nuclear power. The greatly elevated temperatures within the Earth result in a buildup of pressure; some of this heat escapes through fissures and cracks to the surface. Some common examples of these fissures and cracks that you may have heard of are geysers, hydrothermal vents, and hot springs.

More specifically, in the process of geothermal energy production, the naturally heated water and steam from the Earth's interior turn turbines, and this creates electricity. The process involved in creating electricity from geothermal energy is not unlike that of hydroelectric power; instead of water from a dam's reservoirs spinning the turbines and generating electricity, it's water pumped from beneath the Earth that does so. Although surface water from geysers could be used, wells are typically drilled down into the Earth as far as thousands of meters to water that is 300–700 degrees Fahrenheit and then brought to the surface and converted to steam, which powers a turbine. Geothermal energy can also be used directly; in this process, the heated water is piped directly though buildings to heat them—this is a common method for heating homes in Iceland. In a sense, geothermal energy is renewable; however, if the groundwater is used at a faster rate than it is replaced, then this energy source is limited.

Geothermal energy also provides a perfect opportunity for **cogeneration**—using a fuel source to generate both useful heat and electricity. Cogeneration can involve any renewable or nonrenewable energy source, and increases the efficiency of the use of a fuel—all that's required is a system to capture "waste" heat from the energy generation process and put that heat to some useful purpose. In locations such as Iceland where geothermal energy is abundant, using this waste heat is almost always a useful plan.

Geothermal Power Process

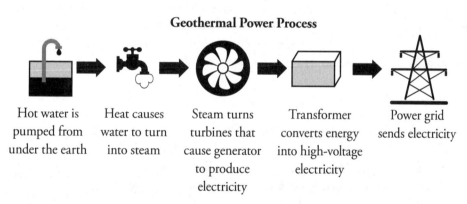

| Hot water is pumped from under the earth | Heat causes water to turn into steam | Steam turns turbines that cause generator to produce electricity | Transformer converts energy into high-voltage electricity | Power grid sends electricity |

Unfortunately, geothermal energy hasn't gained popularity, is expensive to install, and can potentially release toxic gas during the drilling process. The use of geothermal energy is also limited because only a few areas have geothermal sources to tap. Another problem with this renewable energy source is that the salts that are dissolved in the water corrode machinery parts. Additionally, some gases (such as methane, carbon dioxide, hydrogen sulfide, and ammonia) that are trapped in the water may be released as the water is utilized.

Other Sources of Energy

There are two other less widespread renewable energy sources: tidal movement in the ocean (**tidal power**) and **hydrogen cells.** You should be somewhat familiar with both of these energy sources for the exam.

Ocean Tides

The tidal movements of ocean water can be tapped and used as a source of energy. To harvest tidal energy, dams are erected across outlets of tidal basins. Incoming tides are sluiced through the dam, and the outgoing tides pass through the dam, turning turbines and generating electricity. Recently, ocean dams have been developed that allow energy to be harnessed from both the outgoing and incoming tides.

At this time, there is a tidal power plant installed in the East River in New York, NY. It harnesses enough power to power 10,000 homes. There are many different designs of ocean tide power plants in the idea stages. One of these involves having waves push into a chamber of air; the compressed air is then forced through a small hole at the turbine, and turns the turbine as it is released. An experimental prototype of this design has been installed off the coast of Scotland and is nicknamed the LIMPET (Land-Installed-Marine-Powered Energy Transformer).

Hydrogen Cells

Hydrogen is obtained from fossil fuels by a process called reforming. Hydrogen is very difficult to store and not very energy dense, but hydrogen fuel cells are considered by many to be the best, cleanest, and safest fuel source. Free hydrogen is not found on Earth, but it can be released through the process of electrolysis, in which hydrogen atoms are stripped from water, leaving the oxygen atom. Hydrogen can also be obtained from organic molecules, but the use of organic sources can release pollutants—as can the process of electrolysis if a fossil fuel, such as natural gas or coal, is used to drive the process. However, once the free hydrogen is released, it can be stored and then used to generate electricity through the reverse reaction of electrolysis.

One of the major benefits of the use of hydrogen fuel cells is that the only waste from the fuel cell is steam—water vapor. This technology has been used for decades in spacecrafts, but the high cost of the fuel cell and lack of hydrogen fuel stations has limited the technology to just a few test programs. The United States Department of Energy estimates that hydrogen fuel cells large enough to power light trucks and cars in the United States will require the production of 150 megatons of hydrogen per year. As of 2021, the United States is only producing approximately 10 megatons of hydrogen annually.

In order for hydrogen to become a truly viable option as a renewable energy source, an inexpensive and efficient way to produce hydrogen from nonfossil fuel sources must be developed. One of the most promising techniques for this involves the use of photovoltaic cells to harvest sunlight and then power the splitting of the water molecule.

ENERGY CONSERVATION: A FINAL NOTE

When we discuss **energy conservation**, we are basically referring to the practice of reducing our use of fossil fuels and reducing the impact we have on the environment as we produce and use energy.

One important form of energy conservation is the use of alternative fuel cars. They are gaining in popularity and acceptance. **Battery electric vehicles (BEVs)** are slowly becoming more commonplace. These vehicles run on battery power alone and can be charged with electric power. Their downsides include the pollution from the production and disposal of the batteries, expense, and the relative dearth of charging stations. A more popular alternative is the **hybrid-electric vehicle.** Hybrid vehicles are built with two motors: one electric and one gasoline-powered. The electric motor powers the car from 0 to about 35 miles per hour. Above 35 mph, the gasoline engine starts and helps to power the car. At highway speeds, both the electric and gas motors operate. The cars are designed so that when the brakes are applied, some of the energy is transferred from the brakes to recharge the electric motor's battery. Not only do these cars have good gas mileage, but they also produce far less CO_2 pollution than traditional gas-powered vehicles. Several carmakers also make models that use propane or natural gas as fuels, although these are not as common as hybrids. These generate only CO_2 and water as emissions, and they get good gas mileage. A problem is the lack of refueling stations, although devices are available that allow refueling from home. Cars can also be retrofitted with natural gas fuel tanks, so the driver can choose between gasoline or methane fuel.

Another type of alternative fuel is used cooking oils. The oils used in deep-fat fryers can be filtered and then burned in diesel-fueled cars, trucks, and buses. After starting the engine on pure diesel fuel, the driver switches to the **biofuel** to drive. At the end of the trip, the driver runs on pure diesel fuel again for a few minutes before shutting off the engine.

Another front on which steps are being taken to increase energy conservation is building design. Designers are rediscovering and modernizing techniques like passive solar heating, using thermal mass to regulate temperatures, and incorporating photovoltaic cells and smart regulation technology to reduce the amount of energy used from the grid in heating, cooling, lighting, and other energy use. While offices, factories, and other commercial buildings are a prime target of conservation efforts, these ideas can be implemented on a smaller scale in individual homes as well. Residential energy consumers can adjust thermostats to reduce the use of heat and air conditioning, conserve water, use energy-efficient appliances, and use conservation in their landscaping.

It has been argued that finding new fossil fuel sources would serve the same purpose as would reducing our current use of fossil fuels. However, this is not true—this statement does not take into consideration the fact that our use of fossil fuels has numerous negative effects on the environment. Additionally, in the long term it will not help us much to conserve fossil fuel resources—simply because these are not renewable energy sources—so they will eventually be depleted. Therefore, if we are to have dependable, long-term, renewable sources of energy, we must continue to develop, implement, and improve upon current renewable technology and methods.

On the legislative front, the United States has adapted the **CAFE,** or **Corporate Average Fuel Economy**, standards. These standards set mile-per-gallon standards for a fleet of cars. The goal of these standards is to reduce energy consumption by increasing the fuel economy of cars and light trucks. Review the CAFE standards described on page 254.

Finally, do not forget the role of mass transit in reducing pollution. Buses and trains can move many more people than cars. When the amount of pollution made by the vehicle is divided by all the passengers it is carrying, the bus or train generates far less pollution per person than a car.

In order to determine whether or not a resource should be used, citizens, governments, and businesses engage in a process called cost-benefit analysis. Costs and benefits can be either tangible (measurable), or intangible (immeasurable). Consider a corporation interested in clearing a forest for wood. The benefits are jobs and lumber (both tangible), while the costs are the loss of beauty and recreation opportunity (intangible), decreased biodiversity, and carbon dioxide removal (tangible).

In the next chapter, we will further discuss pollution—the effects it has on the Earth and its inhabitants, the types of wastes that currently exist, and how we can manage them.

CHAPTER 8 KEY TERMS

Use some of your renewable brain energy to study these terms.

Types of Energy

energy
potential energy
kinetic energy
radiant energy
convection
conduction
energy quality
net energy yield

Units of Energy

First Law of Thermodynamics
Second Law of Thermodynamics

Nonrenewable Energy

combustion
fossil fuels
seams
exploratory wells
proven reserve
crude oil
tar sands
anthracite
bituminous
subbituminous
lignite
underground mining
subsidence
strip mining
overburden
scrubbers
fly ash
boiler residue
fracking (hydraulic fracturing)
Hubbert peak
fission
breeder reactors
radioactive
half-life
nuclear fusion
boiling water reactors
pressurized water reactors
fuel rods

Renewable Energy

biomass
peat
gasohol
hydroelectric power
fish ladders
solar energy
passive solar energy collection
active collection
photovoltaic cells
wind power
turbine
nacelle
wind farms
geothermal energy
cogeneration
tidal power
hydrogen cells

Energy Conservation: A Final Note

energy conservation
battery electric vehicles (BEVs)
hybrid-electric vehicle
biofuel
CAFE (corporate average fuel economy)

Chapter 8 Drill

Directions: Each of the questions or incomplete statements below is followed by four suggested answers or completions. Select the one that is best in each case. For answers and explanations, see Chapter 13.

1. A fuel's net energy yield is correctly defined as

    (A) how much of the fuel is left in the world

    (B) how much time it takes to extract and transport

    (C) a comparison between the amount of pollution the fuel generates and the amount of useful energy produced

    (D) a comparison between the costs of mining, processing, and transporting a fuel and the amount of useful energy the fuel generates

2. A regular light bulb has an efficiency rating of 3 percent. For every 1.00 joule of energy that bulb uses, the amount of useful energy produced is

    (A) 1.03 joules of light

    (B) 1.03 joules of heat

    (C) 0.97 joules of light

    (D) 0.03 joules of light

3. Methane gas and ethanol are two examples of biogases that are produced in which of the following processes?

    (A) The distillation of oil

    (B) The pressurization of natural gas

    (C) The anaerobic digestion of biomass

    (D) The catalytic reaction of coal and limestone

4. Hybrid car engines have which of the following types of motors?

    (A) Gasoline powered only

    (B) Natural gas powered only

    (C) Electric powered only

    (D) Gasoline- and electric-powered engines

5. All of the following are ways to increase energy efficiency EXCEPT

    (A) using low volume shower spray heads

    (B) insulating your home thoroughly

    (C) switching incandescent light bulbs to fluorescent bulbs

    (D) leaving room lights on

6. A typical coal-burning power plant uses 4,500 tons of coal per day. Each pound of coal produces 5,000 BTUs of electrical energy. How many BTUs are produced each day from this plant?

    (A) 4.5×10^{10}

    (B) 0.45×10^{10}

    (C) 11.5×10^{3}

    (D) 4.5×10^{8}

7. Which of the following produces the least amount of carbon dioxide while generating electricity?

    (A) Oil

    (B) Coal

    (C) Wind turbines

    (D) Wood

8. How much energy, in kWh, is used by a 100-watt computer running for 5 hours?

    (A) 500 kWh

    (B) 200 kWh

    (C) 50 kWh

    (D) 0.5 kWh

9. Photovoltaic cells produce electricity by

    (A) a system of mirrors that focuses sunlight onto a heat collection device

    (B) using the sun's energy to create a flow of electrons in a material such as silicon

    (C) breaking down organic molecules and releasing energy

    (D) warming air, which spins a turbine

10. A sample of radioactive material has a half-life of 20 years. It has an activity of 2 curies. How many years does it take for the material to have an activity level of 0.25 curie?

    (A) 20 years

    (B) 40 years

    (C) 60 years

    (D) 80 years

11. The term *vampire appliances* correctly refers to appliances that

   (A) generate more power than they consume

   (B) consume electricity even when they are not operating

   (C) are EnergyStar rated

   (D) are programmed to turn themselves off at midnight each night

12. All nonrenewable resource power plants use heat to

   (A) make hot air that generates power

   (B) create powerful magnetic fields that make electricity

   (C) create powerful water jets that spin turbines

   (D) produce steam to turn electric generators

13. The acidity of a lake would most likely increase because of

   (A) the construction of a hydroelectric dam in the region

   (B) increased power generation at a local wind farm

   (C) overfishing by commercial fishermen

   (D) the burning of coal by nearby factories

Free-Response Question

1. Nuclear power plants have been described as being part of the solution to the problem of the United States' dependency on foreign energy. Currently, some 20 percent of the electricity produced in the United States is generated by nuclear power.

    (a) **Identify** the key parts of a nuclear power plant. **Describe** the roles of the following: core, fuel rods, coolant, and heat exchanger.

    (b) **Describe** TWO practical methods of dealing with the long-term storage of the highly radioactive wastes produced by a nuclear power plant.

    (c) **Describe** one positive impact that a nuclear power plant might have on air pollution.

    (d) Opponents of nuclear power plants point out the problems caused by thermal pollution of nearby rivers. **Explain** how the thermal pollution occurs and **identify** one method to reduce this problem.

Summary

○ The Laws of Thermodynamics state that energy is neither created nor destroyed and that the universe tends toward entropy (disorder).

○ Nonrenewable energy is the primary source of energy production, using primarily fossil fuels (about 80% of the global demand), including:
- oil
- coal
- natural gas

○ Nuclear energy is a non-fossil fuel, nonrenewable energy source.

○ Renewable energy sources account for approximately 30 percent of the global energy use. These include:
- biofuels
- solar
- hydroelectric
- wind
- geothermal
- hydrogen cells

○ Renewable does not always mean "clean" and nonrenewable does not always mean "dirty"; these terms refer to how long it takes to regenerate the fuel supply (generally less and more, respectively, than 50 years, or a human's lifespan).

○ Reduction in the use of both renewable and nonrenewable energy supplies can be done with greater energy conservation and improved technology efficiency.

Chapter 9
Units 7 and 8: Pollution

In this chapter, we'll review *two* units (Unit 7 and Unit 8) covered on the AP Environmental Science Exam, which AP calls *Atmospheric Pollution* and *Aquatic and Terrestrial Pollution.* Obviously, these two topics are interrelated. According to the College Board, about 7–10% of the test is based directly on each of these topics, so that's a total of about 14–20% of test material devoted to the ideas covered in this chapter. If you are unfamiliar with a topic presented here, consult your textbook for more in-depth information.

We'll begin the chapter with a discussion of what it means for something to be toxic and a review of how toxicity impacts health, along with some related concepts, including common pathogens. We'll move on to discuss toxins in air pollution and then review the major aspects of thermal pollution, water pollution, and the problems that arise as a result of solid waste, finally mentioning hazardous waste and noise pollution. As we go through each type of pollution, we'll also discuss the impact of pollution on the environment and human health, as well as some economic impacts. Let's begin!

TOXICITY AND HEALTH

One way to think about health is that each organism has a range of tolerance for almost any substance. For each substance it might come into contact with, an organism has an optimum range, inside of which it can maintain homeostasis. Outside of this range, the organism may experience physiological stress, limited growth, reduced reproduction, and in extreme cases, death.

A **toxin** is any substance that, when inhaled, ingested, or absorbed at sufficient dosages, damages a living organism, and the **toxicity** of a toxin is the degree to which it is biologically harmful. Almost any substance that is inhaled, ingested, or absorbed by a living organism can be harmful when it is present in large enough quantities—even water!

Substances are usually tested for toxicity using a **dose-response analysis.** In a dose-response analysis, organisms are exposed to a toxin at different concentrations, and the dosage that causes the death of the organism is recorded. The information from a set of organisms is graphed, and the resulting curve is referred to as a **dose-response curve.** The dosage of toxin it takes to kill 50 percent of the test animals is termed LD_{50}, and this value can be determined from the graph. A high LD_{50} indicates that a substance has a low toxicity; a low one indicates high toxicity. A **poison** is any substance that has an LD_{50} of 50 mg or less per kg of body weight.

The **Delaney Clause** is on the list of legislations required for the exam, so make sure to memorize it!

The government regulates certain types of toxins in air, water, and food. The **Food and Drug Administration (FDA)** is the body that regulates food and related products; it was empowered to do so by the Federal Food, Drug, and Cosmetic Act of 1938. One important part of this act is the **Delaney Clause** (part of the Food Additives Amendment of 1958), which specifically bans any food additives found to cause cancer in humans or in animal testing. When the Delaney Clause was included, no one thought it would have a very broad application, but as scientists have identified more and more cancer-causing substances, its relevance has grown.

If just the negative health effects are plotted on the dose-response curve instead of the level of the toxin at which death occurs, the resulting graph indicates the dosage that causes a change in the state of health. This type of graph can be used to find the **ED**$_{50}$—the point at which 50 percent of the test organisms show a negative effect from the toxin. The dosage at which a negative effect occurs is referred to as the **threshold dose**. Two more terms you should know for the test are acute effect and chronic effect. An **acute effect** is an effect caused by a short exposure to a high level of toxin; a snakebite, for example, causes an acute effect. A **chronic effect** is what results from long-term exposure to low levels of toxin; an example of this would be long-term exposure to lead paint in a house.

An **infection** is the result of a pathogen invading the body, and **disease** occurs when the infection causes a change in the state of health. For example, HIV, the virus that causes the disease AIDS, infects the body and typically has a long residence time. When it causes a change in a person's state of health, it has morphed into a disease called AIDS.

As they say, the dose makes the poison. When determining how harmful a substance is, all of the following must be considered:

- dosage amount over a period of time

- number of times of exposure

- size and/or age of the organism that is exposed

- ability of the organism to detoxify that substance

- organism's sensitivity to that substance (due, for example, to genetic predisposition or previous exposure)

- synergistic effect (when more than one substance combines to cause a toxic effect that's greater than any one component)

Pathogens are bacteria, viruses, or other microorganisms that can cause disease. There are five main categories of pathogens.

- Viruses (and other subcellular infectious particles, such as prions)

- Bacteria

- Fungi

- Protozoa

- Parasitic worms

Pathogens can attack directly or via a carrier organism (called a **vector**). One example of a pathogen that relies on a vector is the bacteria that causes Rocky Mountain spotted fever. It lives in the bodies of ticks, and when ticks bite humans, the ticks inject the bacteria, which causes the disease.

Pathogens, being single-celled organisms, are usually able to adapt quickly to take advantage of new opportunities to infect and spread through human populations. Poverty-stricken, low-income areas often lack sanitary waste disposal and have contaminated drinking water supplies, leading to havens and opportunities for the spread of infectious diseases. However, specific pathogens can occur in many environments, regardless of how sanitary conditions may appear to be. The AP exam may test your knowledge of specific pathogens and infectious diseases: on the next page are the ones it specifically tests.

Plague is a disease carried by organisms infected with the plague bacteria. It is transferred to humans via the bite of an infected organism (vector) or through contact with contaminated fluids or tissues. It has distinct forms (bubonic, septicemic, and pneumonic), but all are characterized by fever, chills, headaches, and nausea.

Tuberculosis is a bacterial infection that typically attacks the lungs. It is spread by breathing in the bacteria from the bodily fluids of an infected person. Its symptoms include chronic cough with bloody mucus, fever, night sweats, and weight loss.

Cholera is a bacterial disease that is contracted from infected water. Its main symptom is large amounts of watery diarrhea, which can lead to severe dehydration. Poverty, poor sanitation, and lack of clean drinking water are risk factors for the disease.

Dysentery is a type of gastroenteritis (inflammation of the stomach and small intestine) that results in diarrhea with blood, plus often fever and abdominal pain. It has multiple possible causes, including bacteria, amoebas, chemicals, and parasitic worms. The main risk factor for dysentery is contamination of food and water, usually from untreated sewage in streams and rivers.

Malaria is a parasitic disease caused by bites from infected mosquitoes. It is widespread in the tropical and subtropical regions of the globe, especially in sub-Saharan Africa. It causes fever, tiredness, vomiting, headaches, and, in severe cases, yellowed skin, seizures, coma, and even death. There have been strident efforts worldwide to eradicate this disease, but results have been much more successful in some places than in others.

West Nile virus is transmitted to humans via bites from infected mosquitoes. It is closely related to the viruses that cause zika, dengue, and yellow fever. West Nile virus causes West Nile fever, a disease characterized by fever, headache, vomiting, rash, and in severe cases encephalitis or meningitis.

Zika is a virus caused by bites from infected mosquitoes. It can be transmitted through sexual contact and from a pregnant mother to a fetus. It tends to cause fever, red eyes, joint pain, headache, and rash; it can cause microencephaly and other brain malformations when transferred to a fetus, and in adults it has been linked to Guillain–Barré syndrome, a rapid-onset muscle weakness.

Fast Fact
The top 4 causes of death in the United States are cardiovascular disease, cancer, unintentional injuries, and COPD (chronic obstructive pulmonary disorder). Most cases of these illnesses are attributable to lifestyle choices.

Severe acute respiratory syndrome (SARS) is a viral respiratory disease caused by severe acute respiratory syndrome coronavirus (SARS-CoV-1). There was an outbreak of this disease in 2002–2004. The respiratory symptoms were a form of pneumonia. Another coronavirus, MERS-coronavirus (MERS-CoV) caused an outbreak of **Middle East respiratory syndrome (MERS)**, another viral respiratory illness, in 2012–2013. In 2019, a much more severe outbreak of the related virus strain severe acute respiratory syndrome-coronavirus 2 (SARS-CoV-2) caused the **COVID-19** pandemic. These viruses tend to move from animals to humans and are transferred by inhaling or touching infected fluids.

As you're probably well aware, other things besides pathogens can make people ill, including environmental factors such as tobacco smoke, UV radiation, or asbestos. Also, although you may be exposed to a toxin or an infectious agent and not experience a change in the state of your health, someone else who's exposed to the toxic agent or pathogen could become very ill. It takes careful study to predict the range of likely effects a given toxin or pathogen will have. In fact, it can often be difficult to establish cause and effect between pollutants and other toxins and human health issues, because humans experience exposure to a variety of chemicals and pollutants.

The degree of likelihood that a person will become ill after exposure to a toxin or pathogen is called **risk.** Many environmental, medical, and public health decisions are based on potential risk. Calculating risk is referred to as **risk assessment,** and **risk management** means using strategies to reduce the amount of risk. The U.S. Department of Public Health and Public Services is an organization that makes use of risk assessment and management. For example, the department decides who can receive the flu shot each year. If the risk of getting the flu is high for a particular year, most of the population is encouraged to get the shot; however, if the risk seems small or the predicted flu strains are mild, only older people and the immunocompromised are advised to get the flu shot.

A **pollutant** is any substance (or energy) that, when introduced into its environment, renders the air, soil, water, or some other natural resource harmful, or adversely affects its usefulness. We'll start by examining pollutants that affect the air.

AIR POLLUTION

Substances that are considered contributors to air pollution have two sources: they can be natural releases from the environment or they can be created by humans. The effects of air pollution on humans can range in severity from lethal to simply aggravating. Some natural pollutants include pollen, dust particles, mold spores, forest fire smoke, and volcanic gases. One of the more recently described air pollutants from nature is produced by dinoflagellates, which, you might recall from Chapter 4, are the organisms that cause red tide. The toxins that are produced by these algae are caught in sea spray, in which they can be aerosolized and inhaled by humans, causing respiratory distress.

Although you may think that human-caused pollution is a relatively new phenomenon, people have added pollutants to the air throughout the history of humankind. Early humans' fire created pollutants, and the Romans' lead smelting resulted in air pollution that drifted thousands of miles from the source—and has even been discovered trapped in the ice of Greenland! It is true, however, that the large-scale production of pollutants began with the Industrial Revolution, and this is especially true of air pollution. The beginning of the Industrial Revolution marked the entrance of pollutants from fossil fuel into the atmosphere, for example, which has been environmentally disastrous.

Let's go over some terms used to describe pollution before we get into more specific details. **Primary pollutants** are those that are released directly into the lower atmosphere (remember the troposphere?) and are toxic; one example of a primary pollutant is carbon monoxide (CO). **Secondary pollutants** are those that are formed by the combination of primary pollutants in the atmosphere; an example of a secondary pollutant is acid rain. Acid rain is produced from the combination of **sulfur oxides** (such as SO_2 and SO_3) and water vapor.

> **Fast Fact**
>
> Fossil fuel emissions may also contain sulfur, which reacts during combustion:
>
> $$S + O_2 \rightarrow SO_2$$
>
> $$2SO_2 + O_2 \rightarrow 2SO_3$$
>
> $$SO_3 + H_2O \rightarrow H_2SO_4$$
> (sulfuric acid)

Pollutants can be released by **stationary sources,** such as factories or power plants, or they can be released by **moving sources,** like cars. **Point source pollution** describes a specific location from which pollution is released; an example of a point source location might be a factory or a site where wood is being burned. Pollution that does not have a specific point of release—for example, a combination of many sources, such as a number of cows releasing methane gas within a few square miles—is known as **non-point source pollution.**

The Major Culprits

The Environmental Protection Agency (EPA) has determined that there are six pollutants (familiarly referred to as the dirty half dozen) that do the most harm to human health and welfare, and refers to them as **criteria pollutants.**

The Six Criteria Pollutants

- carbon monoxide, CO
- lead, Pb
- ozone, O_3
- nitrogen dioxide, NO_2
- sulfur dioxide, SO_2
- particulates

In general, gases in the atmosphere are measured in units of parts per million, or ppm, when they are in relative abundance; when they are present in trace (very small) amounts, they are measured in parts per billion (ppb). For example, if in a certain geographic area, the carbon dioxide content of the air is 10 ppm, this would mean that there are 10 molecules of CO_2 per one million molecules of air. These pollutants are monitored with the National Ambient Air Quality Standards (NAAQS): these standards were established by the EPA to protect human health.

Let's go back through the list above. **Carbon monoxide (CO)** is an odorless, colorless gas that's typically released as a by-product of incompletely burned organic material, such as fossil fuels. CO is hazardous to human health because it binds irreversibly to hemoglobin in the blood—the molecule that is responsible for transporting oxygen around the body from the lungs. Hemoglobin has a higher affinity for CO than it does for oxygen, which means that in the presence of both CO and O_2, CO will bind more readily than O_2. In our normal oxygen-rich environments, this competition is not a problem, but in areas where CO is present in large concentrations, it can be deadly. More than 60 percent of the CO released into the atmosphere comes from vehicles that burn fossil fuels.

Lead is an air pollutant that, as you now know, has been around since the time of the Roman smelters. It is generally released into the atmosphere as a particulate (a very small solid particle that can be suspended in the air), but then settles on land and water, where it is incorporated into the food chain and is subject to biomagnification. If it enters the human body, it can cause numerous nervous system disorders, including cognitive and developmental disabilities in children. At one time, lead entered the atmosphere primarily as a result of the burning of leaded gasoline. However, lead gas has been phased out, and now the primary source of lead is industrial smelting. Incidentally, the "lead" in your pencils is not the element lead; in fact, it's the mineral graphite. The graphite in pencils received the name "lead" because of its lead-like color when it's transferred to paper.

We began our discussion of **ozone** in Chapter 6 and have mentioned it several times since. Notice that the ozone the EPA calls one of the dirty half dozen is specifically—and only—the ozone that's formed as a result of human activity. All ozone is O_3 and is the same chemically. However, **stratospheric ozone** (which absorbs UV light from the sun and therefore protects life on our planet) is functionally very different from **tropospheric ozone** (a powerful respiratory irritant and precursor to secondary air pollutants). Up high, ozone helps us; down low, it hurts us. O_3 is a secondary pollutant; it is formed in the troposphere as a result of the interaction of **nitrogen oxides**, heat, sunlight, and volatile organic compounds (VOCs—more on these later). Tropospheric ozone is a major component of what we think of as smog (more on this later).

The next major culprit on the list, **nitrogen dioxide (NO_2),** is one in a family of nitrogen and oxygen gases. NO_2 and the other nitrogen oxides are formed when atmospheric nitrogen and oxygen react as a result of exposure to high temperatures; this type of reaction occurs in combustion engines, for example. In fact, more than half of the nitrogen oxides in the atmosphere are released as a result of combustion engines. Other sources of nitrogen oxides are utilities and industrial combustion. Nitrogen dioxide is also commonly found as a secondary pollutant and is a component of smog and acid precipitation.

Sulfur dioxide (SO_2) is a colorless gas with a penetrating and suffocating odor. It is a powerful respiratory irritant and is typically released into the air through the combustion of coal. As we mentioned in Chapter 8, the use of scrubbers in coal-burning plants has helped reduce the amount of SO_2 released into the atmosphere. However, there are other sources of sulfur dioxide, including metal smelting, paper pulping, and the burning of fossil fuels, especially diesel. Sulfur dioxide can also be a component of indoor pollution as a result of gas heaters, improperly vented gas ranges, and tobacco smoke. In the atmosphere, SO_2 reacts with water vapor to form acid precipitation. Here's one last note to help you with the test: both nitrogen and sulfur can combine with oxygen to make several different molecules. Rather than a list of all the possible molecules, you might see the terms NO_x and SO_x. These terms (O_x) mean that there are several sulfur- and nitrogen-containing compounds mixed together.

Particulate matter is the last on the EPA's list of the dirty half dozen. Like lead, it is not a gas, but exists in the form of small particles of solid or liquid material. These particles are light enough to be carried on air currents, and when humans breathe them in, the particles act as irritants. Examples include soot (black carbon) and sulfate aerosols.

There have been significant decreases in the atmospheric content of both lead and carbon monoxide since the 1970s, mostly because of the phasing out of lead gasoline and the introduction of car engines that burn more cleanly. Many of the changes (in the United States, anyway) are due to the Environmental Protection Agency and the **Clean Air Act (CAA),** first enacted in 1963: a U.S. federal law designed to control air pollution nationally, amended many times—to set standards for controlling automobile emissions, to address acid rain, ozone depletion, and toxic air pollution,

> The **Clean Air Act** is part of the required environmental legislations to know for the exam, so make sure to memorize it!

and to make provisions for investigation and enforcement. However, there are other air pollutants that are of growing concern to environmentalists, including **volatile organic compounds (VOCs),** which are released as a result of various industrial processes including dry cleaning, the use of industrial solvents, and the use of propane. VOCs can react in the atmosphere with other gases to form O_3 and are a major contributor to smog in urban areas. A couple of examples are formaldehyde and gasoline. VOCs evaporate or sublimate at room temperature—hence the title "volatile." Not all of them are human-made pollutants, though: trees also produce some VOCs naturally. You'll have noticed that "contributes to smog" is a characteristic of several pollutants in the dirty half dozen, so the next question is: what exactly is smog?

Smog

As you might be aware, the setting for many of the Sherlock Holmes mysteries is the foggy, smoggy city of London. The smog that covered London throughout the 19th century, and well into the middle of the 20th century, was **industrial smog**—also known as **gray smog** or gray-air smog. As deadly as any of Holmes's adversaries in Sir Arthur Conan Doyle's stories, gray smog killed more than 2,000 people in a prolonged smog incident in 1911. However, the worst pollution-related incident in London occurred in 1952 and led to the death of about 10,000 city dwellers from pneumonia, tuberculosis, heart failure, and bronchitis. It was this disaster, resulting from the burning of large amounts of low-quality coal to heat homes and combat a cold fog, that prompted the Clean Air Act of 1956 in England.

Industrial smog is formed from pollutants that are typically associated with the burning of oil or coal. When CO and CO_2 are released in the process of combustion, they combine with particulate matter in the atmosphere and produce smog. The production of smog can also be aided by weather conditions—air inversions, for example, which trap the pollutants; or fog, which holds the pollutants. As we mentioned above, sulfur dioxide may be another component in gray smog, combining with water vapor to form sulfuric acid that is suspended in the cloud of smog.

Photochemical smog—also known as **brown smog**—is a different type of smog, usually formed on hot, sunny days in urban areas, due to the large numbers of vehicles emitting the specific pollutants that cause it. Many environmental factors, such as topography, amount of sunlight, temperature, humidity, and wind, affect its formation. Photochemical smog hasn't caused the acute disasters industrial smog has, but it can harm human health in several ways, including causing respiratory problems and eye irritation. In photochemical smog, NO_x compounds, VOCs, and ozone all combine to form smog with a brownish hue. It's also known as summer smog because of its tendency to form on warm, sunny days. Nitrogen oxide is produced early in the day. The intensity of sunlight on these days also promotes the formation of ozone from the combination of NO_x compounds. Ozone concentrations peak in the afternoon and are higher in the summer because ozone is produced by chemical reactions between oxygen and sunlight. Los Angeles, California, and Athens, Greece, are two cities that are particularly susceptible to photochemical smog. Athens has enacted mandates that have already reduced the number of cars driven each day in the city and improved the quality of the air. For example, by law, cars with even-numbered license plates can only be driven on even-numbered days—and cars with odd-numbered license plates can only be driven on odd-numbered days!

You should be familiar with the following chemical equations:

Photochemical Smog

$2NO + O_2 \rightarrow 2NO_2$ (causes brownish haze)

$NO_2 + UV\ light \rightarrow NO + O$ followed by:

$O + O_2 \rightarrow O_3$ (O_3 is ozone and is very hazardous to plants, animals, and materials in the troposphere)

hydrocarbons $+ O_2 + NO_2 \rightarrow$ PANs (peroxyacyl nitrates—cause burning eyes and damage vegetation)

Acid Rain

Acid precipitation—in the form of acid rain, acid hail, acid snow, etc.—occurs as a result of pollution (from human-made or, occasionally, natural sources) in the atmosphere, primarily SO_2 and nitrogen oxides. These gases combine with water to form acids (typically nitric acid and sulfuric acid) that are deposited on the Earth through precipitation. Because this acid is highly diluted, acid precipitation isn't acidic enough to burn skin upon contact, but it does have a significant, measurable effect on humans and the environment. How acidic is acid rain? Well, rain usually has a pH of about 5.6, but acid rain can have a pH as low as 2.3. The pH of rain is not 7.0 because of the carbonic acid in rainwater. Also, it's important for you to know that the pH scale is logarithmic, meaning that each whole pH value below 7 is ten times more acidic than the next higher value.

Acid precipitation can be a chronic and significant problem for large urban areas with many vehicles and areas that are downwind of coal-burning plants. While **dry acid particle deposition** occurs two to three days after emission into the atmosphere, **wet deposition** is usually delayed for four to fourteen days after emission; therefore, pollution from wet deposition can travel in air currents to locations that are many miles downwind of the emission source.

Some areas, like those with already acidic soils that were derived from granite, are particularly vulnerable to acid precipitation. Other areas that are particularly vulnerable to acid precipitation are those where the soil has been leached of its natural calcium content. This is because calcium acts as a natural buffer and tempers the effects of acid precipitation. Acid rain is a significant cause of sink holes in Florida due to acid dissolving the limestone rock that much of Florida is made of.

Acid precipitation is responsible for the following effects:

- leaching of some minerals from soil (which alters soil chemistry)

- creating a buildup of sulfur and nitrogen ions in soil

- increasing the aluminum concentration in soil to levels that are toxic for plants

- leaching calcium ions from the needles of conifers

- elevating the aluminum concentration in lakes to levels that are toxic to fish

- lowering the pH of streams, rivers, ponds, and lakes, which may lead to fish kills

- causing human respiratory irritation

- damaging all types of rocks, including statues, monuments, and buildings

In some areas of the world, progress has been made toward controlling acid precipitation. The 1990 amendment to the Clean Air Act has led to significant reductions in the amounts of SO_2 and NO_x that are emitted from industrial plants. Despite **National Ambient Air Quality Standards** (standards established by the United States Environmental Protection Agency), there is still considerable damage being done to soils and lakes in many areas, and these ecosystems will not be able to continue to tolerate significant lowering of their pH.

Motor Vehicles and Air Pollution

Today, all new vehicles sold in the United States must meet the EPA standards (in California, they must meet certain standards set by the state). Due to the Clean Air Act (the CAA) and its amendment (the CAAA), new cars (those produced after the year 1999) emit 75 percent fewer pollutants than cars made before 1970. The most significant device in controlling emissions in cars is the **catalytic converter**. This platinum-coated device oxidizes most of the VOCs and some of the CO that would otherwise be emitted in exhaust, converting them to CO_2. Newer models of catalytic converters also reduce nitrogen oxides, but not very successfully. Another technology that can help reduce the pollution caused by motor vehicles is the **vapor recovery nozzle:** an air pollution control device used on gasoline pumps that prevents fumes from escaping into the atmosphere when fueling a vehicle.

In the Energy Policy and Conservation Act of 1975, the Department of Transportation was given the authority to set what's called **Corporate Average Fuel Economy (CAFE)** for motor vehicles (Chapter 8). CAFE was intended to reduce both fuel consumption and emissions (not surprisingly, because burning less gas creates less air pollution).

Starting in 2011, the CAFE standards were expressed as mathematical functions depending on the vehicle's "footprint" (a measure of vehicle size determined by multiplying the vehicle's wheelbase by its average track width). That complicated 2011 mathematical formula was replaced starting in 2012 with a simpler formula with cut-off values. In July 2011, President Obama announced an agreement with 13 large automakers to increase fuel economy to 54.5 miles per gallon for cars and light-duty trucks by model year 2025. In April 2022, the U.S. Department of Transportation's National Highway Traffic Safety Administration announced new, landmark fuel economy standards for model year 2024–2026 vehicles. The new CAFE standards require an industry-wide fleet average of approximately 49 mpg for passenger cars and light trucks in model year 2026. This is projected to reduce fuel use by more than 200 billion gallons through 2050. All of these new standards will most likely result in higher purchase prices for vehicles, and they have certainly caused an outcry from auto manufacturers and oil refineries. However, the new standards are expected to reduce air pollutants by two million tons per year.

A hydrogen fuel cell vehicle would produce even less pollution than a hybrid vehicle, but don't expect to see them on the market very soon. Mass producing the cells is still not cheap enough to make the cars economically viable, as we mentioned in Chapter 8.

Vehicles of the Future

In 1990, the state of California passed a No-Pollution Vehicle Law mandating that, by 2003, 10 percent of the cars sold in the state would be pollution-free. That law was later rescinded because of problems with the development of the zero-pollution electric car, which looked promising at the time the bill was passed. In the past, electric cars were not widely adopted because they had a limited traveling range, they weighed more than their gasoline-burning counterparts, and they lacked amenities (like air conditioning).

However, a renewed interest as well as new battery technology, which increases energy storage and reduces cost, has motivated car makers to produce a promising generation of fully electric cars. The Nissan Leaf, Toyota Prius, and Tesla Model Y are examples of some zero-pollution electric vehicles.

In addition to new electric cars, hybrid vehicles that run on a mixture of gas and electric power have been gaining popularity over the past few years, with nearly all auto companies releasing hybrid versions of their standard vehicles. Government regulations, incentives, and public acceptance will probably determine how quickly the hybrid car moves into the mainstream vehicle market. One incentive that has been offered through the Inflation Reduction Act is a $7,500 tax credit for all newly purchased, and $4,500 for used purchases, of all-electric, fuel-cell electric, and plug-in hybrid vehicles purchased in 2023 or after. Some individual states also provide incentives to residents who purchase hybrid vehicles.

Aside from the incentives offered through the Inflation Reduction Act, it is highly unlikely that Congress will enact legislation that will provide further incentives for the purchase of hybrid vehicles or other alternatives that would reduce air pollution from vehicles. This is in part due to the fact that lobbying groups representing the oil companies and vehicle manufacturers consistently lobby against these incentives, and part due to the fact that the current congress has entrenched political divisions that preclude the development of such legislation. However, in the future, grassroots organizations that are backed by the voting public may influence legislation.

Indoor Air Pollution

The idea of **indoor air pollution** and the concept of "sick building syndrome" are still relatively new, but it is now widely recognized that air pollutants are usually at a higher concentration indoors than outside. This makes sense if you consider that pollutants that exist outside can also move inside as doors and windows are opened. Once the pollutant is indoors, it remains trapped until air currents move it out the door or windows or through a ventilation system. Additionally, indoor spaces have certain pollutants that are unique to them. The World Health Organization (WHO) estimates that indoor air pollution is responsible for 3.2 million annual deaths world-wide. (That's one death every 8 seconds!) According to the Environmental Protection Agency (the EPA), indoor air pollution is one of the five major environmental risks to human health.

One reason that indoor air pollution has such a great impact on health is because humans spend a significant amount of time indoors. Especially in developed countries, people generally work and live in well-sealed buildings that have little air exchange. In developing countries, however, one of the worst indoor air pollutants is material that's used for fuel. Dung, wood, and crop waste are the primary fuels used by more than half the world's population to heat homes and cook food, and the particulate matter that results from burning these fuels can exceed accept-able levels by hundreds of times.

In developed countries, other pollutants play the biggest roles in the creation of indoor air pollu-tion; the most abundant indoor pollutants are volatile organic compounds (VOCs). VOCs are found in carpet, furniture, plastic, oils, paints, adhesives, pesticides, and cleaning fluids. Even dishwashers are responsible for the creation of VOCs, when chlorine detergent reacts with left-over foods. Another component of pollution in developed countries is carbon monoxide (CO), which is an asphyxiant; CO arises in indoor air as a result of gas leaks or poor gas combustion devices. CO detectors are available for homes, and can prevent CO poisoning.

Two of the most deadly and common indoor pollutants in developed countries are tobacco smoke and radon. **Tobacco smoke** affects not only the health of the smoker, but the health of those around the smoker, as well. Secondhand smoke causes many of the same symptoms in nonsmokers who simply breathe it in as smoking can cause to the smokers themselves. Secondhand smoke, which contains more than 4,000 different chemicals, has been classified by the EPA as a Group A carcinogen (meaning that it causes cancer in humans). It's estimated that secondhand smoke causes approximately 34,000 deaths per year from heart disease, and 7,300 deaths from lung cancer in adults in the United States. In children younger than 18 months, it is responsible for 150,000–300,000 lower respiratory tract infections annually, and it increases the number and severity of asthma attacks in about one million asthmatic children.

Radon is the second leading cause of lung cancer (after smoking) in the United States. Radon is a gas that's emitted by uranium as it undergoes radioactive decay. It seeps up through rocks and soil and enters buildings via basements or cracks in the walls or foundations. It can also sometimes be found in groundwater that enters homes through a well. It is not found everywhere and must be tested for specifically. Homes that were built after 1990 have radon-resistant features.

We have already mentioned smoke from fires and tobacco smoke; other particulates that pose a risk indoors are asbestos and dust. **Dust** is simply a catch-all term for fine particles of solid matter—from various sources such as soil and volcanic eruptions, plant pollen, human and animal hair, textile fibers, paper fibers, etc. In homes, dust is composed of about 50% dead skin cells. It poses health risks to humans because of allergic reactions and the exacerbation of asthma and lung conditions. **Asbestos** is a naturally occurring silicate mineral that was often used as a building material in years past due to its insulating and heat-resisting properties. It has been shown to be dangerous to human health, as the fibers, when inhaled, can cause a host of lung problems. Mesothelioma is a type of cancer caused mainly by exposure to asbestos. Though its use is completely banned in many countries and strictly controlled in others, it remains in many older buildings.

Two more indoor air pollutants worth mentioning are **formaldehyde**, often found in building materials, upholstery, and carpeting, and **lead**, which was often used in paints and can be found in older paint layers (which may become uncovered over time) even when the newer layers are lead-free.

The final indoor pollutants we'll review are actually living: certain living organisms, such as tiny insects, fungi, and bacteria are considered pollutants. Many people are allergic to mold spores, mites, and animal dander, but asthma attacks can also be triggered by these living pollutants. The water tanks for large air-conditioning units are good places for certain types of bacteria to grow, and as air is distributed throughout the house, the bacteria are also distributed. Some bacteria can cause diseases; one example of this is *Legionella pneumophila*, which causes Legionnaires' disease.

Sick Building Syndrome

Sick building syndrome (SBS) is a term that's used when the majority of a building's occupants experience certain symptoms that vary with the amount of time spent in the building and for which no other cause can be identified. SBS is somewhat difficult to diagnose, and specific culprits are very difficult to identify. A condition is referred to as a **building-related illness** when the signs and symptoms can be attributed to a specific infectious organism that resides in the building.

One example of a building-related illness is Legionnaires' disease. Some symptoms of SBS include the following:

- irritation of the eyes, nose, and throat

- neurological symptoms, such as headaches and dizziness; reduction in the ability to concentrate; or memory loss

- skin irritation

- nausea or vomiting

- a change in odor or taste sensitivity

There are many ways in which people can reduce the amounts of indoor pollutants that they're exposed to—for many people, simply quitting smoking or encouraging roommates to quit can make a huge difference. Other precautions that people can take are to limit the amount of exposure they have to certain chemicals, such as pesticides or cleaning fluids. Perhaps the most important step to take is making sure that buildings are as well ventilated as possible.

THERMAL POLLUTION

Urban environments are generally about 20 degrees Fahrenheit warmer than the countryside that surrounds them, and this is due to the heat-absorbing capacity of buildings, concrete, and asphalt, which radiate the heat that they have absorbed. Industrial and domestic machines also directly warm the air. Because of their high temperatures, urban areas are known as **heat islands.** The high temperatures of heat islands increase the rates of photochemical reactions, which adds to photochemical smog.

The temperature profile of an urban area shows peaks and valleys in temperature based on how the land is used. For example, green spaces have lower temperatures than commercial areas, which have lots of parking lots, cars, buildings, and asphalt. Two ways in which the heat island effect can be significantly reduced are: (1) replacing dark, heat-absorbing surfaces (such as roofs) with light-colored heat-reflecting surfaces, and (2) planting trees and adding to green spaces. Trees shade the urban environment from solar radiation; in addition, the process of transpiration (the release of water through plant leaves) creates a cooling effect for the surrounding area. Another reason why urban areas are often less cool than rural areas is because the concrete and asphalt in cities increase water runoff. Runoff leads to increased temperatures because the deep pools of water that are created as a result of runoff are less affected by evaporation than are areas where water is spread out thinly over a larger surface area. Green spaces can reduce runoff by trapping the water and distributing it more evenly across a larger surface area.

One way people are trying to combat thermal pollution is by adding green roofs to city buildings. A green roof, or living roof, is a roof that is fully or partially covered with plants, greenery, gardens, and other vegetation planted over some type of waterproofing material. Not only does this combat the heat island effect, but it also keeps the buildings cool in summer and warm in winter, reduces rainwater runoff, provides habitats for wildlife, and helps to clean the urban air.

Temperature Profile of an Urban Heat Island

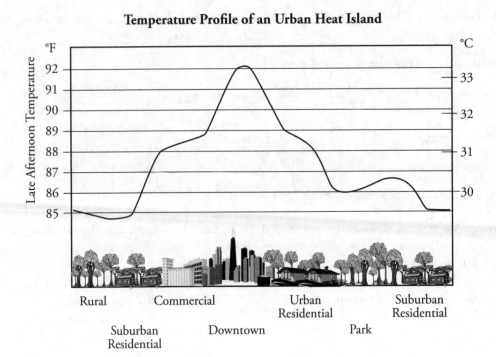

Another type of thermal pollution associated with many urban environments is **temperature inversion**. In this phenomenon, air pollutants become trapped over cities because they are not able to rise into the atmosphere. In normal atmospheric conditions, the warm, polluted air over a city rises into the cooler atmosphere. (Remember that warm air is less dense than the surrounding cool air, and less dense objects float!) In an inversion, the air above the city is warm and blocks the polluted air from rising. The polluted air remains hanging above the city and can cause respiratory problems. Inversions often occur in cities surrounded by mountains or cities bordered by mountains on one side and ocean on the other (for example, Los Angeles and Beirut). But thermal inversions can occur over any city where large masses of warm air can become stalled.

Thermal Inversion

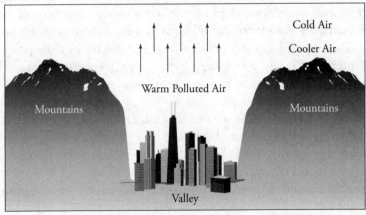

Normal Conditions

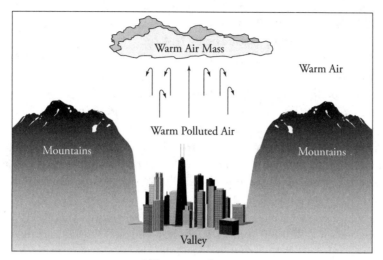

A Temperature Inversion

WATER POLLUTION

When the Cuyahoga River near Cleveland, Ohio, caught fire in 1969, it became a symbol of polluted America. This fire, along with many other problems that began to arise with polluted bodies of water at that time, eventually resulted in the **Clean Water Act (CWA)** of 1972, the primary U.S. federal law governing water pollution, which regulates water quality standards, point-source pollution, and water uses. The CWA had a dramatic effect on the quality of water in the United States. By 2016, 91 percent of community water systems met federal health standards—this number was up from the 79 percent that were considered clean by the government in 1993.

> The **Clean Water Act** is part of the required environmental legislations to know for the exam, so make sure to memorize it!

Experts say that Americans have some of the cleanest drinking (tap) water in the world. From the time of the passage of the CWA to 2017, 60 percent of the stream lengths that were tested were found to be sufficiently clean to allow fishing and swimming, while only 36 percent of the streams that were tested in 1972 were clean enough. Also as a result of the CWA, the annual loss of wetlands has slowed significantly since 1972. The CWA has certainly had a positive effect on our water, but there are still plenty of water issues and bodies of water that need to be cleaned. In 2023, the U.S. Supreme Court narrowed the scope of the Clean Water Act by significantly curtailing the power of the Environmental Protection Agency to regulate the nation's wetlands and waterways. The court decided that the jurisdiction of the Clean Water Act extends only over wetlands that have a "continuous surface connection" with a traditional navigable water body of the United States.

The Clean Water Act needs to be constantly enforced, and the actions of specific citizens and companies need to be monitored. For example, the Flint water crisis represents a breakdown of Americans' assurance in their clean water supply. The crisis began in 2014 when Flint, Michigan, changed its drinking water source from Lake Huron and the Detroit River to the Flint River. The water was not treated properly, and because of this, lead leached from water pipes into the water supply. Residents were exposed to dangerous levels of the toxin.

A federal state of emergency was declared for Flint in 2016, and Flint residents were told not to use the water supply for drinking, cooking, cleaning, or bathing. Water quality was deemed acceptable by 2017, and the lead pipes are being replaced.

One continual problem that contributes to water pollution is that runoff from land carries excess nutrients and pollutants to streams. This can result in large dead zones due to declining levels of dissolved oxygen. To track this, sometimes an **oxygen sag curve** is used—a plot of dissolved oxygen levels versus the distance from a source of pollution (usually excess nutrients and biological refuse). For example, the dead zone in the Gulf of Mexico covers up to 5,000 square miles in the middle of what is the richest area for shellfish in the United States. This dead zone has caused the collapse of the shrimp and shellfish industries in that region.

The dead zone was created because the Mississippi River collects roughly 10,000 pounds of fertilizer and raw sewage pollution from 31 states and some of Canada as it travels south. Then it dumps all of this nutrient-rich water into the Gulf. The warm, nutrient-rich freshwater does not mix well with the colder saltwater, and this results in **eutrophication**—the warm water being overly enriched with minerals and nutrients to the point that excessive growth of algae and other phytoplankton occurs (an **algal bloom**). In turn, the zooplankton that feed on them also experience a population explosion. When the phytoplankton and zooplankton die and sink to the bottom, bacteria metabolize the available dissolved oxygen as they decompose this detritus; the lack of oxygen creates a **hypoxic zone** (or **dead zone**), in which nothing that depends on oxygen can grow. Die-offs of fish and other larger organisms can happen when the aquatic food chain is disrupted. This zone stays in place from May until September, when colder, wetter weather helps to break it up. To save this economically important fishery, a federal-state Hypoxia Task Force was formed in 1997 to reduce the size of the dead zone by two-thirds by 2015, but in 2015 it found that the zone was about the same size it was in 1994. The target date was extended to 2035.

You should review the diagram of the Mississippi dead zone on the next page. The black areas (near the coasts of Louisiana, Mississippi, and Texas) represent areas where the level of dissolved oxygen (DO) is very low. Similar but less severe situations can occur in any waterway in which too much agricultural runoff and/or wastewater accumulates, bringing the nutrient levels to excess. These are known as **eutrophic waterways;** if hypoxia results, they are called **hypoxic waterways.** In contrast, **oligotrophic waterways** have low amounts of nutrients, stable algae populations, and high levels of dissolved oxygen.

Dead Zone

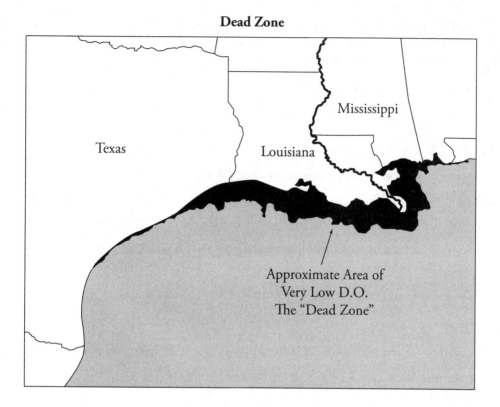

Texas

Louisiana

Mississippi

Approximate Area of
Very Low D.O.
The "Dead Zone"

Sources

Like the terms that are used to describe sources of air pollution, particular sources that are responsible for water pollution, like paper mills, are called point sources, and pollution sources that do not have a definitive source (or result from contributions of many sources) are non-point sources.

Right now, the biggest source of water pollution is agricultural activities; the runners-up are industrial and mining activities. Unfortunately, standing bodies of water, such as ponds, reservoirs, and lakes, do not recover quickly from the addition of pollutants. The lack of water flow prevents the pollutants from being diluted, which means that they accumulate in the water and undergo biomagnification in the food chain. In a similar way, groundwater does not recover well from the addition of pollutants; this is again because there is very little movement of water and therefore very little flushing, mixing, or dilution. Furthermore, groundwater is generally very cold and low in dissolved oxygen, which makes recovery from degradable waste a slow process. The porous rock that surrounds the groundwater absorbs the pollutants, which makes them difficult to remove.

However, flowing streams and rivers can recover from moderate levels of pollution if the pollutants are degradable. As illustrated by the implementation of long sewage pipes that once dumped raw sewage into the ocean off coastal areas, people thought that the ocean was able to dilute and recover from the addition of any amount of pollutants. While oceans can dilute, flush, and decompose large amounts of degradable waste, their capacity for recovery is unknown.

Water pollution is dealt with in two basic ways: (1) reducing or removing the sources of pollution, and (2) treating the water in order to remove pollutants or render them harmless in some way. Here's a list of the major water pollutants:

- excess nutrients (nitrogen, phosphate, etc.)

- organic waste

- toxic waste (pesticides, petroleum products, heavy metals, acids)

- sediments (soil washed with runoff water into streams)

- hot or cold water (hot water discharged from industrial facilities where it was used as a coolant; cold water from dam releases is discharged from the bottom of a reservoir)

- coliform bacteria (bacteria found in the intestines of animals that indicate the presence of fecal matter in water)

- invasive species (zebra mussels, for example)

- thermal pollution

- persistent organic pollutants (POPs)

First, note that thermal pollution, in the context of aquatic environments, refers to what happens when heat released into the water produces negative effects on the organisms in that ecosystem, which can range from habitat disruption or changes in breeding cues to outright death from cooking. In addition, variations in water temperature affect the concentration of dissolved oxygen, because warm water does not contain as much oxygen as cold water. So heat added to aquatic habitats can decrease the amount of this vital resource available to the organisms that live there.

Secondly, what are **persistent organic pollutants (POPs)**? These are specific organic compounds, such as DDT and PCBs, that resist environmental degradation (thus, they are persistent—they're even sometimes called "forever chemicals"). The Stockholm Convention on Persistent Organic Pollutants in 2001 discussed these chemicals, recognized their potential human and environmental toxicity, and made a preliminary list of them, recommending a global ban on the chemicals on its list. Since then, the list has been expanded. One of the reasons POPs can be toxic to organisms is that they are soluble in fat, which allows them to accumulate in organisms' fatty tissues. In addition, because of their persistence, they can travel over long distances via wind and water before being redeposited.

Most of the POPs are also **endocrine disruptors:** chemicals that can interfere with the endocrine systems of animals. Endocrine disruptors can lead to birth defects, developmental disorders, and gender imbalances in fish and other species, including humans. Some that are commonly detected in humans include DDT, polychlorinated biphenyls (PCBs), bisphenol A (BPA), polybrominated diphenyl ethers (PBDEs), other plastics, phytoestrogens, and phthalates.

The accumulation of toxins such as POPs in organisms due to their fat solubility is an example of **bioaccumulation**—the selective absorption and concentration of elements or compounds by cells in a living organism, most commonly fat-soluble compounds. In other words, organisms absorb these substances and have a hard time getting rid of them, so they persist in bodies as well as in the environment. A few substances that you should know bioaccumulate and have negative environmental impacts are PCBs, DDT, and mercury.

You've probably seen lots of news reports about the harmful effects—particularly to babies and very young children—of ingesting seafood that's contaminated with mercury. An EPA study found that one in six women of childbearing age in the United States may have blood mercury levels that could be harmful to a developing fetus. Coal-fired power plants are the major source of mercury pollution in the environment. Airborne mercury can travel hundreds of miles from its source before being deposited in the ground and lakes, streams, and other bodies of water, both directly from rainfall and as a result of runoff. Mercury in water can bioaccumulate in fish, which are then eaten by people. Abandoned metal and coal mines frequently produce **acid mine drainage**, highly acidic water which flows to surrounding areas. In 2012, the EPA issued the Mercury and Air Toxics Standards (MATS) rule that provides emission limits for mercury, particulate matter, hydrogen chloride, and hydrogen fluoride for approximately 600 coal and oil power plants. Bioaccumulation is closely related to the issue of biomagnification.

Biomagnification

Food chains represent the flow of energy in an ecosystem, but other things can flow through food chains, too—including environmental toxins. While the amount of usable energy decreases at every level of the food chain, the concentration of certain toxins increases at each successive level, since most toxins cannot be broken down by organisms.

Biomagnification is the term used to describe the increasing concentration of these toxin molecules at successively higher trophic levels in a food chain. Keep in mind that although really any type of molecule could be described using the terms bioaccumulation and biomagnification, generally these terms are used to describe toxins and heavy metals. And certainly, this is how they'll be used on the test.

> There are also pyramid models for biomass and numbers at each trophic level. They demonstrate the decreasing number of organisms at each level, which is why tertiary consumers are the first to be affected by environmental problems.

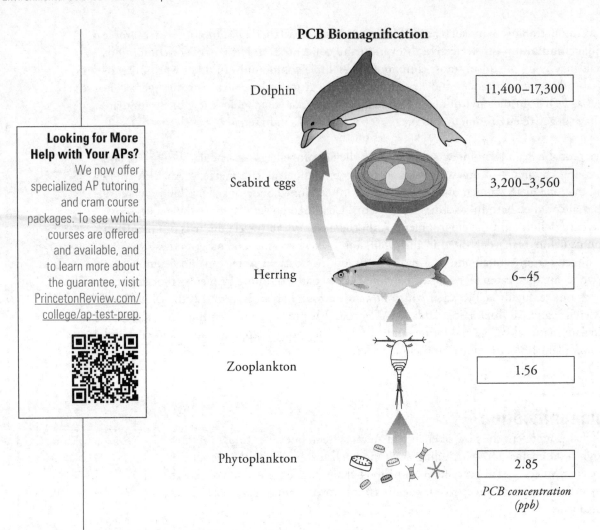

PCB Biomagnification

		PCB concentration (ppb)
Dolphin		11,400–17,300
Seabird eggs		3,200–3,560
Herring		6–45
Zooplankton		1.56
Phytoplankton		2.85

Some effects that can occur in an ecosystem when a persistent substance is biomagnified in a food chain include eggshell thinning and developmental deformities in top carnivores of the higher trophic levels. Humans also experience harmful effects from biomagnification, including issues with the reproductive, nervous, and circulatory systems. Ecologist Sandra Steingraber has pointed out that the production of breastmilk undergoes a final level of bioaccumulation (from the tissues of the mother's body to the breastmilk) and thus infants' food can contain even more of these substances than that of adults.

More generally, most types of water pollution can have many negative effects on ecosystems and on human health. For example, heavy metals from mining operations and from the burning of fossil fuels, once they make their ways into the groundwater, can enter the drinking supply. Diverse effects such as cancers, endocrine disruptions, nervous system damage, congenital disorders, heart and lung problems, and eye and kidney damage have been linked to heavy metals. Mercury in particular has the added danger that when it enters aquatic environments in its elemental state, bacteria in the water quickly convert it to methyl-mercury, which is highly toxic.

Increased sediment of whatever kind in waterways can reduce light infiltration, which affects primary producers and visual predators, disrupting the food chain; furthermore, when the sediment settles it can wreak havoc on riverbed and other habitats. Wetlands and mangroves are particularly sensitive to pollutants from agricultural and industrial waste.

Even in ocean environments, pollutants can cause more immediate problems than the eventual degradation of habitat when pollutants overcome a given area's ability to dilute them: oil spills, for example, can cause organisms to die from the hydrocarbons in oil. Oil that floats on the surface of water can coat the feathers of birds and the fur of marine mammals, drowning them or impeding their everyday survival. Some components of oil sink to the ocean floor, killing bottom-dwellers and ruining habitats there. Even the oil that washes up on beaches can have economic consequences on the fishing and tourism industries and cause problems for shoreline habitats. Sediment runoff is one of the causes of the degradation of the major coral reefs of the world.

Finally, if your AP Environmental Science class has performed water quality tests on water samples, you might have tested for the presence of various chemicals as well as insect larvae, which act as indicator species. Among the most important factors in judging water quality are:

- **pH,** which is a measure of acidity or alkalinity (normal for water is 6–8)

- **hardness,** which is a measure of the concentrations of calcium and magnesium

- **dissolved oxygen**—low levels of dissolved oxygen indicate an inability to sustain life (warm water holds less dissolved oxygen than cool water)

- **turbidity,** or the density of suspended particles in the water

- **biological oxygen demand (BOD),** which is a measure of the rate at which bacteria absorb oxygen from the water

Now, let's talk more specifically about a major water pollutant—wastewater.

Wastewater

Another group of water pollutants that are very dangerous to human health are infectious agents, such as those found in human and animal waste. Fecal waste not only contains the symbiotic bacteria that aid in the human digestive processes, but also contains disease-causing bacteria. Several human diseases, such as cholera and typhoid fever, are caused as a result of human waste entering the water source of a community. In fact, the major reason for the increase in the life span of humans was not modern developments in medicine; it was the introduction of cleaner drinking water and better ways of disposing of wastewater.

The term **wastewater** is used to refer to any water that has been used by humans. This includes human sewage; water drained from showers, tubs, sinks, dishwashers, and washing machines; water from industrial processes; and storm water runoff. Water that is channeled into storm drains, such as storm water, is generally dumped directly into rivers. (This is why storm drain covers in many locations have been stenciled with warnings about not dumping material down them.)

Today in the United States, wastewater that isn't storm water is moved through sewage pipes to a sewage treatment facility, but this was not always the case. Sewage water once was, and in developing countries still is, merely dumped into the nearest river or ocean. While some amounts of sewage can be diluted and broken down in these waters, too much waste poses serious risks to human health and the health of the aquatic ecosystems.

Now in the United States, sewage pipes deliver wastewater to a municipal sewage treatment plant, where it is first filtered through screens (in what's called a **physical treatment**) to remove debris such as stones, sticks, rags, toys, and other objects that were flushed down toilets. This debris is then usually separated and sent to a landfill. The remaining water is passed into a settling tank, where suspended solids settle out as sludge—chemically treated polymers may be added to help the suspended solids separate and settle out. This treatment is known as **primary treatment,** and it removes about 60 percent of the suspended solids and 30 percent of the organic waste that requires oxygen in order to decompose.

Secondary treatment refers to the biological treatment of the wastewater in order to continue to remove biodegradable waste. This treatment can be done using trickling filters, in which aerobic bacteria digest waste as it seeps over bacteria-covered rock beds. Alternately, the wastewater can be pumped into an activated **sludge processor,** which is basically a tank filled with aerobic bacteria. The solids in the water, including the bacteria, are once again left to settle out. The solids remaining are considered **sludge,** which is combined with the sludge from the primary treatment. Sludge used to be dumped into the ocean, but that practice has been banned. Instead, the sludge is further processed with anaerobic bacteria (to break down more organic material). This digestion also produces methane gas that can be used as an alternative fuel to run the treatment plant. After drying, this sludge cake can be processed and sold as fertilizer.

At the end of secondary treatment, 97 percent of the suspended solids; 95–97 percent of the organic waste; 70 percent of the toxic metals, organic chemicals, and phosphates; 50 percent of the nitrogen; and 5 percent of the dissolved salts have been removed from the wastewater. However, almost no persistent organic chemicals, such as pesticides, are removed, nor are radioactive isotopes. Generally, after secondary treatment, the wastewater is chlorinated to remove any remaining living cells and then discharged into a stream, the ocean, or water that's used to water lawns. A negative effect of the final chlorination of the water is that trihalomethanes (potential carcinogens) can be formed when any organic matter left in the water reacts with the chlorine, and this is problematic. Two alternate processes to chlorination—ozonation and UV radiation—have been used to treat secondary-treatment water, but they have not proven to be as effective or long-lasting as chlorine and are also much more expensive.

Some municipal plants deposit wastewater directly into ground water; this is done in San Jose Creek in Los Angeles County, for example. In these places, the water must be further treated by tertiary treatment. **Tertiary treatment** involves passing the secondary treated water through a series of sand and carbon filters and then further chlorination. At the San Jose Creek Plant, the tertiary treated water from the reclamation plants is discharged into percolation basins, where it replenishes groundwater, or it is used for irrigation and for watering lawns, golf courses, and plants in nurseries. Tertiary treatment is expensive, but in arid or semi-arid regions, every gallon that can be reclaimed is one that needs not come from rapidly depleting sources, such as diminished rivers or underground aquifers.

Private wastewater treatment in the form of septic tank systems is hallmarked by some as the most environmentally friendly type of waste disposal. Septic tanks act in a way that's similar to the primary and secondary treatments that take place in municipal treatment plants. The water is then discharged into leachate (drain) fields. In order to install these types of systems, the soil must be able to percolate the water—that is, the water must be made to move from the top of the soil though its various horizons. Some clay soils are not porous enough to allow percolation and thus are unsuitable for a septic field.

Water Quality Legislation

There are many pieces of federal law that cover water quality. The College Board now requires that you are familiar with the following:

Date	Name of Legislation	What It Did
1972	Clean Water Act	Used regulatory and non-regulatory tools to protect all surface waters in the United States. • sharply reduced direct pollutant discharges into waterways • financed municipal wastewater treatment facilities to manage polluted runoff • achieved the broader goal of restoring and maintaining the chemical, physical, and biological integrity of the nation's waters • supported "the protection and propagation of fish, shellfish, and wildlife, and recreation in and on the water"
1974, 1996, 2005, 2011, 2015	Safe Drinking Water Act	Established a federal program to monitor and increase the safety of the drinking water supply. It does not apply to wells that supply fewer than 25 people. Amendments in recent years have led to more stringent regulation of lead and algal toxins in drinking water.

The **Clean Water Act** and the **Safe Drinking Water Act** are both on the list of legislations required for the exam, so make sure to memorize them!

SOLID WASTE (GARBAGE)

Solid waste is any discarded material that is not a liquid or gas. It is generated in the domestic, industrial, business, and agricultural sectors, and it can consist of hazardous waste, industrial solid waste, or municipal waste. Many types of solid waste provide a threat to human health and the environment.

The phrase "reduce, reuse, recycle" might seem simplistic, but it does outline the steps needed to reduce the amount of solid waste that must be dealt with. **Reduce**, of course, refers to the minimizing of disposable waste. For example, there are many types of packaging that are extremely wasteful—if you keep an eye out, you'll see them everywhere. Reducing or eliminating the amount of packaging per item sold is one way disposable waste can be sensibly reduced. **Reuse** applies to products that in some cases are disposable, but can be put to alternative use—sometimes over and over again—before their eventual disposal. For example, refillable bottles and tanks, reusable packing materials, secondhand goods, and cloth shopping bags all represent a reduction in waste when compared to single-use items. Finally, **recycling** is the reuse of materials. Recycling offers the benefit of reducing the global demand for new minerals, but the process is energy-intensive and can be costly. In **primary recycling,** materials such as plastic or aluminum are used to rebuild the same product—an example of this is the use of the aluminum from aluminum cans to produce more aluminum cans. Alternately, in **secondary recycling,** materials are reused to form new products that are usually lower quality goods—examples of this are old tires recycled to form carpet, and plastic bottles recycled to create decking material. Finally, another environmentally important process is composting. **Composting,** the process of allowing organic matter such as food scraps, paper, and yard waste, to decompose, allows the organic material in solid waste to be decomposed and reintroduced into the soil rather than take up space in landfills. The product of this decomposition, known as compost, is useful as fertilizer. Drawbacks to composting include its odor and the attraction of rodents and other pests.

According to the EPA, one of the most effective steps in aiding the environment that occurred in the 20th century was the marked growth in the use of recycling and composting to deal with solid waste. Of the 292.4 million tons of municipal solid waste produced in the United States in 2018, more than 69 million tons were recycled. According to the EPA, the following percent of each of these materials was recycled in the year 2018:

Material	Material Recycled in 2018 (Per 69.1 Million Tons)
Paper and paperboard	66.54%
Metals	12.62%
Glass	4.43%
Plastics	4.47%
Wood	4.49%
Rubber, leather, & textiles	6.05%
Other	1.40%

In order to encourage people to reduce, reuse, and recycle, many communities have established Pay-As-You-Throw (PAYT) programs, which charge municipal customers for the amount of household garbage they throw away. As you can imagine, this has been a strong incentive for people to practice these good habits. An example of an incentive program that has worked beautifully is the bottle redemption bill. Ten states have enacted bottle bills and they have worked fantastically, especially in Michigan, which has a 10-cent redemption fee.

Landfills

In 1987, after it was discovered that landfills on Long Island were contaminating local groundwater, the barge *Mabro* left New York towing 3,186 tons of garbage in search of a dumping ground. However, it was barred from docking in several southern states, followed by the countries of Mexico, Cuba, and Belize. Three months and 6,000 miles later, it returned to New York, where it became a symbol for Americans who were concerned about the status of landfills in the United States. It was also at this time that the term NIMBY (which stands for Not In My Backyard) became popular. It was widely agreed upon that landfills were needed, but no one wanted a landfill close to their home.

Modern landfills are very different from the traditional caricature of a garbage dump filled with heaps of junked cars and rats foraging for food scraps. Federal regulations that protect human health and the environment have paved the way for **sanitary landfills**. For example, federal law prohibits landfills from being located near geological faults, wetlands, or flood plains. Additionally, landfill sites are periodically required to dig large holes in the ground and line them with geomembranes or plastic sheets that are reinforced with two feet of clay on the bottom and sides. Smoothing wet clay is much like making a clay pot; the layer that is created is virtually impermeable. Also, the waste in the landfill must be frequently covered with soil in order to control insects, bacteria, rodents, and odor; and the decomposed material that percolates to the bottom of the landfill (called **leachate**) is piped to the top of the site and collected in leachate ponds, which are closely monitored. The rate at which the material decomposes depends on things like what the trash is composed of and what conditions are present for the microbes responsible for decomposing the waste. Gases from the landfill, like methane, may even be piped up from the site and used to generate electricity. Sometimes the methane is burned in continuously flaming flares to avoid larger fires or explosions. To ensure that landfills do not contaminate the environment, they are required to be positioned at least six feet above the water table, and groundwater at the sites must be tested frequently for quality. When one site (hole) is full, it must be capped with an engineered cover, monitored, and provided with long-term care.

Of course, more and more landfills are needed to keep up with human waste. The question arises of how to best make use of closed landfills. Landfill mitigation strategies range from burning waste for energy to restoring habitat on former landfills for use as parks.

Some countries dispose of their waste by dumping it into the ocean. This practice, along with other sources of plastic, has led to large floating islands of trash in the oceans. Litter that reaches aquatic ecosystems, besides being unsightly, can introduce toxic substances to the food chain. Additionally, wildlife can become entangled in the waste, as well as ingest it; this can create intestinal blockage and choking hazards for wildlife.

Waste may also be burned in municipal incinerators, which are generally capable of sorting out recyclables first. The energy released from the incineration can be used to generate electricity in what's called the **Waste-to-Energy (WTE) program.** This type of system is particularly effective in large municipal areas, where waste only needs to be transported short distances.

Some items are not accepted in sanitary landfills and may be disposed of illegally, leading to environmental problems. One example is used rubber tires, which, when left in piles, can become breeding grounds for mosquitoes that can spread disease. Another category is electronic waste, or **e-waste,** which is composed of discarded electronic devices including televisions, cell phones, and computers. This waste must be disposed of carefully because it can contain potentially hazardous materials such as lead, cadmium, beryllium, mercury, and flame retardants. These can leach from landfills into groundwater if they are not disposed of properly. E-waste can be reduced by recycling and reuse, but even recycling of e-waste must be handled carefully, or recycling workers and their communities may face health risks.

HAZARDOUS WASTE

Hazardous waste is any waste that poses a danger to human health; it must be dealt with in a different way than other types of waste. While the AP exam doesn't explicitly test hazardous waste, it's closely tied to tested information about waste and waste disposal. Hazardous waste includes such common items as batteries, cleaners, paints, solvents, and pesticides. Industry produces the largest amounts of hazardous waste, and most developed countries now regulate the disposal of these wastes. United States law mandates that hazardous materials be tracked "from cradle to grave" thanks to laws like the **Resource Conservation and Recovery Act (RCRA)**. The EPA breaks hazardous wastes down into four categories:

- **corrosive waste:** waste that corrodes metal

- **ignitable waste:** substances such as alcohol or gasoline that can easily catch fire

- **reactive waste:** substances that are chemically unstable or react readily with other compounds, resulting in explosions or causing other problems

- **toxic waste:** waste that creates health risks when inhaled or ingested, or when it comes into contact with skin

Hazardous wastes are disposed of in three main ways: in injection wells, in surface impoundments, and in landfills. Many communities have specific areas in their landfills that are designated for hazardous waste, and the standards for those areas of the landfills are higher than standards for nonhazardous waste areas. **Surface impoundment** is typically used for liquid waste; it involves the creation of shallow, lined pools from which the hazardous liquid evaporates. **Deep well injection** involves drilling a hole in the ground that's below the water table. These wells must reach below the impervious soil layer into porous rock, and waste is injected into the well. All three of these methods have their advantages, but none of them is satisfactory.

As you can probably imagine, radioactive waste must be contained in a different way than other hazardous wastes. For years, the United States has been trying to develop one major site for the disposal of all of our radioactive waste. Yucca Mountain, Nevada, was selected as the location of this site because of its remoteness and because nuclear testing had previously been done at the site. This decision is still controversial, in part because of NIMBY, but also because Nevada has no nuclear power plants. Some argue that because Nevada doesn't benefit from nuclear energy, the state should not be responsible for the spent fuel repository. It's a contentious subject. However, it is critical that some plan for the long-term storage of nuclear waste be made, because in the very near future the nation's nuclear power plants will be at maximum capacity for the storage of spent fuel. The Waste Isolation Pilot Plant in New Mexico is a permanent site for nuclear waste burial. Waste that's left over from the construction of nuclear weapons is known as **transuranic waste**.

Some people define radioactive wastes that produce low levels of ionizing radiation as **low-level radioactive waste** and those that produce high levels of ionizing radiation as **high-level radioactive waste.** However, the EPA categorizes radioactive waste according to its place of origin. Therefore, in the EPA's classification system, some wastes that are considered high level may actually be less radioactive than certain low-level wastes. The EPA puts all radioactive wastes into six categories:

- nuclear reactor waste: high level

- waste from the reprocessing of spent nuclear fuel: high level

- waste from the manufacture of nuclear weapons: high level

- waste from the mining and processing of uranium ore: high level

- radioactive waste from industrial or research industries, including clothing, gloves, tubes, needles, animal carcasses, etc.: low level

- radioactive natural materials: not waste

In general, low-level waste is either stored on-site by licensed facilities until the radioactivity has degraded, or it is shipped to a low-level waste disposal facility. Mixed waste, containing both chemically hazardous waste and radioactive waste, is generally disposed of in the same manner.

In this book, we use the radioactive waste disposal terms used in the EPA classification system. However, on the AP Environmental Science Exam, a discussion of either system of classification would be considered correct—as long as you identify which classification system you're using.

Contaminated Waste Sites

After the 1970s, new regulations for the disposal of hazardous wastes solved many of the problems of how to add new wastes to landfills with minimal impact on the environment. At the same time, the issue lingered of what to do with sites that were already contaminated by hazardous waste or pollutants, known as **brownfield sites**. These sites had to be cleaned up and those who had acted irresponsibly had to be held accountable for the environmental problems they'd caused. For these reasons, the United States legislature created the **Superfund Program**, which was administered by the EPA.

Rocky Flats, Colorado, is a Superfund site where the party responsible for the damage happened to be the United States government. Starting in 1952 and continuing for almost 40 years, components of nuclear weapons, such as plutonium, uranium, beryllium, and stainless steel were all manufactured on this site. Now that the area has been significantly cleaned up, it is home to a variety of plants and animals, including bald eagles, and acts as a wind-power testing site.

Along with the Rocky Flats Plant, you should know the story of **Love Canal** near Niagara Falls, New York. The site was originally a canal built to bring power and employment to the surrounding community. After the canal's failure, the land was purchased by various companies that turned the canal into a landfill. After the town purchased the covered landfill area, 100 homes and a school were built on the site. In 1978, people saw rusting drums full of waste sticking up above ground. They also noticed dead and dying trees and gardens. Homeowners even reported having pools of smelly liquids in their basements; their children reported burning hands and faces after coming in from playing. Environmental Protection Agency employees soon came to the canal area, and by the end of August, 220 families had moved or said they would move out of the area. It was in response to the situation at Love Canal (and other sites in the United States) that laws like the Resource Conservation and Recovery Act (RCRA) and the Comprehensive Environmental Response, Compensation, and Liability Act (CERCLA) were passed.

Laws for Solid and Hazardous Wastes

There are many federal statutes that cover issues concerning solid and hazardous wastes. Make sure you are familiar with the ones in the table below.

Both the **Resource Conservation and Recovery Act (RCRA)** and the **Comprehensive Environmental Response, Compensation, and Liability Act (CERCLA)** are required for the exam, so make sure to memorize them!

Date	Name of Legislation	What It Did
1976	The Resource Conservation and Recovery Act	• The solid waste program encouraged states to develop comprehensive plans to manage nonhazardous industrial solid waste and municipal solid waste, sets criteria for municipal solid waste landfills and other solid waste disposal facilities, and prohibits the open dumping of solid waste. • The hazardous waste program established a system for controlling hazardous waste from the time it is generated until its ultimate disposal—in effect, from "cradle to grave." • The underground storage tank (UST) program regulates underground storage tanks containing hazardous substances and petroleum products.
1980	The Comprehensive Environmental Response, Compensation, and Liability Act (CERCLA), commonly known as Superfund	• Created a tax on the chemical and petroleum industries and provided broad federal authority to respond directly to releases or threatened releases of hazardous substances that may endanger public health or the environment. • Established prohibitions and requirements concerning closed and abandoned hazardous waste sites. • Provided for liability of persons responsible for releases of hazardous waste at these sites. • Established a trust fund to provide for cleanup when no responsible party could be identified.

NOISE POLLUTION

Take your earphones off and think about this for a minute: the EPA considers noise to be a controllable pollutant. The **U.S. Noise Control Act** of 1972 gave the EPA power to set emission standards for major sources of noise, including transportation, machinery, and construction. Occupational Safety and Health Association (OSHA) has also set limits on the amount of noise that people can be exposed to in the workplace. Although the definition of noise pollution can be quite flexible, **noise pollution** in a broad sense is any noise that causes stress or has the potential to damage human health. However, remember that it's not just humans who are affected by noise pollution: it can have harmful effects on whole ecosystems. Some effects of noise pollution on animals in ecological systems include stress, the masking of sounds used to communicate or hunt, damaged hearing, and changes to migratory routes.

These, or equivalent effects, are seen in humans, too. The most acute result of too much noise is that continued exposure to high levels of it can damage hearing. The louder the noise, the shorter the exposure it takes to damage inner ear cells and cause hearing impairment. Unfortunately, certain essential cells in the ear that are involved in hearing do not regenerate, so the loss of hearing is permanent. There are federal laws that regulate noise emissions from some equipment and modes of transportation, and OSHA is responsible for the regulation of noise in the workplace. In local communities, however, noise pollution is usually controlled by state or local laws. The **Quiet Communities Act** provides for the coordination of federal research and activities into noise control. Furthermore, this act authorized the use of Federal Aviation Administration funds for the development of noise abatement plans around airports.

CHAPTER 9 KEY TERMS

Don't *waste* any time; study these words now!

Toxicity and Health

toxin
toxicity
dose-response analysis
dose-response curve
LD_{50}
poison
Food and Drug Administration (FDA)
Delaney Clause
ED_{50}
threshold dose
acute effect
chronic effect
infection
disease
pathogens
vector
plague
tuberculosis
cholera
dysentery
malaria
West Nile virus
zika
severe acute respiratory syndrome (SARS)
Middle East respiratory syndrome (MERS)
COVID-19
risk
risk assessment
risk management
pollutant

Air Pollution

primary pollutants
secondary pollutants
sulfur oxides
stationary sources
moving sources
point source pollution
non-point source pollution
criteria pollutants
carbon monoxide (CO)
lead
ozone
stratospheric ozone

tropospheric ozone
nitrogen oxides
nitrogen dioxide (NO_2)
sulfur dioxide (SO_2)
particulate matter
Clean Air Act (CAA)
volatile organic compounds (VOCs)
industrial smog (gray smog)
photochemical smog (brown smog)
acid precipitation
dry acid particle deposition
wet deposition
National Ambient Air Quality Standards
catalytic converter
vapor recovery nozzle
Corporate Average Fuel Economy (CAFE)
indoor air pollution
radon
dust
asbestos
sick building syndrome (SBS)
building-related illness

Thermal Pollution

heat islands
temperature inversion

Water Pollution

Clean Water Act (CWA)
oxygen sag curve
eutrophication
algal bloom
hypoxic zone (dead zone)
eutrophic waterways
hypoxic waterways
oligotrophic waterways
persistent organic pollutants (POPs)
endocrine disruptors
bioaccumulation
acid mine drainage
biomagnification
pH
hardness
dissolved oxygen
turbidity

biological oxygen demand (BOD)
wastewater
physical treatment
primary treatment
secondary treatment
sludge processor
sludge
tertiary treatment

Solid Waste (Garbage)
solid waste
reduce
reuse
recycling
primary recycling
secondary recycling
composting
sanitary landfills
leachate
Waste-to-Energy (WTE) program
e-waste

Hazardous Waste
hazardous waste
Resource Conservation and Recovery Act
  (RCRA)
corrosive waste
ignitable waste
reactive waste
toxic waste
surface impoundment
deep-well injection
transuranic waste
low-level radioactive waste
high-level radioactive waste
brownfield sites
Comprehensive Environmental Response,
  Compensation, and Liability Act (CERCLA)
Superfund Program
Love Canal

Noise Pollution
U.S. Noise Control Act
noise pollution
Quiet Communities Act

Chapter 9 Drill

Directions: Each of the questions or incomplete statements below is followed by four suggested answers or completions. Select the one that is best in each case. For answers and explanations, see Chapter 13.

1. Which of the following is NOT a direct source of groundwater pollution?

   (A) Automobile exhaust

   (B) Wastewater lagoons

   (C) Underground storage tanks

   (D) Waste injected into deep wells

2. Which of the following cities has the greatest amount of gray-air smog?

   (A) New York, New York

   (B) Beijing, China

   (C) Los Angles, California

   (D) Chicago, Illinois

3. All of the following are true about sanitary landfills EXCEPT that they

   (A) have methods of monitoring leaks in the clay and plastic liners

   (B) pipe generated methane gas to storage tanks

   (C) pump leachate out of the landfill for treatment and disposal

   (D) are built so that trash sits on top of the land

4. Which of the following correctly explains what happens to the level of oxygen dissolved in water when organic waste is put in the water?

   (A) The level remains the same after the waste is added.

   (B) The level increases because of the availability of nutrients to animals that live in the water.

   (C) The level decreases because of the bacteria feeding off the waste and using the oxygen to live.

   (D) The level decreases because of the waste absorbing the oxygen.

5. An abundance of which of the following creatures (and a lack of other creatures) would indicate that water is polluted?

   (A) Trout and other game fish

   (B) Sludge worms, anaerobic bacteria, and fungi

   (C) Carp, gar, and leeches

   (D) Salamanders and turtles

6. Which of the following is the most common way of disposing of municipal solid waste?

   (A) Recycling

   (B) Composting

   (C) Placing in landfills

   (D) Burning

7. Which of the following choices gives the correct order of processing sanitary waste in a sewage treatment plant?

   (A) Disinfection—breakdown of organics by bacteria—solid separation

   (B) Solid separation—breakdown of organics by bacteria—disinfection

   (C) Solid separation—disinfection—breakdown of organics by bacteria

   (D) Breakdown of organics by bacteria—solid separation—disinfection

8. Which of the following is a secondary pollutant?

   (A) CO

   (B) Soot

   (C) VOCs

   (D) PANs

9. The United States proposed building a nuclear waste disposal site in

   (A) Wheeling, West Virginia

   (B) Yucca Mountain, Nevada

   (C) Gallup, New Mexico

   (D) Hudson, New York

10. Oxides of nitrogen create the pollutant

    (A) nitric acid

    (B) nitrogen gas

    (C) sulfuric acid

    (D) carbonic acid

11. In comparison to the surrounding rural areas, cities are

    (A) cooler than the rural areas

    (B) the same temperature as the rural areas

    (C) hotter than the rural areas

    (D) incomparable to the surrounding areas as far as temperatures

12. Which of the following are two of the most deadly and common indoor air pollutants in developed countries?

    (A) Soot and VOCs

    (B) CO_2 and methane

    (C) Nitric acid and sulfuric acid

    (D) Radon and tobacco smoke

13. The dosage at which a negative effect of a toxin occurs is called the

    (A) threshold dose

    (B) dose-response analysis

    (C) LD_{50}

    (D) ED_{50}

Free-Response Question

1. Students from a local high school participated in a study of Hillside Pond. After safely taking samples of some small fish, a fish-eating hawk, some pond water, some zooplankton, and a fish that preys on the small fish, they determined the average concentration of compound "X" in each sample. The table below summarizes their data.

Organism	Compound "X" concentration
Small fish	0.1 ppm
Hawk	3.0 ppm
Pond Water	0.1 ppb
Zooplankton	0.2 ppb
Predatory fish	1.0 ppm

(a) **Identify** one process that would cause compound "X" to contaminate the pond's water.

(b) Use the concentrations of compound "X" to **identify** the correct trophic order of the pond. **Draw** a food chain that illustrates the order and **label** it with the names of the organisms and the concentrations of compound "X" for each part of the chain.

(c) **Describe** a process that would explain the different concentrations of compound "X" in each organism.

(d) **Identify** one real-life example of a substance that behaves like compound "X" in the oceans, and **identify** one negative effect that the substance might have on humans.

Free-Response Question

2. Photochemical smog is one of the most common forms of air pollution today.

   (a) **Identify** TWO primary pollutants that cause photochemical smog. **Describe** how they are produced.

   (b) **Identify** TWO secondary pollutants that make up photochemical smog. **Describe** how they are produced.

   (c) **Give** one reason why photochemical smog is more likely to be found in industrialized nations and gray-air smog is more likely to be found in non-industrialized nations.

   (d) **Identify** one component of photochemical smog that affects human health. **Explain** its consequences.

Summary

- o Toxicity is a measurement of how harmful a substance is to living things. The toxicity of a dosage must take into account the following:
  - frequency and concentration of dosage
  - synergistic effect
  - age/size of organism
  - organism sensitivity

- o Pathogens are bacteria, viruses, or other microorganisms that can cause disease. There are five main categories of pathogens.
  - viruses (and other subcellular infectious particles, such as prions)
  - bacteria
  - fungi
  - protozoa
  - parasitic worms

- o All pollution can either be traced back to a specific location, called point source pollution, or is derived from a combination of many sources, known as non-point source pollution.

Air Pollution

- o The greatest production of air pollution since the Industrial Revolution has been from a drastic increase in the use of fossil fuels.

- o Major criteria air pollutants of concern in the troposphere are:
  - carbon monoxide (CO)
  - lead (Pb)
  - ozone (O_3)
  - nitrogen dioxides (NO_2)
  - sulfur dioxides (SO_2)
  - particulates

- o Acid rain is a secondary pollutant resulting from the combination of primarily SO_2 and nitrogen oxides combining with water vapor to form acids that alter the pH of precipitation.

Thermal Pollution

○ Heat islands occur in urban areas and other locations that are dominated by heat-absorbing materials, like concrete and asphalt. These increased temperatures can cause higher rates of photochemical reactions, which may lead to photochemical smog. Temperature inversions may also cause warm polluted air to not rise out of an urban area when a warm air mass stalls above the city.

Water Pollution

○ The greatest contributor to water pollution is excess nutrients from runoff, which can lead to eutrophication.

○ Major water pollutants include:
  - excess nutrients
  - organic waste
  - toxic waste
  - sediments and suspended particulates
  - thermal pollution
  - coliform bacteria
  - invasive species
  - persistent organic pollutants

○ Bioaccumulation describes how organisms at lower trophic levels accumulate persistent toxins.

○ Biomagnification describes the movement of substances, such as toxins, between the trophic levels.

○ Important parameters for testing water quality are:
  - pH
  - hardness
  - dissolved oxygen (DO)
  - turbidity
  - biological oxygen demand (BOD)

Noise Pollution

o   The U.S. Noise Control Act of 1972 gave the EPA power to set emission standards for major sources of noise, including transportation, machinery, and construction.

o   The Quiet Communities Act provides for the coordination of federal research and activities into noise control.

Waste

o   Solid waste consists of many types that may each be a threat to human health and the environment.

o   Municipal waste (produced by households) and industrial waste (from commercial production) are either recycled, composted, buried in landfills, or incinerated for Waste-to-Energy programs.

o   Hazardous waste poses dangers to human health and must be dealt with specifically via deep well injections, surface impoundments, and landfills, for example.

o   United States legislation created the Superfund program to manage land previously used commercially and polluted with hazardous waste.

Chapter 10
Unit 9: Global Change

In this chapter, we'll review Unit 9 of the AP Environmental Science course, *Global Change*. According to the College Board, about 15–20% of the test is based directly on the ideas covered in this chapter. If you are unfamiliar with a topic presented here, consult your textbook for more in-depth information.

By now, you've probably almost completed your AP Environmental Science course. What's left? In this chapter, we zoom out to the global picture—first, with the biggest environmental problem facing humanity: global climate change. Then we'll talk about the other major global problem: loss of biodiversity. Finally, we'll get around to what steps we as global citizens can take in the face of these problems!

THE GREENHOUSE EFFECT

Remember in Chapter 6 when we discussed the **greenhouse effect** and the greenhouse gases that cause it? It looks something like this: the Sun's rays strike the Earth, and some of the solar radiation is reflected back into space; however, greenhouse gases in the troposphere intercept and absorb a lot of this radiation. This warms the atmosphere and the Earth's surface. Remember: the greenhouse effect results in the surface temperature necessary for life on Earth to exist.

The Greenhouse Effect

The principal greenhouse gases are water vapor (H_2O), carbon dioxide (CO_2), and methane, but several other gases (mainly pollutants), such as nitrous oxide, ozone, chlorofluorocarbons (CFCs), and hydrofluorocarbons (HFCs) have some amount of greenhouse effect as well. As you will have noticed in the previous chapters, several human activities are responsible for the release of these gases in greater quantities than would naturally occur without us.

While the Earth has undergone climate change throughout geologic time, with major shifts in global temperatures causing periods of warming and cooling as recorded with carbon dioxide data and ice cores, we have recently reached the unavoidable consensus that the global climate is undergoing a shift that is almost certainly in large part due to these human contributions: an idea referred to simply as climate change.

CLIMATE CHANGE

Scientists use very sophisticated computer models and make several thousand meteorological observations each day to monitor the daily temperature of the Earth's atmosphere. Over the last several years, their observations have shown that there has been a steady rise in the Earth's average temperature. According to NASA, Earth's average surface temperature in 2022 was the fifth warmest on record. Continuing the planet's long-term warming trend, global temperatures in 2022 were 1.6 degrees Fahrenheit (0.89 degrees Celsius) above the average. The past nine years have been the warmest years since modern recordkeeping began in 1880, meaning that the Earth's surface temperature in 2022 was about 2 degrees Fahrenheit (or about 1.11 degrees Celsius) warmer than the late 19th-century average. Other qualified scientists have carefully documented a decrease in the size of glaciers and ice sheets, a slight rise in the average ocean level, and more severe rainstorms. In response to these phenomena, the Intergovernmental Panel on Climate Change (IPCC), an intergovernmental body of the United Nations, gathered hundreds of scientists from around the world to study these problems. In a 2013 report, the IPCC stated that most of the observed increase in the global average temperature since the mid-20th century is very likely (greater than 95 percent) due to the observed increase in **anthropogenic greenhouse gas** concentrations. And the IPCC 2021 report showed that emissions of greenhouse gases from human activities are responsible for approximately 1.1°C of warming since 1850–1900. The three major gases are carbon dioxide (from pre-industrial levels of 280 ppm to 400 ppm in 2016), methane (from preindustrial levels of 715 ppb to 1,840 ppb in 2016), and nitrous oxide (from preindustrial levels of 270 ppb to 328 ppb in 2016). These gases absorb the infrared heat radiating from the Earth and thus heat the lower atmosphere. This warming is in addition to the normal warming of the atmosphere by the greenhouse effect. The IPCC Sixth Assessment Report, released in 2023, delivered the starkest warnings yet regarding climate change, and found eight key findings:

1. Climate change is happening, it is the result of human activities, and it is a threat to human and natural systems.
2. Global temperatures are rising, glaciers are melting, and sea levels are rising.
3. The frequency and intensity of extreme weather events are increasing.
4. Oceans acidification is accelerating, biodiversity is decreasing, and water is becoming increasingly scarce.
5. Climate change is exacerbating existing inequalities.
6. The global economy is vulnerable to the impacts of climate change.
7. We have options in all sectors to at least halve emissions by 2030.
8. International cooperation is required to tackle climate change.

Average CO₂, 1890–2020

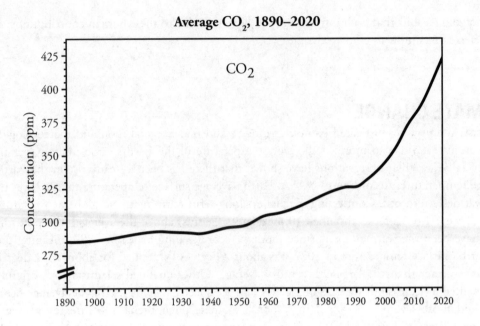

Average Global Temperature, 1880–2020

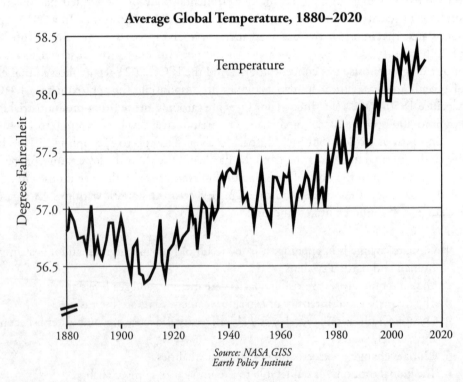

Source: NASA GISS
Earth Policy Institute

The impact of a particular greenhouse gas on global climate change can be thought of in terms of its **global warming potential (GWP)**. Carbon dioxide, which has a GWP of 1, is used as a reference point for the comparison of the impacts of different greenhouse gases on global climate change. Chlorofluorocarbons (CFCs) have the highest GWP, followed by nitrous oxide, and then methane.

It should also be mentioned that while water vapor is a greenhouse gas and even accounts for the largest percentage of the greenhouse effect, it doesn't contribute significantly to global climate change, because it has a short residence time in the atmosphere (an average of about nine days compared to years or centuries for other greenhouse gas molecules), and its levels have remained consistent for some time. However, water *does* respond to and amplify the effects of other greenhouse gases.

The increase in the Earth's temperature will lead to a variety of changes to the Earth. Physical changes on Earth include continued rising temperatures; further melting of glaciers, ice sheets, and permafrost; changes in precipitation patterns (with wet areas getting more precipitation and dry areas getting less precipitation); an increase in the frequency and duration of storms; an increase in the number of hot days; and a decrease in the number of cold days.

While it's clear that rising air and surface temperatures have widespread effects on the biosphere, it's also true that ocean temperatures are increasing due to the increase in greenhouse gases in the atmosphere and the greenhouse effect they produce. The most far-reaching consequence of this is that the thermal expansion of water (warm water is less dense than cooler water, and therefore takes up more space) is one of the factors contributing to rising sea levels (along with the melting of glaciers and ice sheets mentioned above). **Ocean warming** will also likely cause changes in coastlines, ocean currents, sea surface temperatures, tides, the sea floor, and weather.

Complicating the picture of oceanic effects is the phenomenon of **ocean acidification,** the decrease in pH of the oceans that's primarily another effect of the increased carbon dioxide concentrations in the atmosphere due to the burning of fossil fuels, vehicle emissions, and deforestation. As more carbon dioxide is released into the atmosphere, the oceans absorb a large part of that carbon dioxide. Dissolving CO_2 increases the hydrogen ion concentration (as shown below), and causes ocean water to become more acidic.

Ocean Acidification

$$CO_{2\,(aq)} + H_2O \rightleftharpoons H_2CO_3 \rightleftharpoons HCO_3^- + H^+ \rightleftharpoons CO_3^{2-} + 2H^+$$

Climate change will also affect biota. While there will be increased crop yields in cold environments, this is likely to be offset by loss of croplands as other areas suffer droughts and higher temperatures. Cold-tolerant species will need to migrate to cooler climates or they may become extinct. As equatorial-type climate zones spread north and south into what are currently subtropical and temperate climate zones, pathogens, infectious diseases, and any associated vectors will spread into these areas (where these diseases have not previously been known to occur). Entire coastal populations and ecozones will also be displaced by the rising oceans.

Marine ecosystems will also be affected. The change in sea level will have effects: some positive (for example, in newly created habitats on newly-flooded continental shelves) and some negative (for example, in deeper communities that may no longer fall within the photic zone of seawater). But the effects of ocean warming are likely to be even more drastic. Some habitats will be damaged or lost; some species are likely to adapt through metabolic and/or reproductive changes; and, as with sea levels, ocean warming may have positive effects on some habitats and organisms as well.

Ocean acidification has and will continue to have tangible effects on ocean biota as well. Some species are experiencing reproductive and metabolic changes, just as with ocean warming; and the greatest effects of acidification can be seen in organisms that make use of calcification (as in shells): for example, corals.

As we reviewed in Chapter 4, coral reefs are structures found in warm, shallow tropical waters that represent diverse and ecologically crucial ecosystems. Coral reefs are created by small marine animals (called cnidarians), which are involved in mutualistic relationships with photosynthetic algae called zooxanthellae. Reefs provide local populations with a great variety of seafood and are popular recreational areas for humans.

Ocean acidification damages corals by decreasing their ability to calcify (due to loss of calcium carbonate), making it difficult for them to form shells. When you put this together with coral bleaching, the world's coral reefs are under severe threat.

Coral bleaching is what happens when coral polyps expel the symbiotic algae that live in their shells and are vital to their health—the algae provide up to 90 percent of the corals' energy! Once a coral bleaches, it continues to live but begins to starve. Some corals recover, but many don't. The leading cause of coral bleaching is rising water temperatures, but the list of contributing factors also includes:

- increased sunlight exposure
- increased sedimentation (due to silt runoff)
- bacterial infections
- increased or decreased salinity
- herbicides
- exposure (as in extremely low tides)
- mineral dust carried in dust storms caused by drought
- pollutants such as those commonly found in sunscreens
- ocean acidification
- oil and chemical spills
- oxygen starvation caused by increase in zooplankton after overfishing

In many areas of the world, pollution, climate change, and exploitation have led to severe and irreversible damages to these reefs.

Another set of effects we may see due to global climate change is changes to the larger patterns of climate. For example, remember our discussion of winds in Chapter 6—winds generated by atmospheric circulation help transport heat throughout the Earth. Climate change may change the circulation patterns of wind, because the temperature changes may impact Hadley cells and the jet stream. And the winds in turn affect the oceanic currents, which carry heat in the water just as the winds do in the air. When these currents change, it will likely have a great impact on the climate in many places, especially coastal regions.

There is one area (well, two really!) of the globe that has a very specific set of circumstances due to global climate change that you should be familiar with—the polar regions. The Earth's polar regions are showing faster response times to global climate change, because ice and snow in these regions reflect the most energy back out to space, leading to a positive feedback loop. As the Earth warms, this ice and snow melts, meaning less solar energy is radiated back into space and instead is absorbed by the Earth's surface. This in turn causes more warming of the polar regions. In addition, the melting sea ice and thawing tundra release greenhouse gases like methane, which add even more to the global change! And to top it all off, the polar regions represent some of the more delicate ecosystems on the planet, which recover the most slowly from disruptions. One consequence of the loss of ice and snow in polar regions, for example, is the effect on species that depend on the ice for habitat and food. The species that are adapted to the coldest places on Earth may not have anywhere to displace *to* when even those places become warmer.

Climate change will also probably affect soil in many places planetwide, since changes in temperature and rainfall can impact the viability of soil and potentially increase erosion—which can then change the makeup of the soil as well.

Human health will show additional deaths from water- and insect-borne diseases. More frequent heat spells will endanger the very young and old. It is very likely that commerce, transport facilities, and coastal settlements will be disrupted by ocean level changes and stronger, more frequent storms. Marine ecosystem productivity and fishery productivity is also likely to change.

Ozone Depletion

Finally, let's return to a topic we've mentioned in several previous chapters—ozone. Remember when we said that tropospheric ozone (down low) harms us, but stratospheric ozone (up high) helps us? Specifically, we mentioned that while all ozone is O_3 and is the same chemically, stratospheric ozone absorbs UV light from the sun and therefore protects life on our planet, and is thus functionally very different from tropospheric ozone (which is a powerful respiratory irritant and precursor to secondary air pollutants). Ozone in the stratosphere provides us with a much-needed defense against ultraviolet radiation. The ozone layer is responsible for blocking about 95 percent of the sun's ultraviolet radiation (UV), thus protecting surface-dwelling organisms from UV damage. Ozone is naturally created by the interaction of sunlight and atmospheric oxygen. The simplified reaction is

$$O_2 + UV \text{ (sunlight)} \rightarrow O + O$$
$$O + O_2 \rightarrow O_3$$

As early as the mid-1950s, a thinning of the ozone layer above the Antarctic was observed. In the 1970s, atmospheric scientists hypothesized, and later proved, that declining stratospheric ozone levels were due to a group of man-made chemicals known as **chlorofluorocarbons (CFCs)**. Invented in the 1930s, CFCs and many other related compounds (e.g., halons and hydrochlorofluorocarbons) were used in items such as propellants, fire extinguishers, and cans of hairspray.

Once released, CFCs migrate to the stratosphere through atmospheric mixing (they are very stable, which allows them to survive through the rise). In the upper stratosphere, intense UV radiation breaks the CFC molecules apart and releases chlorine atoms that form chlorine monoxide (ClO) while converting O_3 to O_2. Let's take a look at that reaction.

$$Cl + O_3 \rightarrow ClO + O_2$$

During the winter months, chlorine monoxide is concentrated on ice crystals that form in and around the Antarctic polar vortex. In early spring, the returning warmth of the sun frees the chlorine from the chlorine monoxide where it destroys more ozone. The reaction that frees the chlorine from chlorine monoxide is shown here.

$$ClO + O \rightarrow Cl + O_2$$

Ozone loss is greatest in the spring as the chlorine breaks down ozone into O_2. Remember that chlorine acts as a catalyst; it is not changed by its reaction with ozone, and it can help break down another O_3 molecule immediately. As the air continues to warm, the natural production of ozone *increases* as more sunlight catalyzes the combination of oxygen back into ozone. This occurs in January and February (Antarctica's summer).

The Antarctic continent is the area exposed to the greatest amount of UV radiation, but prevailing winds can carry the ozone-depleted air to South America, Australia, and southern Africa. In 2006, the area of ozone thinness was more than 26 million square kilometers. Reduced levels of ozone have been documented over the Arctic and even over some midlatitude regions.

The loss of ozone has serious implications for the Earth's ecosystems as well as for human health. The increased number of UV rays that reach the Earth through the thin ozone layer can kill phytoplankton and other primary producers. The decrease in primary productivity of both marine and terrestrial ecosystems lowers the amount of available fish and crops. Human health issues from increased exposure to UV rays include eye cataracts, skin cancers, and the weakening of our immune systems.

As you can see, there are quite a few different large-scale impacts that humans have on the natural world, and they sometimes overlap, interact, and feed back on each other. Biodiversity and the health of our ecosystems are at stake.

HUMAN IMPACTS ON BIODIVERSITY

Use the acronym **HIPPCO** to memorize the human factors that can cause extinction and decrease biodiversity.

- **H**abitat destruction/fragmentation
- **I**nvasive species
- **P**opulation growth
- **P**ollution
- **C**limate change
- **O**verharvesting/overexploitation

Of these, you'll notice that population growth and pollution have been well covered in their own chapters. Overexploitation of resources has come up in our discussions of land and water use and of energy. In this chapter we've talked about climate change quite a bit already. We've mentioned invasive species, but let's review them, and then we'll take a look at habitat destruction and fragmentation.

Remember from Chapter 4 that **invasive species** are just introduced species (non-native species) that are successful where they're introduced. They can sometimes be beneficial, but they tend to be labeled invasive when their impact alters the environment they are introduced to—by outcompeting native species, using up resources or prey, or physically altering the habitat—in ways that threaten native species. In Chapter 5 we mentioned that invasive species tend to be *r*-selected generalists, while the species most adversely affected by invasives tend to be *K*-selected specialists. The ability of generalists to adapt gives them an advantage in new surroundings, and the characteristics of *r*-selected species (populations below the carrying capacity of their environment, small body size, short lifespans, early maturation and reproduction, many offspring at once) mean there's a good chance that at least some members will survive to establish a foothold there.

Now to tackle habitat destruction and fragmentation. In truth, we've already mentioned numerous causes of habitat loss, including global climate change (via changes in temperature, precipitation, and sea level rise), ocean warming and acidification, pollution, and human development. **Habitat fragmentation** occurs when large habitats are broken into smaller, isolated areas, which can happen as a result of the construction of roads and pipelines, clearing for agriculture or development, and logging, for example. The scale of habitat fragmentation that has an adverse effect on the inhabitants of a given ecosystem will vary from species to species within that ecosystem, so some species may suffer from habitat loss in the very same place where other species are still thriving. Of course, species loss affects food webs, and may eventually be the factor that means the habitat can no longer support other species as well.

A particular mention should be given to wetland ecosystems, as they are some of the most fragile and easily disrupted. They provide a wealth of ecosystem services: wetlands improve water quality by basically acting as giant water filters; they are natural sources of methane; they are a source of biodiversity and a habitat for diverse wildlife; and they provide protection from floods as reservoirs for floodwater to drain to. Given their importance, it's all the more problematic that wetlands are at risk on many fronts. Threats to wetlands include overfishing, pollutants from agriculture and industrial waste, and habitat destruction and fragmentation due to commercial development and dam construction.

It's worth pointing out that habitat loss and fragmentation are not the only ways that biodiversity is being lost. Agriculture—specifically monoculture—has expanded the range of just a few plant species enormously, and diminished that of many others in the balance. Animal species have been affected as well. Some animal species have been somewhat or completely domesticated and many of those are now managed for economic returns (such as with honeybee colonies and domestic livestock). Domestication (whether as pets or livestock) can have a negative impact on the biodiversity of a species—inevitably, humans will breed for characteristics that are desirable or useful to us, which often goes counter to what would naturally increase or maintain high biodiversity.

Threatened and Endangered Species

One of the ways we talk about loss of biodiversity is in labeling species endangered—in other words, in danger of **extinction**. Human activities have caused or contributed to the **extinction** of many species. The International Union for Conservation of Nature (IUCN) evaluates the conservation status of plant and animal species. A species is designated as **critically endangered** if the species is under a very high risk of extinction; **endangered** if the species is likely to become extinct; and **vulnerable** if the species is likely to become endangered if no action is taken. Species assigned to any one of these three categorizations are considered **threatened species.** The IUCN maintains a Red List of Threatened Species that is updated regularly. In 2000, the number of threatened species was just over 10,000. As of 2023, this number approached 45,000. Refer to the figure below.

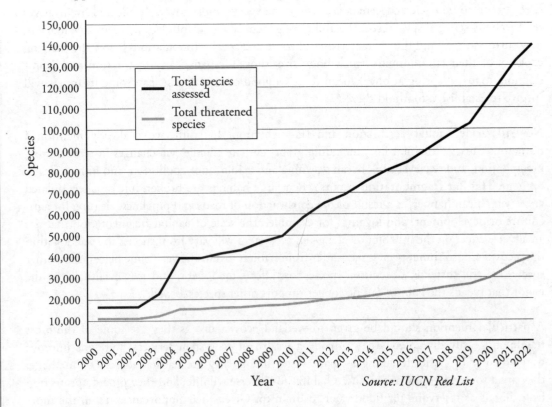

Source: IUCN Red List

During the past 500 million years, there have been five major extinction events, during which a significant percentage of all life on Earth, life inhabiting the oceans included, was exterminated by some catastrophic event. The most famous incident was the asteroid impact 65 million years ago near the present-day Yucatan Peninsula, which eliminated the dinosaurs and other large reptiles, vacating numerous niches that were quickly filled by mammals through radiative adaptation. Radiative adaptation occurs when, through evolutionary forces of natural selection, a species rapidly diversifies into numerous new species to take advantage of newly freed-up resources. The lesson to gain from this information is that life on Earth is very adaptive and capable of surviving great stress. Presently, humans are stressing the planet and may render the Earth uninhabitable for our species. Life, however, will certainly continue.

Besides major extinction events, extinctions have happened throughout the Earth's history. This natural rate of extinction—occurring apart from widespread events—is called the **background extinction rate**. Knowledgeable scientists estimate that the current extinction rate is between 50 and 500 times higher than in the past, likely due to human influence. Extinctions can happen anywhere in the world, but the rates are particularly high in the tropics (mostly on mountains and islands, which are home to isolated small populations that are especially vulnerable to both natural and human-caused changes to their environments).

A variety of factors can lead to a species becoming threatened with extinction, such as being extensively hunted, having a limited diet that becomes less available, being outcompeted by invasive species, or having specific and limited habitat requirements in an area where those are getting harder to fulfill. Not all species will be in danger of extinction when exposed to the same changes in their ecosystem. Species that are able to adapt to changes in their environments—or that are able to move to new environments—are much less likely to face extinction.

Thus, the species that are most endangered have several factors in common: they require large ranges of habitat to survive, have low reproductive rates, have specialized feeding habits, have some sort of value to humans (medicinal or food), and have low population numbers.

Humans play a major role in the extinction of species because of our destruction of animal and plant habitats. Poverty and rapid population growth cause people to use destructive practices, such as slash-and-burn farming, that destroy species' habitats. When we build roads or cities, habitats are lost or fragmented; this fragmentation may prevent the free movement of a species to find mates or escape danger, or it may reduce the area a species has for all the activities of its life cycle below a critical threshold. Finally, we cause habitat degradation by adding pollutants to the environment. Other factors that can contribute to extinction are invasive species and the direct hunting or overexploitation of a species or desired resources. Dr. Norman Myers coined the term **biodiversity hot spot** to describe a highly diverse region that faces severe threats and has already lost 70 percent of its original natural habitat by area.

WHAT CAN WE DO ABOUT IT?

> Conservation is a great moral issue, for it involves the patriotic duty of ensuring the safety and continuance of the nation.
>
> —Theodore Roosevelt

The human population is in the process of becoming aware of the looming climate and biodiversity catastrophes, and we are beginning to address these issues. Let's talk about what's being done, and what can be done.

In Chapter 8 we covered many types of renewable energy. Switching from dependence on fossil fuels, the production and use of which release greenhouse gases and other kinds of pollution, to cleaner energy sources, is one of the biggest steps humanity can take to diminish the global crisis.

In general, what is called for is a switch to **sustainability** as the model underlying our relationship with the Earth and its natural resources. Remember that idea from Chapter 7? Sustainability just means humans using resources in such a way as to not deplete those resources for future generations, following all the environmental indicators we've been discussing, such as biological diversity, food production levels, average global surface temperatures and CO_2 concentrations, human population, and resource depletion.

> **Did you know?**
>
> **Carbon Sequestration**
> is the process of capturing and storing atmospheric carbon dioxide.

In addition, humans will need to adapt to the changes that are already happening. In terms of climate change, adaptations to the warmer climate will need to occur at many levels of society. Technological improvements like **carbon sequestration** and the reduction of emissions from engines, behavioral changes such as turning off lights to conserve electricity, and policy changes such as enacting new treaties (like the **Kyoto Protocol**) and legislation will all be necessary in the next few decades. The promotion of sustainable growth will enhance the abilities of all societies to adapt to the new climate.

One area in which efforts to fix environmental problems have already met with some measure of success is in reducing ozone depletion. The loss of ozone was driven by increasing CFCs and related chemicals released during human use and rising to the stratosphere. But fortunately, there are several methods to manage the amounts of CFCs we release. In 1987, the **Montreal Protocol** was signed by more than 146 nations. The protocol calls for the worldwide end of CFC production. The United States stopped production in 1995. Since the institution of the Montreal Protocol, the release of ozone-depleting chemicals has been reduced by 95 percent. There are many nations that still rely on CFCs, though work is being conducted to develop safe and effective substitutes. Hydrofluorocarbons (HFCs) are one such replacement, but some are strong greenhouse gases. Though the problem still exists, a global recovery is underway. Ozone levels are now beginning to increase, as indicated in Stage 2 in the diagram below.

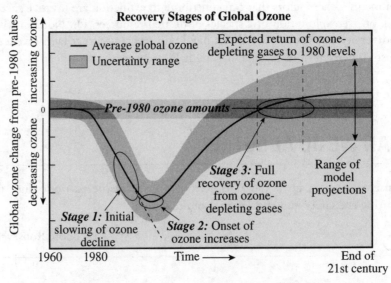

There are several international agreements that cover pollution issues, but the following are the ones you need to know for the exam:

Date	Name of Agreement	What It Did
1987	Montreal Protocol	Cut the emissions of CFCs that damage the ozone layer. This was amended in Copenhagen (1992) to include other key ozone-depleting chemicals.
1997	Kyoto Protocol	Required the participating 38 developed countries to cut their greenhouse gas emissions back to 5% below 1990 levels. While the United States signed the agreement, it did not ratify the agreement. As a result, the United States is not bound to abide by the Kyoto Protocol.

> The **Kyoto Protocol** and the **Montreal Protocol** are both on the list of legislations required for the exam, so make sure to memorize them!

When it comes to the loss of biodiversity, there are several ways humans can mitigate its impact. First, we can conserve habitats by requiring that large tracts of land be set aside and protected from human activity. In these protected habitats, organisms can find their niches and survive without risk of human interference. National parks and animal sanctuaries are two examples of protected habitats. This protection will slow the loss of species, and in some cases, allow species to recover and gain numbers. Making it illegal to trade in specific organisms means that those organisms will not be hunted or collected. Obviously, legislation is an important tool, and the **Endangered Species Act,** which requires its agency to designate which species are threatened and endangered and codifies the protection of those species and their habitats, has been key to this effort.

There are many United States laws that have been passed to reduce the rates of extinctions and protect specific organisms. The following are relevant examples:

Date	Name of Legislation	What It Did
1973	**Endangered Species Act** Program for the protection of threatened plants and animals and their habitats	The act prohibited the commerce of those species considered to be endangered or threatened.
1973	Convention on International Trade in Endangered Species of Wild Flora and Fauna (**CITES**)	This international treaty bans the capture, exportation, or sale of endangered and threatened species.

> **Endangered Species Act** and **CITES** are part of the required environmental legislations to know for the exam, so make sure to memorize those two!

Living sustainably and conserving resources help lower the demand that destroys habitats. Sustainable land use practices, including green building practices, sustainable agriculture, and new ideas to make roads and other infrastructure more environmentally friendly, will have a huge impact, since land use is a driving factor in habitat destruction and fragmentation. From simple techniques such as rainwater collection and home composting to more ambitious projects such as permeable pavement for roads and parking, organic farming, and **habitat corridors**—areas of habitat connecting wildlife populations separated by human activities or structures (such as roads, development, or logging)—these practices can help humans integrate better into our ecosystems and minimize the damage our daily existence does.

Finally, it is important that we not stop at protection, and work to actively restore lost habitats. Given that human intervention has been causing so many wildlife populations to diminish, it makes sense that some intervention is required to preserve the functioning of the biosphere. Restoring habitats involves a deep understanding of the conditions a habitat requires, and careful management of soil conditions and leaf litter accumulation, for example, along with the control of invasive species, can be essential. We can also help organisms with a species-by-species approach. Zoos and other institutions have captive breeding programs in which endangered species are bred under human control until their populations are high enough to be reintroduced into the wild. Wildlife management is a complex field, and careful management of conditions for breeding and protection of vulnerable young within preserves can also be components of a restoration strategy.

When it comes to controlling invasive species, a variety of human interventions are employed, depending on the species involved and their relationships to their environment. For example, pest animal species, especially those in contained environments such as islands, can sometimes be successfully eradicated through extermination. Invasivorism is a movement that explores the idea of eating invasive species in order to control, reduce, or eliminate their populations—chefs from around the world have begun seeking out and using invasive species as alternative ingredients. Genetic manipulation through breeding can be used to fight invasive pathogens, as in the example of the reintroduction of blight-resistant chestnut trees to fight the chestnut blight in the U.S.

It's important to recognize that these efforts are complicated and can go wrong. Sometimes, an invasive species becomes a primary food source or other necessary resource for another part of the food web, and disturbing it will then damage the ecosystem. Sometimes a good idea has unforeseen consequences and environmental damage results even from well-intentioned interventions. Sometimes the only solutions are compromises. For example, non-native species can be introduced to fill an ecological role that previously was performed by a native species now extinct (this is known as **taxon substitution**).

Public Policy

Public policy is a subject that will help you understand the human response to the global changes we've been talking about in this chapter, though the specifics of policy and other avenues to environmental progress aren't tested explicitly on the exam, except for the specific laws we've mentioned so far. The exploitation of public resources has been the motivation behind environmental policy at the international, national, state, and local level for as long as public policies have been made. Strictly speaking, **policy** is defined as a plan or course of action—as of a government, political party, or business—intended to influence and determine decisions, actions, and other matters.

While policies that we make as a nation are usually fairly easy to enforce—because they often have our collective best interests as a nation in mind—international policy, as is established through the United Nations (UN), is only achievable and realistic if the affected countries all cooperate with decisions that are made collectively. For example, in the 1994 International Conference on Population and Development (which was sponsored by the UN), one of the goals agreed on by the participants at the conference was to enroll 90 percent of all children in primary school by 2010. However, this policy can be put into action only if the countries that signed the agreement are willing to carry out the necessary steps.

There are ways in which the UN can attempt to force countries to follow mandates that are agreed on by the majority. These include withholding borrowing power through the World Bank, trade rules, and withholding aid. However, there are often certain environmentally significant countries that don't belong to the UN, didn't sign whatever agreement is at issue, or just don't have the infrastructure to enforce the objective—however worthy. Additionally, international agencies often don't have the power to control what happens inside a particular country.

Much more effective are international policies that are put into effect through treaties that the countries involved have all agreed to—their governments have all ratified the treaty. Obviously, policies that countries agree to are most often ones that benefit these countries in some way, so it isn't too surprising that they are more readily enforced. In other words, international laws that are not agreeable are not usually followed because they don't provide countries with incentives. Moreover, it is not possible to punish countries that don't follow these policies.

As we touched upon above, it's understandably much easier to enforce laws and policies in the United States than it is for us to police the other nations of the world. In the United States, state and local laws have an effect on the environment, but if there is a conflict between state or local law and federal law, most times federal law will take precedence. However, in some cases states have legislated controls that are even stricter than those the federal law requires. In these cases, the state laws are the ones enforced, rather than the more lenient federal law. Additionally, some laws are passed and enforced regionally because of particular geographic needs; one example of this is the difference in water laws to the east and west of the Mississippi River.

East of the Mississippi River, water laws are based on the principle that the upstream consumers control the water but, by law, cannot impede or reduce its flow or change its quality. A number of lawsuits based on this premise have been filed. One example is the diversion of water from the Apalachicola-Chattahoochee-Flint (ACF) and Alabama-Coosa-Tallapoosa (ACT) river basins. The state of Georgia would like to divert water from these basins to supply the growing needs of the urban area of Atlanta. The states of Alabama, Tennessee, and Florida, which are downstream from the diversion, are concerned about the flow and quality of water that will reach them if this diversion project is carried out. In April 2021, the U.S. Supreme Court dismissed Florida's lawsuit against Georgia over water use in the Apalachicola-Chattahoochee-Flint River Basin.

While we're on the topic, Atlanta is a good example of a city whose ecological footprint is far larger than the resources that are available on the land it occupies. (Remember that an ecological footprint is the amount of resources available to support a population and absorb its wastes.) City lawmakers are rightfully concerned about water shortages in the near future.

On the other hand, water laws west of the Mississippi are based on **water rights**. West of the Mississippi, it's held that the person who first files a claim on a water resource has rights to the use of the water. The amount of water an individual with a water right has claim to each year is determined by water flow that year (and also by how much water the individual wants!). But, regardless of the amount of water present or the place where the right was claimed, the first person who made the water claim gets to use their share of water before anyone else can partake. Obviously, water rights in the West do not require sharing, as they do in the East.

Environmental Policy in the United States—A Short History

Although the first laws of the United States, such as those contained in the Constitution, do not mention the environment specifically, the Bill of Rights includes the Fifth Amendment, which prohibits the taking of private property for public use without just compensation. This has been interpreted to include the "preventing of serious public harm" by those who wish to take private property and do something on it that will affect the environment or those around them in a negative way. Basically, this means that your neighbor cannot decide to build a small nuclear power plant on their residential property because this would violate zoning laws.

Early laws did not mention the environment because when these laws were written, there was so much land and so many resources in the United States that it was unimaginable that they could ever be in danger of being used up. In fact, many early laws, such as the Homestead Act of 1862, encouraged the settlement and exploitation of western lands. Others, such as the Mineral Lands Act of 1866, encouraged the use of resources, and, unfortunately, this exploitative act is still in effect for many mining regulations. A few years later, the General Mining Act of 1872, a federal law, was created to systematically oversee and control prospecting and mining for economic minerals, such as gold, platinum, and silver, on federal public lands.

Shortly after the Civil War, as people continued to migrate to the West, it was realized that the United States did not have an endless supply of land or resources. In fact, in order to preserve some of the lands in the West that were being very quickly settled, the first national park, Yellowstone National Park, was established in 1872. Further legislative action in 1891 created the forest reserves, which made these lands off-limits to logging in order to protect the land from being overharvested and to maintain the existing watersheds. This legislation marked the beginning of the federal government assuming an environmentally protective role.

Political and Cultural Activism

During this time period there were several men who stood out as early environmental activists, including Henry David Thoreau (1817–1862). Thoreau's book, *Walden*, describes his retreat from society and the quiet years that he spent living on Walden Pond studying nature.

More Great Books

Walden is a great book to read if you are interested in philosophy or environmentalism. If you live in the Boston area, you can even go for a swim in Thoreau's beloved pond!

Another important writer and scientist of this time period was George Perkins Marsh (1801–1882), whose book *Man and Nature* helped the American public understand that there are limits to natural resources. His plan for the conservation of resources is the basis for many of the resource conservation principles that we try to adhere to today. Another early environmental advocate was John Muir, a nature preservationist who founded the Sierra Club in 1892. He led a campaign for the protection of lands from human exploitation and advocated low-impact recreational activities such as hiking and camping. These ideas did not become popular until the 1960s.

As far as political leaders, arguably the most environmentally active president in the history of the United States was Theodore Roosevelt (1858–1919). Roosevelt was interested from an early age in the workings of the environment and even began his own natural history museum as a child. Interestingly enough, that collection became a part of the founding collection for New York's American Museum of Natural History.

Roosevelt's term as president has been called the **Golden Age of Conservation** because of the many environmentally friendly laws and policies he put into effect. During his presidency (1901–1909), he increased the area of national forest lands by 400 percent (up to 194,000,000 acres), establishing 150 new national forests and adding area to others. He established the first 51 bird reserves, signing the first one into existence by asking his advisors, "Is there any law which prohibits me from declaring this island a bird refuge?" When his advisors determined that there was not, he signed the bill with gusto, announcing, "Very well, then, I so declare it." Additionally, he established five national parks, including the Grand Canyon, four national game preserves, 18 national monuments (established under the 1906 Antiquities Act), 24 reclamation projects, and seven conservation commissions.

Roosevelt also appointed the first chief of the United States Forest Service in the history of the United States, Gifford Pinchot (1865–1946). Pinchot applied the principles of sustainable harvest and multiple-use to wildlife protection, recreation, and resource extraction.

As you're probably well aware, the 1960s were a turbulent time in U.S. history, when the baby boomers born after World War II began to come of age and express their opinions to the world. The book *Silent Spring*, written by Rachel Carson in 1962, awoke in many Americans an awareness of the state of the environment. The air was dirty, the water was polluted, and hazardous wastes were collecting in landfills all over the country. Also at this time, the Apollo space missions allowed Americans to see planet Earth from afar for the first time, and this popularized the term "spaceship Earth." Paul Ehrlich's 1968 book, *The Population Bomb,* warned of the myriad problems that would arise along with the quickly increasing human population, and an entertainer named John Deutschendorf took the stage name of John Denver and began to popularize the environmental movement through song.

More Great Books
Head to your local bookstore or library and look over a copy of Rachel Carson's famous book—you'll be glad you did!

A multitude of environmental laws and policies were initiated during the 1970s. For example, the first Earth Day was celebrated on April 22, 1970. Also in 1970, Richard Nixon signed into law the National Environmental Policy Act (NEPA); this act created the Council on Environmental Quality and required the submission of an environmental impact statement before any major federal action could be taken. One of Nixon's major environmental contributions was to consolidate two agencies that had environmental responsibilities into a bureau called the Environmental Protection Agency (EPA). Finally, two major legislative actions were enacted in this new era of environmental awareness in the United States. The first was the Clean Air Act of 1963, which we have mentioned many times in these pages, and the second was the Clean Water Act (introduced in 1972).

This is just a brief history of environmental activity and activists in the history of the United States. Throughout this book, we've highlighted some other important environmental laws that you should be aware of for this exam. The ones in bold are those that have had a particularly significant impact. Make flashcards of all of these, as well as the Clean Air Act of 1963 and the Clean Water Act, so that you know them cold for test day!

Some Relevant Environmental Policy Acts

As you just learned, a few pieces of legislation helped form the environmental policy of the United States. The table below shows some of the legislation relevant to U.S. environmental policy, but are not required by the College Board. If you want to see required laws that deal with particular problems like endangered species, clear water, or clean air, go back to those chapters!

Date	Name of Legislation	What It Did
1970	National Environmental Policy Act	Created the Council on Environmental Quality that resulted in the creation of the Environmental Protection Agency (EPA) from the consolidation of various environmental agencies. It also mandates that federal agencies prepare environmental impact statements.
1990	Pollution Prevention Act	Designed to promote source reduction (stop pollution from being produced).

What Have We Done for Us Lately?

One of the most significant pieces of climate legislation in U.S. history is the Inflation Reduction Act of 2022. The bill includes nearly $370 billion over the next decade to be used in clean energy and climate investments, including low-carbon technologies, scaling-up renewable energy production, and tax credits to promote electric vehicle sales. However, there is a distinct anti-environmental movement in the United States, and large, established environmental groups, such as the Sierra Club, have declined in membership, which is an interesting litmus test of environmentalism in America. However, hope lies in the fact that new grassroots environmental organizations are currently growing throughout the United States. **Non-governmental organizations (NGOs)** like Greenpeace and the World Wildlife Fund also play a role in protecting the environment. Six environmental issues that are expected to take center stage in the 21st century include:

- climate change
- water shortages and water supplies
- population growth
- loss of biodiversity
- air and chemical pollution
- ocean acidification and pollution

Keep your eye out for discussions of these topics in the news, and listen critically to campaigning politicians to see where they stand on these issues and how they voted by visiting Congress.gov. Such issues will prove to impact you personally in more and more ways as time goes by. The discussion of how to best reduce greenhouse gas emissions, for example, currently revolves around what's called **cap-and-trade policy**, an approach that provides economic incentives for limiting emissions of pollutants.

AVENUES TO ENVIRONMENTAL PROGRESS

Most often, the United States government has approached environmental issues by passing "command and control" laws. These laws set limits on factors, such as the amounts of pollution that are allowable from various sources, and they establish penalties for those that go over the limits. Wildlife has always been protected by similar types of legislation. There is no doubt that these laws have led to cleaner air and water as well as the conservation of soil and other natural resources. Endangered and threatened species have also both been protected, and some extremely endangered species have even been able to recover somewhat.

However, there have always been problems with the "command and control" approach. For example, consider the Endangered Species Act. Red-cockaded woodpeckers are endangered because the open forests with big, old pine trees have been replaced by forests with younger, smaller pines. Also, periodic natural fires, which historically kept the pinewoods open, have been suppressed because humans have settled in these areas. Periodic fires are needed to control the brushy understory and keep the pinewoods open. Creating yet another problem for the endangered bird, timber owners have been known to kill them in order to avoid preserving their habitat. However, it's very hard to prove what happens to these birds—are they being exterminated by landowners, or are they simply migrating elsewhere or declining in number for other reasons?

Green Taxes

There are other approaches that are more successfully used to continue environmental improvement without forcing the enactment of other types of command and control. Over time, it has become clear that the act of punishing actions that hurt the environment is not nearly as effective as rewarding actions that help the environment. Since the 1970s, the United States has substantially increased taxes on labor and modestly increased taxes on income, while allowing actions that create pollution and cause resource depletion to remain largely untaxed. The result is that the tax system of the United States encourages resource depletion and discourages investments in machinery and labor. A worldwide discussion is taking place about how to move away from taxing "goods," such as investments and employment (activities we should be encouraging), and toward taxing "bads," like pollution, which we would like to discourage. Pollution taxes have now been embraced by a growing number of mainstream economists and policy makers and are just one of a new group of taxes called **green taxes**.

A green tax shift is a fiscal policy that lowers taxes on income, including wages and profit, and raises taxes on consumption, particularly the unsustainable consumption of nonrenewable resources. Some taxes that could be lowered by the implementation of a green tax shift are payroll and income taxes, and the following is a list of taxes that could be implemented or, if currently in existence, increased:

- carbon taxes on the use of fossil fuels
- taxes on the extraction of mineral, energy, and forestry products
- license fees for fishing and hunting
- taxes on technologies and products that are associated with substantial negative externalities
- garbage disposal taxes
- taxes on effluents, emissions, and other hazardous wastes

Additionally, taxes on certain forms of consumption may occur through the "feebate" approach, in which additional fees are imposed on less sustainable products—such as sport-utility vehicles—and then pooled to fund rebates on more sustainable alternatives, such as hybrid electric vehicles.

In this scheme, taxes serve as policy tools as well as a way to protect the environment.

The three main goals of green taxes are:

- to generate revenue to correct past pollution damage and reduce future pollution

- to change behavior

- to use the funds received from pollution taxes for restoration

Market permits are also being used somewhat successfully to encourage reduction in pollutants. Market permits are cap-and-trade permits that work in this way: companies are allowed to buy permits that allow them to discharge a certain amount of substances into certain environmental outlets. If they can reduce their discharge, they are allowed to sell the remaining portion of their permit to another company. Economically speaking, it is to a company's advantage to reduce its discharge and sell the remainder of its allowable discharge to another company. But perhaps a better idea is for the government to buy back the unused permits rather than have them sold to another industry; this would reduce the overall discharge.

Many people think that subsidies (which are giveaways or tax breaks on certain resources to encourage their use) are hurting the environment more than they're helping it; detractors think that subsidies only encourage the use of unsustainable products.

Note that all policies, treaties, and laws are important to our environment. However, for purposes of the AP Environmental Science Exam, you will probably only be asked questions concerning United States federal laws, which is why we suggested that you commit the different acts you saw throughout this book to memory. International treaties, summits, and policies such as those that are directed from the United Nations or one of its agencies will probably come up on future exams, but you don't have to worry about that right now.

Globalization

As you can imagine, our world is becoming more and more interconnected. Aircraft can fly around the world in about 24 hours; we have instant communication worldwide via phones, television, and the Internet. This is called **globalization,** and it affects society, the economy, and the environment. Positive effects can be seen in new economic opportunities, our expanded access to information, and the interactions of many societies. For example, grapes can be grown in Chile, shipped north, and be sold in your supermarket in less than a week. There are also several negative impacts of globalization. In certain parts of China, large piles of unusable electronic components have been creating water pollution problems as rainwater leaches out heavy metals. The rapid spread of emerging diseases, increased levels of air pollution and hazardous waste, and the loss of marine fish stocks are just a few more examples of globalization's negative impacts. Not to mention those grapes from Chile required an enormous amount of energy (and fossil fuel) and probably added to the pollution during their trip north!

Remember when you read about the Commons, resources owned by no one but accessible by everyone? This concept is important when we consider global access to those resources. Fresh water, clean air, ample supplies of fish, and access to fertile croplands are all examples of the global Commons. It is important to use these resources sustainably because they are the foundations for economic and social development.

Poverty and greed can cause people to use resources in an unsustainable manner and damage the environment. Cutting down important rainforest habitats to raise crops and accepting companies that generate a lot of harmful pollutants are two examples of how people's hunger for money can lead to unsustainable practices. Unfortunately, the economically disadvantaged people who allow unsustainable practices to continue are also the ones most susceptible to environmental issues brought about by climate change and have the least amount of resources to combat the health and environmental problems that result.

International organizations such as the World Bank and the United Nations are two examples of institutions that are trying to ameliorate the poverty issue. The World Bank uses loans to reduce poverty and to help foster improvements in biodiversity, environmental policies, land management, pollution management, and water resources management. The United Nations, through its environmental program, seeks to promote international cooperation, develop regional programs to promote sustainability, and to assess global, regional, and national environmental trends.

Now you're ready for the key terms review and the following drills. Remember to use our techniques as you go through them.

CHAPTER 10 KEY TERMS

Study these terms and you will be sure to write excellent essays.

The Greenhouse Effect

greenhouse effect
climate change

Climate Change

anthropogenic greenhouse gas
global warming potential (GWP)
ocean warming
ocean acidification
coral bleaching
chlorofluorocarbons (CFCs)
ozone loss

Human Impacts on Biodiversity

invasive species
habitat fragmentation
extinction
critically endangered
endangered
vulnerable
threatened species
background extinction rate
biodiversity hot spot

What Can We Do About It?

sustainability
carbon sequestration
Kyoto Protocol
Montreal Protocol
Endangered Species Act
CITES
habitat corridors
taxon substitution
policy
water rights
Golden Age of Conservation
non-governmental organizations (NGOs)
cap-and-trade policy

Avenues To Environmental Progress

green taxes
market permits
globalization

Chapter 10 Drill

Directions: Each of the questions or incomplete statements below is followed by four suggested answers or completions. Select the one that is best in each case. For answers and explanations, see Chapter 13.

1. Overexploitation of a species can happen by all of the following EXCEPT

   (A) excessive hunting

   (B) use of a species for food

   (C) use of species as a pet

   (D) habitat conservation

2. Which of the following gases involved in global climate change is increasing in the atmosphere at the fastest rate?

   (A) H_2O

   (B) Methane

   (C) Chlorofluorocarbons

   (D) CO_2

3. Which of the following is NOT a consequence of stratospheric ozone depletion?

   (A) UV damage that kills primary producers in the food chain

   (B) Fewer crops and fish

   (C) Human and animal respiratory irritation

   (D) Weaker human immune systems

4. The release of CFCs was banned under an international treaty written in which of the following cities?

   (A) Montreal

   (B) New York

   (C) New Delhi

   (D) Kyoto

5. The listing of threatened species and the purchase of land to protect their habitats is legislated in which of the following?

   (A) Federal Noxious Weed Act

   (B) Endangered Species Act

   (C) Convention on Biological Diversity

   (D) Fish and Wildlife Conservation Act

6. A cap-and-trade policy might be effective in controlling which of the following types of pollutants?

   (A) Thermal pollution in rivers

   (B) Organic waste pollution in oceans

   (C) Underground water pollutants

   (D) Carbon dioxide in the atmosphere

7. Which of the following can be classified as an NGO?

   (A) World Wildlife Fund

   (B) Bureau of Land Management

   (C) Fish and Wildlife Commission

   (D) International Trade Commission

8. In which region are response times to global climate change the fastest?

   (A) Equator

   (B) Tropics

   (C) Temperate zones

   (D) Polar regions

9. Which of the following does NOT directly contribute to coral bleaching?

   (A) Ocean warming

   (B) Herbicide exposure

   (C) Ocean acidification

   (D) Change in ocean currents

10. Which of the following is NOT directly a threat to biodiversity?

    (A) Climate change

    (B) Ozone recovery

    (C) Habitat fragmentation and loss

    (D) Monoculture

Free-Response Question

1. The town council of Hilltop Valley has just received a proposal from its advisory panel to build a new coal-fired power plant on the western edge of the county where Hilltop Valley is located. The plant will be located along the back of the county's major river so that water can easily be transported to the plant.

   (a) **Describe** TWO parts of the Clean Air Act that will impact the building of this plant. **Identify** ONE example of a compound that would be impacted under the National Ambient Air Quality Standards (NAAQS) and **discuss** the regulation of hazardous air pollutant levels.

   (b) **Describe** TWO positive and TWO negative impacts that the plant might have on the region.

   (c) Opponents of the plant say that they will use the Endangered Species Act (ESA) to prevent the plant's construction. **Describe** how the ESA could be successfully used to stop construction.

   (d) Some residents of Hilltop Valley are lobbying to build a power plant that uses renewable energy rather than coal. **Make a claim** about one type of plant that could be built in Hilltop Valley and **identify** one drawback your choice of energy source would bring.

Summary

- o The greenhouse effect results in the surface temperature necessary for life to exist on Earth.

- o Greenhouse gases (GHGs) are molecules that absorb heat and warm the lower atmosphere.
  The major GHGs are:
  - water (H_2O)
  - carbon dioxide (CO_2)
  - nitrous oxide (N_2O)
  - methane (CH_4)

- o The increase in anthropogenic greenhouse gas concentrations is driving an increase in the global average temperature. This and several other effects (such as ocean acidification) are collectively referred to as climate change.

- o Global climate change has a range of detrimental effects on weather, ocean temperatures, circulation patterns of air and water, soil, and the biosphere.

- o The effect of climate change on the polar regions is especially severe due to the positive feedback loop involved in its effects.

- o Ozone depletion is a very specific climatic effect of one specific class of pollutants that has varied effects (including adding to climate change). It fortunately has easier solutions than many climate change factors, and steps are being taken to control it.

o   Use the acronym HIPPCO to memorize the human factors that can cause extinction and decrease biodiversity:
  • Habitat destruction/fragmentation
  • Invasive species
  • Population growth
  • Pollution
  • Climate change
  • Overharvesting/overexploitation

o   Steps being taken or that can be taken to reduce the impacts of global climate change and loss of biodiversity include eliminating dependence on fossil fuels and other pollutants; using sustainability as a model; conserving and protecting habitats and wildlife; implementing sustainable land use practices; and active restoration of habitats and wildlife. Legislation is a powerful tool in this process.

Chapter 11
AP Environmental
Science in the Lab

The College Board requires the AP Environmental Science course to have a laboratory and field investigation component that will complement students' learning about the environment. Although there is no formal lab manual for the course, teachers are expected to provide lab experiences that ultimately incorporate test concepts and principles learned throughout the course. These experiences must also help you design experiments, collect data, apply mathematical routines and methods, and refine testable explanations and predictions.

While the College Board allows a wide range of lab and field activities that can be conducted in the course, they all should encompass these elements:

- connect to major concepts in science and to one or more areas of the course

- allow students to have direct experience with an organism or system in the environment

- involve observing systems, collecting data/information, and communicating observations or results

They should also encourage you to do the following:

- generate questions for investigation

- choose which variables to investigate

- design and conduct experiments

- design your own experimental procedures

- collect, analyze, interpret, and display data

- determine how to present your conclusions

The variety of different labs that can be performed gives students opportunities to explore various topics in environmental science; this also allows instructors the freedom to adapt their course laboratories to their geographic locations, which can make the course more interesting for students.

In addition, the AP Environmental Science course includes an emphasis on Science Practices. These describe what you should be able to do while exploring the concepts of the course and should be a focus in lab investigations. The skills involved in each of these practices form the basis of the tasks on the AP Environmental Science Exam.

The Science Practices and their component skills are as follows:

Practice 1: *Concept Explanation* Explain environmental concepts, processes, and models presented in written format.
    Skills: Describe and explain environmental concepts and processes, including models, in both theoretical and applied contexts.

Practice 2: *Visual Representations* Analyze visual representations of environmental concepts and processes.
    Skills: Use visual representations of environmental concepts, processes, or models to describe characteristics, explain relationships, and relate ideas to broader issues, in both theoretical and applied contexts.

Practice 3: *Text Analysis* Analyze sources of information about environmental issues.
    Skills: Identify authors' claims, perspectives, assumptions, and reasoning, and evaluate the credibility of sources and the validity of conclusions.

Practice 4: *Scientific Experiments* Analyze research studies that test environmental principles.
    Skills: Identify testable hypotheses or questions; identify and describe research methods, designs, and measures used in experiments; make observations and collect data from laboratory setups; and explain modifications to experimental procedures that will alter results.

Practice 5: *Data Analysis* Analyze and interpret quantitative data represented in tables, charts, and graphs.
    Skills: Describe patterns or trends in data and relationships among variables; explain patterns and trends to draw conclusions; interpret experimental data and results in relation to hypotheses; and explain what the data implies or illustrates about environmental issues.

Practice 6: *Mathematical Routines* Apply quantitative methods to address environmental concepts.
    Skills: Determine an approach or method to use to solve a given problem; apply appropriate mathematical relationships to solve problems (with work shown); and calculate accurate numeric answers with appropriate units.

Practice 7: *Environmental Solutions* Propose and justify solutions to environmental problems.
    Skills: Describe environmental problems, potential responses or approaches to them, and disadvantages, advantages, or unintended consequences of potential solutions; use data and evidence to support potential solutions; make claims proposing solutions to environmental problems in applied contexts; and justify proposed solutions by explaining potential advantages.

These practices and skills are not only centrally important to your course learning, but are also reflected in the structure of your exam: each multiple-choice question tests one of these practices along with some specific aspect of course content, and every free-response question tests multiple Science Practices.

On the AP Environmental Science Exam, you could be asked a number of different types of questions in which the lab component from your class will be helpful. However, you won't be able to get full credit simply by remembering the details of what you did in a particular lab. You will have to use your critical thinking skills to answer these questions.

A SAMPLE FREE-RESPONSE QUESTION INVOLVING A LAB

A lab-related free-response question on the exam might look a lot like the one below.

The county range management office has just received a federal grant to study the breeding success of red-tailed hawks in your area and you—their student intern for the summer—have been chosen to design and perform the research. The range office wants to know how the hawk population at the state prairie reserve compares with the hawk population on the federal grazing land and two private ranches nearby.

The office provides you with the following population data for the red-tailed hawks collected over four years, along with the land management policies in effect at the four locations.

Your supervisor points out to you that while the numbers of hawks counted during migration season is decreasing, the summer numbers are relatively steady. She offers the explanation that it's thought that the hawks' migration patterns are changing, with more of the hawks venturing farther south beyond the counting area.

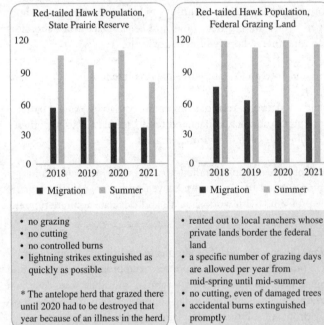

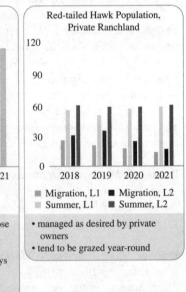

- no grazing
- no cutting
- no controlled burns
- lightning strikes extinguished as quickly as possible

* The antelope herd that grazed there until 2020 had to be destroyed that year because of an illness in the herd.

- rented out to local ranchers whose private lands border the federal land
- a specific number of grazing days are allowed per year from mid-spring until mid-summer
- no cutting, even of damaged trees
- accidental burns extinguished promptly

- managed as desired by private owners
- tend to be grazed year-round

(a) **Identify** and **explain** one possible environmental reason for the shifting migration patterns indicated by the discrepancy between the migration and summer population trends for all the locations.

(b) **Explain** the likely reason why the summer population in the state prairie reserve dropped in 2021.

(c) The study you have been chosen to perform is on the hawks' breeding success (number of eggs laid vs. number of offspring that survive to reproductive age). Design a plan of action for determining the hawks' breeding success in the four areas. **Explain** how you will carry it out, including your plans for setup and for data collection. **Identify** what your control will be.

(d) **Describe** how the land management plans of each of the four areas likely affect the success of the hawk population.

(e) **Explain** why a predator, such as the hawk, might provide evidence of a successful land management plan.

Look at part (a). Your supervisor explains, in the question stem, that it's thought that the hawks' migration patterns are changing, with more of the hawks venturing farther south beyond the counting area. What could cause the hawks to venture farther south? The simplest explanation is probably either climate change or some more local human environmental destruction—the hawks' normal migration habitat is no longer suitable, so they must venture farther to find one that is. Since the question asks you to **identify** *and* **explain**, make sure you do both: name the source of the change (identify) and explain why it means the hawks migration patterns have changed. Each of the tasks here is most likely worth one point.

In part (b), you're asked to explain the likely reason why the summer population in the state prairie reserve dropped in 2021. Look for information you are given that only pertains to that year. In the land management section of the figure, there is a note under the state prairie reserve that the antelope herd that grazed there until 2020 had to be destroyed that year because of an illness in the herd. While it specifies 2020 rather than 2021, the effects of this cull likely became visible in the following year. A large change in the local ecosystem, such as the elimination of a population, would certainly have a ripple effect throughout the food chain. With less grazing, the grass is likely to have overgrown, limiting the diversity of the plant population on the prairie reserve and thus making a more challenging environment for herbivores. Losses in the herbivore populations will have made it more difficult for the hawks and other predators to survive, bringing their numbers down. This explanation is probably worth one point.

In part (c), you're asked to design a plan of action. The **explain** portion asks for your plans for setup and data collection, and the **identify** portion asks what your control will be. This looks like it's worth 3 points.

You know what your dependent variable is—the breeding success of the hawk population—and you know how you are going to judge this success—by the number of eggs laid (which you can count) and the number of offspring that survive. You should make this second number more measurable by approximating it with something easy to count, such as the number of fledglings that leave the nest.

What is the independent variable in this experiment? It's the way in which the land is managed in each of the four areas. Remember that you cannot change the management practices; you have been given those.

What is the control in the experiment? This is a little less clear-cut. You're given four management plans, none of which operates on land that's left untouched. Therefore, you could set up a control at one of the reserves, but it might be difficult to attract hawks and start with an adequate base control number. Another option is to do a **library control**. In a library control, you would determine the maximum number of eggs and fledglings that could be expected based on data collected in previous years. Either one of these types of controls would be acceptable.

Let's summarize the plan of action that you might propose.

You will grid off four random plots, each 10 acres in area, in each of the four management areas. You will then observe the hawks in these plots and determine the locations of their nests. You will examine each nest (using binoculars, because the nests are usually high in trees) and determine the number of eggs in each clutch. You will continue to make observations throughout the summer to determine how many birds are hatched and how many live to the fledgling stage and leave the nest. You will statistically determine the percent success rate of each nest and the overall area to determine the success of the hawks. Then you will compare your data from the four areas.

So, the plan of action that you write in your exam booklet for answering part (c) of the question might include the following points:

Plan of Action

- Using library resources, determine the maximum success rate of offspring production in hawks (control).

- Plot out four random 10-acre plots on each of the four management areas.

- Survey plots for hawks and nests within each grid and record the locations of the nests.

- Count the number of eggs in each nest.

- Observe the nests and hatchlings and count the number of hatched birds. Band baby birds for tracking purposes.

- Count the number of fledglings that survive and leave the nest.

- Perform statistical analysis to determine the success of the hawks on each of the four management areas.

The plan of action is your longest answer in this section, but hopefully you can see that it wasn't too difficult. You took the information you learned this year and applied it to the situation, but you didn't need to know much about hawks.

Unfortunately, however, the next parts will require you to know a little more, rather than just asking you to use your imagination. For example, part (d) asks you to describe how the land management plan of each of the four areas likely affects the success of the hawk population.

Let's go through them one by one. The state land would be heavy in undergrowth and might have damaged trees located on it. The diversity of the plant population in this area would be relatively low because the grass would predominate—it would be able to grow unchecked. New grass would grow, but probably not many other plant species. This would be more true now that the antelope herd is not there to graze.

On the other hand, the federal land is grazed, so in this area there will be abundant growth of new grass. Also, due to this natural thinning of the grass, there will probably be some diversity in the vegetation, but without fall grazing to clean up the dying grass and continue to thin it, the federal land might not be as diverse as the private land, which is presumably grazed in the fall.

The private land can be grazed at any time. Controlled fires can also be used on this land to remove dead grass, add nutrients back into the soil, and remove some of the plant species that compete with grass, such as cactus or sage brush. For these reasons, the private land may have more diversity than any of the other land management areas. You could also theorize that the private land could be overgrazed and thus exhibit low diversity.

Really, you could make either presumption and be counted as correct. The two ranches may also differ in their policies and their results. However, since you are given data on the fire practices and grazing on the rest of the land, the first choice might be best. Why is the vegetation diversity important? The more vegetation diversity there is, the more choices for the herbivores in food and habitats—and probably the greater the success of the herbivores. In any event, the more successful the plants are, the more successful the herbivores in the area will be. Because hawks are carnivores, the more successful the herbivores, the more successful the hawks will be. See how this line of reasoning naturally leads to your answer to part (d)?

You're almost done; let's think about part (e). The success of the management plan would be demonstrated by the success of the hawk because the hawk is the tertiary consumer in the food chain. It cannot be successful if the rest of the food chain isn't strong.

You're done, and you really did not need to know much about the prairie or hawks. What you did need to know was how to interpret patterns in the data you were given, how to set up a field experiment, the concept of the food chain, and something about land management plans. Even the management plans could have been deduced with some critical thinking.

SOME COMMON AP ENVIRONMENTAL SCIENCE LABS

In the following pages, you will find some of the common labs performed during an AP Environmental Science course, a summary of the procedures you might follow, and the take-home message of each.

Remember that each AP Environmental Science course is different; you may have performed some of these labs, but you probably have not completed all of them. It's a good idea to review all of these labs and understand their basic workings as well as their intent.

Soil Analysis Lab

- **Soil Testing Laboratory:** In this lab, soil is tested for physical traits and chemical properties, which provide information about the soil's condition and suitability for crops, septic fields, or other purposes. All of the factors tested are listed below.

Chemical Properties

1. pH—Clay soil requires more lime (calcium oxide) or alum (aluminum sulfate) to alter its pH than do sandy or loam soils. Iron necessary for plant growth is unavailable when the soil becomes alkaline. Gymnosperms (pine, fir, etc.) grow better in mildly acidic soil.

2. Nitrogen—common plant nutrient component

3. Phosphorus—common plant nutrient component

4. Potash—common name for a compound that contains one of the potassium oxides

Physical Characteristics

1. Soil type—Sand, silt, clay. Use mesh screen, cheese cloth, and soil settling in water tubes to determine the percent of each type of particle in the sample.

2. Water-holding capacity—Because of the small pores between clay particles, water moves very slowly through clay. Therefore, clay has a greater holding capacity than silt or sand.

3. Permeability—the movement of gas or liquid through the soil

4. Friability—Good soil is rich, light, and easily worked with fingers, which is good for plant growth because roots can easily grow through it.

5. Percent humus—A measure of soluble organic constituents of soil; the higher the number, the better. Organic soil has qualities of both sand and clay. The small particles of organic soil come together to form larger clusters. Water can be retained inside a cluster, but can move between clusters to percolate. Organic material is also high in nutrients.

6. Buffering capacity—Resilience of different types of bedrock, such as marble, granite, and basalt, when exposed to acid. Marble has high calcium content and is a better buffer than other rocks.

Water Analysis Lab

- **Water Chemical and Physical Analysis:** Water can be tested from many different sources. Sample kits have tools for many different tests—for example, the LaMotte kit and the spectrophotometer type kit. Below are some commonly performed tests and some of the expected results. It is probably not necessary for you to memorize all the standards, but you should be familiar with them.

1. pH—Normal pH of water is between 6.5–8.5.

2. DO—Measure of dissolved oxygen in water. Warm-water fish require a minimum of 4 ppm and cold-water fish require 5 ppm.

3. Turbidity—Measurement of water clarity. Higher turbidity means there will be low clarity, and little sunlight will be able to penetrate the water. A Secchi disk may be used to measure turbidity, but a more accurate measure can be made with a turbidity unit.

4. Phosphate—An important plant nutrient, typically found in fertilizer and runoff from agricultural lands. Too much leads to eutrophication of water, high BOD, and low DO levels. Levels should not exceed 0.025 mg/L in still water and 0.05 mg/L in flowing water.

5. Alkalinity—Measure of compounds that shift the pH toward the alkaline. There are no EPA standards, but normal is between 100–250 ppm.

6. BOD—Biological oxygen demand, which is required for the aerobic organisms in a body of water. Unpolluted natural waters have a concentration of 5 mg/L or less. High nutrient levels or lots of biological material ready for decomposition are associated with high BOD, and low levels with low BOD.

7. Chlorine—EPA standards dictate that Cl cannot exceed 250 mg/L. NaCl is applied to roads and parking lots and can run off into streams. Other sources of excess Cl are animal waste, potash fertilizer (KCl), and septic tank effluent. Chlorine is associated with limestone deposits but is not common in other soils, rocks, or minerals.

8. Hardness—A measure of salts composed of calcium, magnesium, or iron. Most water-testing kits test for $CaCl_2$. Water with more than 120 ppm is classified as hard, and water with less than 60 ppm is classified as soft. Water between 60 and 120 ppm is classified as "moderately hard."

9. Iron—Normal range is 0.1–0.5 ppm.

10. Nitrates—As an important plant nutrient, nitrogen is typically found in fertilizer and is a component of runoff from agricultural lands. Too much leads to eutrophication of water, high BOD, and low DO levels. Over 0.10 mg/L is considered elevated, and the EPA limit is 10 mg/L.

11. Total solids—Weight of the suspended solids and dissolved solids. All natural waters have some suspended solids, but problem solids are sewage, industrial waste, or excess amounts of algae.

12. Total dissolved solids—Naturally occurring in water, but may cause an objectionable taste in drinking water. They are also unsuitable for irrigation because they leave a salt residue on the soil. EPA standard is 500 mg/L, but dissolved solids may range from 20–2,000 mg/L.

13. Fecal coliform—Any bacteria that ferments lactose and produces gas when grown in lactose broth. New tests for this are performed by adding a water sample to a specialized media and observing color changes. Drinking water should show no colonies of growth from the water sample.

Air Quality Labs

- **Air Quality**: Air quality can be assessed using various methods.

- **Particulates**: Sticky paper can be used to collect air particulates from various sources, and then the paper can be examined under a microscope. It is not possible to see the smallest particulates, but they do color the white paper.

- **Ozone**: In this lab, an eco badge or a homemade potassium iodide gel sampler is hung or worn in order to collect data on tropospheric ozone. The badge or KI sample changes color in the presence of ozone and becomes more intensely colored as the amount of ozone increases.

- **Carbon Dioxide**: In this lab, a commercial sampling device is used to determine the amount of carbon dioxide in an air sample. Car exhaust, tobacco smoke, and other pollutants can also be sampled.

Other Labs

- **Lichen**: A lichen survey can be used to judge air quality. Lichens are sensitive to air pollution, particularly sulfur dioxide. The most sensitive lichens are the fruticose types, followed by the foliose, and then the crustose.

- **Scrubber Model**: A model of a scrubber can be constructed to attempt to remove sulfur contaminates from burning coal. A calcium compound can be used to try to wash the contaminant from the air column.

- **Biodiversity of Invertebrates**: Insects can be counted in an area and then plotted to assess biodiversity. Traps such as fall traps or sticky traps can be set, and bait such as tuna or sugars can be used. The number of different insects captured is counted, and this number divided into the total number captured would give an idea of the biodiversity of the area. A taxonomic key can be used to determine the number of species and their taxonomic names. A similar setup can be used to determine the impact of an invasive species. In this process, native bugs can be trapped and set up in terrariums; the invasive species is then introduced and the effects documented.

- **Field Trips**: Field trips can be taken to various areas of interest, such as power plants, landfills, or municipal waste treatment plants. The possibilities are almost endless. Think about what you learned from these experiences and how they can help you on this exam.

- **Energy Audit**: In this lab, students are asked to use their homes as laboratories and perform an energy audit, examining the amount of electricity used by their families over a set period of time and then using appliance standards to determine which is the largest energy consumer. A result of this lab is that students suggest how their families' electrical energy needs could be reduced.

- **Food Chain**: In this lab, students observe a natural ecosystem and examine the food chain. They identify the organisms that are producers, primary consumers, and secondary consumers and determine how many levels make up the food chain, what organisms act as decomposers, and the presence of symbiotic relationships.

- **Model Building**: In this lab, models are built of land formations, coastlines, tectonic plates, mining operations, or any number of other physical formations. In modeling tectonic plates, for example, students can slide or bump together plate models that are lying on top of a viscous substance representing the magma. From this model, students see how subduction zones and volcanoes work. Other models also allow students to construct a representation of a physical structure in order to better understand that physical structure.

- **Mining**: Students can construct a model of a mine, representing the rock layers and the mineral deposits.

- **LC$_{50}$ (or LD$_{50}$)**: In this lab, students use a kit or a lab procedure to test the concentration (LC) or dose (LD) that would cause the death of 50 percent of a test organism. An example of this procedure is to test the effects of copper (common algaecide and fungicide ingredient) on Daphnia (a tiny species of freshwater crustacea). Daphnia are exposed to a range of Cu levels and then fed fluorescently tagged sugar. Healthy Daphnia show up blue under UV light.

- **Population Density and Biomass**: This experiment can be performed in a number of different ways. One method is to mark off two plots and then remove all the vegetation from Plot 1. Plot 2 is left to grow for an additional period of time. This investigation may also be performed using grass squares in the lab or bottles of algae. The vegetation from Plot 1 is dried and weighed. This provides the baseline data for the amount of biomass present at the beginning of the experiment. After the time period of growth for the second plot, the vegetation is removed from the second plot and treated as the first was. This provides information about the increase in biomass for the experimental time period. Another way this can be expressed is as the productivity of vegetation. By subtracting the initial growth (Plot 1) from the experimental growth (Plot 2), productivity is calculated. This would constitute the net productivity for the plot. The grass in the second plot produces X amount of glucose (photosynthate). However, some of this photosynthate must be used to support the needs of the plant, so the number you calculate is the net, not the gross, productivity.

- **Gross and Net Productivity**: If an experiment is performed using bottles of growing algae or duckweed, the gross productivity can also be determined. One of the bottles in this experiment contains the starting plant material, but instead of being exposed to light, it is covered with foil to prevent light from entering the bottle. In this bottle, the plants are only able to perform respiration. Therefore, at the end of the experimental time, this bottle should have less biomass than the initial plant material because the plants have to use stored sugar for metabolic process. This provides you with the metabolic rate information, which, when added to the net productivity, gives you the gross productivity.

- **Other Uses for Vegetation Plots**: Sample plots may be used to examine the biodiversity of vegetation or the patterns of plant growth or dispersal. For example, one species of plant can be marked in a plot, and unless a garden was used as the sample plot, the target plant will probably be dispersed in clumps. Remember that random dispersal is not typical in nature, and even distribution typically only occurs in plots planted by humans.

- **Population Growth**: Population growth experiments can involve fast-growing populations such as bacteria, duckweed (*Lemma minor*), roly-polies (sowbugs), or fruit flies (*Drosophila*) and can involve graphing, as does the analysis of human population data. Population growth can be graphed, with time along the *x*-axis and population growth on the *y*-axis. Initially, the curve is a J shape, but it becomes S-shaped. Bacterial growth curves have an extended hook on the J due to the lag time in growth as the bacteria acclimate to the new media.

- **Variables**: In population growth experiments, often other variables are added to samples of bacteria to compare the growth of a normal population to one that has been altered in some way. This can be used to assess the effect of extra nutrients, such as nitrogen or phosphorus, or the negative effects of certain substances on the organisms.

- **Turbidity and Bacteria**: When studying bacteria, turbidity is commonly observed and recorded. The more turbid (cloudy) the tube, the more growth has taken place. Turbidity can be observed with a spectrophotometer. When the sample tube is inserted into the spectrophotometer, the instrument passes light through the sample tube and measures the amount of light that passed through or was absorbed.

- **Population Size**: In this type of lab, the size of a population of species such as the spongy moth, caterpillar, or other insect is studied. Remember that a population is a group of individuals of the same species located in a given area. The experiment could, for example, involve a collection box and colored stickers that attract caterpillars. The first day, caterpillars are captured and marked. On the second day, the caterpillars are captured and the new and recaptured individuals are marked. By dividing the number of recaptured insects by the size of the population on day one, an estimate of the population that was originally captured can be obtained. On day three, the total number of caterpillars captured is counted. The total number of caterpillars captured on day three is divided by the calculated number from day two—this gives an estimate of the total population.

- **Salinization**: In this lab, students determine whether and how salt levels slow seed germination. Students use various dilutions of salts (a variety of salt types may be used) to saturate a growing surface. Seeds are sown to determine levels of salt that inhibit germination. This lab mimics what happens in irrigated farm land with salinization.

Note: If you are asked to graph data on this exam, make sure you use reasonable units! Count the graph blocks you're given; this may help you determine an appropriate scale. Also, remember that time is usually plotted on the *x*-axis.

Well, this marks the end of the content review for this exam—congratulations, you've finished! Study the words in the Glossary, which follows, and then take Practice Tests 2 and 3.

Good luck!

Chapter 12
Glossary

Scan the QR code and register your book to access the digital flashcards of these terms.

GLOSSARY

These are all AP Environmental Science terms you should know for the exam, so view the digital flashcards in your Student Tools or write down these words in a notebook, and study them any chance you get. Do whatever you have to do to commit them to memory before exam day!

Chapter 4: The Living World: Ecosystems and Biodiversity

10% Rule—the rule that in a food chain, only about 10% of the energy is transferred from one level to the next

abiotic—related to factors or things that are separate and independent from living things; nonliving

abyssal zone—the deepest region of the ocean, marked by extremely cold temperatures and low levels of dissolved oxygen, but high levels of nutrients because of the decaying plant and animal matter that sinks down from the zones above

ammonification—the production of ammonia or ammonium compounds in the decomposition of organic matter, especially through the action of bacteria

anaerobic—without oxygen

aquatic life zones—ecosystems in aqueous environments

assimilation—the process in which plants absorb ammonium (NH_3), ammonia ions (NH_4^+), and nitrate ions (NO_3) through their roots

autotrophs—producers; organisms that can produce their own organic compounds from inorganic compounds; they use energy from the Sun or from the oxidation of inorganic substances

barrier island—a long, relatively narrow island running parallel to the mainland, built up by the action of waves and currents and serving to protect the coast from erosion by surf and tidal surges

bathyal zone—the middle region of the ocean, characterized by less density of organisms because it does not receive enough light to support photosynthesis

benthic zone—the surface and sub-surface layers of the river-, lake-, pond-, or streambed, characterized by very low temperatures and low oxygen levels

biodiversity—the number and variety of organisms found within a specified geographic region, or ecosystem, or the variability among living organisms, including the variability within and between species and within and between ecosystems

biogeochemical cycles—the complex cycles through which nutrients such as carbon, oxygen, nitrogen, phosphorus, sulfur, and water move through the environment

biological extinction—true extermination of a species; no individuals of this species are left on the planet

biomes—ecosystems based on land

biotic—living or derived from living things

chaparral—scrub forest or shrubland; a biome characterized by moderate precipitation, shallow or infertile soil, small trees with large, hard evergreen leaves, and spiny shrubs

chemotrophs—autotrophic bacteria that use chemosynthesis to produce energy in anaerobic environments

climax community—a stable, mature community in a successive series that has reached equilibrium after having evolved through stages and adapted to its environment

coastal zone—ocean zone consisting of the ocean water closest to land, usually defined as between the shore and the end of the continental shelf (the edge of the tectonic plate); characterized by abundant sunlight and oxygen

combusted—burned

commensalism—symbiotic relationship in which one organism benefits while the other is neither helped nor hurt

commercial (economic) extinction—a few individuals exist but the effort needed to locate and harvest them is not worth the expense

community—formed from populations of different species occupying the same geographic area

competition—the relationship that exists when two individuals—of the same species or of different species—compete for resources in the same environment

competitive exclusion—the process that occurs when two different species in a region compete and the better adapted species wins

coniferous forest (taiga)—a biome characterized by moderate precipitation, acidic soil, and coniferous trees

consumers—organisms that must obtain food energy from secondary sources, for example, by eating plant or animal matter

coral reef—an erosion-resistant marine ridge or mound consisting chiefly of compacted coral together with algal material and biochemically deposited magnesium and calcium carbonates

cultural services—use of nature for science and education, therapeutic and recreational uses, and spiritual and cultural uses

deciduous forest—a biome characterized by adequate precipitation, rich soil with high organic content, and hardwood trees

decomposers—organisms that consume dead plant and animal material—the process of decomposition returns nutrients to the environment

denitrification—the process by which specialized bacteria (mostly anaerobic bacteria) convert ammonia to NO_3, NO_2, and N_2, which are released back into the atmosphere

deserts—biomes characterized by extremely low precipitation, coarse sandy soil, and cactus and other low-water adapted plants

detritivores—organisms that derive energy from consuming nonliving organic matter, such as dead animals or fallen leaves

ecological extinction—the condition in which there are so few individuals of a species that the species can no longer perform its ecological function

ecological succession—the transition in species composition of a biological community, often following ecological disturbance of the community; the establishment of a biological community in any area virtually barren of life

economic extinction—see commercial extinction

ecoregions—see ecozones

ecosystem—a system of interconnected elements: a community of living organisms and its environment

ecosystem services—benefits that humans receive from the ecosystems in nature when they function properly

ecotones—regions where different biomes overlap

ecozones (ecoregions)—smaller regions within ecosystems that share similar physical features

edge effect—the condition in which there is greater species diversity and biological density at ecosystem boundaries than there is in the heart of ecological communities

energy pyramid—the structure obtained if we organize the amount of energy contained in producers and consumers in an ecosystem by kilocalories per square meter, from largest to smallest

epilimnion—the uppermost and thus the most oxygenated layer of freshwater

estuary—the part of the wide lower course of a river where its current is met by the tides

euphotic zone—the photic, upper layers of ocean water; the euphotic zone is the warmest region of ocean water and has the highest levels of dissolved oxygen

eutrophication—warm water becoming overly enriched with minerals and nutrients to the point that excessive growth of algae and other phytoplankton occurs (an algal bloom)

evaporation—conversion of a liquid into a gas (vapor)

evolution—change in the genetic composition of a population during successive generations as a result of natural selection acting on the genetic variation among individuals and resulting in the development of new species

evolutionary fitness—the better-adaptedness of individual organisms for their environment that allows them to live and reproduce, ensuring that their genes are part of their population's next generation

exchange pool—a site where a nutrient sits for only a short period of time

extinction—the death of an entire species; permanent inactivity

food chain—a succession of organisms in an ecological community that constitutes a continuation of food energy from one organism to another as each consumes a lower member and, in turn, is preyed upon by a higher member

food web—a complex of interrelated food chains in an ecological community

fundamental niche—the niche a species would have if there were no competition

Gause's principle—states that no two species can occupy the same niche at the same time, and that the species that is less fit to live in the environment will either relocate, die out, or occupy a smaller niche

gene pool—the total genetic makeup of a population

generalist—a species that has a broad niche, is highly adaptable, and can live in varied habitats

genetic drift—the random fluctuations in the frequency of the appearance of a gene in a small, isolated population, presumably owing to chance, rather than natural selection

grasslands—a biome characterized by moderate precipitation, rich soil, and sod-forming grasses

Gross Primary Productivity—the amount of sugar that the plants produce in photosynthesis

groundwater—any water that comes from below the ground (from wells or from aquifers)

habitat—the area or environment where an organism or ecological community normally lives or occurs

habitat fragmentation—when the size of an organism's natural habitat is reduced, or when development occurs that isolates a habitat or part of one

heterotroph—an organism that cannot synthesize its own food and is dependent on complex organic substances for nutrition

hypolimnion—the lower, colder, and denser layer of freshwater

indicator species—species that are used as a standard to evaluate the health of an ecosystem

indigenous species—species that originate and live, or occur naturally, in an area or environment

interspecific competition—competition between individuals of different species

intraspecific competition—competition between individuals of the same species

invasive species—introduced, non-native species

keystone species—a species whose very presence contributes to an ecosystem's diversity and whose extinction would consequently lead to the extinction of other forms of life

Law of Conservation of Matter—states that matter can neither be created nor destroyed

Law of the Minimum—states that living organisms will continue to live, consuming available materials until the supply of these materials is exhausted

Law of Tolerance—describes the degree to which living organisms are capable of tolerating changes in their environment

limiting factor—any factor that controls a population's growth

limnetic zone—the surface of open ocean water; the region that extends to the depth that sunlight can penetrate

littoral zone—ocean zone beginning with the very shallow water at the shoreline and extending to the depth at which rooted plants stop growing

macroevolution—large-scale patterns of evolution within biological organisms over long periods of time

mangrove swamps—coastal wetlands (areas of land covered in freshwater, saltwater, or a combination of both) found in tropical and subtropical regions

microevolution—small-scale evolutionary changes over relatively short periods of time

mutualism—a symbiotic relationship in which both species benefit

natural selection—the process by which, according to Darwin's Theory of Evolution, only the organisms best adapted to their environment tend to survive and transmit their genetic characteristics in increasing numbers to succeeding generations, while those less adapted tend to be eliminated

Net Primary Productivity (NPP)—the amount of energy that plants pass on to the community of herbivores in an ecosystem; the amount of sugar that the plants produce in photosynthesis minus the amount of energy the plants need for growth, maintenance, repair, and reproduction

niche—the total sum of a species' use of the biotic and abiotic resources in its environment

nitrification—the process in which soil bacteria convert ammonium (NH_4^+) to a form that can be used by plants: nitrate, or NO_3^-

nitrogen fixation—the conversion of atmospheric nitrogen into compounds, such as ammonia, by natural agencies or various industrial processes

parasitism—a symbiotic relationship in which one member is helped by the association and the other is harmed

phosphorus cycle—the cycle through which phosphorus moves through the environment in different forms

photosynthesis—the process in green plants and certain other organisms by which carbohydrates are synthesized from carbon dioxide and water using light as an energy source: most forms of photosynthesis release oxygen as a byproduct

phylogenetic tree—a branching diagram used to model evolution and describe the evolutionary relationships between species

pioneer species—the organisms that take root and adapt to the conditions of a habitat in the first stages of ecological succession

population—a group of organisms of the same species that live in the same area

precipitation—water that condenses in the atmosphere and falls to the Earth's surface in liquid or solid form

predation—when one species feeds on another

predator—a species that feeds on another species

prey—a species that's subject to predation by another species

primary consumers—organisms that consume producers (plants and algae)

primary succession—when ecological succession begins in a virtually lifeless area, such as the area behind a moving glacier

producer—an organism that is capable of converting radiant energy or chemical energy into carbohydrates

profundal zone—the depths; ocean water that is too deep for sunlight to penetrate. Because the profundal zone is aphotic (a place where light cannot reach), photosynthesizing plants and animals cannot live in this region

provisioning services—water, food, medicinal resources, raw materials, energy, and ornaments provided to humans by functioning ecosystems

realized niche—the compromised niche a species occupies—smaller than the niche it would occupy in the absence of competition

red tide—a bloom of dinoflagellates that causes reddish discoloration of coastal ocean waters; certain dinoflagellates of the genus *Gonyamlax* produce toxins that kill fish and contaminate shellfish

regulating services—waste decomposition and detoxification, purification of water and air, pest and disease control and regulation of prey populations through predation, and carbon sequestration

reservoir—a place where a large quantity of a resource sits for a long period of time

residency time—the amount of time a resource spends in a reservoir or an exchange pool

resource partitioning—when different species rely on the same resource but use slightly different parts of the habitat to avoid direct competition

respiration—the process by which animals (and plants!) breathe and give off carbon dioxide from cellular metabolism

runoff—that portion of rainfall that runs into streams as surface water rather than being absorbed into groundwater or evaporating

saprotrophs—decomposers that use enzymes to break down dead organisms and absorb the nutrients

savanna—a biome characterized by low rainfall, porous soil with only a thin layer of humus, and grasses with widely spaced trees

secondary consumers—organisms that consume primary consumers

secondary succession—ecological succession that takes place where an existing community has been cleared (by disturbance events such as fire, tornado, or human impact), but the soil has been left intact

selective pressure—any cause that reduces reproductive success (fitness) in a portion of the population (these are what drive natural selection)

specialist—a species that has a narrow niche and can only live in a certain habitat

speciation—formation of new species through evolution

species—a group of organisms that are capable of breeding with one another and incapable of breeding with other species

species richness—the number of different species found in an ecosystem

sulfur cycle—the cycle through which sulfur moves through the environment in different forms

supporting services—primary production, nutrient recycling, soil formation, and pollination (the category of ecosystem services that makes other ecosystem services possible)

symbiotic relationships—close, prolonged associations between two or more different organisms of different species that may, but do not necessarily, benefit the members

taiga—see **coniferous forest**

temperate rainforest—a biome characterized by abundant rain, moderately rich soil, and coniferous and broadleaf trees, epiphytes, mosses, ferns, and shrubs

terrestrial cycle—the portion of a biogeochemical cycle that takes place on land

tertiary consumers—organisms that consume secondary consumers or other tertiary consumers

theory of island biogeography—a field that studies species richness and diversification in isolated communities: oceanic islands, and also other isolated ecosystems

thermocline—a layer in a large body of water, such as a lake, that sharply separates regions differing in temperature, so that the temperature gradient across the layer is abrupt

transpiration—the act or process of transpiring, or releasing water vapor, especially through the stomata of plant tissue or the pores of the skin

trophic level—one of the feeding levels in a food chain

tropical rainforest—a biome characterized by abundant rain, poor quality soil, and tall trees with few lower limbs, vines, epiphytes, and plants adapted to low light intensity

tundra—a biome characterized by very low precipitation, permafrost, and herbaceous plants

upwelling—a process in which cold, often nutrient-rich, waters from the ocean depths rise to the surface

wetlands—lowland areas, such as marshes or swamps, that are saturated with moisture, especially when regarded as the natural habitat of wildlife

Chapter 5: Populations

Actual Growth Rate—a simple formula to calculate population change over time (excluding the effects of immigration and emigration): (birth rate – death rate) / 10

age-structure diagrams—see **age-structure pyramids**

age-structure pyramids (age-structure diagrams)—diagrams that provide a graphical representation of the total numbers or percentages of individuals of given ages in populations

biotic potential—the amount that the population would grow if there were unlimited resources in its environment

boom-and-bust cycle—population cycle characterized by rapid increases and equally rapid drop-offs

carrying capacity (K)—the maximum population size that can be supported by the available resources in a region

clumped—the most common dispersion pattern for populations, in which individuals flock together in clumps

cohort—the group of individuals in a population born at a given time or a group of individuals in the same stage of life

crude birth rate—the number of live births per 1,000 members of the population in a year

crude death rate—the number of deaths per 1,000 members of the population in a year

demographic transition—the movement of a population from one state to the next in the demographic transition model

demographic transition model—a model that is used to predict population trends based on the birth and death rates as well as economic status of a population

density-dependent factors—population-limiting factors that are a result of the size of the population

density-independent factors—population-limiting factors that operate independently of population size

doubling time—the time it takes for a population to double

emigration—the movement of individuals out of a population

exponential population growth—the growth pattern of a population well below the size dictated by the carrying capacity of its habitat; the curve of population over time makes a J shape

food deserts—neighborhoods in which the majority of residents are low-income and access to fresh, healthy food is difficult; residents rely on low-quality processed foods for subsistence

genetically modified organisms (GMOs)—organisms grown or bred by humans, in which strands of DNA that code for desired characteristics have been inserted

hunger—state which occurs when insufficient calories are taken in to replace those being expended

immigration—the movement of individuals into a population

industrial state—third stage of the demographic transition model, in which population growth is still fairly high, but the birth rate drops, becoming similar to the death rate

infant mortality rate—the number of deaths of children under 1 year old per 1,000 live births

K-**selected organisms**—organisms that reproduce later in life, produce fewer offspring, and devote significant time and energy to the nurturing of their offspring

logistic population growth—the condition in which a population is well below the size dictated by the carrying capacity of its habitat, so it will grow exponentially; but as it approaches the carrying capacity, its growth rate will decrease and the size of the population will eventually become stable

macronutrients—nutrients that are needed in large amounts, including proteins, carbohydrates, and fats

malnutrition—poor nutrition that results from an insufficient or poorly balanced diet

Malthusian catastrophe—the idea that humans might overshoot the carrying capacity of the Earth as a whole and suffer population dieback

micronutrients—nutrients that are needed in smaller amounts, including vitamins, iron, and minerals such as calcium

overshoot—when a population exceeds its carrying capacity

population—a group of organisms of the same species

population density—the number of individuals of a population that inhabit a certain unit of land or water area

population dispersion—how individuals of a population are spaced within a region

population momentum—the condition in which a large proportion of women are in child-bearing years, so that a country's population will continue to grow even if the fertility rate falls to replacement levels

postindustrial state—the final stage of the demographic transition model, in which the population approaches and reaches a zero growth rate or even drops below it

post-reproductive—the age 45 and older cohort in an age-structure diagram

preindustrial state—first stage of the demographic transition model, in which the population exhibits a slow rate of growth and has a high birth rate and high death rate because of harsh living conditions

pre-reproductive—the age 0–14 cohort in an age-structure diagram

random—dispersion pattern in which the position of each individual is not determined or influenced by the positions of the other members of the population

replacement birth rate—the number of children a couple must have in order to replace themselves in a population

reproductive—the age 15–44 cohort in an age-structure diagram

r-selected organisms—organisms that reproduce early in life and often and have a high capacity for reproductive growth

Rule of 70—says that the time it takes (in years) for a population to double can be approximated by dividing 70 by the current growth rate of the population

survivorship curve—curve representing the number of individuals in a population born at a given time that remain alive as time goes on

total fertility rate (TFR)—the number of children an average woman will bear during her lifetime, an estimate based on an analysis of data from preceding years in the population in question

transitional state—second stage of the demographic transition model, in which birth rates are high, but due to better food, water, and health care, death rates are lower, which allows for rapid population growth

type I—survivorship curve with a convex shape, which indicates that most individuals in the population survive into adulthood, with a sharp increase in mortality as the population approaches the species' maximum age

type II—survivorship curve with a straight-line pattern indicating that mortality and survival rates are fairly constant throughout life

type III—survivorship curve with a concave shape, which indicates that most offspring die young, but if they live to a certain age, they will live a longer life

undernourished—having not been provided with sufficient quantity or quality of nourishment to sustain proper health and growth

uniform—dispersion pattern in which the members of the population are uniformly spaced throughout their habitat

Chapter 6: Earth Systems and Resources

A horizon—soil horizon (layer) below the O horizon, formed of weathered rock with some organic material; often referred to as topsoil

active volcanoes—volcanoes that are currently erupting or have erupted within recorded history (that is, within the last 10,000 years)

additions—soil-forming process involving bringing in material from off-site

albedo—the fraction of solar energy that is reflected back into space

asthenosphere—the part of the mantle that lies just below the lithosphere

atmosphere—the gaseous mass or envelope surrounding a celestial body—especially the one surrounding the Earth—that is retained by the celestial body's gravitational field

axis—the line about which a rotating body, such as the Earth, turns

B horizon—soil horizon (layer) below the E horizon that receives the minerals and organic materials that are leached out of the A horizon; called the subsoil or zone of illuviation

biological weathering—any weathering that's caused by the activities of living organisms

biosphere—the part of the Earth and its atmosphere where living organisms exist or that is capable of supporting life

C horizon—parent material; soil horizon (layer) below the B horizon made up of unconsolidated material, loose enough to be dug up with a shovel

chemical weathering—weathering that occurs as a result of chemical reactions of rock with water, air, or dissolved minerals

cinder volcanoes—volcanoes with small, short, and steeply sloped cones that form when molten lava erupts and cools quickly in the air, hardening into porous rocks (called cinders or scoria) that fracture as they hit the Earth's surface

clay—the finest soil, made up of particles that are less than 0.002 mm in diameter

climate—weather conditions, especially temperature and precipitation, that remain constant over 30 years or more

climate—one of the six soil-forming factors: climate influences soil formation because it involves differences in temperature and precipitation across the globe, and both heat and water facilitate chemical and biochemical reactions; seasonal fluctuation of heat and moisture affects processes such as freeze-thaw cycles that weather rock; and climate also helps to determine what organisms grow in a particular location

Cl-O-R-P-T-H—the six soil-forming factors: climate, organisms, relief, parent material, time, and human influence

composite volcanoes—tall volcanoes with broad bases and steeper slopes that are formed at subduction zones and are associated with violent eruptions that eject lava, water, and gases as superheated ash and stones

composition—the mixture of particle sizes in a given soil

convection cell—a rotating section of atmosphere that owes its movement to the evaporation, condensation, and precipitation of water and the convection of air

convection currents—vertical airflow near the Earth's surface due to radiative heating and the convection of air

convergent boundary—a plate boundary where two plates are moving toward each other

Coriolis effect—the observed effect of the Coriolis force, especially the deflection of an object moving above the Earth, rightward in the Northern Hemisphere, and leftward in the Southern Hemisphere, as it moves away from the equator

crust—the solid surface of the Earth

cyclone—see hurricane

deltas—landforms made of deposited sediments created where rivers drop most of their sedimentary load as they meet the ocean

dew point—the temperature at which water vapor condenses into liquid water

divergent boundary—a plate boundary at which plates are moving away from each other; causes an upwelling of magma from the mantle to cool and form new crust

doldrums—a region of the ocean near the equator where the trade winds converge, characterized by constant rising air, calms, light winds, or squalls; intertropical convergence zone (ITCZ)

dormant volcanoes—volcanoes that have not been known to erupt during the last 10,000 years

E horizon—soil horizon (layer) below the A horizon found mainly in forest soils; it is light in color and coarse in texture because no organic matter has traveled down from the A horizon, while clays and minerals like iron and aluminum oxides have been washed out by leaching and eluviation; the eluviated horizon

earthquakes—the result of vibrations that release energy from within the Earth; they often occur due to sudden plate movements as two plates slide past each other at a transform boundary

El Niño—a climate variation that takes place in the tropical Pacific about every three to seven years, for a duration of about one year, during which the normal trade winds are weakened or reversed because of a reversal of the high and low pressure regions on either side of the tropical Pacific and the process of upwelling slows or stops, meaning the water off the coast becomes warmer and contains fewer nutrients; widespread changes in seasonal weather in coastal regions results

eluviation—the movement of waterborne minerals, humus, and other materials from higher soil layers to lower soil layers due to the downward movement of water via gravity

ENSO—El Niño/Southern Oscillation: the alternations of atmospheric conditions that lead to El Niños and La Niñas

epicenter—the initial surface location of an earthquake

erosion—the process of soil particles being carried away by wind or water; moves the smaller particles first and hence degrades the soil to a coarser, sandier, stonier texture

exosphere—the furthest layer of the atmosphere, extending to 10,000 km (6,200 miles) or more above the Earth, although the upper limit of this layer is not definitively settled

extinct volcanoes—volcanoes that are thought not to be likely to erupt ever again

fertility—how well plants can grow in a given soil

focus—the location at which an earthquake begins within the Earth

freshwater—water that contains only minimal quantities of dissolved salts, especially sodium chloride

greenhouse effect—the phenomenon whereby the Earth's atmosphere traps solar radiation, caused by the presence in the atmosphere of gases such as carbon dioxide, water vapor, and methane that allow incoming sunlight to pass through but absorb heat radiated back from the Earth's surface

Gulf Stream—a steady ocean current that carries sun-warmed water northward along the east coast of the United States and across the Atlantic Ocean as far as Great Britain

Hadley cell—a system of vertical and horizontal air circulation that creates major weather patterns, predominately in tropical and subtropical regions

horizons—distinct layers of soil

horizontal airflow—flowing air along the Earth's surface that results from large masses of warm, moist air rising and leaving space to be occupied by cooler air that flows in this pattern; this is one way surface winds are created

horse latitudes—the regions between about 30 degrees to 35 degrees north and 30 degrees to 35 degrees south of the equator where subsiding dry air and high pressure result in very weak winds; the subtropical high

hot spots—regions of volcanic activity that do not form at plate boundaries, but rather are found in the middle of tectonic plates, where columns of unusually hot magma melt through the mantle and weaken the Earth's crust

human influence—one of the six soil-forming factors: use of fertilizer, pollution, and acid rain alter soil chemistry on a broad scale; construction activities such as digging and plowing tend to mix soils and blur the distinctions between horizons; traffic and machinery lead to compaction; removal of stabilizing vegetation leads to erosion; and irrigation and groundwater depletion lead to salinization

humus—the dark, crumbly, nutrient-rich material that results from the decomposition of organic material, which is also a product of composting organic waste

hurricane (typhoon, cyclone)—a severe tropical storm originating in the equatorial regions of the Atlantic Ocean or Caribbean Sea or eastern regions of the Pacific Ocean, that travels north, northwest, or northeast from its point of origin and usually involves high-speed winds and heavy rains; they are referred to as *hurricanes* in the Atlantic Ocean, but they are called *typhoons* or *cyclones* when they occur in the Pacific Ocean

hydrosphere—the Earth's oceans and freshwater bodies

igneous—type of rock that results when rock is melted (by heat and pressure below the crust) into a liquid and then resolidifies when cooled

illuviation—the deposition of waterborne minerals, humus, and other materials from higher soil layers in a lower soil horizon

insolation—incoming solar radiation

ionosphere—not a distinct layer of the Earth's atmosphere, but rather regions dispersed throughout the upper mesosphere, thermosphere, and lower exosphere: it comprises regions of ionized gases that absorb most of the energetic charged particles from the Sun (the protons and electrons of the solar wind); it also reflects radio waves, making long-distance radio communication possible

jet stream—a high-speed, meandering wind current that often has a large influence on local weather patterns, generally moving from a westerly direction at speeds often exceeding 400 km (250 miles) per hour at altitudes of 15 to 25 km (10 to 15 miles) (in the tropopause)

La Niña—a cooling of the ocean surface off the western coast of Central and South America (the reverse of El Niño), occurring periodically every 4 to 12 years and affecting Pacific and other weather patterns

lake effect—a small-scale monsoon effect seen on the shores of large lakes or bays, in which breezes or winds are generated by the difference between how fast the land and water warm during the day and cool during the night

lava domes—small and short volcanoes with steep slopes and rounded tops, formed from lava that is too viscous to travel far that instead hardens into a dome shape

leaching—the loss of dissolved (not suspended) organic and chemical compounds from the soil profile by draining into the groundwater

lithosphere—the outer part of the Earth, consisting of the crust and upper mantle, approximately 100 km (62 miles) thick

loam—soil composed of a mixture of sand, clay, silt, and organic matter that has a proportion of the three size classes considered optimal for plant growth: 7–27% clay, 28–50% silt, and less than 52% sand

losses—soil-forming process involving loss of material through erosion, leaching, or biological activity (plant uptake)

magma—molten rock beneath the Earth's crust

mantle—the layer of the Earth between the crust and the core

mechanical weathering—see **physical weathering**

mesosphere—layer of the atmosphere extending from about 50 km (31 miles) to about 80 km (50 miles) above the Earth's surface; the area where meteors usually burn up

metamorphic—type of rock formed as a great deal of pressure and heat produces physical and/or chemical changes in existing rock, which can happen as sedimentary rocks sink deeper into the Earth and are heated by the high temperatures found in the Earth's mantle

meteorologists—scientists who study weather and climate

monsoons—seasonal winds in coastal areas that are usually accompanied by very heavy rainfall, which occur when land heats up and cools down more quickly than water does

O horizon—topmost soil horizon (layer), which is made up of organic matter at various stages of decomposition, including animal waste, leaves and other plant tissues (such as dead roots), and the decomposing bodies of organisms

ocean conveyor belt—a major ocean current that moves cold water in the depths of the Pacific Ocean while creating major upwellings in other areas of the Pacific

organisms—one of the six soil-forming factors: different local conditions support different organisms, which influence the soil in a multitude of ways

ozone layer—a thin band of ozone in the lower half of the stratosphere that traps the high-energy radiation of the Sun, holding some of the heat and protecting the troposphere and the Earth's surface from this radiation

Pangaea—a supercontinent theorized to exist during the Paleozoic and Mesozoic Eras when the continents were joined together

parent material—one of the six soil-forming factors: the starting point for soil development; as parent material varies from location to location, so will the soil that develops at each location; its mineral properties, hardness, and topographical form affect how it is weathered into soil

pedosphere—the Earth's soil

permeability—how easily fluids such as water and air move through a given soil

physical (mechanical) weathering—any process that breaks rock down into smaller pieces without changing the chemistry of the rock; typically wind and water

plate boundaries—the edges of tectonic plates

polar easterlies—winds between latitudes 60 degrees and the North Pole that blow from the north and east, and winds between 60 degrees and the South Pole that blow from the south and east

porosity—the number of "holes" or pockets in a given soil

precipitation—water that condenses in the atmosphere and falls to the Earth's surface in liquid or solid form

prevailing winds—belts of air that distribute heat and moisture unevenly around the globe, resulting from the Coriolis effect

R horizon—bedrock; the soil horizon (layer) below the C horizon which constitutes a layer of consolidated (cemented), unweathered rock (not strictly part of the soil)

rain shadow effect—the low-rainfall region that exists on the leeward (downwind) side of a mountain range; the result of the mountain range's causing precipitation on the windward side

relief—one of the six soil-forming factors: topographical relief affects where water moves on the landscape, the depth of the water table in a given location, erosion (which locations are likely to lose surface material through the action of wind and rain, and which locations are likely to accumulate eroded material), how much sun a location receives, and which organisms grow in a particular location

resource—any substance, capability (such as work performed by humans or animals), or other asset that is available in a supply that can be accessed and drawn on as needed, to be utilized for the effective functioning of an organism or community, often for economic gain

revolution—a single cycle of a celestial body (such as the Earth) around a central point (such as the Sun)

Richter scale—scale that measures the amplitude of the highest S-wave of an earthquake

rift valleys—the areas of new crust formation at divergent boundaries, usually between two oceanic plates

rock cycle—the biogeochemical cycle that recycles rocks, in which time, pressure, and the Earth's heat interact to transform rocks into other rocks

rotation—turning on an axis

> **Movie Reference**
> Raise your hand if you're a *Star Wars* fan!

sand—the coarsest soil, with particles 0.05–2.0 mm in diameter; "It's coarse and rough and irritating, and it gets everywhere" —Anakin Skywalker, *Star Wars*

sea water—the saltwater in the world's oceans

sedimentary—type of rock formed as sediment (eroded rocks and the remains of plants and animals) and dissolved minerals build up (usually underwater, on a stream bed or ocean floor) and are compressed as more material is deposited and then cemented together

seismograph—instrument used to measure the magnitude of earthquakes

shear wave—see **S-wave**

shield volcanoes—tall volcanoes with broad bases and gentle slopes that generally form over oceanic hot spots and usually have mild eruptions with slow lava flow, with occasional pyroclastic flows when water enters the vent

silt—soil type with medium coarseness, with particles 0.002–0.05 mm in diameter

soil pH—the pH of the soil's water component

solid Earth—the Earth's solid, rocky outer shell

Southern Oscillation—the atmospheric pressure conditions corresponding to the periodic warming of El Niño and cooling of La Niña

stratosphere—layer of the atmosphere above the tropopause that extends about 20–50 km (7.5–31 miles) above the Earth's surface; its lower half contains the ozone layer; gases in the stratosphere are not well mixed and temperatures increase with distance from the Earth

subduction—the result of a converging ocean-ocean or ocean-continent boundary, in which a heavy ocean plate is pushed below the other plate and melts as it encounters the hot mantle

subduction zone—the site of a convergent boundary between oceanic and continental plates, or sometimes between two oceanic plates, at which the subducting plate is recycled into new magma, which rises through the overlying plate to create volcanoes inland

S-wave (shear wave)—a seismic body wave that shakes the ground up and down or side to side, perpendicular to the direction the wave is moving

tectonic plates—large pieces of lithosphere that move slowly over the mantle of the Earth

thermosphere—layer of the atmosphere above the mesosphere that extends from 80 to around 500 km above the Earth (50–435 miles), in which gases are very thin (rare) and auroras occur

time—one of the six soil-forming factors: more time equals more change, though parent material and relief affect how much time changes take

trade winds—the more or less constant winds blowing in horizontal directions over the Earth's surface as part of Hadley cells

transform fault boundary—a plate boundary at which two plates slide against each other in opposite directions

transformations—soil-forming processes involving the changing of materials in place by weathering or chemical activity

translocations—soil-forming processes involving vertical movement from one soil horizon to another

tropopause—atmospheric layer that acts as a buffer between the troposphere and the stratosphere, where the jet streams travel

troposphere—layer of the atmosphere that lies closest to the Earth and contains about 75–80% of the Earth's atmosphere by mass: it extends from the Earth's surface to about 12 km (7.5 miles) at the poles and 20 km (12.4 miles) at the equator; it is where all the weather that we experience takes place, and contains 99% of the atmosphere's water vapor and clouds

tsunamis—very large and potentially destructive ocean waves, or chains of waves, caused by the movement of the Earth during an earthquake or volcanic eruption

typhoon—see **hurricane**

water holding capacity—the total amount of water a given soil can hold

watershed—the region draining into a river system or other body of water

weather—the day-to-day variations in temperature, air pressure, wind, humidity, and precipitation mediated by the atmosphere in a given region

weathering—the gradual breakdown of rock into smaller and smaller particles, caused by natural chemical, physical, and biological factors

westerly—moving air mass (named for the direction from which it originates) caused by the Ferrel cell, which travels north and east in the Northern Hemisphere and south and east in the Southern Hemisphere

wind—air that's moving as a result of the unequal heating of the Earth's atmosphere

Chapter 7: Land and Water Use

aggregates—clumps of soil formed and held together by such substances as clay particles and organic matter

agroforestry—the planting together of trees and crops, creating a mutualistic symbiotic relationship between them

aquaculture (fish farming)—the raising of fish and other aquatic species in captivity for harvest

aquifers—an underground layer of porous rock, sand, or other material that allows the movement of water between layers of nonporous rock or clay; frequently tapped for wells

arable—land that is fit to be cultivated

bottom trawling—a fishing technique in which the ocean floor is scraped by heavy nets that smash everything in their path

brownfields—urban areas that contain abandoned factories or former residential sites where the possibility that the soil and water are contaminated hinders redevelopment

by-catch—any species of fish, mammals, or birds that are caught during fishing that are not the target organism

capture fisheries—fish production in which fish are caught in the wild and not raised in captivity for consumption

clear-cutting—the removal of all of the trees in an area

compacted—condition in which the mineral grains making up an aquifer collapse on each other and the material is unable to hold as much water

concentrated animal feeding operations (CAFOs) (feedlots)—new meat production industry operations, in which animals are confined and concentrated into smaller spaces in order to keep costs down and quickly get livestock ready for slaughter

confined aquifer—aquifer with boundaries that don't readily transport water

conservation—the management or regulation of a resource so that its use does not exceed the capacity of the resource to regenerate itself

consumption—the day-to-day use of environmental resources such as food, clothing, and housing

contour plowing—a process in which rows of crops are plowed across the hillside; this prevents the erosion that can occur when rows are cut up and down on a slope

controlled burns (prescribed burns)—small fires started when the conditions are just right and which lower the amounts of fuel in forests so as to avoid more dangerous fires

cost-benefit analysis—weighing the benefits against the costs in order to make a decision on resource use

crop rotation—the practice of alternating the crops grown on a piece of land to replenish soil nutrients; e.g., corn one year, legumes for two years, and then back to corn

crown fires—forest fires that start on the ground or in the canopies of forests that have not experienced recent surface fires, spread quickly and are characterized by high temperatures because they consume underbrush and dead material on the forest floor (these fires are a huge threat to wildlife, human life, and property)

deforestation—the removal of trees for agricultural purposes or purposes of exportation

drift nets—nets that drift free in the water and indiscriminately catch everything in their path

drip irrigation—irrigation technique that allots an area only as much water as is necessary and delivers the water directly to the roots using perforated hoses that release small amounts of water

Dust Bowl—dusty drought conditions that occurred in the Great Plains in the 1930s due to a bad drought and farming practices used at that time

ecological footprint—the amount of the Earth's surface required to supply the needs of and dispose of the waste from a particular population

ecosystem capital—the economic value of natural resources to the humans who use them

externalities—unwanted or unanticipated consequences of using resources

feedlots—see **concentrated animal feeding operations (CAFOs)**

fish farming—see **aquaculture**

fishery—the industry or occupation devoted to the catching, processing, or selling of fish, shellfish, or other aquatic animals

flood irrigation—irrigation technique that involves flooding a field with water—water then flows via gravity to all parts of the field

free-range grazing—animal feeding in which animals are able to move about outdoors and eat the foods they are adapted to eat

furrow irrigation—irrigation technique that involves cutting furrows between crop rows and filling them with water

gangue—the waste material that results from mining

genetically modified organisms (GMOs)—organisms grown or bred by humans, in which strands of DNA that code for desired characteristics have been inserted

green manure—soil enricher made by leaving plants (uprooted or simply sown) to wither and then serve as mulch, and then plowing them under and incorporating them into the soil before they can rot, providing valuable nutrients

Green Revolution—the time after the Industrial Revolution when farming became mechanized and crop yields in industrialized nations boomed as farmers began using large amounts of chemical fertilizers and pesticides

greenbelts—open or forested areas built at the outer edges of cities

ground fires—smoldering fires that take place in bogs or swamps and can burn underground for days or weeks; originating from surface fires, ground fires are difficult to detect and extinguish

groundwater—any water that comes from below the ground—that is, from wells or from aquifers

intangible—immeasurable; not having a physical value

integrated pest management (IPM)—an environmentally sensitive approach to pest control that uses a combination of several methods to try to keep the pest population down to an economically viable level, including introducing natural insect predators to the area, intercropping, using mulch to control weeds, diversifying crops, crop rotation, releasing pheromone or hormone interrupters, using traps, and constructing barriers, saving chemical pesticides for only the worst-case scenarios

interbasin transfer—way of dealing with potential water shortages in communities in which water is transported very long distances from its source, through aqueducts or pipelines

intercropping (strip cropping)—the practice of planting bands of different crops across a hillside or field

IPAT model—a mathematical model to describe the impact that humans have on the environment: $I = P \times A \times T$, where I = the total impact, P = population size, A = affluence, and T = level of technology

land degradation—deterioration of land quality (topsoil, organisms, vegetation, water quality), usually caused by exploitation

loamy—soils composed of a balanced mixture of the three particle sizes (clay, silt, and sand) as well as organic matter

long lining—in fishing, the use of long lines with baited hooks, which will be taken by numerous aquatic organisms

marginal benefits—added benefits of a particular step in a process

marginal costs—added costs of a particular step in a process

megacities—cities with populations in excess of 10 million people

metallic minerals—minerals mined for their metals (for example, zinc), which can be extracted through smelting and used for various purposes

mine restoration—projects through which air, land, and water harmed by mining can be reclaimed

mineral deposit—an area where a particular mineral is concentrated

mining—the excavation of earth for the purpose of extracting ore or minerals

monoculture—the cultivation of a single crop on a farm or in a region or country; a single, homogeneous culture without diversity or variety

nonmetallic minerals—minerals mined to be used in their natural state

nonrenewable resources—resources that are often formed by very slow geologic processes, and therefore considered incapable of being regenerated within the realm of human existence

no-till methods—agricultural methods in which farmers plant seeds without using a plow to turn the soil

old growth forest—a forest that has never been cut; it has not been seriously disturbed for several hundred years

ore—a rock or mineral from which a valuable substance can be extracted at a profit

overburden—the rocks and earth that are removed when strip mining for a commercially valuable mineral resource

overgrazed—when grass is consumed by animals at a faster rate than it can regrow

perennial crops—crops that grow back without replanting each year

photosynthate—plants' photosynthetic products

plantation farming—a type of industrialized agriculture in which a monoculture cash crop such as bananas, coffee, or vegetables, is grown and then exported to developed nations

plantations (tree farms)—planted and managed tracts of trees of the same age (because they were planted by humans at the same time) that are harvested for commercial use

polyculture—the practice of planting several crops on the same plot of land simultaneously to increase biodiversity and sustainability

prescribed burns—see **controlled burns**

preservation—the maintenance of a species or ecosystem in order to ensure its perpetuation, with no concern as to its potential monetary value

prior appropriation—the giving of water rights to those who have historically used the water in a certain area

production—the use of environmental resources for profit

renewable resources—resources, such as plants and animals, which can be regenerated quickly (within the time it takes to draw down their supply) if harvested at sustainable yields

riparian—of, on, or relating to the banks of a natural course of flowing water

riparian right—the right, as to fishing or to the use of a riverbed, of one who owns or is allowed to use riparian land

rotational grazing—a method of pasturing involving moving grazing animals to different paddocks and allowing the ones not in use to rest and recover

salinization—a condition that occurs when soil becomes waterlogged from excess irrigation and then dries out; as the water evaporates, the salt crystallizes and forms a layer on the soil surface, which prevents the growth of plants

second growth forests—forest areas where cutting has occurred and a new, younger forest has arisen

selective cutting—the removal of select trees in an area, leaving the majority of the habitat in place and therefore having less of an impact on the ecosystem

shaft mining—subsurface mining in which vertical tunnels are built to access and then excavate minerals that are underground and otherwise unreachable

shelter-wood cutting—a method of harvesting trees in which mature trees are removed, generally over a period of 10–20 years, and a portion of the forest canopy is left intact to provide shelter and protection for regeneration

silviculture—the management of forest plantations for the purpose of harvesting timber

slash-and-burn—a process in which an area of vegetation is cut down and burned before being planted with crops

soil structure—the extent to which soil aggregates or clumps

spray irrigation—irrigation technique that involves pumping water into spray nozzles and spraying fields

strip cropping—see **intercropping**

strip mining—surface mining that involves the removal of the Earth's surface all the way down to the level of the mineral seam

suburbs—satellite communities on the outskirts of cities

surface fires—fires that typically burn only the forest's underbrush and do little damage to mature trees; these fires actually serve to protect the forest from more harmful fires by removing underbrush and dead materials that would burn quickly and at high temperatures

sustainability—humans using resources in such a way as to not deplete those resources for future generations, following environmental indicators such as biological diversity, food production levels, average global surface temperatures and CO_2 concentrations, human population, and resource depletion

sustainable yield—the amount of a renewable resource that can be taken without reducing the available supply

tailings—piles of gangue, which is the waste material that results from mining

tangible—measurable; having a physical value

terracing—creating flat platforms in the hillside that provide a level planting surface, which reduces soil runoff from the slope

traditional subsistence agriculture—a process in which each family in a community grows crops for itself and relies on animal and human labor to plant and harvest crops

Tragedy of Free Access—the depletion of marine fisheries worldwide (seen as an example of the Tragedy of the Commons on an international scale)

Tragedy of the Commons—a situation in which a common resource is used by many people and then becomes depleted as these people do not regulate their consumption of the resource (based on an example in which a piece of open land, a commons, was used collectively by the townspeople for grazing their cattle: each townsperson who used the land continued to add one cow or ox at a time until the commons was overgrazed)

tree farms—see **plantations**

unconfined aquifer—aquifer in which water is free to flow both vertically and horizontally

uneven-aged management—the broad category under which selective cutting and shelter-wood cutting fall; selective deforestation

urban sprawl—the emigration of people out of the city and into the suburbs

water-scarce—countries that have a renewable annual water supply of less than 1,000 m^3 per person

water-stressed—countries that have a renewable annual water supply of about 1,000–2,000 m^3 per person

windbreak—one or more rows of trees or shrubs planted near crops in such a way as to provide shelter from eroding winds

Chapter 8: Energy Resources and Consumption

active collection—the use of devices, such as solar panels, to collect, focus, transport, or store solar energy

anthracite—the cleanest-burning coal; almost pure carbon

battery electric vehicles (BEVs)—vehicles that run on battery power alone and can be charged with electric power

biofuel—filtered cooking oil that can be burned in diesel engines

biomass—burning plant or animal matter, including wood, charcoal, food waste, crops, animal waste products, etc., for energy production

bituminous—the second purest form of coal

boiler residue—a waste product produced by the burning of coal

boiling water reactors—nuclear reactors that use the heat of the reactor core to boil water into steam, which is piped directly to the turbines and spins them to generate electricity

breeder reactor—a nuclear reactor that generates more fissionable material than it consumes

CAFE (corporate average fuel economy)—mile-per-gallon standards meant to reduce energy consumption by increasing the fuel economy of cars and light trucks

cogeneration—using a fuel source to generate both useful heat and electricity

combustion—the process of burning

conduction—the transfer of energy through matter from particle to particle

convection—the transfer of heat by the movement of the heated matter

crude oil—the form petroleum takes when in the ground

energy—the capacity to do work

energy conservation—the practice of reducing the use of fossil fuels and reducing the impact on the environment from the production and use of energy

energy quality—the characteristics of different energy forms in terms of their heat value, versatility, and environmental performance

exploratory well—a well drilled to sample a particular area for fossil fuel reserves (if an exploratory well hits a fossil fuel reserve, it can provide an estimate of the amount of fuel that can be obtained from that area)

First Law of Thermodynamics—law that states that energy can neither be created nor destroyed; it can only be transferred and transformed

fish ladders—structures on or around artificial and natural barriers (such as dams, locks, and waterfalls) that facilitate fishes' natural migration and movements by enabling fish to pass around the barriers by swimming and often leaping up a series of relatively low steps

fission—a nuclear reaction in which an atomic nucleus, especially a heavy nucleus such as an isotope of uranium, splits into fragments, usually two fragments of comparable mass, releasing from 100 million to several hundred million electron volts of energy

fly ash—a waste product produced by the burning of coal

fossil fuel—a hydrocarbon deposit, such as petroleum, coal, or natural gas, derived from living matter of a previous geologic time and used for fuel

fracking (hydraulic fracturing)—a process by which natural gas and oil are extracted from rock that lies deep underground by drilling a deep well and then injecting millions of gallons of fracking fluid—a mix of water, sand, and chemicals—at a high enough pressure to fracture the rock and release the oil or gas

fuel rods—cylindrical tubes filled with pellets of nuclear fuel

gasohol—a gasoline extender made from a mixture of 90 percent gasoline and 10 percent ethanol; it has higher octane than gasoline and burns more slowly, coolly, and completely, thus resulting in reduced emissions of some pollutants

geothermal energy—energy that's produced by harnessing the Earth's internal heat

half-life—the amount of time it takes for half of a radioactive sample to degrade

Hubbert peak—an influential theory that concerns the long-term rate of conventional oil (and other fossil fuel) extraction and depletion; it predicts that future world oil production will soon reach a peak and then rapidly decline

hybrid-electric vehicle—a vehicle built with two motors: one electric and one gasoline-powered

hydraulic fracturing—see **fracking**

hydroelectric power—power generated via the systematic placement of river dams that spin turbines to produce electricity or, on a smaller scale, by using turbines placed directly in small rivers to generate power

hydrogen cells—technology in which free hydrogen is stored and then used to generate electricity through electrolysis

kinetic energy—the energy of motion

lignite—the least pure coal

nacelle—the base (containing the generator and gearbox) of a wind turbine

net energy yield—the amount of useful energy derived from a fuel less the energy cost of extraction, processing, and transportation

nuclear fusion—a nuclear reaction in which two nuclei are fused to form one or more different atomic nuclei (and subatomic particles)

overburden—the rocks and earth that are removed when strip mining

passive solar energy collection—the use of building materials, building placement, and design to passively collect solar energy that can be used to keep a building warm or cool

peat—partially decomposed organic material harvested from peatlands or bogs and burned as fuel

photovoltaic cell—a semiconductor device that converts the energy of sunlight into electric energy

potential energy—energy at rest, or stored energy

pressurized water reactors—nuclear reactors in which one water supply is kept under high pressure to prevent the water from boiling and then passed through the reactor heat exchanger, where heat from the first water supply is transferred to the second water supply, which forms steam to spin the turbines and generate electricity

proven reserve—an estimate of the amount of fossil fuel that can be obtained from a reserve

radiant energy—the energy of electromagnetic and gravitational radiation (for example, sunlight)

radioactive—having unstable nuclei which emit energy over time as particles or photons

scrubbers—devices containing alkaline substances that precipitate out much of the sulfur dioxide from industrial plants' air effluent

seams—long, continuous deposits of coal or minerals

Second Law of Thermodynamics—law that states that the entropy (disorder) of the universe is increasing; one corollary of the Second Law of Thermodynamics is that, in most energy transformations, a significant fraction of energy is lost to the universe as heat

solar energy—the energy of (or energy derived from) the electromagnetic radiation of the Sun

strip mining—surface mining that involves the removal of the Earth's surface all the way down to the level of the mineral seam

subbituminous—the third purest form of coal

subsidence—the sudden sinking or gradual downward settling of the ground's surface with little or no horizontal motion, such as is caused by groundwater withdrawal, mine cave-ins, etc.

tar sands—oil reserves found in surface sands: they are a combination of clay, sand, water, and bitumen

tidal power—energy harnessed from the tidal movements of ocean water

turbine—a rotary mechanical device that extracts energy from a fluid flow and converts it into useful work, which can then be used for generating electrical power

underground mining—mining that involves the sinking of shafts to reach underground deposits; networks of tunnels are dug or blasted, and humans enter these tunnels in order to manually retrieve the coal

wind farms—groups of modern wind turbines used to generate wind energy

wind power—energy harnessed from wind

Chapter 9: Pollution

acid mine drainage—highly acidic water that results from water seeping through mines and carrying off sulfur-containing compounds: the chemical conversion of sulfur-bearing minerals occurs through a combination of biological (bacterial) and inorganic chemical reactions, and the result is the buildup of extremely acidic compounds in the soil surrounding the deposit, which are carried by water as it flows to surrounding areas and can severely harm local stream ecosystems

acid precipitation—acid rain, acid hail, acid snow, etc.; all of which occur as a result of pollution in the atmosphere, primarily SO_2 and nitrogen oxides, which combine with water to form acids (typically nitric acid and sulfuric acid) that are deposited on the Earth through precipitation

acute effect—the effect caused by a short exposure to a high level of toxin

algal bloom—exponential growth in the population of single-cell algae, which form blooms of color

asbestos—a naturally occurring silicate mineral that was often used as a building material in years past due to its insulating and heat-resisting properties, and which has been shown to be dangerous to human health (the fibers, when inhaled, can cause a host of lung problems)

bioaccumulation—the selective absorption and concentration of elements or compounds by cells in a living organism, most commonly fat-soluble compounds

biological oxygen demand (BOD)—a measure of the rate at which bacteria absorb oxygen from water (an important factor in judging water quality)

biomagnification—the increasing concentration of any molecules (normally toxins or heavy metals) at successively higher trophic levels in a food chain

brown smog—see **photochemical smog**

brownfield sites—industrial or commercial sites that are idle or underused because they are contaminated by hazardous waste or pollutants

building-related illness—a condition in which the signs and symptoms of an illness can be attributed to a specific infectious organism that resides in the building

carbon monoxide (CO)—an odorless, colorless gas that's typically released as a by-product of incompletely burned organic material, such as fossil fuels, and is hazardous to human health because it binds irreversibly to hemoglobin in the blood (one of the EPA's six criteria pollutants)

catalytic converter—a platinum-coated device that oxidizes most of the VOCs and some of the CO that would otherwise be emitted in automobile exhaust, converting them to CO_2

cholera—a bacterial disease that is contracted from infected water; its main symptom is large amounts of watery diarrhea

chronic effect—an effect that results from long-term exposure to low levels of toxin

Clean Air Act (CAA)—a U.S. federal law enacted in 1963 and designed to control air pollution nationally, amended many times: to set standards for controlling automobile emissions, to address acid rain, ozone depletion, and toxic air pollution, and to make provisions for investigation and enforcement

Clean Water Act (CWA)—the primary U.S. federal law governing water pollution, enacted in 1972, governing water quality standards, regulation of point-source pollution, and water uses

composting—the process of allowing organic matter such as food scraps, paper, and yard waste to decompose, allowing the organic material in solid waste to be reintroduced into the soil as fertilizer

Corporate Average Fuel Economy (CAFE)—standards that set mile-per-gallon standards for automotive vehicles, with the goal to reduce energy consumption by increasing the fuel economy of cars

corrosive waste—waste that corrodes metal

COVID-19—a 2019–2022 pandemic caused by the virus strain severe acute respiratory syndrome-coronavirus 2 (SARS-CoV-2)

criteria pollutants—the six pollutants that the EPA has determined do the most harm to human health and welfare: carbon monoxide, lead, ozone, nitrogen dioxide, sulfur dioxide, and particulates

dead zone—see **hypoxic zone**

deep-well injection—hazardous waste disposal method that involves drilling a hole in the ground that reaches below the water table and the impervious soil layer into porous rock, and injecting waste into the well

Delaney Clause—a part of the Food Additives Amendment of 1958 (an amendment to the Federal Food, Drug, and Cosmetic Act of 1938) that specifically bans any food additives found to cause cancer in humans or in animal testing

disease—occurs when the infection from a pathogen in the body causes a change in the state of health

dissolved oxygen—amount of oxygen dissolved in a given amount of water (warm water holds less dissolved oxygen than cool water); an important factor in judging water quality, since low levels of dissolved oxygen indicate an inability to sustain life

dose-response analysis—a process in which organisms are exposed to a toxin at different concentrations, and the dosage that causes the death of the organism is recorded

dose-response curve—the curve that results from graphing the information obtained in a dose-response analysis

dry acid particle deposition—acid deposition that occurs in the absence of precipitation, when particles and gases stick to the ground, plants, or other surfaces

dust—a catch-all term for fine particles of solid matter (from various sources such as soil and volcanic eruptions, plant pollen, human and animal hair, textile fibers, paper fibers, etc.) that pose health risks to humans because of allergic reactions and the exacerbation of asthma and lung conditions

dysentery—a type of gastroenteritis (inflammation of the stomach and small intestine) that results in diarrhea with blood, plus often fever and abdominal pain, with multiple possible causes, including bacteria, amoebas, chemicals, and parasitic worms

ED$_{50}$—the dosage at which 50 percent of the test organisms show a negative effect from a toxin

endocrine disruptors—chemicals that can interfere with the endocrine systems of animals

eutrophic waterways—any waterways in which too much agricultural runoff and/or waste-water accumulates, bringing the nutrient levels to excess

eutrophication—water becoming overly enriched with minerals and nutrients to the point that excessive growth of algae and other phytoplankton occurs

e-waste—discarded electronic devices including televisions, cell phones, and computers

Food and Drug Administration (FDA)—the governmental body in the United States that regulates food and related products, empowered by the Federal Food, Drug, and Cosmetic Act of 1938

gray smog—see **industrial smog**

hardness—a measure of the concentrations of calcium and magnesium in water (an important factor in judging water quality)

hazardous waste—any waste that poses a danger to human health and therefore must be dealt with in a different way from other types of waste

heat islands—urban areas that heat up more quickly and retain heat better than nonurban areas

high-level radioactive waste—radioactive wastes that produce high levels of ionizing radiation

hypoxic waterways—any waterways in which hypoxia results from decomposers using up the available oxygen due to eutrophication and rapid growth in the populations of phytoplankton and zooplankton

hypoxic zone (dead zone)—zone in which nothing that depends on oxygen can grow; created when nutrient-rich water (containing runoff from land carrying fertilizer, raw sewage, and/or pollutants) gets dumped into oceans via rivers and streams: because the warm, nutrient-rich freshwater does not mix well with the colder saltwater, eutrophication results and allows phytoplankton and the zooplankton that feed on them to grow almost uncontrollably; when these die and sink to the bottom, bacteria metabolize the available dissolved oxygen as they decompose this detritus, using up the available oxygen

ignitable waste—hazardous waste containing substances such as alcohol or gasoline that can easily catch fire

indoor air pollution—pollutants that move indoors from outside and/or are produced indoors from sources such as carpet, furniture, plastic, oils, paints, adhesives, pesticides, cleaning fluids, gas leaks, tobacco smoke, dust, asbestos, and living organisms

industrial smog (gray smog)—smog resulting from emissions from industry and other sources of gases produced by the burning of fossil fuels, especially coal, from which CO and CO_2 combine with particulates in the atmosphere (sometimes aided by air inversions or fog and sometimes also containing sulfuric acid formed from water vapor and sulfur dioxide)

infection—the result of a pathogen invading the body

LD_{50}—the dosage at which 50 percent of the test organisms die from a toxin

leachate—the decomposed material that percolates to the bottom of a landfill

lead—air pollutant that is generally released into the atmosphere as a particulate, but then settles on land and water, where it is incorporated into the food chain and is subject to biomagnification; if it enters the human body, it can cause numerous nervous system disorders, including cognitive and developmental disabilities in children (one of the EPA's six criteria pollutants)

Love Canal—site in Niagara Falls, New York, that had been the location of a landfill and only after becoming a residential neighborhood in the 1970s was realized to be a brownfield site and environmental disaster; the health of hundreds of residents was harmed, resulting in the passing of CERCLA (the Superfund Act)

low-level radioactive waste—radioactive wastes that produce low levels of ionizing radiation

malaria—a parasitic disease caused by bites from infected mosquitoes that is widespread in the tropical and subtropical regions of the globe (especially in sub-Saharan Africa) and causes fever, tiredness, vomiting, headaches, and, in severe cases, yellowed skin, seizures, coma, and even death

Middle East respiratory syndrome (MERS)—a viral respiratory illness caused by a coronavirus, MERS-coronavirus (MERS-CoV), of which there was an outbreak in 2012–2013

moving sources—non-stationary sources of pollution, such as cars

National Ambient Air Quality Standards—a set of maximum permissible levels for pollutants established by the Environmental Protection Agency to protect human health

nitrogen dioxide (NO$_2$)—one in a family of nitrogen and oxygen gases (nitrogen oxides) formed when atmospheric nitrogen and oxygen react as a result of exposure to high temperatures such as those in combustion engines, utilities, and industrial combustion; also commonly found as a secondary pollutant and a component of smog and acid precipitation (one of the EPA's six criteria pollutants)

nitrogen oxides—a family of nitrogen and oxygen gases formed when atmospheric nitrogen and oxygen react as a result of exposure to high temperatures such as those in combustion engines, utilities, and industrial combustion; also commonly found as secondary pollutants and components of smog and acid precipitation

noise pollution—any noise that causes stress or has the potential to damage human health

non-point source pollution—pollution that does not have a specific point of release

oligotrophic waterways—waterways that have low amounts of nutrients, stable algae populations, and high levels of dissolved oxygen

oxygen sag curve—a plot of dissolved oxygen levels versus the distance from a source of pollution (usually excess nutrients and biological refuse)

ozone—O$_3$, which has very different effects depending on where it's found: stratospheric ozone absorbs UV light from the sun and protects life on our planet in the ozone layer; while tropospheric ozone—the ozone that's formed as a result of human activity—is a powerful respiratory irritant and a secondary pollutant formed in the troposphere as a result of the interaction of nitrogen oxides, heat, sunlight, and volatile organic compounds, and is a major component of smog (tropospheric ozone is one of the EPA's six criteria pollutants)

particulate matter—small particles of solid or liquid pollutant (such as soot or sulfate aerosols) light enough to be carried on air currents (and when humans breathe them in, the particles act as irritants) (one of the EPA's six criteria pollutants)

pathogens—microorganisms that can cause disease (viruses and other subcellular infectious particles, bacteria, fungi, protozoa, or parasitic worms)

persistent organic pollutants (POPs)—specific organic compounds, such as DDT and PCBs, that resist environmental degradation and are soluble in fat, which allows them to accumulate in organisms' fatty tissues, making them fairly toxic to many organisms; in addition, because of their persistence, they can travel over long distances via wind and water before being redeposited

pH—a scale used to specify the acidity or alkalinity of an aqueous solution: the scale is logarithmic and inversely indicates the concentration of hydrogen ions in the solution (an important factor in judging water quality)

photochemical smog (brown smog)—a type of smog usually formed on hot, sunny days in urban areas from a combination of NO$_x$ compounds, VOCs, and ozone; can harm human health in several ways, including causing respiratory problems and eye irritation

physical treatment—in a sewage treatment plant, the initial filtration that is done to remove debris such as stones, sticks, rags, toys, and other objects that were flushed down the toilet

plague—a disease carried by organisms infected with the plague bacteria, transferred to humans via the bite of an infected organism (vector) or through contact with contaminated fluids or tissues; it has distinct forms (bubonic, septicemic, and pneumonic), but all are characterized by fever, chills, headaches, and nausea

point source pollution—pollution released from a specific location (e.g., a factory where wood is being burned)

poison—any substance that has an LD_{50} of 50 mg or less per kg of body weight

pollutant—any substance (or energy) that, when introduced into its environment, renders the air, soil, water, or some other natural resource harmful, or adversely affects its usefulness

primary pollutants—those pollutants that are released directly into the lower atmosphere

primary recycling—type of recycling in which materials such as plastic or aluminum are used to rebuild the same product

primary treatment—a process during which physically treated sewage water is passed into a settling tank, wherein suspended solids settle out as sludge; chemically treated polymers may be added to help the suspended solids separate and settle out

Quiet Communities Act—1978 U.S. federal legislation that provides for the coordination of federal research and activities into noise control and authorized the use of Federal Aviation Administration funds for the development of noise abatement plans around airports

radon—a gas that's emitted by uranium as it undergoes radioactive decay; it seeps up through rocks and soil and enters buildings via basements or cracks in the walls or foundations, and can also sometimes be found in groundwater that enters homes through a well; the second leading cause of lung cancer (after smoking) in the United States

reactive waste—hazardous waste containing substances that are chemically unstable or react readily with other compounds, resulting in explosions or causing other problems

recycling—the reuse of materials to make new products

reduce—minimize disposable waste in order to lessen its environmental impact

Resource Conservation and Recovery Act (RCRA)—U.S. Federal law enacted in 1976 and governing the disposal of solid and hazardous waste

reuse—the continued use of products that are disposable but can be put to alternative use—sometimes over and over again—before their eventual disposal, in order to reduce the consumption of new items and thus lessen the waste produced

risk—degree of likelihood that a person will become ill after exposure to a toxin or pathogen

risk assessment—calculating risk

risk management—using strategies to reduce the amount of risk

sanitary landfills—modern landfills that follow federal regulations: large holes in the ground lined with geomembranes or plastic sheets that are reinforced with two feet of clay on the bottom and sides, creating a virtually impermeable containment layer; the waste in the landfill is frequently covered with soil in order to control insects, bacteria, rodents, and odor; and the leachate is piped to the top of the site and collected and monitored in leachate ponds; in addition, they are required to be positioned at least six feet above the water table and may not be located near geological faults, wetlands, or flood plains; and groundwater at the sites is tested frequently; when one site is full, it is capped, monitored, and provided with long-term care

secondary pollutants—pollutants that are formed by the combination of primary pollutants in the atmosphere

secondary recycling—type of recycling in which materials are reused to form new (different) products that are usually lower quality goods

secondary treatment—the biological treatment of wastewater in order to continue to remove biodegradable waste, using trickling filters (in which aerobic bacteria digest waste as it seeps over bacteria-covered rock beds) or an activated sludge processor (which is basically a tank filled with aerobic bacteria in which the solids are left to settle out)

severe acute respiratory syndrome (SARS)—a viral respiratory disease caused by severe acute respiratory syndrome coronavirus (SARS-CoV-1), of which there was an outbreak in 2002–2004; the respiratory symptoms were a form of pneumonia

sick building syndrome (SBS)—a condition in which the majority of a building's occupants experience certain symptoms that vary with the amount of time spent in the building, without being able to identify a specified cause or illness

sludge—the solids that remain after the secondary treatment of sewage

sludge processor—a tank filled with aerobic bacteria that's used to treat sewage

solid waste—any discarded material that is not a liquid or gas; it is generated in the domestic, industrial, business, and agricultural sectors, and can consist of hazardous waste, industrial solid waste, or municipal waste

stationary sources—non-moving sources of pollution, such as factories

stratospheric ozone—ozone (O_3) found in the stratosphere, where it absorbs UV light from the sun and protects life on our planet as the ozone layer

sulfur dioxide (SO_2)—a colorless gas with a penetrating and suffocating odor; a powerful respiratory irritant typically released into the air through the combustion of coal, and also during metal smelting, paper pulping, and the burning of fossil fuels; can also be a component of indoor pollution as a result of gas heaters, improperly vented gas ranges, and tobacco smoke; in the atmosphere, SO_2 reacts with water vapor to form acid precipitation (one of the EPA's six criteria pollutants)

sulfur oxides—a family of sulfur and oxygen gases related to sulfur dioxide; pollutants contributing to acid precipitation and industrial smog

Superfund Program—a program funded by the federal government and a trust funded by taxes on chemicals; identifies pollutants and cleans up hazardous waste sites

surface impoundment—hazardous waste disposal method typically used for liquid waste; it involves the creation of shallow, lined pools from which the hazardous liquid evaporates

temperature inversion—phenomenon in which the warm, polluted air over a city is blocked from rising into the cooler atmosphere by warm air above it and remains hanging above the city

tertiary treatment—wastewater treatment step undertaken when treated water is to be deposited directly into groundwater, in which water is further treated by passing secondary-treated water through a series of sand and carbon filters and then further chlorination

threshold dose—the dosage level of a toxin at which a negative effect occurs

toxic waste—hazardous waste that creates health risks when inhaled or ingested, or when it comes into contact with skin

toxicity—the degree to which a substance is biologically harmful

toxin—any substance that, when inhaled, ingested, or absorbed at sufficient dosages, damages a living organism

transuranic waste—waste that's left over from the construction of nuclear weapons

tropospheric ozone—ozone (O_3) found in the troposphere (the ozone that's formed as a result of human activity); it's a powerful respiratory irritant and a secondary pollutant formed as a result of the interaction of nitrogen oxides, heat, sunlight, and volatile organic compounds, and is a major component of smog (one of the EPA's six criteria pollutants)

tuberculosis—a bacterial infection that typically attacks the lungs, is spread by breathing in the bacteria from the bodily fluids of an infected person, and has symptoms including chronic cough with bloody mucus, fever, night sweats, and weight loss

turbidity—the density of suspended particles in the water; water clarity/cloudiness (an important factor in judging water quality)

U.S. Noise Control Act—U.S. federal legislation enacted in 1972 that gave the EPA power to set emission standards for major sources of noise, including transportation, machinery, and construction

vapor recovery nozzle—an air pollution control device used on gasoline pumps that prevents fumes from escaping into the atmosphere when fueling a vehicle

vector—the carrier organism through which pathogens can attack, such as a tick, flea, or mosquito

volatile organic compounds (VOCs)—pollutants released as a result of various industrial processes including dry cleaning, the use of industrial solvents, and the use of propane; sublimate at room temperature, can react in the atmosphere with other gases to form O_3, and are a major contributor to smog in urban areas

Waste-to-Energy (WTE) program—program that uses energy released from municipal waste incineration to generate electricity

wastewater—any water that has been used by humans; this includes human sewage, water drained from showers, tubs, sinks, dishwashers, washing machines, water from industrial processes, and storm water runoff

West Nile virus—a virus transmitted to humans via bites from infected mosquitoes that is closely related to the viruses that cause zika, dengue, and yellow fever; it causes West Nile fever, a disease characterized by fever, headache, vomiting, rash, and in severe cases, encephalitis or meningitis

wet deposition—acid precipitation

zika—a virus caused by bites from infected mosquitoes that can also be transmitted through sexual contact and from a pregnant mother to a fetus: it tends to cause fever, red eyes, joint pain, headache, and rash; it can cause microencephaly and other brain malformations when transferred to a fetus, and in adults it has been linked to Guillain–Barré syndrome, a rapid-onset muscle weakness

Chapter 10: Global Change

anthropogenic greenhouse gas—atmospheric greenhouse gases, the sources of which are human activity, such as burning of fossil fuels, deforestation, agriculture and livestock, land-fill emissions, manufacturing processes, and industrial chemicals

background extinction rate—natural rate of extinction occurring apart from widespread events

biodiversity hot spot—a highly diverse region that faces severe threats and has already lost 70 percent of its original natural habitat by area

cap-and-trade policy—approach to the reduction of greenhouse gas emissions that provides economic incentives for limiting emissions of pollutants

carbon sequestration—the process of capturing and storing atmospheric carbon dioxide

chlorofluorocarbons (CFCs)—a group of man-made chemicals implicated in declining stratospheric ozone levels, invented in the 1930s and used in items such as propellants, fire extinguishers, and cans of hairspray; once released, CFCs migrate to the stratosphere through atmospheric mixing (they are very stable, which allows them to survive through the rise), and then in the upper stratosphere, intense UV radiation breaks the CFC molecules apart and releases chlorine atoms that form chlorine monoxide (ClO) while converting O_3 to O_2

CITES—treaty (the Convention on International Trade in Endangered Species of Wild Fauna and Flora) that bans the capture, exportation, or sale of endangered and threatened species

climate change—the shift the global climate is undergoing that is, by consensus, almost certainly in large part due to human contributions

coral bleaching—what happens when coral polyps expel the symbiotic algae that live in their shells and are vital to their health (once a coral bleaches, it continues to live but begins to starve; some corals recover, but many don't); caused by rising water temperatures, among a multitude of other factors

critically endangered—designation given if a species is under a very high risk of extinction

endangered—designation given if a species is likely to become extinct

Endangered Species Act—1973 U.S. legislation that prohibited the commerce of those species considered to be endangered or threatened

extinction—the death of an entire species

global warming potential (GWP)—a measure of the impact of a particular greenhouse gas on global climate change that uses carbon dioxide (which has a GWP of 1) as a reference point

globalization—the increasing interconnectedness of the world in modern times

Golden Age of Conservation— Theodore Roosevelt's presidency (1901–1909), during which many environmentally friendly laws and policies were put into effect

green taxes—taxes (such as on pollution or certain forms of consumption) that aim to generate revenue to correct past pollution damage and reduce future pollution and provide an incentive to change behavior to practices more protective of the environment

greenhouse effect—the phenomenon whereby the Earth's atmosphere traps solar radiation, caused by the presence in the atmosphere of gases such as carbon dioxide, water vapor, and methane, that allow incoming sunlight to pass through but absorb heat radiated back from the Earth's surface

habitat corridors—areas of habitat connecting wildlife populations separated by human activities or structures (such as roads, development, or logging)

habitat fragmentation—when the size of an organism's natural habitat is reduced, or when development occurs that isolates a habitat or part of one

invasive species—introduced, non-native species

Kyoto Protocol—international treaty that required the participating 38 developed countries to cut their greenhouse gas emissions back to 5% below 1990 levels

market permits—cap-and-trade permits that allow companies that buy them a certain amount of discharge of substances into certain environmental outlets; if companies can reduce their amount of discharge, they are allowed to sell the remaining portion of their permits to other companies

Montreal Protocol—international agreement that addressed the problem of ozone depletion by mandating cutting the emissions of CFCs that damage the ozone layer (was amended in 1992 to include other key ozone-depleting chemicals)

non-governmental organizations (NGOs)—organizations founded by citizens that are independent of government involvement, such as Greenpeace and the World Wildlife Fund

ocean acidification—decrease in pH of the oceans that's primarily an effect of the increased carbon dioxide concentrations in the atmosphere due to the burning of fossil fuels, vehicle emissions, and deforestation

ocean warming—increase in ocean temperatures due to the increase in greenhouse gases in the atmosphere and the greenhouse effect they produce

ozone loss—a thinning of the ozone layer (declining stratospheric ozone levels) due to the effect of chlorofluorocarbons (CFCs)

policy—a plan or course of action—as of a government, political party, or business—intended to influence and determine decisions, actions, and other matters

sustainability—humans using resources in such a way as to not deplete those resources for future generations, following environmental indicators such as biological diversity, food production levels, average global surface temperatures and CO_2 concentrations, human population, and resource depletion

taxon substitution—when non-native species are introduced to fill an ecological role that previously was performed by a native species now extinct

threatened species—species assigned to any one of three categorizations: critically endangered, endangered, or vulnerable

vulnerable—designation given if a species is likely to become endangered if no action is taken

water rights—system of U.S. water laws in use west of the Mississippi River, in which the person who first files a claim on a water resource has rights to the use of the water

Chapter 13
Chapter Drills:
Answers and
Explanations

CHAPTER 1 DRILL

Multiple Choice

1. **C** Population dispersion refers to how individuals of a population are spaced within a region. The three main ways in which populations of species can be dispersed are uniform, in which members of a population are evenly spaced throughout a geographic region; random, in which the position of members of a population are not influenced by the positions of other members of the population; and clumped, in which individuals are grouped together in a region. Accordingly, the correct answer is (C).

2. **D** Physical weathering is the process that breaks rock down into smaller pieces without changing the chemistry; eliminate answer choice (A). Ecological weathering is not one of the weathering processes; eliminate answer choice (B). Chemical weathering occurs as a result of chemical reactions of rock with water, air, or dissolved minerals; eliminate answer choice (C). Biological weathering takes place as a result of the activities of living organisms, which may act through biological or chemical means. Accordingly, the correct answer is (D).

3. **A** As the question notes, the city of Houston has experienced tremendous growth over the past few decades, which has led to the establishment of communities outside of Houston. As noted, this growth has led to difficulty in knowing where the city of Houston ends and suburban communities like Katy, Texas, and Galveston, Texas, begin. The scenario presented exemplifies urban sprawl, which refers to the emigration of people out of the city and into the suburbs. The suburban revolution is not a concept discussed, so answer choice (B) can be eliminated. Gentrification is the process whereby the character of a poor urban area is changed by wealthier people moving in, improving housing, attracting new businesses, and typically displacing current inhabitants in the process; eliminate answer choice (C). Brownfields are urban areas that cannot be efficiently redeveloped as they contain abandoned factories or former residential sites where there's the possibility that the soil and water are contaminated; eliminate answer choice (D). Thus, the correct answer is (A).

4. **C** Acid precipitation occurs due to pollution in the atmosphere—primarily the SO_2 and nitrogen oxides. Due to the high dilution rate of acid precipitation, it will not burn skin upon contact, but, nonetheless, acid precipitation has a significant impact on humans and the environment. Indeed, acid precipitation is responsible for leaching minerals from soil, which alters the soil chemistry, and damages rock, including statues, monuments, and buildings. While acid precipitation does cause human respiratory irritation, severe acute respiratory syndrome (SARS) is a viral respiratory disease caused by the SARS coronavirus, which is spread through the inhalation or absorption of infected fluids. Accordingly, only statements I and III are correct, and the correct answer is (C).

5. **B** Humans have a tremendous impact on biodiversity. In order to remember the human factors that can cause extinction and decrease biodiversity, you should memorize the acronym HIPPCO, which stands for Habitat destruction, Invasive species, Population growth, Pollution, Climate change, and Overharvesting. Accordingly, infectious disease is not one of the human factors that can cause extinction and decrease biodiversity, so the correct answer is (B).

6. **D** In order to answer this question, you must be aware of the types of ecosystems, their world locations, and the major vegetation associated with each biome. Chapparal is located in western North America and the Mediterranean region and has both spiny shrubs and small trees with large, hard evergreen leaves as its major vegetation; eliminate answer choice (A). Coniferous forests are populated with coniferous trees and located in northern North America and northern Europe; eliminate answer choice (B). Deciduous

forests have hardwood trees as the major vegetation and are located in North America, Europe, Australia, and Eastern Asia; eliminate answer choice (C). Finally, tundra is located in regions that are just south of the ice caps of the Arctic and extend across North America, Europe, and Siberia in Asia, and herbaceous plants are its major vegetation; the correct answer is (D).

7. **B** In order to answer this question, you must be familiar with the major environmental legislation passed in the United States. The Soil and Water Conservation Act set up criteria to evaluate the conditions of American soil, water, and related resources, and also established soil and water conservation programs to aid landowners and users; eliminate answer choice (A). The Safe Drinking Water Act established a federal program to monitor and increase the safety of the drinking water supply; the correct answer is (B). The Clean Water Act used regulatory and nonregulatory tools to protect all surface waters in the United States; eliminate (C). The Comprehensive Environmental Response, Compensation, and Liability Act, commonly known as Superfund, created a tax on the chemical and petroleum industries and provided broad federal authority to respond directly to releases of hazardous substances that may endanger public health or the environment; eliminate answer choice (D).

8. **C** In order to answer this question, you need to know the correct actual growth rate formula: (birth rate – death rate / 10). Accordingly, the actual growth rate for this country is 3.5%. You also need to know the Rule of 70, which states that the time it takes (in years) for a population to double can be approximated by dividing 70 by the current growth rate of the population. Therefore, the population will double in approximately $70 \div 3.5 = 20$ years. Thus, the population will double, reaching 40 million individuals, in 20 years or in 2042; therefore, the correct answer is (C).

9. **A** This question requires you to have knowledge about the demographic transition model (DMT), which is used to predict population trends based on the birth and death rates of a population. In the DMT, birth rates exceed death rates in all states of the demographic transition model except in the preindustrial state, in which the population exhibits a slow rate of growth and has both high birth and high death rates because of harsh living conditions. Accordingly, (A) is the correct answer.

10. **B** Ground fires are smoldering fires that occur in bogs or swamps, originate from surface fires, and are difficult to detect and extinguish; eliminate answer choice (A). Surface fires typically only burn the underbrush of a forest and serve to protect the forest from more harmful fires by removing underbrush and debris; eliminate answer choice (C). Controlled burns, or prescribed burns, are controlled fires that are intentionally started to reduce the fuel buildup and, in turn, decrease the likelihood of more severe fires; eliminate answer choice (D). Crown fires may start on either the ground or forest canopies that have not experienced recent surface fires, spread quickly, and are characterized by high temperatures. Thus, the correct answer is choice (B).

11. **B** Make sure you know about renewable and nonrenewable resources for the AP Exam! Renewable resources are those that can be regenerated quickly or occur perpetually in nature, such as plants, animals, wind, and solar, while nonrenewable resources are those that do not regenerate quickly, such as minerals and fossil fuels. The time necessary for hardwood trees to mature, which is about 50 years, is widely considered the crossover point from renewable to nonrenewable resources. Accordingly, all of the answer choices except (B) are true statements, so the correct answer is (B).

CHAPTER 2 DRILL

Free Response

1. (a) This is a Rule of 70 calculation question. The population has doubled twice from 1972 to 2012. Therefore, the doubling period is 20 years. Using the Rule of 70 formula, the correct calculation is 70 ÷ 20 = 3.5 percent.

   (b) This is an endangered species question, so consult your HIPPCO acronym. Prairie dogs declined due to:
   - **H**abitat loss to livestock and agriculture competing for land
   - **P**ollution to their habitat through a deliberate poisoning eradication program, because farmers thought the tunnels posed threats to horse and cow legs
   - **C**limate—massive drought (Dust Bowl fell between 1880 and 1972)

   (c) This is a keystone species question. The burrow colonies developed by the prairie dogs contribute to the success of other species by providing shelter for animals such as burrowing owls, ferrets, and snakes; the colonies also provide a food source for predators; and their burrowing churns the soil to enable the Earth to better sustain plant life and infiltrate water.

   (d) This is an economic/ecological pros question.

Economic

For	Against
• Provides food source for economically important species	• Produces holes that can injure livestock
• Eradication plans are expensive	• Dangerous or destructive to property, infrastructure, and development
• Eco-tourism to see organisms that rely on the prairie dog	

Ecological

For	Against
• Provides food source for other species	• Overpopulation of prairie dogs could alter ecological dynamics
• Provides burrows for other species	
• Churned soil is better for plants	

   (e) This is a policy question. You should consider any policy that is used to protect an endangered species or its habitat.
   - Endangered Species Act (EPA)
   - Convention on International Trade in Endangered Species (CITES)
   - Marine Mammals Protection Act (MMA)
   - Migratory Bird Conservation Act

2. (a) This question requires you to read the dissolved oxygen (DO) and biochemical oxygen demand (BOD) graphs of the oxygen sag curve, which show that the biological oxygen demand is greater than the oxygen available. This is because the bacteria and organisms that gravitate to the waste and begin to consume it require great amounts of oxygen and deplete the supply. Therefore, the limiting factor in the second, third, and fourth zones is oxygen. Organisms that require greater levels of oxygen will not be able to survive in these areas, but organisms with wide ranges of tolerance for oxygen availability will survive here.

(b) Again, this question asks you to read the diagram for information. The health of the stream diminishes right after the point source pollution because the pollution eliminates what species are normally found there. You may also reference specific indicator species as those require specific levels of water quality being absent or present in a zone.

(c) This question asks you to explore your knowledge of the wastewater treatment process. There are two different ways to think about how both wastewater treatment plants and wetlands improve water quality: (1) removing solids in primary treatment, and (2) biological treatment during secondary treatment. The physical removal of solids can be done through processes such as filtering, screening, skimming, and settling out suspended solids. The biological treatment is using bacteria and microbes to digest and decompose any waste in the water.

(d) This question also asks you to explore your knowledge of the final stage of the wastewater treatment process. Possible contaminates are microbes/bacteria used during secondary treatment, pathogens, *E.coli*, coliform bacteria, *Giardia*, cholera. Methods for disinfection are chlorine, UV light, lime, or microfiltration.

3. This question is designed to have you consider the many issues associated with waste management. This is also a computational question.

(a) Be aware of the fact that this question is only looking for options that have an environmental impact. If you choose *sanitary landfill,* acceptable environmental reasons include the following:
   • Sanitary landfills produce lower air pollution and little odor.
   • After retirement, methane can be recovered from sanitary landfills to be used as a fuel source.
   • Sanitary landfills still encourage waste reduction and encourage recycling.

   If you choose *incineration,* options for acceptable environmental reasons include the following:
   • Incineration requires less land space.
   • Trash volume is reduced.
   • Incineration reduces the chance of water pollution.

(b) Be aware of the fact that this question is only looking for one economic impact. If you choose *sanitary landfill,* acceptable economic reasons include the following:
   • Sanitary landfills can be built quickly and cheaply, especially if the city has a lot of space.
   • After retirement, methane can be recovered from sanitary landfills to be used as a fuel source and/or sold.

   If you choose *incineration,* acceptable economic reasons include the following:
   • The process of incineration produces electricity, which may produce a profit for the city.
   • Incineration frees up land that can then be used for other businesses.

(c) This is a calculations question.

    i. A $350/week profit means you are earning $50/day. You have two major costs: gas and the driver's salary. Gas = 10 gals/day × $2/gal = $20/day. Driver = $100/day. Total cost per day = $120/day. You'll need to make $170/day for a $50/day profit, and therefore you'll need to make $1,190/week to earn a $350 profit.

    ii. You have already determined from (i) that you'll need to earn $170/day. Since each can is worth 2 cents, or $0.02, you can solve for x in $0.02x = \$170$, so $x = 8,500$ cans per day.

(d) There are several methods that could be described here.

- Educate people on the concept that trash doesn't simply just "go away" as well as on consumption patterns.

- Promote Pay-as-You-Throw programs.

- Develop a compostable waste collection program.

- Limit the amount people are allowed to throw away, or provide disincentives with fees.

- Encourage manufacturers to produce less packaging.

- Outlaw plastic shopping bags or charge a fee for usage.

- Encourage low-waste societies.

CHAPTER 4 DRILL

Multiple Choice

1. **D** A population is said to be undergoing natural selection when a habitat selects certain organisms to live and reproduce, while allowing others to die. Any cause that reduces reproductive success in a portion of the population is a selective pressure, and such selective pressures drive natural selection. Furthermore, natural selection acts upon a whole population rather than an individual organism during its lifespan. Accordingly, statements (I), (II), and (III) are correct, and the correct answer is choice (D).

2. **A** In order to answer this question, you must be aware of the types of ecosystems, their world locations, and the major vegetation associated with each biome. Savannas receive between 10–30 cm of rain, have porous soil that contains only a thin layer of humus and vegetation consisting of grasses with widely spaced trees, and are in Australia, South America, India, and parts of Africa; keep (A). Grasslands receive between 10–60 cm of water, have rich soil and vegetation comprised of sod-forming grasses, and are in the North American plains, Russian steppes, South African velds, and Argentinian pampas; eliminate answer choice (B). Chapparal is located in western North America and the Mediterranean region and has both spiny shrubs and small trees with large, hard evergreen leaves as its major vegetation; eliminate answer choice (C). Deserts receive less than 25 cm of water, have soil characterized by a coarse, sandy texture, contain vegetation that consists of cacti and other low-water adapted plants, and are located 30 degrees north and south of the equator; eliminate answer choice (D).

3. **B** In order to answer this question, you need to know the different types of interspecies interactions and symbiotic relationships. Predation occurs when a predator feeds on prey, driving change in the population size, as seen in the relationship between a coyote and a rabbit; eliminate answer choice (A). Mutualism is a type of symbiotic relationship in which both species benefit, as seen in the relationship between a clown fish and a sea anemone, in which the clown fish protects the anemone from some of its predators and the stinging cells of the anemone protect the clown fish; keep answer choice (B). Commensalism is a type of symbiotic relationship in which one organism benefits while the other is neither helped nor hurt, as exemplified by the relationship between trees and epiphytes; eliminate answer choice (C). Finally, parasitism is a symbiotic relationship in which one species is harmed while the other benefits, such as the relationship between a cat and its fleas; eliminate answer choice (D). Thus, the best option is answer choice (B).

4. **A** In order to answer this question, you need to know the layers of freshwater bodies. The littoral zone is characterized by shallow water at the shoreline, and is home to turtles, frogs, and other species that travel back and forth from water to land; keep answer choice (A). The limnetic zone is the region that extends to the depth that sunlight can penetrate and is home to short-lived organisms that rely on sunlight, such as photosynthesizing plankton; eliminate answer choice (B). The profundal zone consists of water that is too deep for sunlight to penetrate and is home to organisms that have adapted to low-light, low-temperature, and low-oxygen environments; eliminate answer choice (C). Finally, the benthic zone consists of the surface and sub-surface layers, and is home to organisms such as bottom-feeders, scavengers, and decomposers; eliminate answer choice (D). Thus, the correct answer is choice (A).

5. **C** In order to answer this question, you need to know how the ocean zones are divided. The coastal zone, which is closest to land, is characterized by abundant sunlight and oxygen due to its proximity to land; eliminate answer choice (A). The euphotic zone is the warmest region of ocean water and has the highest levels of dissolved oxygen; eliminate answer choice (B). The bathyal zone is the middle region of the ocean, characterized by little sunlight and cool waters; keep answer choice (C). Finally, the abyssal zone is the deepest region of the ocean and is characterized by cold waters, low levels of dissolved oxygen, and no sunlight; eliminate answer choice (D). Accordingly, the best answer choice is (C).

6. **B** This question requires that you know the different phases of the water cycle. Precipitation occurs when water condenses from a gaseous state to form a liquid or solid, becoming dense enough to fall to the Earth; eliminate answer choice (A). Evaporation occurs when water is returned to the atmosphere from both Earth's surface and living organisms; eliminate answer choice (C). Infiltration occurs when water soaks into the soil from the ground level, moving between soil and rocks, to be soaked up by roots to encourage plant growth; eliminate answer choice (D). Transpiration is the process of releasing water vapor into the air, especially through the stomata of plant tissue or the pores of the skin. Accordingly, the best answer choice is (B).

7. **D** While all of the elements listed are essential for successful plant growth, phosphorous is often a limiting factor for plant growth. Indeed, plants that have access to little phosphorous are often stunted. Thus, the correct answer is (D).

8. **C** This question requires knowledge regarding the different types of consumers. An autotroph is an organism that can produce its own organic compounds from inorganic chemicals; eliminate answer choice (A). A decomposer is an organism that consumes dead plant and animal material; eliminate answer choice (B). A detritivore is an organism that gains energy from consuming nonliving organic matter, such as dead animals and fallen leaves; keep answer choice (C). A saprotroph is a decomposer, such as fungi, that uses enzymes to break down dead organisms and absorb their nutrients; eliminate answer choice (D). Accordingly, an earthworm that consumes nonliving organic matter would be best classified as a detritivore, and the correct answer is choice (C).

9. **B** This question requires you to recognize the different ways of depicting relationships among living organisms. The image provided depicts an energy pyramid, in which the amount of energy available at each trophic level is organized from greatest to least. As dictated by the 10% Rule, only about 10% of the energy is transferred from one level to the next. Accordingly, the correct answer is (B).

10. **A** Primary consumers include herbivores that only consume producers, such as plants and algae. Caterpillars, bees, and rabbits are all primary consumers that only consume such producers, while frogs are secondary consumers that also ingest primary consumers, such as insects. Therefore, the correct answer is (A).

11. **C** A number of factors affect the total fertility rates, and therefore birth rate, of a population. The primary factors are the availability of birth control, the demand for children in the labor force, the base level of education for women, the existence of public and/or private retirement systems, and the population's religious beliefs, culture, and traditions. Accordingly, the standard hygiene practices of a population do not have an impact on the population's birth rate, and the correct answer is (C).

Free Response

1. You might need some information from the chapter to answer this question. This will give you an opportunity to see how prepared you are for this exam before you review all of the major topics you'll need to know. How does your answer compare with ours?

(a) The runoff volume would be much greater in the clear-cut plot than the forested plot. In the forested area, the trees help hold water in the soil. Also, the leaves slow down the speed of the raindrops, lessening their impact on the soil. The roots also help bind the soil and hold it in place, so there is no erosion. In the cut area, the water stays closer to the surface, so there is more water for runoff. The rain can fall onto the soil with full force and erosion can take place. The soil is not held together, and the particles have a greater likelihood of moving.

(2 points maximum—1 point for the correct answer and 1 point for a correct explanation)

(b) The phosphate levels would be much higher in the runoff from the cut plot. The phosphate would be leached out of the soil in the clear-cut plot because of all the water running off the soil. Another explanation would be that the trees absorb most of the phosphorus out of the soil. There would be less phosphate in the soil of the forested land, so the runoff would contain less.

(2 points maximum—1 point for the correct answer and 1 point for a correct explanation)

(c) While several negative effects are possible, you only need to provide one here! First, the added nutrients could cause an algal bloom in the stream. The bloom might make the stream less habitable for fish or insect larvae. As the algae decompose, the amount of dissolved oxygen would go down. The sediment might increase the water's turbidity, making it cloudier, and thus lowering the ability of producers to live in the stream. Also, the increased water volume might cause more erosion or possibly flooding farther downstream. Finally, the lack of shade would increase the water's temperature. This increase would lower the dissolved oxygen (DO) levels.

(2 points maximum—1 point for the correct answer and 1 point for a correct explanation)

(d) Because of the very rapid rate of decay and high metabolism of the living plants, there is little organic material in rain forest soil. Anything that falls to the forest floor is quickly decomposed, and the remains are rapidly absorbed by plants. When the forest is cut down, this soil is directly exposed to the large amounts of rain that falls in these forests. The rain quickly washes away the remaining organic matter, leaving even fewer nutrients in the soil.

(4 points maximum—2 points for the correct explanation of why there are few nutrients and 2 points for the correct explanation of the rapid loss of the remaining nutrients)

CHAPTER 5 DRILL

Multiple Choice

1. **D** For this question, you must be aware of the three types of population dispersion: uniform, random, and clumped. Random dispersion occurs when the positions of members of a population are not influenced by the positions of other members of the population; groups of fish are not random occurrences, so eliminate answer choice (A). Uniform dispersion occurs when members of a population are evenly spaced throughout a geographic region; groups, or schools of fish, are neither uniform in population size nor uniformly distributed in an ocean. Answer choice (B) can be eliminated. Linear dispersion is not one of the types of population dispersion, so eliminate answer choice (C). Clumped dispersion occurs when individuals of a population are grouped together in a region; schools or groups of fish that swim together to avoid predation exhibit clumped dispersion. Therefore, the correct answer is (D).

2. **B** Biotic potential refers to the amount a population would grow if there were unlimited resources in its environment; eliminate answer choice (A). Carrying capacity (K) refers to the maximum population size of a certain species in a certain environment that can be sustainably supported by the environment's available resources; keep answer choice (B). Logistic population density is not a term used in population growth; eliminate answer choice (C). Population overshoot occurs when a population exceeds its carrying capacity; eliminate answer choice (D). Thus, the best answer is (B).

3. **C** This question requires that you know the difference between K-selected and r-selected organisms. K-selected organisms have populations that live in stable environments where competition for resources is relatively high, tend to be large, have long life spans, reproduce later in life, have fewer offspring, and experience growth that is limited by the carrying capacity of the environment. Organisms that are r-selected tend to be small and have short lifespans, and they have populations that live in environments where competition for resources is relatively low, reproduce early in life and have many offspring, provide little care to their offspring, and experience population growth that is limited only by the species' own biological limits. Accordingly, humans, lions, and horses would be classified as K-selected organisms, while frogs would be classified as an r-selected organism. Thus, the correct answer is (C).

4. **D** To answer this question, you need to be familiar with r-selected organisms and survivorship curves. Organisms that are r-selected tend to be small and have short lifespans, and they have populations that live in environments where competition for resources is relatively low, reproduce early in life and have many offspring, provide little care to their offspring, and experience population growth that is limited only by the species' own biological limits. Survivorship curves represent the number of individuals in a population born at a given time that remain alive as time goes on; r-selected species tend to follow a Type III curve, while K-selected species tend to follow either a Type I or Type II curve. Accordingly, statements I and III are correct, r-selected species follow a Type III curve and provide little care to their offspring, while statement II is incorrect, as r-selected species produce *high* numbers of offspring that encounter few bottlenecks to survival. Thus, answer choice (D) is correct.

5. **A** In order to answer this question, you must know the formula to calculate actual growth rate: actual growth rate = (birth rate – death rate) ÷ 10. You are told that country Z has a birth rate of 32.4 and a death rate of 9.8. Accordingly, the actual growth rate of country Z is (32.4 – 9.8) ÷ 10 = 22.6 ÷ 10 = 2.26%. The correct answer is (A).

6. **B** The Tragedy of the Commons is a situation in which a common resource is used by populations and then becomes depleted as the populations do not regulate their consumption of the resource; eliminate answer choice (A). The idea that humans might overshoot the carrying capacity of the Earth and suffer some sort of devastation is referred to as the Malthusian catastrophe; keep answer choice (B). The Tragedy of Free Access is the depletion of marine fisheries worldwide; eliminate answer choice (C). The Hadley effect is not an AP Environmental Science term, but a Hadley cell is a system of vertical and horizontal air circulation that creates major weather patterns, predominately in tropical and subtropical regions; eliminate answer choice (D). Thus, the best answer choice is (B).

7. **C** In order to answer this question, you must be familiar with age-structure pyramids, the three categories of age, pre-productive (0–14), reproductive (15–44), and post-reproductive (45+), and how to identify the growth rate of a population given the age-structure pyramid. Given the age-structure pyramid in the question, the population is experiencing greater birth rates than death rates, the population *does* have fewer individuals in the reproductive stage than in the pre-reproductive stage, and the population *does* have more individuals in the reproductive stage than in the post-reproductive stage. However, the population is not experiencing negative growth, which occurs when the post-reproductive population outnumbers both the pre-reproductive and reproductive populations. Accordingly, the correct answer is (C).

8. **D** This question, like many on the AP Environmental Science Exam, requires that you know the demographic transition model. In the preindustrial state, populations exhibit slow growth rates due to harsh living conditions; eliminate answer choice (A). In the transitional state, populations exhibit high growth rates, due to improved environmental conditions; eliminate answer choice (B). In the industrial state, population growth remains fairly high; eliminate answer choice (C). In the postindustrial state, the population reaches zero growth rate and may even experience a negative growth rate. Accordingly, countries like Germany and Russia, which are experiencing zero or negative birth rates, are in the postindustrial state of the demographic transition model; the correct answer is (D).

9. **C** Another question that requires knowledge of the demographic transition model—make sure you really understand the model before test day! In the preindustrial state, a population has a slow rate of growth due to high birth and high death rates; eliminate answer choice (A). In the transitional state, a population experiences increased growth due to high birth rates, but lower death rates due to advances in living conditions; eliminate answer choice (B). In the industrial state, a population experiences a high rate of growth, but the birth rate drops, becoming similar to its death rate; keep answer choice (C). In the postindustrial state, the population approaches and reaches zero growth, even experiencing negative growth, at times; eliminate answer choice (D).

10. **A** Macronutrients, such as proteins, carbohydrates, and fats, are needed in large amounts to keep the human body healthy and resist disease. Micronutrients, such as vitamins, iron, and calcium, are other nutrients needed to maintain one's health. Accordingly, iron is a micronutrient, and the correct answer is (A).

11. **C** Hunger occurs when insufficient calories are taken in to replace those being expanded; eliminate answer choice (A). Undernourishment occurs when people have not been provided with sufficient quantity or quality of nourishment to sustain proper health and growth; eliminate answer choice (B). Malnutrition, which often affects those whose diets lack essential vitamins and other components, is poor nutrition

that results from an insufficient or poorly balanced diet; keep answer choice (C). Starvation occurs when the body experiences a severe lack of food for a prolonged period of time, and the body can no longer maintain itself; eliminate answer choice (D). Thus, the correct answer is (C).

Free Response

1. (a) The carrying capacity is the maximum number of individuals that a habitat can sustain for a long period of time. If a population exceeds the carrying capacity, there will be a die-off of individuals until the population dips below the carrying capacity. When the population is lower than the carrying capacity, the population can begin to increase. Factors that can limit population in a habitat are physical factors (temperature, nutrient availability, amount of light, amount of dissolved oxygen, or pH) and biotic factors (parasites, predators, competitors).

(4 points maximum—2 for the definition and 1 for each correct example)

(b) One example of how nature limits consumption is competition that occurs between two populations for the same habitat. For example, if two different species of animals prey on the same species in the same habitat, they are said to be in direct competition. Sooner or later, the population of prey will be small enough that one predator in the population would not have enough resources. This might cause them to become extinct, to leave that area, or to switch to another food source, thus ending the competition. Some examples of competition are two raptor birds that compete for mice or fish; hunting cats like cheetahs and lions, which compete for grazing animals; or two species of birds that compete for insects.

(4 points maximum—2 points for correctly explaining how consumption can be controlled and 1 point each for 2 correct examples)

(c) Human activities that violate the limits of population growth can include examples of how we harvest natural resources to help grow food; this affects the habitat of certain plant and animal species. Examples of harvesting more natural resources might include irrigation to increase water availability; fertilizers to overcome a lack of certain minerals in the soil; or turning the natural biome into farmland to raise more food crops. We eliminate competitors for our food supply and we use medicines to kill parasites. There are numerous possible correct answers to this question.

(2 points maximum—1 point for each correct explanation of how humans remove competition and exploit resources)

CHAPTER 6 DRILL

Multiple Choice

1. **D** In order to answer this question, you must be aware of the physical spheres that regulate life on Earth. The atmosphere consists of the envelope of gases that surrounds the Earth; eliminate answer choice (A). The hydrosphere consists of the Earth's oceans and freshwater bodies; eliminate answer choice (B). The upper shell of the solid Earth is called the lithosphere, which interacts the most with other spheres; eliminate answer choice (C). The pedosphere consists of the soil mantle that exists on Earth. Accordingly, the correct answer is (D).

2. **C** Scientists believe the Earth to be between 4.5 and 4.8 billion years old. Accordingly, the best answer is choice (C).

3. **C** This question requires that you know the different types of tectonic plate boundary interactions. An interaction at a transform fault boundary occurs when two plates slide against each other in opposite directions; eliminate answer choice (A). A friction boundary is not one of the types of boundary interactions tested on the AP Environmental Science Exam; eliminate answer choice (B). A divergent boundary interaction occurs when two plates move away from each other, creating a gap between plates that may be filled with rising magma that forms a new crust when cool; keep answer choice (C). A convergent boundary interaction occurs when two plates are pushed toward and into each other, causing one of the plates to be pushed deep into the mantle; eliminate answer choice (D). Accordingly, the correct answer is choice (C).

4. **B** For this question, you need to be aware of the four types of volcanoes and their properties. Shield volcanoes generally form over oceanic hot spots and usually have mild eruptions with slow lava flow; eliminate answer choice (A). Cinder volcanoes form at subduction zones and are associated with violent eruptions of lava, water, and gases; keep answer choice (B). Composite volcanoes form when molten lava erupts and cools quickly in the air, hardening into porous rocks that fracture as they hit the Earth; eliminate answer choice (C). Lava domes form from lava that is too viscous to travel far, but instead hardens into a dome shape; eliminate answer choice (D). Accordingly, the correct answer is (B).

5. **A** Make sure you are familiar with the rock cycle for the exam! In the rock cycle, time, pressure, and the Earth's heat interact to create three basic types of rocks: igneous, sedimentary, and metamorphic. Igneous rocks are formed when magma cools and solidifies; basalt is a type of igneous rock. Sedimentary rock is formed as sediment builds up and is compressed under water; limestone is an example of a sedimentary rock. Metamorphic rock is formed when high pressure and heat produce physical and/or chemical changes in existing rock; slate is an example of a metamorphic rock. Accordingly, the first statement is false, as time, pressure, and the *water* do not interact to create igneous, sedimentary, and metamorphic rock. Thus, the correct answer is (A).

6. **D** Here, you need to know about the basic characteristics, types, and properties of soil. Soil plays the pivotal role of protecting water quality by filtering and cleaning the water that moves through it. Soil types can be categorized according to size: clay, which has particles less than 0.002 mm in diameter, silt, with particles 0.002–0.05 mm in diameter, and sand, with particles 0.05–2.0 mm in diameter. Thus, statement I is true. Based on the size of the different types of soil particles, soil that consists mainly of sand particles would *not* be highly compact; eliminate statement II. Finally, the composition of a given soil affects its porosity, permeability, and fertility; thus, statement III is true. Accordingly, the correct answer is (D).

7. **A** The A horizon, which is the topsoil and most intensively weathered soil layer, has a dark color due to the accumulation of organic matter and is known as the zone of leaching; keep answer choice (A). The B horizon, or subsoil, is called the zone of illuviation; eliminate answer choice (B). The E horizon, known as the eluviated horizon, contains no organic matter, and is mostly found in soils developed under forests; eliminate answer choice (C). The O horizon is composed of organic matter at various stages of decomposition, which results in a stable residue known as humus; eliminate answer choice (D). Accordingly, the correct answer is (A).

8. **D** Water, wind, temperature, and living organisms are major agents of weathering, which can be categorized as physical, chemical, or biological. Physical weathering is any process that breaks rock down into smaller pieces without changing the chemistry of the rock. Chemical weathering occurs as a result of chemical reactions of rock with water, air, or dissolved minerals. Biological weathering takes place as the result of the activities of living organisms, which may act through physical or chemical means. Accordingly, temporal weathering is not a type of weathering process that creates soil, and the correct answer is (D).

9. **C** Make sure that you know the order and characteristics of the layers of Earth's atmosphere for the exam! The thermosphere extends between 80–500 km above the Earth and is where colorful auroras like the northern borealis take place; eliminate answer choice (A). The stratosphere extends 20–50 km above the Earth's surface and contains the ozone layer in its lower half; eliminate answer choice (B). The mesosphere is the layer of the Earth's atmosphere that exists between 50–80 km above Earth's surface and is where meteors usually burn up; keep answer choice (C). The troposphere extends from the Earth's surface to 20 km above the surface and is where all the weather we experience takes place; eliminate answer choice (D). Thus, the correct answer is (C).

10. **B** The alternations of atmospheric conditions that lead to El Niño and La Niña are referred to as ENSO (El Niño/Southern Oscillation) events. ELLA events are not a term tested on the AP Environmental Science Exam; eliminate answer choice (A). Typhoons and cyclones are the names of hurricanes when they occur in the Pacific Ocean; eliminate answer choices (C) and (D). Accordingly, the correct answer is (B).

11. **D** Albedo is the percentage of incoming solar radiation reflected by a surface. The lower the surface albedo, the more solar radiation is absorbed. An albedo value of 0 corresponds to zero reflection and absorption of all radiation, whereas an albedo value of 1 corresponds to reflection of all incoming radiation. Snow and ice have high albedo values, and the correct answer is (D).

Free Response

1. (a) Solar radiation comes in through the atmosphere and, when it reaches the surface, some is reflected while most is absorbed, warming the Earth and giving plants the energy they use to make food. The Earth's surface then reemits some of the energy as heat. While some of that heat is lost to space, the greenhouse gases are those that absorb and trap heat in the atmosphere and emit it back toward the surface. More greenhouse gases mean more heat trapped.

(2 points maximum—1 for correct explanation of greenhouse effect and 1 for explanation of the intensification of it)

(b) Reasons for the seasons:
- The Earth's tilt means that sunlight hits more directly on the hemisphere facing the Sun, meaning that (1) there is more sunlight per surface area to be absorbed and (2) the sunlight passes through a shorter distance of atmosphere on its way down, decreasing how much of its energy is absorbed by the atmosphere and increasing how much is left to warm the Earth's surface.

- (3) The axial tilt of the planet also means that the length of daytime is longer during a hemisphere's summer, which increases the amount of time per day that the surface receives solar energy.

More incoming solar energy at a given place and time means higher temperatures, which can change patterns of air movement, affect precipitation, and increase evaporation of water.

(4 points maximum—3 for factors explaining the seasons and 1 for exploration of what hotter temperature means for weather)

(c) The ice in the polar regions has high albedo and reflects the most sunlight back into space. As the icecaps melt, albedo decreases and more solar energy is absorbed. The other effect of melting ice that intensifies the greenhouse effect is that the melting ice releases greenhouse gases!

(2 points maximum—1 for each factor explaining the effects of polar melting on the greenhouse effect)

(d) A hurricane (cyclone, typhoon) is a rotating tropical storm with wind speeds greater than 120 km/hr, that develops when trade winds blow over very warm ocean water, causing the air to warm. An intense low-pressure system forms and the air picks up more and more water vapor from the ocean surface. Wind circles the low-pressure system due to the Coriolis effect, and as the water vapor condenses, the heat energy released increases the wind speed and intensifies the storm.

(2 points maximum—1 for description of what a hurricane is and 1 for description of how it forms)

CHAPTER 7 DRILL

Multiple Choice

1. **D** Approximately 75 percent of the Earth's surface is covered by water; this includes both saltwater and freshwater.

2. **D** Clear-cutting is the removal of all trees from an area at the same time. This is typically done in areas that support fast-growing trees, like pines.

3. **A** Moderate irrigation with groundwater over a long period of time can cause serious problems, including a significant buildup of salts on the soil's surface, which makes the land unusable for crops. This condition is known as salinization.

4. **D** One of the major motivations for cutting down rainforests is to increase the amount of grazing land for cattle and other farm animals. All of the other options are problems associated with deforestation, but (D) benefits humans.

5. **A** Greenbelts are used in urban planning in order to increase green space and control the growth of cities. They are open or forested areas built at the outer edge of a city. Because no growth is permitted in them, they can increase the quality of life for people living near them. Sometimes satellite towns are built outside the greenbelts and connected to the city by highways and mass transportation methods.

6. **C** The Bureau of Land Management is responsible for the management of federal rangeland.

7. **C** The amount of copper ore in the Earth is limited. It is considered a nonrenewable resource. Choices (A), (B), and (D) are considered to be renewable because a renewable resource is one that will be available as long as humans use it in a sustainable manner.

8. **B** In the Tragedy of the Commons, a common resource is used by many people and then becomes depleted as these people do not regulate their consumption of the resource. Some sources say that 75 percent of the world's commercially usable fish are either overfished or are being fished at their maximum sustainable yield.

9. **A** Traditional industrialized agriculture consumes large amounts of energy and other resources. When all aspects are considered (growing crops, processing, and transportation), 17 percent of the United States' total commercial energy use goes into food production. Typically, for every unit of food energy eaten, it takes 10 units of fossil fuel energy to prepare and deliver the food.

10. **B** Pesticide resistance is the only answer choice that represents a problem that results from the use of pesticides. Because pesticides have been used in large amounts, some species of pests have evolved traits that allow them to resist the action of pesticides.

11. **B** Sulfuric acid forms as water seeps through mines and carries off sulfur-containing compounds. The chemical conversion of sulfur-bearing minerals occurs through a combination of biological (bacterial) and inorganic chemical reactions.

12. **C** Sustainability depends on the long-term utilization of resources; this is described in (C). The other options could cause resources to be used up more rapidly.

Free Response

1. Use the checklists below to determine whether your responses are correct. We will use checklists like these when there are many different ways that you could have answered the question. Remember that your answers should be in paragraph form!

(a) Aspects

Positive Aspects	Negative Aspects
Low cost to implement this way of watering	Lots of water lost to evaporation
Little technology and training necessary	Not all areas well-adapted to this technique
Low cost to maintain this way of watering	Delivers more water than plants need
	Water distribution is uneven
	Promotes weed growth along with crops

(2 points maximum—1 point for a correct positive aspect and 1 point for a correct negative aspect)

(b) Alternatives

- Lining canals (Positive: reduces water lost to infiltration; Negative: expensive to do and uses resources)

- Leveling fields (Positive: water flows where needed; Negative: very expensive to level fields)

- Irrigating at night (Positive: avoids evaporation; Negative: requires careful planning and training)

- Irrigating only when necessary (Positive: less waste; Negative: difficult to time)

- Using drip irrigation (Positive: water drops right to roots; Negative: uses resources of plastic)

- Using center pivot (Positive: low waste as water is sprayed directly onto plants; Negative: equipment is very costly and runs on fossil fuels)

  (3 points maximum—1 point for identifying the process, 1 point for a correct positive impact, and 1 point for a correct negative impact)

(c) Possible negative impacts

- Saltwater intrusion: As the aquifer diminishes, nearby ocean water can migrate underground.

- Diminished water for domestic or industrial use

- Subsidence: As water is withdrawn, the soil settles and sinkholes can develop; these can damage buildings and destroy ecosystems.

(2 points maximum—1 point for naming the impact and 1 point for the explanation)

(d) Possible positive and negative effects

Positive Effects	Negative Effects
Production of low-cost electricity	Costly to build
Reservoirs used for many recreational activities	Negative impact on local ecology
Provides flood control	Prevents silt recharging of floodplain
Irrigation water can be controlled	Decrease in fish migration and spawning
Durable	Great danger if breached

(3 points maximum—1 point for each positive and negative impact)

CHAPTER 8 DRILL

Multiple Choice

1. **D** Net energy yield is a comparison between cost of extraction, processing, and transportation and the amount of useful energy derived from the fuel. For example, the net energy ration of natural gas used to heat homes is 4.9 and the net energy yield for electrical heating is 0.3. When the two are compared, the net yield from gas is much higher because the cost of getting the gas to your home is very small compared to the costs of running a nuclear power plant.

2. **D** This question asks you to calculate the amount of light produced by a bulb. You know the amount of energy going into the bulb (1.00 joule) and its efficiency (3 percent). So, 3 percent of 1.00 is 0.03. Choice (D) is the correct answer because the useful energy produced is light, not heat.

3. **C** Both methane and ethanol are created as bacteria break down biomass. An example of this is seen when farmers produce methane from decomposing manure to heat their barns. The manure is placed in underground pits, where bacteria break it down and release methane.

4. **D** The most common hybrid engines are gasoline-electric hybrids. The two engines work together to provide acceleration and power. When the car is driving slowly, less than 56 kph (35 mph), the electric engine powers the car.

5. **D** The only answer choice that describes a waste of energy is leaving room lights on. Most incandescent light bulbs have an energy efficiency rating of 3–5 percent. So, for every 100 units of energy, we only get 5 units of useful light.

6. **A** The correct formula for this problem is 4,500 tons × (2,000 lb/1 ton) × (5,000 BTU/lb) = 4.5×10^{10}. It is much easier to use scientific notation here, so $(4.5 \times 10^3) \times (2 \times 10^3) \times (5 \times 10^3) = 45 \times 10^9$ or 4.5×10^{10}.

7. **C** CO_2 emissions are high in any process that releases energy by burning material. Therefore, burning coal, oil, diesel, and wood all generate CO_2. The processing and reprocessing of uranium into fuel rods generates a moderate amount of CO_2. The only non-combustion generation occurs by wind turbines.

8. **D** WATTS × TIME = kWh. 1,000 watts = 1 kWh, so $\dfrac{500}{1,000} = 0.5 \, kWh$.

9. **B** Photovoltaic cells are constructed from silicon and boron. When sunlight strikes the cells, electrons are energized. These electrons can then flow freely, producing an electric current. By placing wires in the correct positions, this current can be used to power devices. Because the cells are expensive to make, the cost per kilowatt hour is high, but they are useful for applications in which there is no other source of electricity.

10. **C** Radioactive half-life is the amount of time it takes for half of a radioactive sample to degrade. In this question, there are 3 half-lives: the first is 2 curies → 1 curie, the second is 1 curie → 0.5 curie, and the third is 0.5 curie → 0.25 curie. According to the statement, each half-life lasts 20 years. So, 3 × 20 = 60 years.

11. **B** Appliances like microwaves do not have an off switch. They continuously draw power to remain in an "instant on" mode.

12. **D** Coal, oil, and natural gas are burned to generate heat and a nuclear reactor generates heat by radioactive decay. The heat turns water to steam and then steam spins a turbine that in turn spins a generator.

13. **D** The burning of coal causes air pollution in the form of sulfur and nitrogen oxides, which combine with atmospheric water to form acid rain. Acid rain is a leading cause of environmental harm to forests and lakes.

Free Response

1. (a) The core is the place in a nuclear reactor where nuclear reactions occur. It is made of very high-strength steel and other high-performance materials. Inside the core are the coolant, fuel rods, and moderators that control the reaction. The fuel rods are made of enriched uranium U-235. In the reactor, a chain reaction generates the heat, which is converted into steam to drive the electricity-producing turbines. The coolant prevents the melting of the uranium, or core, by removing the heat generated by the chain reaction. Some of the heat is used to turn water to steam that spins turbines. In most reactors, the heat produced in the core does not come in contact with the water that is vaporized; instead, a device called a heat exchanger takes heat from the core (carried by water or molten sodium) and transfers it to water, which vaporizes.

    (2 points maximum—½ point for each correct description)

    (b) Currently, all highly radioactive waste is stored on the grounds of the power plant in deep pools of water. Some materials are stored in steel drums that are housed inside storage containers with walls made of lead and concrete. The United States is currently investigating other possible deep underground storage sites. Low-level radioactive materials are also stored in radioactive landfills.

    (4 points maximum—2 points each for how and where materials are stored)

(c) Because nuclear power plants do not use the combustion of fossil fuels to generate the heat to produce steam, there are no by-products of combustion. There is no particulate produced, no production of oxides of nitrogen or sulfur, and no release of heavy metals. Some CO_2 is generated, but this occurs primarily in the processing of the fuel rods.

(2 points maximum—1 point for describing the lowered emissions and 1 point for a correct example)

(d) Thermal pollution is generated from the turbines, which are powered by steam from the heat exchanger. These hot turbines must be cooled by water piped in from nearby oceans or rivers. This problem is mitigated by allowing the turbine coolant water to flow up a series of pipes built inside cooling towers. As cool air enters the bottom of the tower, it rises by convection and removes the heat from the water circulating in the pipes. The cooler water is then released.

(2 points maximum—1 point for describing how thermal pollution occurs and 1 point for a method to reduce it)

CHAPTER 9 DRILL

Multiple Choice

1. **A** Choices (B), (C), and (D) all contribute directly to the pollution of groundwater. Choice (A) can contribute indirectly if the exhaust is converted into a secondary pollutant (such as acid precipitation).

2. **B** Gray-air smog is a result of the burning of large volumes of coal to generate electricity and provide heat to homes. China uses a great deal of coal to meet its energy needs. The other cities listed have reasonably effective controls on their generation stations, and people in those cities do not use much coal to heat their homes.

3. **D** Sanitary landfills are designed to completely isolate garbage. These landfills contain clay and plastic liners that hold back and collect water as it passes through. The trash is compacted into the smallest area possible and covered with soil and more plastic and clay. As the trash decomposes, the methane generated is collected to be used for other purposes.

4. **C** This process is called eutrophication, and it occurs when organic material (or nutrients) is added to water. The waste supplies food to the bacteria, which thrive using the oxygen for metabolism. This lowers the level of oxygen in the water, which deprives fish and other aquatic animals.

5. **B** The presence of anaerobic organisms indicate that there is little oxygen in a body of water. The presence of trout, perch, and insect larvae would mean that the water was not very polluted. The absence of these indicator species would indicate that the water is polluted.

6. **C** Each person in the United States produces about 725 kgs (1,600 lbs) of waste each year. 54 percent of this waste is put into landfills, 34 percent is recycled, and most of the rest is burned (12 percent).

7. **B** In the correct order, solids are removed by a series of screens first. Next, the water is passed into tanks that contain bacteria. They remove 97 percent of the organic waste. Finally, chlorine is added to kill bacteria and denature viruses.

8. **D** PANs (peroxyacyl nitrates) are secondary pollutants. They are produced from the reaction of hydrocarbons, oxygen, and nitrogen dioxide. All the remaining options are primary pollutants, which are produced from burning fossil fuels.

9. **B** The U.S. government wanted the site in Yucca Mountain, Nevada. President Obama delayed the storage of any radioactive material there, and the Biden administration is presently looking into other options.

10. **A** When nitrogen oxides react with water vapor in the atmosphere, the result is nitric acid, HNO_3.

11. **C** Urban heat islands are created because of the presence of buildings, highways, factories, and automobiles, and the use of lights and machines that warm the surrounding air. This can cause the formation of clouds and can trap pollution near the Earth and prevent it from being diluted.

12. **D** Two of the most common AND most deadly indoor air pollutants in developed countries are radon and tobacco smoke. Both are leading causes of lung cancer. While carbon monoxide, VOCs, and mold spores are also important indoor air pollutants in developed countries, soot from fuel is much more common in developing countries. CO_2 and methane are involved in atmospheric pollution, while nitric and sulfuric acids are components of acid rain.

13. **A** In a dose-response analysis, organisms are exposed to a toxin at different concentrations and the dosage that causes the death of the organism is recorded. The dosage of toxin it takes to kill 50 percent of the test animals is termed LD_{50}. If just the negative health effects are considered, ED_{50} is the point at which 50 percent of the test organisms show a negative effect from the toxin. The threshold dose is the dosage at which any negative effect occurs.

Free Response

1. (a) Compound "X" can enter the pond from surface water runoff that carries the compound. It could also enter the pond by being carried in by rain, snow, or dust that falls into the water. The substance might get carried in by groundwater from a polluted aquifer.

   (2 points maximum—1 point for the correct name and 1 point for a correct description of the process)

   (b) Water (0.1 ppb) → Zooplankton (0.2 ppb) → Small fish (0.1 ppm) → Predatory fish (1.0 ppm) → Hawk (3.0 ppm)

   (2 points maximum—1 point for the correct order and 1 point for the concentrations)

   (c) Bioaccumulation and biomagnification are two of the most important processes to know for the test. In bioaccumulation, fat-soluble molecules accumulate and stay in the fatty tissues of animals since they cannot dissolve in water. Biomagnification occurs when compounds are passed from prey to predator. Since a predator needs to eat a lot of prey, each of the prey organisms gives some of the compound to the predator. The compounds accumulate, and the concentration becomes much higher than you would expect to be in the environment.

   (4 points maximum—2 points for a correct description of each process)

   (d) Mercury and PCBs (polychlorinated biphenyl) are two very common molecules that bioaccumulate and biomagnify. Negative effects of PCBs include skin conditions such as chloracne and rashes, changes in blood and urine that may indicate liver damage, dermal and ocular lesions, irregular menstrual cycles, lowered immune response, fatigue, headache, cough, and poor cognitive development in children. Negative effects of mercury include itching, burning or pain, skin discoloration (pink cheeks, fingertips, and toes), swelling, desquamation (shedding of skin), sweating, tachycardia, increased salivation, and hypertension. Affected children may show red cheeks, nose, and lips; loss of hair, teeth, and nails; transient rashes; muscle weakness; and increased sensitivity to light. Other symptoms may include kidney dysfunction, emotional lability, memory impairment, and insomnia.

(2 points maximum—1 point for the compound and 1 point for a correct symptom associated with the compound)

Use the following lists to check your answers, but remember that you must write all of your answers in paragraph (essay) form.

2. (a)   Primary pollutants include:

Oxides of carbon—CO and CO_2

Oxides of nitrogen—NO and NO_2

Oxides of sulfur—SO_2

Unburned hydrocarbons

All of these primary pollutants are produced by the burning of fossil fuels.

(3 points maximum—1 point for each correct pollutant and 1 point for the correct origin)

(b)   Secondary pollutants include:

Sulfur trioxide—SO_3

Nitric acid—HNO_3

Sulfuric acid—H_2SO_4

Hydrogen peroxide—H_2O_2

Ozone—O_3

Peroxyacyl nitrates or PANs—$R\text{-}C(O)OONO_2$

Aldehydes—R-COH

All of these secondary pollutants are formed in the atmosphere. When primary pollutants combine with water vapor and sunlight energy, the reactions that take place produce the products above.

(3 points maximum—1 point for each correct pollutant and 1 point for a correct explanation of how they are produced)

(c)   Photochemical smog is more likely to be found in industrialized nations because of their extensive use of fossil fuels (mostly coal and oil). Gray-air smog is found in areas where coal is the dominant fossil fuel.

(2 points maximum—1 point for photochemical smog explanation and 1 point for gray-air smog explanation)

(d)   Ozone causes breathing problems, eye irritation, coughing, and reduced immune response. It also aggravates chronic diseases. CO can reduce the oxygen-carrying capacity of blood and cause dizziness and nausea. It can also slow fetal development. NO_2, SO_2, and PANs cause breathing problems, aggravate existing breathing problems, and increase susceptibility to disease.

(2 points maximum—1 point for the irritant and 1 point for the human effect)

CHAPTER 10 DRILL

Multiple Choice

1. **D** Habitat conservation is the only one of these factors that promotes species growth. All the other factors cause a population to decline.

2. **D** The concentration of CO_2 in the atmosphere is increasing more rapidly than that of any other gas. Carbon dioxide is produced during the combustion of fossil fuels, and levels of CO_2 are expected to increase rapidly in the next 20 years. Water vapor also contributes to the greenhouse effect, but its levels have remained consistent for hundreds of years.

3. **C** Because the ozone depletion that has occurred in recent times has, as a primary effect, the allowance of more UV radiation through to the Earth's surface, damage caused by excess UV kills primary producers such as phytoplankton. In turn, this effect ripples up the food chain, decreasing numbers of crops and fish. The effects on human health include eye cataracts, skin cancers, and weakened immune systems.

4. **A** The Montreal Protocol was signed in 1987 by 36 nations concerned with the depletion of the ozone layer by chlorofluorocarbons (CFCs). The pact stated that CFCs had to be reduced by 35 percent between 1989 and 2000. There was a further refinement of the agreement in Copenhagen, Denmark, in 1992.

5. **B** The Endangered Species Act was passed by Congress in 1973 to provide protection for species that are threatened with extinction. This act authorizes federal agencies to undertake conservation programs to protect species listed as endangered or threatened and to purchase land to protect habitats.

6. **D** A cap-and-trade policy limits carbon dioxide in the atmosphere by implementing caps and offering incentives for reducing emissions.

7. **A** WWF is an organization that is not associated with any branch of government.

8. **D** The Earth's polar regions are showing the fastest response times to global climate change, because ice and snow in these regions reflect the most energy back out to space, leading to a positive feedback loop.

9. **D** Ocean warming is the main cause of coral bleaching, but ocean acidification and herbicides are among the several other contributing causes. While changing ocean currents may bring warmer water or pollutants to coral reefs, by themselves they don't contribute to bleaching.

10. **B** The most significant factors that cause extinctions and pose threats to biodiversity are those listed in the acronym HIPPCO: habitat destruction/fragmentation, invasive species, population growth, pollution, climate change, and overharvesting/overexploitation. But other threats, such as monoculture and ozone depletion, also have an impact. Ozone recovery, however, is the gradual replenishment of stratospheric ozone that's projected now that CFC use has been severely curtailed: it should, if anything, have an opposite effect.

Free Response

1. (a) The CAA sets two standards: National Ambient Air Quality Standards for six outdoor pollutants and the national emissions standards for more than 100 toxic pollutants. The NAAQS sets limits on pollutants that people or the environment may be exposed to over a certain period of time. Carbon monoxide, ozone, sulfur dioxide, and nitrogen dioxide are compounds of concern to people near the plant. The second set of standards describes specific limits of hazardous air pollutants. Examples include lead, mercury, and radionuclides.

(2 points maximum—1 to describe the goal of each standard [NAAQS and hazardous air pollutant] and 1 for a correct example of each)

(b)   Positive results include more employment, reliable electricity, lower cost electricity, less dependence on foreign energy sources, greater tax reviews, and a positive impact on the local economy. Negative impacts include more air pollution, more water pollution (thermal is an example), the issue of ash disposal, the need to build infrastructure to support the plant (railroads for coal, roads to and from the plant), and the loss of the natural beauty of the area.

(4 points maximum—1 point for each positive and each negative impact)

(c)   ESA could be used if a threatened or endangered species is found in or near the area where the plant is to be built. The act allows for the conservation of ecosystems that support the endangered or threatened species. It is possible that the plant would not be built in order to conserve the ecosystem in which the species lives.

(2 points maximum—1 point for defining the goal of the ESA and 1 point for stating that the plant could not be built)

(d)   Renewable energy sources and drawbacks

Sources	Drawback
Solar	energy use/pollution involved in manufacture of PV cells
	need for space to array cells
	non-constant source
	cost of setup
Hydroelectric	construction cost
	environmental impact of damming river, including flooding and downriver changes
Wind	need for space for wind farm
	non-constant source
	construction cost
Geothermal	may not be possible in the area
	may release gases
	cost of drilling and upkeep

(2 points maximum—1 for choosing a reasonable source of renewable energy for the plant and 1 for a drawback of that choice)

Part VI
Practice Tests

Practice Test 2

AP® Environmental Science Exam

SECTION I: Multiple-Choice Questions

DO NOT OPEN THIS BOOKLET UNTIL YOU ARE TOLD TO DO SO.

Instructions

At a Glance

Total Time
1 hour and 30 minutes
Number of Questions
80
Percent of Total Grade
60%
Writing Instrument
Pencil required

Section I of this examination contains 80 multiple-choice questions. Fill in only the ovals for numbers 1 through 80 on your answer sheet. A blank answer sheet is located at the back of this book or can be downloaded and printed from your Student Tools.

Indicate all of your answers to the multiple-choice questions on the answer sheet. No credit will be given for anything written in this exam booklet, but you may use the booklet for notes or scratch work. After you have decided which of the suggested answers is best, completely fill in the corresponding oval on the answer sheet. Give only one answer to each question. If you change an answer, be sure that the previous mark is erased completely. Here is a sample question and answer.

Sample Question   Sample Answer

Chicago is a

 (A) state

 (B) city

 (C) country

 (D) continent

Use your time effectively, working as quickly as you can without losing accuracy. Do not spend too much time on any one question. Go on to other questions and come back to the ones you have not answered if you have time. It is not expected that everyone will know the answers to all the multiple-choice questions.

About Guessing

Many candidates wonder whether or not to guess the answers to questions about which they are not certain. Multiple-choice scores are based on the number of questions answered correctly. Points are not deducted for incorrect answers, and no points are awarded for unanswered questions. Because points are not deducted for incorrect answers, you are encouraged to answer all multiple-choice questions. On any questions you do not know the answer to, you should eliminate as many choices as you can, and then select the best answer among the remaining choices.

GO ON TO THE NEXT PAGE.

ENVIRONMENTAL SCIENCE
Section I
Time—1 hour and 30 minutes
80 Questions

Directions: Each of the questions or incomplete statements below is followed by four suggested answers or completions. Select the one that is best in each case and then fill in the corresponding oval on the answer sheet.

Questions 1 and 2 refer to the following map.

The map below shows the major plate boundaries of the world and their boundaries and movement.

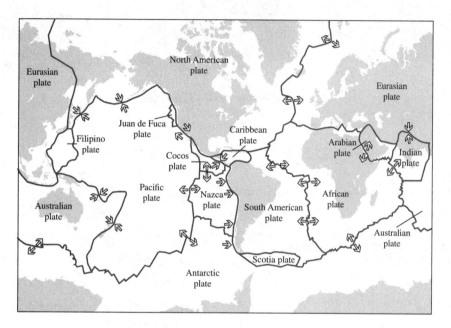

1. Which of the following plate boundaries most likely is characterized by seafloor spreading and rift valleys?

    (A) Nazca–South American boundary

    (B) Australian–Pacific boundary

    (C) Antarctic–Pacific boundary

    (D) Indian-Eurasian boundary

2. Which type of plate boundary typically results in mountain creation, island arcs, earthquakes, and/or volcanoes?

    (A) Convergent boundary

    (B) Divergent boundary

    (C) Transform boundary

    (D) Triple junction

GO ON TO THE NEXT PAGE.

Questions 3 and 4 refer to the following information and diagram.

A ring species is a connected series of neighboring populations, each of which can interbreed with the populations that neighbor it directly, but for which there are at least two end populations that are too distantly related to interbreed.

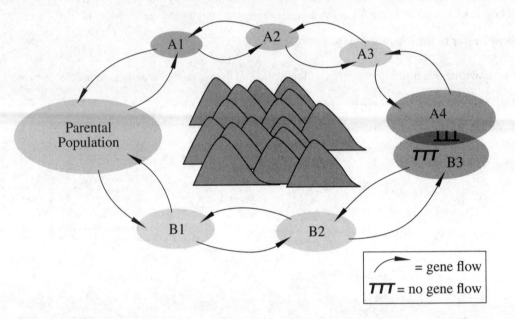

3. For which evolutionary process do ring species provide evidence most directly?

   (A) Speciation

   (B) Natural selection

   (C) Sexual selection

   (D) Extinction

4. Which two populations in the diagram above represent the end populations?

   (A) Parental population and A4

   (B) Parental population and B3

   (C) A1 and B1

   (D) A4 and B3

GO ON TO THE NEXT PAGE.

Questions 5 and 6 refer to the following chart.

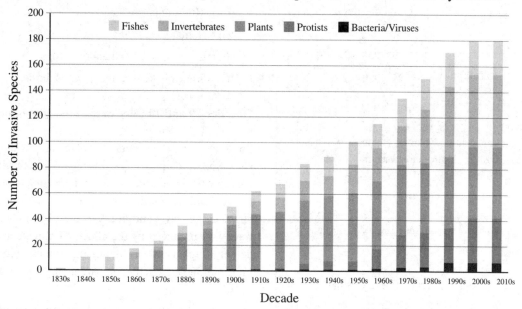

Cumulative Number of Aquatic Invasive Species in the Great Lakes by Decade

5. Based on the chart, which category of invasive species in the Great Lakes showed the greatest increase from the 1950s to the 1960s?

   (A) Invertebrates

   (B) Plants

   (C) Protists

   (D) Bacteria/Viruses

6. Which of the following represents the approximate percentage increase in the total number of invasive species in the Great Lakes between the 1980s and the 2000s?

   (A) 13%

   (B) 20%

   (C) 30%

   (D) 80%

GO ON TO THE NEXT PAGE.

Questions 7 and 8 refer to the following model.

The model below shows an example of primary ecological succession resulting in a deciduous forest.

Ecological Succession

Bare rock
↓
Lichen, Algae, Mosses, Bacteria
(Break down rock and leave organic debris that together form soil)
↓
Grasses
(Add organic matter to soil and anchor it in place)
↓
Small herbaceous plants
(Continue to add organic matter to soil)
↓
Small bushes
(Add shelter and shade for other plants)
↓
Conifers
(Create additional habitats)
↓
Short-lived hardwoods such as dogwood and red maple
(Can tolerate shade of conifers but are short-lived and vulnerable to damage)
↓
Long-lived hardwoods
(More specialized, hardier hardwoods such as oak and hickory)

7. What type of event might result in a bare area where this type of ecological succession could occur?

(A) Fire

(B) Tornado

(C) Clear-cutting

(D) Glacial retreat

8. Which of the following animal species is likely to exist in this area during the second stage (Lichen, Algae, Mosses, Bacteria)?

(A) Mites, ants, spiders

(B) Nematodes, flying insects

(C) Lugs, snails, salamanders, frogs

(D) Squirrels, foxes, mice, moles, birds

Questions 9 and 10 refer to the following information.

Soil is a complex, ancient material teeming with living organisms. Soil development is an intricate dance that involves four basic processes and six soil-forming factors. It takes hundreds to thousands to millions of years for a soil to develop its characteristic layers or profile. Any soil you see is a dynamic formation produced by the effects of climate and biological activity (organisms), as modified by topography (relief) and human influences, acting on parent materials over time.

9. All of the following could be components of the influence of human activity on soil formation EXCEPT

(A) alteration of soil chemistry

(B) compaction

(C) decomposition

(D) salinization

10. Which of the following is true about soils used by humans?

(A) Loamy soils composed of a balanced mixture of clay, silt, and sand are not ideal for plant growth.

(B) Soil is considered a nonrenewable resource.

(C) Monoculture practices in modern agriculture tend to enhance soil quality.

(D) The structure of soil (the extent to which it aggregates or clumps) is unimportant with respect to its arability.

GO ON TO THE NEXT PAGE.

Questions 11–13 refer to the following information.

One effect of global climate change is ocean acidification: the ongoing decrease in the pH of the oceans caused by the increase in atmospheric CO_2, which seawater takes up and dissolves. About 30–40% of the carbon dioxide that's released into the atmosphere due to human activities ends up in seawater and lakes and rivers. CO_2 reacts with water to form carbonic acid (H_2CO_3), and some of the carbonic acid molecules separate into bicarbonate ions (HCO_3^-) and hydrogen ions (H^+). The increase in hydrogen ions is interpreted as an increase in acidity. This increase can upset the balance of marine ecosystems by interfering with metabolism and immune response in some organisms, making it more difficult for organisms like coral and plankton to form calcium carbonate shells, and (along with ocean warming) contributing to coral bleaching and reef die-offs.

11. Which of the following is NOT a possible effect of ocean acidification that may impact human life and industry?

    (A) Coral bleaching

    (B) Disruption of marine food webs

    (C) Release of more CO_2 into the atmosphere

    (D) Putting endangered species more at risk due to habitat loss

12. Which is likely to be the best long-term solution to the problem of ocean acidification?

    (A) Reducing CO_2 emissions

    (B) Reducing overfishing and water pollution

    (C) Reforestation to add more oxygen to the atmosphere

    (D) Feeding iron to phytoplankton to speed up photosynthesis

13. The surface pH of the world's oceans has decreased since preindustrial times by about 0.11 pH units, which indicates an increase of almost 30% in the concentration of H^+ ions. If ocean surface pH is expected to drop by a further 0.3 to 0.5 pH units by 2100, approximately what range of possible increase does that represent in terms of H^+ ion concentration?

    (A) 82%–136%

    (B) 100%–216%

    (C) 173%–355%

    (D) 400%–600%

14. The goal of the second stage of a wastewater treatment plant is to

    (A) remove the large solid material

    (B) aerate the water

    (C) remove chemicals such as DDT or PCBs

    (D) lower the amount of organic material in the water

15. Which of the following organisms are the first to be adversely affected by thermal pollution in a stream?

    (A) Trees along the bank

    (B) Insect larvae in the water

    (C) Large fish migrating upstream

    (D) Bacteria in the water

GO ON TO THE NEXT PAGE.

Questions 16–19 refer to the following information and graph.

A scientist placed 100 fish eggs into each of seven solutions with different pH values. After 96 hours, the number of survivors was counted and converted into a percentage. The survival percentages are given in the graph below.

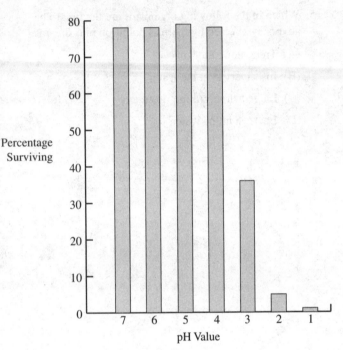

16. Which of the values below best represents the LD$_{50}$ in this experiment?

    (A) 4.0

    (B) 4.5

    (C) 3.0

    (D) 2.5

17. Which of the following, if added to the solutions, would lead to an increase in pH?

    (A) Limestone

    (B) Carbon dioxide

    (C) Nitrogen dioxide

    (D) Sulfur dioxide

18. Which of the following best describes the goal of the experiment?

    (A) To observe how many fish would hatch at different pH values

    (B) To find out how many fish live in streams with different pH values

    (C) To understand how acid rain affects life in streams

    (D) To see what chemical is best at changing the pH of water

19. The pH value is a measure of the

    (A) amount of heavy metals in the water

    (B) biochemical oxygen demand (BOD) of the water

    (C) concentration of oxygen in the water

    (D) concentration of hydrogen ions in the water

20. Which of the following laws created the Superfund program?

    (A) Comprehensive Environmental Response, Compensation, and Liability Act

    (B) Resource Conservation and Recovery Act

    (C) Clean Air Act

    (D) National Environmental Policy Act

21. Late fall frosts and the northward migration of some tree and plant species may indicate which of the following global changes?

    (A) Increased global temperatures

    (B) The effects of more ultraviolet light from the sun

    (C) A reduction in the volume of ice at the North and South Poles

    (D) Changes in global precipitation patterns

22. High infant mortality rates are likely to exist in countries that have

    (A) a strong and stable economy

    (B) high levels of education for adults

    (C) a stable food supply

    (D) high levels of infectious diseases

GO ON TO THE NEXT PAGE.

23. All of the following statements are true EXCEPT

    (A) energy can be converted from one form to another

    (B) energy input always equals energy output

    (C) energy and matter can generally be converted into each other

    (D) at each step of an energy transformation, some energy is lost to heat

24. Oxygen-depleted zones of the oceans, such as the one at the mouth of the Mississippi River, are most likely caused by

    (A) a reduction in the plant life in rivers that empty into the ocean near the dead zone

    (B) excessive fertilizers carried into the ocean, which cause algal blooms that lower the oxygen levels

    (C) thermal pollution in the ocean

    (D) acid precipitation falling onto the ocean

25. One potential benefit of using genetically modified foods is

    (A) the release of genes to other plants or animals

    (B) the resistance of crops against pesticides

    (C) their growth in monoculture leads to a reduction in biodiversity in the area

    (D) unknown effects on the ecosystem into which they are released

26. Which of the following compounds would probably supply the greatest amount of useful energy to humans?

    (A) The exhaust from a car

    (B) Unrefined aluminum ore

    (C) A glass bottle

    (D) A liter of gasoline

27. Which of the following choices gives the geologic eras in the correct sequence, from the oldest to the most recent?

    (A) Cenozoic—Mesozoic—Paleozoic—Precambrian

    (B) Precambrian—Paleozoic—Mesozoic—Cenozoic

    (C) Paleozoic—Precambrian—Cenozoic—Mesozoic

    (D) Paleozoic—Cenozoic—Precambrian—Mesozoic

28. Approximately what percentage of the world's solid waste does the United States produce?

    (A) 50 percent

    (B) 33 percent

    (C) 10 percent

    (D) 5 percent

29. Which of the following correctly describes conservation easement?

    (A) A process that conserves soil from erosion

    (B) A binding agreement that preserves land from further development in exchange for tax write-offs

    (C) An agreement that allows a developer to add new land to a housing project with little input from neighbors

    (D) A practice that prevents the breakdown of stream banks

30. The highest priority of the Clean Water Act is to provide

    (A) guidance in toxic chemical disposal

    (B) funds to reclaim old strip mines

    (C) policies to lessen the number of oil spills in the ocean

    (D) policies to attain fishable and swimmable waters in the United States

31. Which of the following best describes changes in the genetic composition of a population over many generations?

    (A) Evolution

    (B) Mutation

    (C) Natural selection

    (D) Biomagnification

32. Women have fewer and healthier children when all of the following are true EXCEPT

    (A) they have little education

    (B) they live where their rights are not suppressed

    (C) they have access to medicine and health care

    (D) the cost of a child's education is high

GO ON TO THE NEXT PAGE.

33. Which of the following is a true statement?

    (A) The population size of organisms following a logical population growth is not limited by a carrying capacity.

    (B) The population size of organisms following an exponential growth is limited by a carrying capacity.

    (C) Increasing the death rate of a population can lower the carrying capacity.

    (D) Increasing the rate of food production can increase the carrying capacity.

34. An increase in the amount of UV light striking the Earth as a result of ozone loss will cause which of the following?

    (A) Global climate change

    (B) Increased skin cancer rates in humans

    (C) Lowering of ocean water levels

    (D) An increase in CO_2 in the atmosphere

35. Ozone in the troposphere can result in all of the following EXCEPT

    (A) eye irritation

    (B) lung cancer

    (C) bronchitis

    (D) headache

36. Which of the following describes the amount of energy that plants pass on to herbivores?

    (A) The amount of solar energy in a biome

    (B) The First Law of Thermodynamics

    (C) The Net Primary Productivity (NPP) of an area

    (D) The Second Law of Thermodynamics

37. The Second Law of Thermodynamics relates to living organisms because it explains why

    (A) matter is never destroyed but it can change shape

    (B) plants need sunlight in order to survive

    (C) all living things must have a constant supply of energy in the form of food

    (D) the amount of energy flowing into an ecosystem is the same as the amount flowing out of that system

38. Acid deposition most severely affects amphibian species because amphibians

    (A) do not care for their young

    (B) are not mammals

    (C) need to live in both terrestrial and aquatic habitats

    (D) seldom reproduce

39. All of the following are internal costs of an automobile EXCEPT

    (A) car insurance

    (B) fuel

    (C) pollution and health care costs

    (D) raw materials and labor

40. Scrubbers are devices installed in smokestacks to

    (A) reduce the amount of materials such as SO_2 in the smoke they discharge

    (B) clean out the stack so smoke can move rapidly upward

    (C) reduce the amount of sulfur in coal before it is burned

    (D) reduce the amount of toxic ash produced

41. After ore is mined, the unusable part that remains is placed in piles called

    (A) overburden

    (B) seam waste

    (C) leachate

    (D) tailings

42. All of the following are examples of externalities EXCEPT

    (A) a construction worker purchasing an automobile that reduces his commute time to work

    (B) a factory producing air pollution that leads to acid rain in the neighboring forest area

    (C) the construction of a new football stadium leading to increased income for local businesses

    (D) driving electric and hybrid vehicles reducing the amount of greenhouse gas emissions

GO ON TO THE NEXT PAGE.

43. Which fuel contains the greatest amount of sulfur?

    (A) Wood

    (B) Natural gas

    (C) Nuclear reactor fuel rods

    (D) Coal

44. Biological reserves are areas that allow countries to

    (A) concentrate agricultural production in one area

    (B) set aside critical habitats to ensure the survival of species

    (C) control the flow of rivers and storm waters

    (D) provide grazing land in order to ensure economic growth

45. Which of the following countries has the shortest population doubling time?

    (A) Denmark

    (B) Australia

    (C) United States

    (D) Kenya

46. Which of the following processes leads to an increase in atmospheric water content?

    (A) Precipitation

    (B) Transpiration

    (C) Infiltration

    (D) Condensation

47. Full cost pricing of a refrigerator would include

    (A) adding the cost of employee salaries to the total cost

    (B) the refrigerator's total impact on the environment

    (C) the cost of transporting the refrigerator to the retail store

    (D) the value of the refrigerator if it was donated to a nonprofit group

48. Since 2000, the atmospheric concentration of which of the following has decreased as a result of anthropogenic activity?

    (A) Methane

    (B) Carbon dioxide

    (C) Chlorofluorocarbons

    (D) Nitrous oxide

49. During a society's postindustrial state, the population will exhibit

    (A) rapid growth with a low birth rate and high death rate

    (B) slow growth with a slowing birth rate and a low death rate

    (C) rapid growth with a high birth rate and low death rate

    (D) zero growth with a low birth rate and low death rate

50. Which of the following is NOT a disadvantage of old-style landfills?

    (A) They generate gases that can be recovered and used as fuel.

    (B) Bad odors come from these landfills.

    (C) Toxic wastes leach into ground water.

    (D) Subsidence of the land occurs after the landfill is filled.

51. The international treaty concerning endangered species (CITES) has tried to protect endangered species by taking which of the following steps?

    (A) Making more countries keep these species in zoos

    (B) Paying the debts of member countries in order to relieve the pressure to sell endangered species

    (C) Developing a list of endangered species and prohibiting trade in those species

    (D) Restoring endangered habitats

52. In 2000, the population of Country A was 50,000. If Country A has a constant birth rate of 33 per 1,000 and a constant death rate of 13 per 1,000, when will the population of Country A equal 200,000?

    (A) 2035

    (B) 2050

    (C) 2070

    (D) 2105

GO ON TO THE NEXT PAGE.

53. Salt intrusion into freshwater aquifers, beach erosion, and the disruption of coastal fisheries are all possible results of which of the following?

(A) Rising ocean levels as a result of global warming

(B) More solar ultraviolet radiation on the Earth

(C) More chlorofluorocarbons in the atmosphere

(D) Reduced rates of photosynthesis

54. The chemical actions that produce compost would best be described as

(A) photosynthesis

(B) augmentation

(C) respiration

(D) decomposition

55. Which of the sources below would produce non-point source pollution?

(A) The smokestack of a factory

(B) A volcano

(C) A pipe leading into a river from a sewage treatment plant

(D) A large area of farmland near a river

56. A nation's gross domestic product represents

(A) the ability to provide health care

(B) the amount of goods it imports

(C) the amount of its economic output

(D) the quality of its environment

57. Which of the following mining operations requires people and machinery to operate underground?

(A) Mountain top removal

(B) Contour stripping

(C) Dredging

(D) Shaft sinking

58. A country's total fertility rate (TFR) indicates which of the following?

(A) The life expectancy of women in the country

(B) The average number of babies born to women between the ages of 14 and 45

(C) The number of babies under one year of age who die per 1,000

(D) The total use of contraceptives in the country

59. The wastes stored in Love Canal contaminated the surrounding area by all of the following methods EXCEPT

(A) leaching into the ground water

(B) fumes from burning the wastes

(C) runoff into a nearby stream

(D) spilled drums of waste

60. All of the following are currently used by humans to produce energy EXCEPT

(A) nuclear fusion

(B) nuclear fission

(C) harnessing solar energy

(D) harnessing the Earth's internal heat

61. In sea water, carbon is mostly found in the form of

(A) phosphoric acid

(B) carbon disulfide

(C) bicarbonate ions

(D) methane gas

62. Acid rain and snow harm some areas more than others because certain areas

(A) have more bacteria in the soil than others

(B) have a lesser ability to neutralize the acids

(C) are at a higher elevation than the unaffected areas

(D) are closer to lakes than are the unaffected areas

63. Which one of the following does NOT store a great deal of phosphorus?

(A) Rocks

(B) Water

(C) Atmosphere

(D) Living organisms

64. The addition of oxygen to the early Earth's atmosphere most likely occurred through the process of

(A) volcanic outgassing

(B) photosynthesis

(C) meteorite impact

(D) respiration by animals

GO ON TO THE NEXT PAGE.

65. Scientists use which of the following to estimate environmental risks to humans?

  I.   Animal studies
  II. Epidemiological studies
  III. Statistical probabilities

  (A) II only

  (B) I and II only

  (C) I and III only

  (D) I, II, and III

Questions 66–69 refer to the following graph.

A group of students did a Biological Oxygen Demand (BOD) study along a 30-mile section of a stream. The data they obtained is given in the graph below.

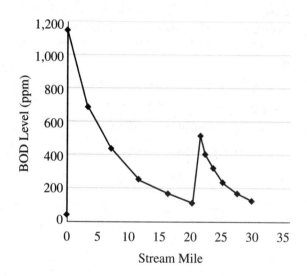

66. Which of the following best describes the type of pollution at mile 0?

  (A) Point source

  (B) Acid deposition

  (C) Secondary pollutant

  (D) Deep well

67. The BOD at mile 12 is approximately

  (A) 700 ppm

  (B) 220 ppm

  (C) 200 ppm

  (D) 175 ppm

68. The BOD test is designed to directly measure

  (A) how much light can pass to the bottom of the stream

  (B) the amount of nitrates in the water

  (C) the amounts of coliform bacteria

  (D) the rate at which oxygen is being consumed by microorganisms

69. Anaerobic bacteria, sludge worms, and fungi are most likely to be found in which part of this stream?

  (A) 0 to 5 miles

  (B) 10 to 15 miles

  (C) 15 to 20 miles

  (D) 25 to 30 miles

70. Riparian zones are important parts of lands because they are

  (A) the area where most cattle feed when they graze

  (B) an area of diverse habitats along the banks of rivers

  (C) important buffers against wind

  (D) areas where varying amounts of light cause different layers of plant growth

71. Which of the following is a disadvantage of fish farming?

  (A) It can allow for genetic engineering, which leads to bigger yields.

  (B) It is very profitable.

  (C) It can lead to large die-offs due to disease.

  (D) It can reduce the pressure to harvest wild species.

72. The form of nitrogen that plants can use directly is

  (A) nitrates

  (B) nitrites

  (C) N_2 gas

  (D) methane

GO ON TO THE NEXT PAGE.

73. Which of the following best describes the effects of a thermal inversion?

    (A) Cold ocean water moves to the surface and warm water sinks.

    (B) Warm, polluted air rises and mixes with cool upper air, and pollutants escape.

    (C) Warm river water cools when it enters the ocean.

    (D) Polluted air at the surface cannot rise because it is blocked by warm air above it.

74. Shifting taxes to tax pollution and waste rather than taxing the cost of products will allow people to

    (A) maximize profit

    (B) increase the tax base in a city

    (C) use tax money for local schools

    (D) shift to a pattern of more sustainable development

75. Which of the following molecules is most damaging to stratospheric ozone?

    (A) H_2O

    (B) CO_2

    (C) Chlorofluorocarbons

    (D) N_2O

76. Which of the following ocean zones has the highest levels of nutrients?

    (A) Coastal zone

    (B) Euphotic zone

    (C) Bathyal zone

    (D) Abyssal zone

77. Samples of atmospheric gases from past eras can most easily be obtained from which of the following sources?

    (A) Methane gas trapped in oil reserves

    (B) Different types of sedimentary rock

    (C) Gases trapped in polar ice caps

    (D) Mud samples from eutrophic lakes

78. Acid deposition on soil kills beneficial decomposers. Which of the following cycles would be most affected by the loss of decomposers?

    (A) Sulfur cycling

    (B) Phosphorus cycling

    (C) Hydrologic cycling

    (D) Nitrogen cycling

79. Which of the following is a trace element necessary for plant growth?

    (A) Carbon

    (B) Nitrogen

    (C) Phosphorous

    (D) Magnesium

80. Concerns that people of color and poor people are disproportionately exposed to environmental pollution are most likely to be addressed by people who believe in the

    (A) Earth stewardship view

    (B) planetary manager view

    (C) environmental justice movement

    (D) sustainability point of view

END OF SECTION I

ENVIRONMENTAL SCIENCE
SECTION II
Time—1 hour and 10 minutes
3 Questions

Directions: Answer all three questions, which are weighted equally; the suggested time is about 23 minutes for answering each question. Where calculations are required, clearly show how you arrived at your answer. Where explanation or discussion is required, support your answers with relevant information and/or specific examples.

1. The following editorial is excerpted from a recent edition of the Hilltop Express:

Hilltop Express

New Pests Invade Farm

A new species of corn-infesting insect has recently been discovered in a local farmer's field. Bill Jones stated: "Last week a section of my corn field was covered in small black beetles. They can fly from plant to plant, and they eat large holes in the leaves. I called the county extension agent Sarah Smith and she came out and identified them. I'm going to start spraying tomorrow morning." In a telephone interview with Sarah, she stated that this species was new to the county and has the potential for causing real damage to the corn crop. She stated that the adults do most of the damage to growing leaves.

The grubs live near the base of the plant and feed on bacteria and other organisms living in the soil. She added that the beetle was resistant to the most common pesticide, NOGrub. NOGrub, she commented, had been tried in another county and was not found to be effective. The editors of the Hilltop Express realize the potential dangers to the county's most important cash crop. We urge the county agents to recommend a series of new pesticide treatments to control this new menace to our livelihood.

(a) **Explain** how the beetle might have become resistant to NOGrub. Assume that NOGrub had been applied to a population of beetles in another county.

(b) If the county agents do not have information on which pesticide is most effective against the beetle, the county plans to investigate by trying several pesticides on controlled sections of beetle-infested crop.
   i. **Identify** the independent and dependent variables in such a trial.
   ii. **Describe** an effective control group and the experimental groups for this experiment.
   iii. **Identify** TWO environmental factors that must be controlled to keep the experiment from producing skewed results.

(c) **Identify** TWO negative impacts of using chemical pesticides on the surrounding ecosystem.

(d) One strategy for combating pests is Integrated Pest Management (IPM).
   i. **Describe** IPM.
   ii. **Identify** one benefit and one difficulty the county would likely encounter in using IPM to control this outbreak.

GO ON TO THE NEXT PAGE.

2. The map below shows two cities: City *X* and City *Z*, separated by several kilometers.

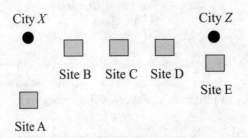

Students from a high school in between the two cities studied soil pH values at the sites labeled A through E on the map. The results of the pH study are given in the following table:

Site	pH value
A	6.2
B	5.6
C	5.0
D	4.5
E	4.3

(a) Refer to the table and diagram above to answer the following questions.
    i.   **Describe** one point source for the pollution that caused the change in the soil's pH as shown.
    ii.   In the description, **identify** a fuel that could create the pollution.

(b) Assume that the fuel identified in (a)(ii) is the source of the pollution.
    i.   **Identify** one primary and one secondary pollutant that can cause the change in the soil's pH.
    ii.   **Describe** the process that causes the change in the pH.

(c) **Propose** one possible method to reduce the air pollutants that are causing the pH change.

(d) **Identify** and **describe** one provision of the Clean Air Act of 1990 that could be used to control and reduce the emissions.

GO ON TO THE NEXT PAGE.

3. According to the United States Energy Information Administration, the consumption of natural gas by the United States increases at 8 percent per year. The U.S. receives its supplies from a variety of international and domestic locations. Natural gas is used in the home, for industry, and for power generation.

   (a) **Calculate** the approximate number of years it would take to double the consumption of natural gas in the United States. **Show** all your work.

   (b) **Identify** one method by which natural gas is recovered and transported.

   (c) **Describe** TWO benefits to the environment that would occur if the United States switched from coal to natural gas-fired electric power generation.

   (d) Some people advocate increasing the use of coal versus natural gas for the production of electricity. Give one argument that the proponents of coal might use to **justify** their position.

   (e) Others advocate for non-hydrocarbon fuel alternatives.
      i.   **Identify** one non-hydrocarbon fuel alternative.
      ii. **Describe** one drawback of the alternative identified in (e)(i).

STOP

END OF EXAM

_____

Practice Test 2: Answers and Explanations

PRACTICE TEST 2 ANSWER KEY

1.	C	21.	A	41.	D	61.	C
2.	A	22.	D	42.	A	62.	B
3.	A	23.	C	43.	D	63.	C
4.	D	24.	B	44.	B	64.	B
5.	C	25.	B	45.	D	65.	D
6.	B	26.	D	46.	B	66.	A
7.	D	27.	B	47.	B	67.	B
8.	A	28.	B	48.	C	68.	D
9.	C	29.	B	49.	D	69.	A
10.	.B	30.	D	50.	A	70.	B
11.	C	31.	A	51.	C	71.	C
12.	A	32.	A	52.	C	72.	A
13.	B	33.	D	53.	A	73.	D
14.	D	34.	B	54.	D	74.	D
15.	B	35.	B	55.	D	75.	C
16.	C	36.	C	56.	C	76.	D
17.	A	37.	C	57.	D	77.	C
18.	A	38.	C	58.	B	78.	D
19.	D	39.	C	59.	B	79.	D
20.	A	40.	A	60.	A	80.	C

PRACTICE TEST 2: ANSWERS AND EXPLANATIONS

Section I—Multiple-Choice Questions

1. **C** The map shows whether plates at each boundary are moving apart (divergent) or toward each other (convergent). A boundary characterized by seafloor spreading and rift valleys should be a divergent boundary. Since (A), (B), and (D) are convergent boundaries, only (C) can be correct.

2. **A** Mountain creation is a result of orogenic belts (also called collision boundaries), which are convergent boundaries between two continental plates. Subduction zones, or convergent boundaries between oceanic and continental plates, usually cause volcanoes to form, along with deep trenches. Convergent boundaries between two oceanic plates can form island chains. All of these types can result in earthquakes due to the compression and tension between colliding plates.

3. **A** Speciation is the process by which populations evolve to become distinct species. A ring species is an example of a species that is "caught in the act" of speciation: the end populations. If looked at in the absence of the other populations, they would be considered distinct species because of their inability to interbreed. However, the connectedness of the ring shows that the parent population was originally one species.

4. **D** The passage states that the end populations are those *that are too distantly related to interbreed*. Looking at the key, you can see that there is gene flow between each set of neighboring populations except for A4 and B3. There's no evidence in the diagram as to whether the parental population can interbreed with other populations besides those it neighbors.

5. **C** Looking at the chart, you can see that for most types of organisms there was no significant increase between the bars for the 1950s and the 1960s. However, the protists group just about doubled. Therefore, protists showed the greatest increase between those decades.

6. **B** To find a percentage change, divide the difference by the original amount, and then multiply by a factor of 100 to get the value into percentage form. The total number of invasive species in the Great Lakes in the 1980s was about 150, while in the 2000s it was about 180. The difference is 180 − 150 = 30. So, divide: 30 ÷ 150 = 0.2. This is equivalent to 20%.

7. **D** Primary succession begins in virtually lifeless areas, such as those that result from lava flows or glacial retreat. Secondary succession takes place where an existing community has been cleared by a disturbance event (such as a fire, tornado, or human impact) but the soil has been left intact. Since the description tells you this example is of primary ecological succession (and the model begins with bare rock, not soil), only (D) can be correct.

8. **A** While ecological succession theory was developed mostly by botanists, the stages include animal species as well as plants. The most likely species to survive in an early stage such as that dominated by lichens, algae, mosses, and bacteria are the smallest: tiny insects that can feed on these primary producers and live in the cracks between them. Therefore, (A) is the best choice.

9. **C** The effects of human activity must increasingly be acknowledged as a factor in soil development. Human use of fertilizers and pesticides, and pollution and its side effects such as acid rain, can alter soil chemistry. Roads with vehicle traffic and heavy machinery for agriculture and construction can compact soil, and irrigation and depletion of groundwater can cause salinization. Decomposition, on the other hand, is the province of other organisms such as bacteria and fungi that live within the soil, so (C) does NOT represent a human influence.

10. **B** The question asks which statement is true: only (B) is correct. Soil is considered nonrenewable because it requires a great length of time to form. Loamy soils are considered ideal for plant growth; monocultural agriculture tends to decrease soil quality by decreasing biodiversity and leaching specific nutrients, and soil needs good structure to be arable.

11. **C** Coral bleaching, (A), affects whole ecosystems built around coral reefs, which can impact human fishing (as well as leisure activities such as snorkeling). That's one example of the larger category of problems that (B) represents: any marine ecosystem that is disrupted by acidification can become imbalanced and this affects its food webs, which directly impacts human fishing. Choice (D) affects humans less directly, but habitat loss causing further endangerment to already-threatened species may cause extinctions that could have big repercussions to ecosystems and food chains. However, release of more CO_2 into the atmosphere is not a potential problematic effect of ocean acidification. CO_2 emissions cause ocean acidification, not the other way around, and acidification actually removes CO_2 from the atmosphere ((C) is correct).

12. **A** The only effective solution currently considered for the problem of ocean acidification is the reduction of CO_2 emissions on a global scale, (A). This makes sense, since the problem is a direct result of these emissions. Reducing overfishing and water pollution, (B), will certainly help with other problems and help increase resilience of marine organisms to acidification, but doesn't actually address the problem itself. Reforestation, (C), will help with air quality but won't directly impact CO_2 emissions or ocean acidification. Finally, iron fertilization, (D), has been considered as a possible mitigation approach to counter the effects of acidification, but hasn't been shown to be effective or free from unwanted side effects. If you don't know anything about iron fertilization, you can still tell that (A) is the best choice because you know it directly addresses the problem at hand.

13. **B** Keep in mind that pH is a logarithmic scale. If a decrease of 0.11 pH units corresponds to an increase of almost 30% in H^+ ions, that means that $\log_{10} x - \log_{10} y \approx 0.11$. Remember that $\log_b x - \log_b y = \log_b \frac{x}{y}$, so $\log_{10} \frac{x}{y} \approx 0.11$; in other words, $10^{0.11} \approx \frac{x}{y}$. So $\frac{x}{y} \approx 1.28824....$ and you can see that that is equivalent to almost a 30% increase. So use the same process to find equivalent percentage increases for the range of pH decreases given: if the pH drops by 0.3 units, then $10^{0.3} \approx \frac{x}{y}$, so $\frac{x}{y} \approx 1.9953$, which is approximately a 100% increase. And if the pH drops by 0.5 units, then $10^{0.5} \approx \frac{x}{y}$, so $\frac{x}{y} \approx 3.1623$, which is approximately a 216% increase. Choice (B) is correct.

14. **D** Wastewater treatment plants use a three-step process in removing waste from water. The main goals of sewage treatment are to remove the solid waste and reduce the biological oxygen demand (BOD). This BOD is a measure of how much organic material is in the water and how much bacteria can live in the

water. The first stage of wastewater treatment is the mechanical removal of solid objects, and screens of various sizes are used to perform this task. In the second stage, the water is sprayed on a bed of rocks that harbor billions of bacteria. The bacteria break down the organic molecules and lower the BOD. After the remaining liquid is treated, chlorine is put into the water to control bacteria populations.

15. **B** Insect larvae are the organisms most vulnerable to changes in abiotic factors like thermal pollution. As is true with the young of most species, the larval forms constitute the most vulnerable stage of an insect's life cycle. Species whose populations rise and fall with changes in abiotic factors are called *indicator species*. The presence of an indicator species in a certain ecosystem tells ecologists that the stream water is ideal to live in—if a fragile indicator species inhabits the region, then it must be safe for all species. If the indicator species is not there, then the water is not optimally clean.

16. **C** The LD_{50} is the value at which 50% of a population dies. If you read the graph, you can see that that value falls closest to the solution with the pH of 3.0. LD_{50} values are important because they help us understand the health risks of certain materials. As you might imagine, if a chemical is very toxic, it has a very low LD_{50}. For example, the nerve agent VX has an LD_{50} of 0.14 mg/kg body weight! It is extremely harmful to humans: this LD_{50} value tells you that it only takes 10.2 mg of nerve agent VX to kill half of the test population of 155-pound males! For the test, just remember that the smaller the LD_{50} number, the more toxic a chemical is.

17. **A** Choices (B), (C), and (D) are all pollutants involved in the production of acid precipitation that would lead to a decrease in the pH, not an increase. Therefore, these choices are incorrect. Limestone, (A), is a base and would increase the pH of the solution. Choice (A) is correct.

18. **A** The purpose of this test is to observe how many fish hatch at different pH values, plain and simple. If you chose any other answer, then you might have been reading too much into the experiment. Remember not to make any inferences—you must answer the question only on the basis of the information you're given. In this experiment, the survival percentage is the dependent variable, and the pH is the independent variable. The data collected relate to hatching survival rates. pH is one of the most important abiotic factors for aquatic organisms. If the pH value shifts out of the ideal range, then the essential proteins in the cells that constitute fish eggs can denature. If that happens, the young will not survive.

19. **D** Only (D) correctly defines pH. pH is the measure of the number of hydrogen ions in a solution. Chemically speaking, it is the negative logarithm (base 10) of the hydrogen ion concentration. So, if a solution has 10^{-7} hydrogen ions in it, then $-\log 10^{-7} = -(-7) = 7$. The pH of the solution is 7; the solution is neutral. The pH scale runs from 0–14; acidic solutions have a pH less than 7, while basic solutions have a pH greater than 7. Logarithms are used to keep the numbers simple because the range of hydrogen ion concentrations can be as high as 100 trillion!

20. **A** The Comprehensive Environmental Response, Compensation, and Liability Act (CERCLA) is a law that was passed to establish the "Superfund." This law created a tax on the chemical and petroleum industries and provides broad federal authority to respond directly to releases or threatened releases of hazardous substances that may endanger public health or the environment. Choice (B) is incorrect:

RCRA established a system for managing non-hazardous and hazardous solid wastes in an environmentally sound manner. Specifically, it provides for the management of hazardous wastes from their point of origin to their point of final disposal ("cradle to grave"). RCRA also promotes resource recovery and waste minimization. Choice (C) is responsible for air pollution control. Finally, (D) mandates that environmental impact studies be done before major construction projects are started.

21. **A** Increased global temperatures are responsible for the phenomena in most of the other answer choices! Global warming is more formally referred to as tropospheric warming. You learned that the atmosphere contains much H_2O vapor as well as trace amounts of CO_2, CH_4, and NO_2. These gases reflect infrared radiation that is, itself, reflected off the Earth's surface as it is warmed by the sun. This process of tropospheric warming is called the "greenhouse effect," and it is a natural phenomenon. However, as humans burn huge amounts of fossil fuels, the excess CO_2 and NO_2 enter the atmosphere. These gases increase the heat-storing capacity of the atmosphere, which causes average global temperatures to increase. This, in turn, can cause late fall frosts and the northward migration of tree and plant species.

22. **D** This is an easy one. All of the answer choices are "positive" answer choices—which you would not think would cause high infant mortality—except (D). In fact, (D) is one strong indicator of countries that exhibit high infant mortality. Generally, high infant mortality rates occur in nations where the Gross Domestic Income (GDI) is low (less than $4,000). These nations cannot offer infant support services such as medicine, clean drinking water, sewage removal, and food—and this leads to high rates of infectious disease. Choices (A), (B), and (C) all occur in countries that have low infant mortality rates.

23. **C** Choice (C) is the only false statement among the answer choices. This question requires that you be familiar with the first two laws of thermodynamics. The first law states that energy cannot be created or destroyed. That is, the total energy of a system will always remain the same. The second law states that at each energy transformation, some energy is lost to the surroundings in an unusable form (usually as heat). For example, as a flashlight runs using battery power, some of the energy from the batteries goes toward warming the light bulb, which is not useful work.

24. **B** The area of the Gulf of Mexico that extends from just south of the Mississippi River mouth to Texas receives nutrient-rich water from both the Mississippi and Atchafalaya Rivers. The water from these rivers carries runoff fertilizers from farms that lie in or near the watersheds for the rivers. These nutrients promote a rapid rise in the algae population in the Gulf waters, which in turn causes eutrophication. When the algal cells die, bacteria living in the water decompose them, or break them down. These decomposers use oxygen for metabolism, and eventually the water becomes extremely oxygen poor. Low levels of oxygen often make these bodies of water uninhabitable for other aquatic organisms.

25. **B** Choice (B) is a positive effect, while all of the other answer choices cite negative effects of genetically modified foods. Genetically modified foods (GMFs) are crop plants to which genes from other species have been added. For example, the "Flavr Savr" tomato, sold from 1994 to 1997, contained select genes from fish! These genes were inserted into the tomato's genes so that the tomatoes would not freeze in a frost. Additional positive aspects for GMFs include increased yields, increased resistance to pests, resistance against pesticides, and the potential to grow in habitats that previously did not support the plants.

26. **D** "Energy usefulness" is defined as how helpful an energy source is to humans. A source is more useful if it costs little to harvest, transport, and use. Useful energy is most commonly in a concentrated form. For example, coal is highly useful compared to wind energy; (A), (B), and (C) are all examples of low-quality energy and matter.

27. **B** Review Chapter 6 if you have trouble remembering the basics of the geologic time scale. You will not be expected to memorize this chart, but you should be familiar with the most recent era and the most talked-about ones. Choice (B) correctly starts with the earliest era, the Precambrian (600 million years ago), then the Paleozoic (500 to 250 million years ago), then the Mesozoic (250 to 65 million years ago), and finally, the Cenozoic (65 million years ago to today)!

28. **B** Although the United States holds only 4.6 percent of the world's population, it produces 33 percent of the world's solid waste.

29. **B** A conservation easement is a voluntary agreement that allows a landowner to limit the type or amount of development on their property, while retaining private ownership of the land. The easement is signed by the landowner, who is the easement donor, and the conservancy, who is the party receiving the easement. The conservancy accepts the easement with understanding that it must enforce the terms of the easement in perpetuity. After the easement is signed, it is recorded with the County Register of Deeds and applies to all future owners of the land. In conservation easements, the owner usually receives a tax break for the promise that they will not build on that land.

30. **D** Choice (D) is the defining policy statement of the CWA. This law governs the discharge of pollutants into the waters of the United States. It gives the Environmental Protection Agency the authority to implement pollution control programs and sets wastewater standards for industry. The Clean Water Act also continues to set water quality standards for all contaminants in surface waters. The act makes it unlawful for any person to discharge any pollutant from a point source into navigable waters unless a permit was obtained under its provisions. It also funds the construction of sewage treatment plants under the construction grants program, and recognizes the need to address the critical problems posed by non-point source pollution.

31. **A** Evolution is the change in the genetic makeup of a population over time. For this exam, the key is to remember that individuals do not evolve and that only populations can evolve. Choice (B) is not a good answer because a mutation is a change in DNA. Choice (C) is also incorrect—evolution can occur in ways other than natural selection. Finally, we can eliminate (D) because it refers to an increase, or magnification, of toxins in a food chain.

32. **A** This question requires you to understand the factors that determine the number of offspring that women are expected to have from a statistical standpoint. Generally speaking, prosperity, high income, and a high level of education are all factors that lower a woman's total fertility rate (the number of offspring she will have). Statistically, the less wealthy and less educated a woman is, the higher her expected total fertility rate will be. Choices (B) through (D) all list factors that lower the total fertility rate, so the correct answer is (A).

33. **D** The carrying capacity is the maximum number of organisms that can live sustainably in a habitat. Increasing the rate of food production can increase the carrying capacity, so (D) is the correct answer. Organisms with a carrying capacity have a logistical population growth, so (A) and (B) are incorrect. Choice (C) is also incorrect because carrying capacity is not affected by the death rate.

34. **B** As chlorine atoms (from CFCs) catalyze the breakdown of ozone molecules in the stratosphere, more ultraviolet light is expected to reach the Earth's surface. These UV rays will break down DNA molecules in human skin cells, which can lead to cancers, cataracts, and other illnesses. Of course, other animals and plants can also be affected. While you may have been tempted to choose (A), (B) is a better option because it is more specific—(A) is much too general to be correct. Choice (C) is an opposite and (D) is a result of global warming.

35. **B** Ozone is generated as a secondary pollutant in photochemical smog. This is an EXCEPT/NOT/LEAST question, so remember that you're looking for the incorrect answer! The exposure to tropospheric ozone can cause all of the symptoms given in (A), (C), and (D). However, (B), lung cancer, is primarily caused by smoking cigarettes.

36. **C** The Net Primary Productivity (NPP) is the rate at which all the plants in an ecosystem produce net useful chemical energy. It is equal to the difference between the rate at which the plants in an ecosystem produce useful chemical energy and the rate at which they use some of that energy through cellular respiration. It is calculated by taking the Gross Primary Productivity (the amount of sugar that the plants produce in photosynthesis) and subtracting from that the amount of energy the plants need for growth, maintenance, repair, and reproduction. NPP is measured in kilocalories per square meter per year ($kcal/m^2/y$). Choice (A) is incorrect because it denotes the amount of energy available for photosynthesis. Choice (B) is no good; it merely states that energy cannot be created or destroyed. Choice (D) can also be eliminated; it states that at each energy transfer, some energy is lost to the environment.

37. **C** Choice (C) is the only option that relates to the Second Law of Thermodynamics. Recall that the second law states that at each step of energy transformation, a large amount of energy is lost as heat to the environment. When we eat food, our bodies convert the energy in the food's chemical bonds to chemical bonds in our molecules. During this conversion, most energy is lost as heat (which is why our bodies are warm!).

38. **C** Acid deposition affects species by damaging habitats and by reducing or contaminating food sources through uptake of toxic levels of metals. Species such as amphibians, which require both aquatic and terrestrial environments, are most at risk. For example, in the acid-sensitive areas of eastern Canada, 16 of the 17 amphibian species have more than 50 percent of their ranges affected by acidic deposition. Monitoring amphibian populations may provide a biological indication of changes in acid deposition.

39. **C** Internal costs are all the costs that are incurred by the buyer and seller of the item in question. For example, the cost of gasoline and the purchase price of the car are both buyer costs. The costs of metal, plastics, employee salaries, and plant maintenance are internal costs of the seller. Choice (C) is the correct answer because it is not an internal cost, but an external one; it is a cost paid by society. Health care costs that arise as a result of air pollution and the cost of building roads are both external costs.

40. **A** Scrubbers are devices that are installed in smokestacks in order to reduce the amount of pollutants that rise up the stack. They extract the pollutants in different ways: by trapping dust in filter material, using cyclonic air motion to concentrate dust in one area, or spraying water in the stack to trap both dust and gases that are water-soluble. Choice (B) is simply not the best answer; (C) occurs before smoke is produced; and (D) is simply untrue—in fact, the amount of ash is greater when scrubbers are in place because scrubbers trap materials.

41. **D** After ore is mined, the unusable parts that remain are placed into piles called tailings. This is a question that you'd either know the answer to or you wouldn't. Remember that if you come to a question like this and have absolutely no idea which choice is correct, pick your favorite letter and move on. You can also circle it and come back to it in the next pass—maybe you'll remember the answer if you give your subconscious mind a chance to do its thing. Let's look at the other answer choices: (A), overburden, is the rock that's removed from the ground in the surface mining of coal, so it's wrong. We made up (B), we hope you didn't choose it! Choice (C), leachate, is the liquid that percolates through a mine or landfill.

42. **A** An externality is an unanticipated consequence experienced by unrelated third parties. Choices (B), (C), and (D) are all examples of externalities. Choice (A) is not an externality because the effect is experienced by a related party (the construction worker who purchased the automobile).

43. **D** The fossil fuel coal is sedimentary rock that's derived from decayed and fossilized plant and animal materials. During the fossilization process, sulfur is incorporated into the rocks; this sulfur comes from bacteria that perform the decomposition of the material, as well as from the decaying material itself. None of the other fuels contains as much of the element sulfur (S).

44. **B** Choice (B) is the best definition of a biological reserve. These reserves are areas of habitat that are crucial to the survival of species. Strictly speaking, a biological reserve is an area of land and/or water designated as having protected status for purposes of preserving certain biological features. Reserves are managed primarily to safeguard these features and provide opportunities for research into the problems underlying the management of natural sites and of vegetation and animal populations. Regulations are usually imposed that control public access and prevent people from disturbing these areas. All other answer choices would lower the biodiversity of an area.

45. **D** The doubling time is the amount of time it takes for the population of a country to double. The country with the shortest doubling time is the country with the most rapid population growth. Among the answer choices, Kenya is the only one with a rapid population growth, so (D) is the correct answer. Denmark is experiencing zero growth, so (A) is wrong. The United States and Australia are experiencing slow growth, so (B) and (C) are also incorrect.

46. **B** Transpiration is the evaporation of water from plants into the air. This process increases the atmospheric water content. Choice (B) is the correct answer.

47. **B** Choice (B) is the definition of full cost pricing: full cost includes the environmental effects. All the other answer choices are part of the external costs paid by the buyer or seller.

48. **C** In 1987, the Montreal Protocol was signed. The protocol called for the worldwide end of chlorofluoro-carbon (CFC) production. As the result of the protocol, the atmospheric concentration of CFCs has decreased, allowing for the ozone hole to recover. Choice (C) is correct. Choices (A), (B), and (D) are all greenhouse gases that have been continually *increasing* in atmospheric concentration, not decreasing.

49. **D** In the postindustrial state, the population has reached its carrying capacity and will exhibit a zero growth rate, or even drop below a zero growth rate.

50. **A** Methane is a useful biofuel that is produced from the decomposition of organic molecules in a landfill. This gas can be trapped, purified, and used to power vehicles. All of the other effects that are listed are undesirable results of the existence of old landfills.

51. **C** The CITES treaty (the Convention on International Trade in Endangered Species of Wild Fauna and Flora) entered into force in 1975. It was first discussed in the late 1960s when people recognized that there was no way to internationally protect endangered or threatened species. CITES is international in scope and is designed to halt the killing of endangered or threatened species for food, collection, or medicinal purposes by penalizing those who collect, trade, or buy those species.

52. **C** The growth rate of Country A is equal to its birth rate subtracted by its death rate: 33 births per 1,000 − 13 deaths per 1,000 = an increase of 20 persons per 1,000, or a growth rate of 2%. The doubling time can be calculated using the Rule of 70: 70 divided by 2 is equal to 35. This means that it takes 35 years for the population of Country A to double. In order for the population of Country A to reach 200,000, the population needs to double twice. This would take 70 years, so the answer is 2070, which is (C).

53. **A** Increased ocean levels is the most logical cause of the phenomena listed in the question. The increase is due to two main factors: the first factor is that as water warms, it expands. This is known as thermal expansion. Secondly, as the average global temperature increases, glaciers and other ice formations will melt; this will increase the volume of the world's oceans. All of the answer choices represent events that could be caused by an increase in the volume of the world's oceans.

54. **D** Compost is formed from decaying plants and other organic material. In the composting process, both bacteria and fungi decompose large organic molecules into smaller molecules. Compost can then be used to fortify and condition soil. While it might have been tempting, (C) is too general an answer, because all living organisms undergo respiration.

55. **D** Non-point pollution sources are defined as pollution that comes from a broad, ill-defined area, such as a large area of farmland. Since most farmers in the area presumably fertilize their crops, you cannot point to one specific area as the only source. Choices (A) through (C) are all point sources of pollution.

56. **C** Choice (C) represents the standard definition of GDP. It is calculated by totaling the value of all the goods and services in the country for a specific period and is often considered a measure of the standard of living in a country.

57. **D** Choice (D) is the correct answer; all other answer choices are open-surface mining practices.

58. **B** The total fertility rate in a country represents the average number of children a woman will have during her reproductive lifetime (between ages 14 to 45). The total fertility rate is affected by many factors, including level of education, culture, and the country's standard of living.

59. **B** Choice (B), or the fumes produced from the burning of wastes, is the only process that does not directly contribute to the contamination of the surrounding soil. When the EPA started to clean up the wastes from Love Canal, they had to deal with wastes that contaminated the surrounding land by all the other methods listed in the answer choices.

60. **A** Humans have not yet managed to make nuclear fusion viable. Choice (A) is correct. The other choices are all different methods currently used by humans to produce energy.

61. **C** The bicarbonate ion (HCO_3^-) is produced when atmospheric carbon dioxide reacts with ocean water. Bicarbonate ions act as a buffer in seawater and allow the pH to be relatively stable, ranging between 7.5 and 8.5. Choice (A), phosphoric acid, contains no carbon atoms. Neither methane nor carbon disulfide exist in large quantities, so (B) and (D) are wrong as well.

62. **B** Choice (B) is correct because areas rich in limestone or other basic minerals can neutralize the acid deposition products. Limestone is composed mostly of calcium carbonate ($CaCO_3$). This dissolves in water to form bicarbonate ions (HCO_3^-), which act as the buffering agent.

63. **C** Atmosphere (C) is the only answer choice that does not store significant amounts of phosphorus. Phosphorus is rarely present in gases. Additionally, it is relatively insoluble in water. It is found most frequently in rocks and minerals.

64. **B** The current hypothesis for the formation of large volumes of oxygen in the atmosphere is called the autotrophic hypothesis. According to the hypothesis, the early atmosphere had almost no O_2. It was not until the evolution of photosynthesis that the levels of O_2 increased. As for the other answer choices, volcanic outgas mostly consists of oxides of carbon, nitrogen, and hydrogen sulfide; meteorites could not carry gases effectively; and animals produce CO_2 not O_2.

65. **D** "Animal studies" refers to tests conducted on animals in laboratories, under controlled conditions. For example, this type of study is used to determine the dosage of a chemical that will kill 50 percent of a test population (LD_{50}). Epidemiological studies are done by examining human populations exposed to certain risks (for example, smoking). Statistical probabilities quantify the likelihood that a given event will occur.

66. **A** This is a single dose of the pollutant from a point source. You can determine this because the BOD drops off, meaning that the pollutant is being diluted by unpolluted water. Choice (B) is incorrect because it is a secondary air pollutant. Choice (C) is a class of pollutants that result from the mixing of combustion products in the atmosphere. Choice (D), deep well, is a location where toxic materials are stored deep underground. It is unlikely that it would pollute the stream.

67. **B** This question tests your ability to estimate based on the graph. Look carefully at the axes and scale. It is helpful to use your pencil to draw a line from the 12-mile mark in the x-axis up to the line, and then

another line from that point to the *y*-axis. You should be able to estimate 220 ppm (parts per million). Choice (A) is too high an estimate. Choices (C) and (D) are estimates that are too low.

68. **D**   The BOD, or Biological Oxygen Demand, test determines the rate at which microorganisms take oxygen out of the water. This test is an estimate of the amount of biodegradable organic matter in the water and is an indirect indicator of water quality. Choice (A) is the turbidity test—the number of particles that scatter light in the water. Choice (B), nitrates, are important plant nutrients and are measured by tests that can quantitatively determine the concentration of the chemical ions. Choice (C), coliform bacteria, live in the intestinal tracts of animals and are an indication that fecal matter is in the water. This is not a BOD test because the coliform test measures the number of living cells, not their biological activity.

69. **A**   You should recognize that the three organisms live in conditions with very low oxygen content. Looking carefully at the graph, you will see that the BOD is highest from 0 to 5 miles. Because the BOD is highest here, the amount of oxygen in the water is the lowest. Remember, high demand means low amounts of oxygen in the water! Choices (B), (C), and (D) are areas where the BOD is low—so there is more oxygen in the water. The three listed organisms do not live in areas of high oxygen content.

70. **B**   Choice (B) is a correct definition of a riparian zone. It is an ecotone between a river or stream and land. Riparian zones act as buffer areas, absorbing excess water and pollutants that travel from the land to the water. Choice (A) defines rangeland; (C) is the role of a tree line; (D) is the understory of a forest.

71. **C**   Choice (C) is the only answer that is a disadvantage. Choices (A), (B), and (D) are all advantages of fish farming. When a fish (or any animal) population is dense, parasites and diseases can flourish.

72. **A**   There is a little saying that might help you remember which form of nitrogen is utilized by plants. A mnemonic is "The plants ate the nitrate." Notice that nitrate has the same last three letters as *ate*. Choices (B) and (C) are parts of the nitrogen cycle, but are not absorbed by plants. Choice (D) is not a nitrogen-containing compound.

73. **D**   Thermal inversions occur when a warm layer of air prevents polluted air from rising up over a city. Normally, cool upper air allows the warm air to rise, and the pollutants are dispersed along with the warm air. When an inversion occurs, the upper air is warmer and the polluted air cannot rise. This traps the pollution close to the Earth's surface. Choices (A) and (C) both deal with water; thermal inversions are atmospheric phenomena. Choice (B) is the pattern in a normal atmosphere.

74. **D**   Choice (D) is the correct answer, based on the principles of an environmentally sustainable economy. This type of economy is defined as a low-waste society that uses conservation, recycling, and reduction to keep its use of nonrenewable resources at sustainable levels. In a sustainable society the focus is on preserving that society for future generations.

75. **C**   Chlorofluorocarbons (CFCs) contain the element chlorine, which catalyzes the breakdown of ozone in the stratosphere. The energy for this reaction comes from the sun's ultraviolet radiation. It is estimated that one atom of chlorine will break down 100,000 molecules of ozone during the time it remains in the atmosphere. Choice (A) is a normal component of the atmosphere; (B) is a greenhouse gas; (D) causes acid deposition.

76. **D**   The abyssal zone is the deepest region of the ocean and the zone with the highest levels of nutrients due to decaying plant and animal matter that sinks down from the zones above. Choice (D) is correct.

77. **C**   The ice cap that lies over much of Greenland is more than two miles thick and has trapped gases from the atmosphere from hundreds of thousands of years ago. Other samples—(B) and (D)—do not list an actual sample; methane is produced in the oil-making process, among other sources, so (A) is wrong as well.

78. **D**   Bacteria are vital for the nitrogen cycle. They convert nitrogen from one form to another, and eventually into nitrates, which plants can use. Choices (A), (B), and (C) are not dependent on bacteria or fungi in the soil.

79. **D**   By definition, trace elements are needed in small amounts. Choices (B) and (C) are typical components of fertilizers and are needed in large quantities; (A) is supplied by the atmosphere in large quantities.

80. **C**   Choice (C) states a fundamental tenet of the environmental justice movement. Choice (A), the Earth stewardship view, is a system of beliefs that includes the belief that people can learn to live in natural harmony with the planet. Those who have the viewpoint in (B) believe that humans are the best planet managers and that our technology and understanding of systems will help us make the correct decision; and sustainability, (D), is the point of view that all people must live in ways that allow the Earth to be sustainable.

Section II—Free-Response Questions

Question 1
Question 1 refers to the editorial on page 393.

(a) **Explain** how the beetle might have become resistant to NOGrub. Assume that NOGrub had been applied to a population of beetles in another county.

In a sexually reproducing population, there is natural variation in the genetic makeup of the population. There are large numbers of beetles, so some of them have a genetic makeup that allows for their survival in an environment with NOGrub in it. When the pesticide is applied, the susceptible beetles die and the resistant ones survive. The survivors reproduce in a habitat of lessened competition; therefore, their genes are passed along.

(1 point maximum—for the idea of natural selection/selective survival, and growth of resistant populations)

(b) If the county agents do not have information on which pesticide is most effective against the beetle, the county plans to investigate by trying several pesticides on controlled sections of beetle-infested crop.

i. **Identify** the independent and dependent variables in such a trial.

The independent variable in the trial is which pesticide is used. The dependent variable is its effect on the beetles: for example, what percentage of beetles die.

ii. **Describe** an effective control group and the experimental groups for this experiment.

A control group for the experiment would be a section of beetle-infested crop on which no pesticide was used. The experimental groups would be separate sections of beetle-infested crop, each treated with one of the pesticides under consideration.

iii. **Identify** TWO environmental factors that must be controlled to keep the experiment from producing skewed results.

Environmental factors:

- Density and health of beetles' food source (the crop): the control and experimental groups should be similar in terms of what the beetles have to eat.
- Presence of predators: the control and experimental groups should not differ in terms of likelihood that beetles will be preyed upon.
- Disturbance: crop sections should not be disturbed, or steps should be taken to ensure the amount of disturbance is equal between groups.
- Irrigation: all crop sections should receive the same amount of water.
- Environmental conditions: the sections should be similar in terms of shade, shelter, and landscape features.

(4 points maximum—1 for correct identification of the independent and dependent variables, 1 for correct description of the control and experimental groups, and 1 for each factor to control for—maximum of 2)

(c) **Identify** TWO negative impacts of using chemical pesticides on the surrounding ecosystem.

Negative impacts of using chemical pesticides on the surrounding ecosystem include the following:

- Pesticides can migrate to other habitats, thus harming other organisms.
- Pesticides can kill beneficial insects, which act as predators of destructive insects.
- Pesticides can combine with other chemicals and create other toxic chemicals.
- Pesticides can be accidentally ingested by humans and cause illness.

(2 points maximum—1 for each correct impact)

(d) One strategy for combating pests is Integrated Pest Management (IPM).

i. **Describe** IPM.

- IPM employs a combination of biological, chemical, and physical (any two of the three) methods for pest control.
- Goal of IPM is to curtail (or eliminate) pesticide use.
- IPM generally does not eradicate the pest population, but brings it to a tolerable level.

(1 point for showing understanding of one of the listed aspects of IPM)

ii. **Identify** one benefit and one difficulty in using IPM to control this outbreak.

Benefits of Using Integrated Pest Management	Difficulties in Using Integrated Pest Management
Lowers costs for pesticides	Needs expert knowledge of insect life cycle
Reduces amounts of pesticides in environment	Takes time to implement management practices
Several weaknesses of insects can be exploited	Initial costs might be higher
Reduces use of fertilizers	Governments subsidize pesticide use
Improves crop yields	Training takes a long time
Lowers genetic resistance issues	Methods for one area might not apply in other areas

(2 points maximum—1 for a correct benefit, and 1 for a correct difficulty)

Question 2

The map shows two cities: City *X* and City *Z*, separated by several kilometers.

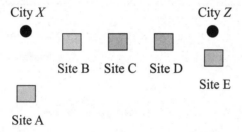

Students from a high school in between the two cities studied soil pH values at the sites labeled A through E on the map. The results of the pH study are given in the following table:

Site	pH value
A	6.2
B	5.6
C	5.0
D	4.5
E	4.3

(a) Refer to the table and diagram above to answer the following questions.

i. **Describe** one point source for the pollution that caused the change in the soil's pH as shown.

ii. In the description, **identify** a fuel that could create the pollution.

Point sources could include the following:

- Electricity generation plants

- Industry sites

- Cars or trucks

- Homes

The fuels involved would be coal or oil. (Natural gas and nuclear would not be accepted as correct.)

(2 points maximum—1 for the point source and 1 for the fuel)

(b) Assume that the fuel identified in (a)(ii) is the source of the pollution.

   i. **Identify** one primary and one secondary pollutant that can cause the change in the soil's pH.

   ii. **Describe** the process that causes the change in the pH.

Primary pollutants are either SO_x or NO_x (saying just "sulfur" or "nitrogen" is not correct, as they do not directly produce the acid reaction). Also, suspended particulate matter can carry acidifying chemicals.

As the fuel is burned, SO_x or NO_x are produced; these enter the atmosphere and react with water vapor to create sulfuric acid or nitric acid. Farther downwind, these chemicals fall as wet or dry deposition and acidify the soil.

(4 points maximum—1 for identifying the primary pollutant and 3 for describing the process of acid formation and deposition)

(c) **Propose** one possible method to reduce the air pollutants that are causing the pH change.

Methods of reducing the air pollutants that are causing the pH change include:

- Reducing factory emissions of SO_x or NO_x by using scrubbers to remove chemicals from smoke

- Using cleaner-burning fuels (natural gas, low-sulfur coal, etc.)

- Reducing demand for electricity, which lowers production

- Adding catalytic converters to lower pollution from cars and trucks

(2 points maximum—1 for the method and 1 for the correct description)

(d) **Identify** and **describe** one provision of the Clean Air Act of 1990 that could be used to control and reduce the emissions.

Provisions of the Clean Air Act of 1990 that could be used to control and reduce the emissions include:

- National Ambient Air Quality standards—a set of maximum permissible levels for pollutants

- Emissions trading policy—each year, certain factories are given "rights" or "credits" to release set amounts of pollutants; these credits can be bought, sold, or traded to reduce a company's liability under the CAA

- National emission standards for hundreds of toxic pollutants (e.g., mercury)

(2 points maximum—1 for the policy and 1 for the correct description of the policy)

Question 3

According to the United States Energy Information Administration, the consumption of natural gas by the United States increases at 8 percent per year. The United States receives its supplies from a variety of international and domestic locations. Natural gas is used in the home, for industry, and for power generation.

(a) **Calculate** the approximate number of years it would take to double the consumption of natural gas in the United States. **Show** all your work.

Apply the Rule of 70 to calculate the doubling time. The formula is $\frac{70}{8} = 8.75$. In other words, it would take about nine years to double the consumption. (Either an exact or rounded figure would be considered correct.)

(2 points maximum—1 point for the correct setup, 1 point for the answer)

(b) **Identify** one method by which natural gas is recovered and transported.

Natural gas is recovered through a series of pipes and valves fitted over a drilled hole. These are often associated with oil wells, coal beds, or natural gas pockets. The gas can then be transmitted through piping or cooled and compressed into Liquefied Natural Gas (LNG) and transported by specially fitted trucks, trains, or ships. Some natural gas is collected as a product of the anaerobic decay of organic material in landfills.

(2 points maximum—1 for a correct description of a collection method, 1 point for a correct transport method)

(c) **Describe** TWO benefits to the environment that would occur if the United States switched from coal to natural gas-fired electric power generation.

The environment would benefit from the following:

- Less habitat destruction from the mining of coal

- Less acid mine drainage due to fewer mines being dug

- Less aesthetic damage done to the environment because fewer mines are dug

- Reduced SO_2 emissions, which will reduce the levels of acid deposition

- Reduced CO_2 emissions, which will reduce the levels of climate-changing gases

- Reduced soot emissions, which will reduce the formation of industrial (gray) smog

- Less fly ash produced

- Less mercury or radioactive materials released by the burning of coal

- Longer boiler life because gas produces fewer products that harm the boilers

- Less cost in building new generation plants, because gas combustion produces fewer dangerous by-products

(2 points maximum—1 point for benefit, 1 point for description. Remember, you receive credit only for your first two answers.)

(d) Some people advocate increasing the use of coal versus natural gas for the production of electricity. Give one argument that the proponents of coal might use to **justify** their position.

Coal-use proponents could make the following arguments:

- There is a larger supply of domestic coal than natural gas, so the United States will be less dependent on foreign energy sources.

- Coal is safer to use. In an accident, natural gas can explode, harming people or facilities.

- The infrastructure for transporting coal already is in place. We would have to do little to increase the production and transportation of coal.

- It might be cheaper to refit existing coal plants with scrubbers to clean up toxic emissions than to build a new, cleaner-burning gas plant.

(2 points maximum—1 point for argument, 1 point for description)

(e) Others advocate for non-hydrocarbon fuel alternatives.

i. **Identify** one non-hydrocarbon fuel alternative.

ii. **Describe** one drawback of the alternative identified in (e)(i).

Possible answers are given in the table on the next page.

Energy Source	Drawbacks
Nuclear	• Safety concerns (such as potential for radiation leaks) • Waste disposal concerns (such as radioactive or toxic waste disposal) • Raw materials storage (safe storage of radioactive materials) • Thermal pollution (very hot reactor cooling fluids)
Hydroelectric	• Water loss via evaporation at dam site • Dam may hinder fish and other aquatic migration • Silting • Habitat destruction/alteration (via flooding upstream, or alteration of water quality downstream) • Potential for displacing populations or wildlife living in dam's flood zone • Potential for seismic activity below reservoir • Potential for plants killed by flooding to produce methane gas (a greenhouse gas)
Solar	• Initial costs are high • Material for solar panels uses hydrocarbon fuels • Solar panels lack aesthetic appeal • Disposal of toxic chemicals (from rechargeable batteries)
Wind	• Initial costs are high • Turbines lack aesthetic appeal • Sound and vibration produced are annoying • Turbines may affect bird migration patterns
Geothermal	• Geographical limitations (There are only certain places in the world where geothermal energy can be harvested.) • Thermal pollution • Toxic chemicals (potential for releasing H_2S into environment)

(2 points maximum—1 point for listing energy source, 1 point for listing related drawback)

HOW TO SCORE PRACTICE TEST 2

Section I: Multiple-Choice

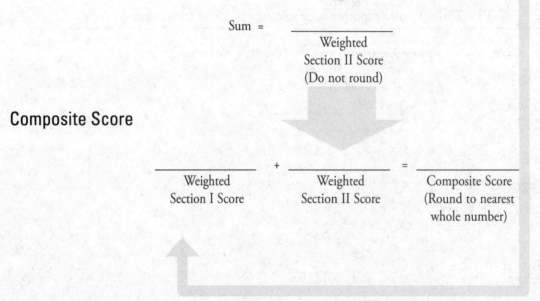

_____ × 1.125 = _____
Number Correct   Weighted
(out of 80)   Section I Score
                            (Do not round)

Section II: Free-Response

Question 1   _____   × 2 =   _____
                  (out of 10)   (Do not round)

Question 2   _____   × 2 =   _____
                  (out of 10)   (Do not round)

Question 3   _____   × 2 =   _____
                  (out of 10)   (Do not round)

Sum = _____
        Weighted
        Section II Score
        (Do not round)

AP Score Conversion Chart Environmental Science

Composite Score Range	AP Score
107–150	5
90–106	4
73–89	3
56–72	2
0–55	1

Composite Score

_____ + _____ = _____
   Weighted   Weighted   Composite Score
Section I Score   Section II Score   (Round to nearest
                                          whole number)

Note: This score sheet is to help you estimate your approximate score for the official exam, not your actual score.

Practice Test 3

AP® Environmental Science Exam

DO NOT OPEN THIS BOOKLET UNTIL YOU ARE TOLD TO DO SO.

At a Glance

Total Time
1 hour and 30 minutes
Number of Questions
80
Percent of Total Grade
60%
Writing Instrument
Pencil required

Instructions

Section I of this examination contains 80 multiple-choice questions. Fill in only the ovals for numbers 1 through 80 on your answer sheet. A blank answer sheet is located at the back of this book or can be downloaded and printed from your Student Tools.

Indicate all of your answers to the multiple-choice questions on the answer sheet. No credit will be given for anything written in this exam booklet, but you may use the booklet for notes or scratch work. After you have decided which of the suggested answers is best, completely fill in the corresponding oval on the answer sheet. Give only one answer to each question. If you change an answer, be sure that the previous mark is erased completely. Here is a sample question and answer.

Sample Question   Sample Answer

Chicago is a

(A) state

(B) city

(C) country

(D) continent

Use your time effectively, working as quickly as you can without losing accuracy. Do not spend too much time on any one question. Go on to other questions and come back to the ones you have not answered if you have time. It is not expected that everyone will know the answers to all the multiple-choice questions.

About Guessing

Many candidates wonder whether or not to guess the answers to questions about which they are not certain. Multiple-choice scores are based on the number of questions answered correctly. Points are not deducted for incorrect answers, and no points are awarded for unanswered questions. Because points are not deducted for incorrect answers, you are encouraged to answer all multiple-choice questions. On any questions you do not know the answer to, you should eliminate as many choices as you can, and then select the best answer among the remaining choices.

GO ON TO THE NEXT PAGE.

ENVIRONMENTAL SCIENCE
Section I
Time—1 hour and 30 minutes
80 Questions

Directions: Each of the questions or incomplete statements below is followed by four suggested answers or completions. Select the one that is best in each case and then fill in the corresponding oval on the answer sheet.

Questions 1 and 2 refer to the following graph.

A survivorship curve graph showing populations of humans, chickens, and a tree species is shown below.

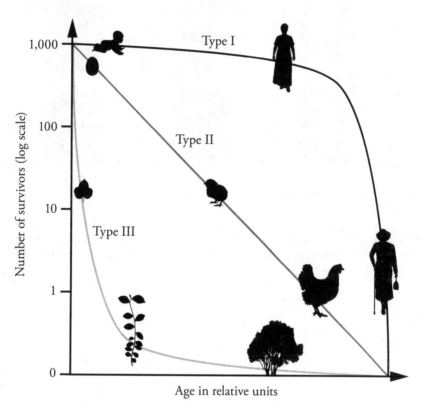

Survivorship indicates the probability that a given organism will live to a certain age.

1. The survivorship curve for the tree species above is an example of a Type III curve. Which of the following species would show a similar curve?

    (A) Moose

    (B) Songbirds

    (C) Sea turtles

    (D) Oysters

2. Which of the following is an accurate description of a Type II curve?

    (A) High survival probability in early and middle life, followed by a rapid decline in later life

    (B) Roughly constant survival probability regardless of age

    (C) High mortality in early life, followed by relatively high survival probability for those surviving the bottleneck

    (D) Relatively high survival probability in early and late life, with a bottleneck of high mortality in the middle of the lifespan

GO ON TO THE NEXT PAGE.

Questions 3 and 4 refer to the following chart.

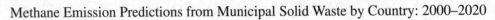

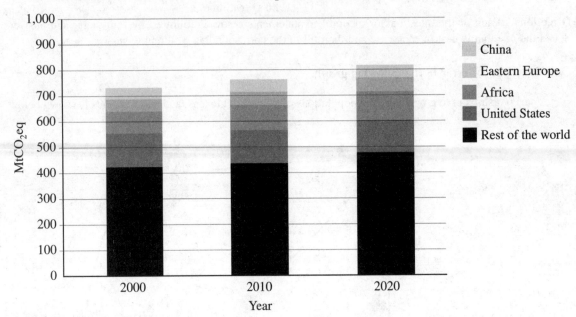

Methane Emission Predictions from Municipal Solid Waste by Country: 2000–2020

3. According to the chart, approximately what percentage of global methane emissions from municipal solid waste in 2010 were predicted to come from the United States?

(A) 7%

(B) 17%

(C) 57%

(D) 73%

4. The amount of emissions of methane from municipal solid waste was predicted to remain the most stable in which region over the period given?

(A) China

(B) Africa

(C) United States

(D) Rest of the world

GO ON TO THE NEXT PAGE.

Questions 5 and 6 refer to the following map.

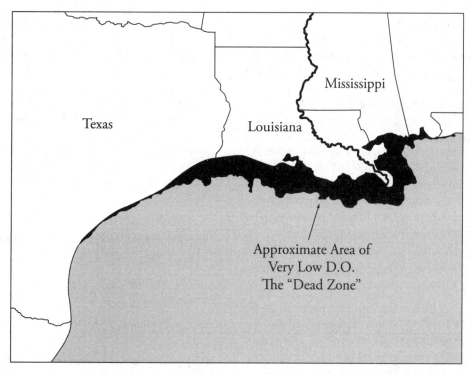

Approximate Area of
Very Low D.O.
The "Dead Zone"

5. The map shows an area of eutrophication ("dead zone") in the Gulf of Mexico. Which of the following accurately describes the chain of events leading to its formation?

(A) Mississippi River carries fertilizer and sewage into the Gulf—phytoplankton and zooplankton experience major population decrease—detritus from dead phytoplankton and zooplankton feeds bacteria at sea floor—bacteria metabolize available oxygen while decomposing this food source

(B) Mississippi River carries fertilizer and sewage into the Gulf—phytoplankton and zooplankton experience population explosion—less detritus from dead phytoplankton and zooplankton to feed bacteria at sea floor—bacteria produce less oxygen while decomposing this food source

(C) Mississippi River carries fertilizer and sewage into the Gulf—phytoplankton and zooplankton experience major population decrease—less detritus from dead phytoplankton and zooplankton to feed bacteria at sea floor—bacteria produce less oxygen while decomposing this food source

(D) Mississippi River carries fertilizer and sewage into the Gulf—phytoplankton and zooplankton experience population explosion—detritus from dead phytoplankton and zooplankton feeds bacteria at sea floor—bacteria metabolize available oxygen while decomposing this food source

6. Which of the following is NOT an effect of the presence of this "dead zone"?

(A) Collapse of the shrimp and shellfish industries in the area

(B) Nothing that depends on oxygen can grow in the zone from May to September

(C) Reproductive problems in fish involving decreased size of reproductive organs, low egg counts, and lack of spawning

(D) Mass migration of fish to other areas of the Gulf

GO ON TO THE NEXT PAGE.

Questions 7 and 8 refer to the following information.

Ecosystem services are benefits that humans receive from the ecosystems in nature when they function properly. There are four categories: **provisioning services**: providing humans with water, food, medicinal resources, raw materials, energy, and ornaments; **regulating services**: waste decomposition and detoxification, purification of water and air, pest and disease control and regulation of prey populations through predation, and carbon sequestration; **cultural services**: use of nature for science and education, therapeutic and recreational uses, and spiritual and cultural uses; and **supporting services** (the ones that make other services possible): primary production, nutrient recycling, soil formation, and pollination.

7. Which of the following is an example of a regulating service?

(A) In New York City, authorities worked to restore the polluted Catskill Watershed, restoring soil absorption and filtration of chemicals via soil and its microbiota, improving water quality.

(B) Restoration of wild bee populations to an area relieves large-scale farms from having to import non-native honeybees to pollinate crops.

(C) In the Roman Empire, water-powered mills produced flour from grain, and were also utilized for power to saw timber and cut stone.

(D) In modern ecotherapy, patients are encouraged to spend time in nature to enhance their physical, mental, and emotional well-being.

8. All of the following are negative consequences to humans of the disruption of ecosystem services EXCEPT

(A) lower crop yields due to depletion of soil nutrients and microbiota

(B) outbreaks of diseases carried by insect vectors due to habitat loss in predator species

(C) higher costs of mining due to depletion of mineral resources

(D) greater availability of habitats in different stages of ecological succession for scientific study due to human development

Questions 9–11 refer to the following information.

The greenhouse effect is a natural process that warms the Earth's surface and the layer of the atmosphere closest to the Earth's surface by reducing radiative loss from the Earth's surface to space. In other words, the sun's energy warms the Earth's surface, and some is radiated back toward space; certain gases in the troposphere absorb some heat and reradiate it downward again, effectively trapping heat that would otherwise be lost. The most important greenhouse gases are carbon dioxide, methane, water vapor, nitrous oxide, and chlorofluorocarbons (CFCs).

9. Which of the following greenhouse gases is NOT a significant contributor to global climate change?

(A) CFCs

(B) Methane

(C) Water vapor

(D) Carbon dioxide

10. The enhanced, or anthropogenic, greenhouse effect refers to a strengthening of the greenhouse effect due to human activities that are contributing to global climate change. Which of the following is NOT true of this anthropogenic effect?

(A) Depletion of the stratospheric ozone layer is a contributing factor.

(B) It is mainly due to increases in emissions of greenhouse gases and pollutants, as well as to changes in land use.

(C) Some of its effects include sea level rise, changes in ocean properties, and an increase in extreme weather events.

(D) Possible effects on the biosphere include changes in biomes, mass migrations and extinctions, spread of disease vectors, and changes in population dynamics.

11. Global warming potential (GWP) is a measure of how much heat a greenhouse gas can trap in the atmosphere, relative to the reference point of carbon dioxide. Which of the following greenhouse gases has the highest GWP?

(A) Chlorofluorocarbons

(B) Nitrous oxide

(C) Carbon dioxide

(D) Methane

GO ON TO THE NEXT PAGE.

12. Exposure to which of the following noises would cause the most damage to a person's hearing?

    (A) A vacuum cleaner

    (B) A chain saw

    (C) A factory

    (D) The firing of a rifle

13. The phrase that best defines population density is

    (A) the number of individuals in a certain geographic area

    (B) the rate at which a population increases

    (C) the maximum number of individuals that a habitat can sustain

    (D) the time it takes for a population to increase to carrying capacity

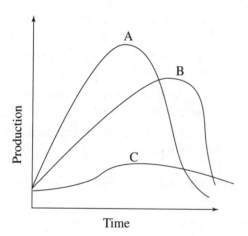

14. The graph above represents possible depletion curves of a nonrenewable resource. Curve C best describes the resource as it

    (A) has just been discovered and no technology exists to use the resource

    (B) is quickly used up and is not recycled

    (C) is newly discovered and in high demand

    (D) has expanding reserves and consumption is reduced

15. The shrinking of the Aral Sea and the ecological disaster that followed was mainly caused by

    (A) the diversion of the sea's two feeder rivers for agricultural use

    (B) withdrawing groundwater from the area

    (C) a major earthquake that hit the region

    (D) the massive use of pesticides

16. Which of the following is a negative result of overfishing a particular species of edible fish?

    I. Loss of so many fish that there is no longer a breeding stock
    II. The removal of non-target species
    III. The reduction of other species that rely on the edible species as food

    (A) II only

    (B) I and II only

    (C) II and III only

    (D) I, II, and III

17. Which type of irrigation results in the greatest amount of water lost to evaporation and runoff?

    (A) Flood irrigation

    (B) Drip irrigation

    (C) Furrow irrigation

    (D) Spray irrigation

GO ON TO THE NEXT PAGE.

Questions 18–21 refer to the following graph.

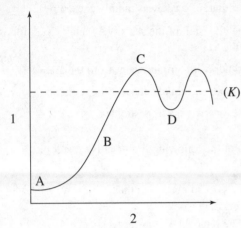

18. The population is growing at its highest rate at which letter?

    (A) A

    (B) D

    (C) B

    (D) C

19. Which of the following phrases best describes line *K* in the graph above?

    (A) The rate of population growth

    (B) The carrying capacity of the environment

    (C) The time it takes for the population to double in size

    (D) The birth rate of the population

20. The best label for the axis labeled 2 in the graph above is

    (A) time

    (B) population number

    (C) logistic growth rate

    (D) environmental resistance

21. Which of the following statements is true concerning the events occurring at point D in the graph above?

    (A) Environmental resistance is high.

    (B) Environmental resistance is low.

    (C) The population will continue to fall.

    (D) The population is above its carrying capacity.

22. The long-term storage of phosphorus and sulfur is in which of the following forms?

    (A) Rocks

    (B) Water

    (C) Plants

    (D) Atmosphere

23. The movement of sections of the Earth's lithosphere is known as

    (A) mass depletion

    (B) plate tectonics

    (C) background extinction

    (D) migration

24. Which of the following best defines the Green Revolution?

    (A) An international effort to stop the construction of nuclear power plants

    (B) A group whose goal is to improve how nations affect the environment

    (C) Increasing the yield of farmland by using more fertilizer, better irrigation, and faster growing crops

    (D) A method that makes a viable soil conditioner by using household waste

25. Which of the following best defines the term "infant mortality rate"?

    (A) How many children live in each square hectare

    (B) The number of births in a population

    (C) The number of infant deaths per 1,000 people aged zero to one

    (D) The difference between the birth rate and the death rate in a population

26. The release of chemicals from underground storage tanks is most likely to pollute which of the following?

    (A) A landfill

    (B) The atmosphere

    (C) The ecotone

    (D) Aquifers

GO ON TO THE NEXT PAGE.

27. The United States Congress failed to ratify which of the following international agreements that is designed to control the release of carbon dioxide?

    (A) CITES agreement

    (B) Kyoto Protocol

    (C) Montreal Protocol

    (D) Clean Air Act

28. Which of the following is the most sustainable way to ensure sufficient energy for the future?

    (A) Find more fossil fuels

    (B) Develop more effective solar power generators

    (C) Build more nuclear reactors

    (D) Reduce waste and inefficiency in electricity use and transmission

29. Which of the following best describes the goals of the CAFE standards?

    (A) Reduce pollution by coal-fired power plants

    (B) Improve the quality of air around cities

    (C) Protect certain endangered species

    (D) Improve the fuel efficiency of automobiles in the United States

30. Which of the following best describes the use of DDT?

    (A) It supplies needed nitrogen to plants.

    (B) It kills weeds and unwanted plants.

    (C) It decreases the amount of pollution from car exhaust.

    (D) It is an insecticide.

31. Which of the following best explains why pesticides become ineffective over time?

    (A) Pesticide manufacturers learn how to cut corners and sell an inferior product.

    (B) Pests acquire resistance to the pesticide during exposure to the pesticide.

    (C) Pests resistant to the pesticide survived and passed on their resistance to the next generation.

    (D) Pests acquired resistance to the pesticide and passed on their resistance to the next generation.

32. For every ton of plastic recycled, 0.3 fewer tons of plastic is required to be manufactured. If 150 million tons of plastic are produced and 50 million tons of plastic are recycled each year in North America, how many tons of plastic need to be newly manufactured each year?

    (A) 1.5 million tons

    (B) 15 million tons

    (C) 115 million tons

    (D) 135 million tons

33. All of the following statements about the degradation of soil by climate change are true EXCEPT

    (A) increased flooding in coastal areas leads to soil desalinization

    (B) increased global temperature produces more deserts

    (C) increased evaporation of water leads to soil salinization

    (D) increased global temperature leads to increased loss of organic material in soil

Freshwater Usage in the United States, 2000

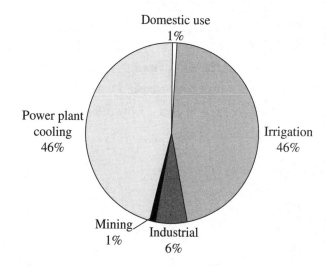

34. According to the diagram above, cooking, showering, and using toilets accounted for approximately what percentage of total water use?

    (A) 92 percent

    (B) 46 percent

    (C) 6 percent

    (D) 1 percent

GO ON TO THE NEXT PAGE.

35. One result of increased troposphere temperatures that is observed today is

    (A) an increase in skin cancers in people

    (B) an increase of 10 to 20 cm in the average global sea level

    (C) more radon seepage into people's homes

    (D) a deeper permafrost in Arctic regions

36. All of the following are major components of air pollution EXCEPT

    (A) sulfur dioxide

    (B) lead

    (C) arsenic

    (D) ozone

37. Which of the following is the root cause of habitat loss, especially in less developed nations?

    (A) Road building

    (B) Poverty

    (C) Conversion of forest to farmland

    (D) Capturing exotic animals for resale

38. All of the following are causes of urban sprawl EXCEPT

    (A) increased number of roads

    (B) higher crime rates in urban areas

    (C) higher levels of air and noise pollution in cities

    (D) increased oil prices

39. The motion of tectonic plates accounts for most of the Earth's

    (A) CO_2 emissions

    (B) river formation

    (C) changes of season

    (D) volcanic activity

40. "K" and "r" are used to describe which of the following aspects of populations?

    (A) The place in a habitat where these organisms live

    (B) The number of males and females in the population

    (C) The reproductive tactics used by the populations

    (D) The time it takes a population to double

41. Convectional heating and cooling of the atmosphere transfers which of the following to other parts of the Earth?

    I. Heat
    II. Moisture
    III. Nutrients

    (A) I only

    (B) II only

    (C) I and II only

    (D) I, II, and III

GO ON TO THE NEXT PAGE.

Questions 42 and 43 refer to the following diagram.

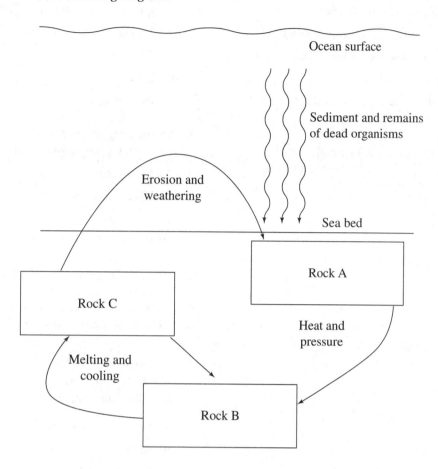

Ocean surface

Sediment and remains of dead organisms

Erosion and weathering

Sea bed

Rock A

Rock C

Heat and pressure

Melting and cooling

Rock B

42. Which of the following gives the correct sequence in the rock cycle shown above?

(A) Rock A—Metamorphic/Rock B—Igneous/ Rock C—Sedimentary

(B) Rock A—Sedimentary/Rock B—Metamorphic/ Rock C—Igneous

(C) Rock A—Igneous/Rock B—Sedimentary/ Rock C—Metamorphic

(D) Rock A—Sedimentary/Rock B—Igneous/ Rock C—Metamorphic

43. Which of the following rock types would contain the greatest number of fossils?

(A) Igneous rock only

(B) Metamorphic rock only

(C) Sedimentary rock only

(D) Sedimentary and igneous rock

GO ON TO THE NEXT PAGE.

44. The North Atlantic Current provides which of the following for Europe and North America?

    (A) Fish to feed predators such as killer whales

    (B) Warm water that moderates land temperatures

    (C) Large amounts of CO_2 to promote photosynthesis

    (D) Mineral-rich waters to reduce depleted mineral reserves

45. Which of the following indoor air pollutants is composed of microscopic mineral fibers that can produce lung cancer in humans?

    (A) Nitrogen oxides

    (B) Asbestos

    (C) Radon

    (D) Formaldehyde

46. Which of the following is an example of commensalism?

    (A) Cheetahs and antelope

    (B) Bees and flowers

    (C) Humans and tapeworms

    (D) Barnacles and whales

47. Which two countries together are responsible for 40% of global greenhouse gas emissions?

    (A) United States and Great Britain

    (B) United States and China

    (C) China and India

    (D) Great Britain and Russia

48. Nuclear reactors use which of the following to absorb neutrons in the reactor core?

    (A) A steam condenser

    (B) Control rods

    (C) A heat exchanger

    (D) Fuel rods

49. Riparian areas are vital to the preservation of high-quality

    (A) mountain slopes

    (B) grazing land

    (C) rivers and streams

    (D) ocean beaches

50. Which of the following fishing techniques is most damaging to ocean bottom ecosystems?

    (A) Trawling

    (B) Drift nets

    (C) Long lines

    (D) Purse seine

51. Which of the following treaties is responsible for lower levels of CFC production worldwide?

    (A) Montreal Protocol

    (B) Kyoto Protocol

    (C) Clean Air Act

    (D) Rio Earth Summit of 1972

52. DDT is an insecticide sprayed to control insects. Years after it was introduced, DDT was found in large predatory birds, such as the osprey. Which of the following processes caused the DDT to be found in the osprey?

    I. Biomagnification
    II. Bioremediation
    III. Bioaccumulation

    (A) I only

    (B) III only

    (C) I and III only

    (D) I, II, and III

53. Doing which of the following could most cost effectively reduce acid rain and acid deposition?

    (A) Reducing the use and waste of electricity

    (B) Making taller smokestacks

    (C) Adding lime to acidified lakes

    (D) Moving power plants to desert areas

54. Which of the following forms of radiation is most harmful to humans?

    (A) Alpha

    (B) Gamma

    (C) Beta

    (D) Infrared

GO ON TO THE NEXT PAGE.

55. Which of the following ecosystems does NOT use solar energy as its ultimate energy source?

    (A) Pond

    (B) Deep-sea hydrothermal vent

    (C) Rain forest

    (D) Tundra

56. All of the following are true about CO_2 sequestering EXCEPT

    (A) it can be accomplished by pumping CO_2 into carbonated beverages

    (B) it can be accomplished by pumping CO_2 into crop lands

    (C) it can be accomplished by pumping CO_2 deep under the ocean floor

    (D) it can be accomplished by pumping CO_2 deep underground into dried-up oil wells

57. Which of the following is NOT true concerning invasive species?

    (A) They can outcompete native species in a habitat.

    (B) They are highly specialized and have narrow niches.

    (C) They alter the biodiversity of the area they are invading.

    (D) They are introduced into a habitat and are not native.

58. The Second Law of Thermodynamics is best exemplified by which of the following?

    (A) The amount of solar radiation going into an eco-system is equal to the total amount of energy going out of that system.

    (B) The amount of carbon in the atmosphere has in-creased due to the combustion of fossil fuels.

    (C) As electricity is transmitted through wires, some of the power is lost to the environment as heat.

    (D) Wind-generated electricity has more power than electricity generated at a hydropower plant.

59. Salinization of soil can be caused by all of the following EXCEPT

    (A) flooding in coastal areas

    (B) rising temperatures

    (C) excessive irrigation

    (D) drip irrigation

60. Which of the following is true about early loss populations, such as fish?

    (A) The chances of an adult dying are about the same as a child dying.

    (B) The maturation process is slow.

    (C) The populations are close to the carrying capacity.

    (D) Many individuals die at an early age.

GO ON TO THE NEXT PAGE.

Questions 61 and 62 refer to the illustration of succession below.

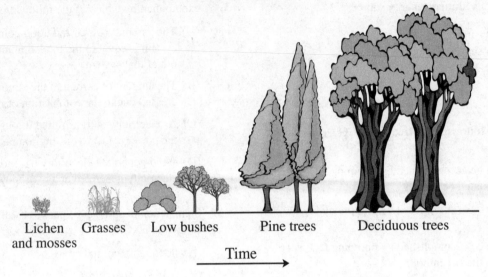

Lichen and mosses   Grasses   Low bushes   Pine trees   Deciduous trees

Time

61. A farmer stops farming a certain tract of land, and small bushes soon grow there. The land then progresses to the deciduous tree stage. This process is known as

(A) pioneer succession

(B) wetland succession

(C) secondary succession

(D) primary succession

62. According to the diagram, low species diversity and small-sized plants are characteristics of which stage of succession?

  I. Late-stage succession

 II. Midstage succession

III. Early-stage succession

(A) I only

(B) II only

(C) III only

(D) I and III only

GO ON TO THE NEXT PAGE.

63. Ozone depletion is occurring most rapidly in the Earth's polar regions because

    (A) the atmosphere is thicker at the poles, so ozone destruction is easier to observe

    (B) large amounts of chlorofluorocarbons (CFCs) can accumulate on ice crystals formed in the cold atmosphere

    (C) the upper atmosphere winds form a pattern of high- and low-pressure systems that can cause the destruction of ozone

    (D) the solar UV radiation is stronger at the poles, promoting the breakdown of ozone

64. Smaller forest fires are beneficial to forests for all of the following reasons EXCEPT

    (A) combustion of dried leaves or needles, which reduces the threat of large fires

    (B) burning the crowns of trees

    (C) germinating seeds of certain plant species

    (D) making burned matter available as a nutrient

65. What are the negative impacts of dams on ecosystems?

    I. Loss of silt in the river downstream from the dam
    II. Generation of low pollution electricity
    III. Loss of terrestrial biodiversity in areas surrounding the dam

    (A) I only

    (B) III only

    (C) I and III only

    (D) I, II, and III

66. Which of the following best illustrates the process of evolution?

    (A) A parasite population becomes resistant to a drug

    (B) Rabbits can have brown fur in summer and white fur in winter

    (C) Frogs burrow deep into the mud during winter

    (D) A baby is born and has a different color hair than his or her parents

67. Which of the following is a renewable energy source?

    (A) Crude oil

    (B) Coal

    (C) Natural gas

    (D) Hydrogen cells

68. The energy necessary to produce stratospheric ozone comes from which of the following?

    (A) Sunlight

    (B) Radioactive decay

    (C) Magma

    (D) Wind

69. Which of the following chemicals can cause lung irritation in the troposphere but is very helpful to humans in the stratosphere?

    (A) O_2

    (B) O_3

    (C) Chlorofluorocarbons

    (D) H_2SO_4

70. Which of the following is true of primary and secondary pollutants?

    (A) Primary pollutants rise up the smokestack before secondary pollutants are formed.

    (B) Primary pollutants are formed from secondary pollutants interacting in the water.

    (C) Secondary pollutants are formed from primary pollutants interacting in the atmosphere.

    (D) Secondary pollutants are directly created by the burning of coal and primary pollutants from the burning of oil.

71. A sample of radioactive iodine-131 is found to have an activity level of 4×10^{-6} curies and a half-life of 8 days. How much time must pass before the activity level of the radioactive waste drops to 2.5×10^{-5} curies?

    (A) 4 days

    (B) 8 days

    (C) 16 days

    (D) 32 days

GO ON TO THE NEXT PAGE.

72. Coal, oil, and natural gas were all formed as a result of

    (A) the decay of organic matter

    (B) the movement of magma in volcanoes

    (C) sedimentary rock turning into metamorphic rock

    (D) the radioactive decay occurring inside the Earth

73. All of the following are negative impacts of food production EXCEPT

    (A) increased erosion

    (B) air pollution from fossil fuels

    (C) bioaccumulation of pesticides

    (D) lower death rates

74. Soils found in mid-latitude grasslands would be most accurately described as having

    (A) a high acid content with little organic matter

    (B) a deep layer of humus and decayed plant material

    (C) a layer of permafrost right below the O-horizon

    (D) a high content of iron oxides and very little moisture

75. All of the following are useful methods for reducing domestic water use EXCEPT

    (A) using low-flow shower heads

    (B) using low-flush-volume toilets

    (C) fixing leaks as soon as they start

    (D) lowering the temperature of the water heater

76. Biodiversity is a direct result of which of the following?

    (A) Deforestation

    (B) Respiration

    (C) Erosion

    (D) Evolution

77. Students studying a river found high levels of fecal coliform bacteria. They concluded that

    (A) this water is fit to swim in

    (B) a nearby treatment plant added chlorine to the waste water

    (C) they can safely drink the water

    (D) untreated animal waste was put into the water

78. During an El Niño-Southern Oscillation, weather events change in which of the following areas?

    (A) The Pacific and Indian Oceans

    (B) The Atlantic and Indian Oceans

    (C) The Arctic Sea

    (D) The Indian and Antarctic Oceans

79. Which of the following pairs correctly matches the source of gray water with its most frequent use in the home?

    (A) Dishwasher and sink water used to flush toilets

    (B) Flushed toilet water used to irrigate garden plants

    (C) Dishwasher and sink water used to irrigate garden plants

    (D) Water collected from rainfall used to flush toilets

80. The Clean Water Act did all of the following EXCEPT

    (A) set water quality standards for all contaminants in surface waters

    (B) make it unlawful for any person to discharge any pollutant from a point source into navigable waters

    (C) demand that an environmental impact statement be prepared for any major development

    (D) fund the construction of sewage treatment plants

END OF SECTION I

ENVIRONMENTAL SCIENCE
Section II
Time—1 hour and 10 minutes
3 Questions

Directions: Answer all three questions, which are weighted equally; the suggested time is about 23 minutes for answering each question. Where calculations are required, clearly show how you arrived at your answer. Where explanation or discussion is required, support your answers with relevant information and/or specific examples.

1. A class wished to determine the LD_{50} of a particular herbicide, Chemical X, on seedlings of the most common type of pest plant at a local tree farm. Using standard laboratory apparatus and glassware, they accurately made the following dilutions: 1.0M, 10^{-1}M, 10^{-2}M, 10^{-3}M, 10^{-4}M, and 10^{-5}M. They grew the seedlings under standard conditions, varying only the concentrations of Chemical X. Finally, they determined the percentage of seedlings that germinated at each concentration.

   (a) Use the description of the experiment above to answer the following questions.
       i. **Identify** a reasonable hypothesis that this experiment could test.
       ii. **Describe** the experimental control group and offer one method for performing repeated trials.

   (b) **Identify** a set of hypothetical results.
       i. Using the axes below, graph your results. Properly **identify** and label the axes and provide a title for the graph.
       ii. **Calculate** and identify the LD_{50} concentration on your graph. **Show** your work.

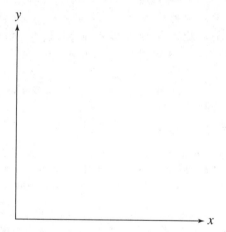

   (c) **Describe** one positive outcome and TWO negative outcomes of using herbicides in the environment.

   (d) Given the possible negative effects of herbicide use, **propose** one alternative strategy the tree farm could employ to control this pest plant.

GO ON TO THE NEXT PAGE.

2. The diagram below illustrates the demographic transition model of the relationship between economic status and population.

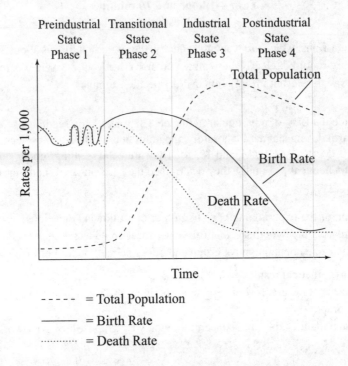

(a) In Phases 2 and 3, there is a large difference between the birth rate and the death rate. **Describe** the effects on the overall population as a result of this difference. **Explain** why the population doubling time during these phases is short.

(b) Choose one of the four phases and **identify** an economic factor that would account for the differences between birth rate and death rate.

(c) **Describe** one biological method of birth control.

(d) Population experts have reported that in some developing countries, the population is experiencing a reverse transition from Phase 2 to Phase 1. **Describe** what would happen to a country's population during such a reverse transition and **make a claim** about one event that could cause this reverse transition.

GO ON TO THE NEXT PAGE.

3. Under certain conditions, an internal combustion car engine produces approximately 3 grams of NO_x per kilometer driven. In Country C, there are 300 million cars, and each car is driven only 20,000 km per year.

   (a) **Calculate** the number of metric tons of NO_x produced by the cars in Country C under these conditions in one year. 1 metric ton = 1,000,000 g.

   (b) **Identify** a secondary pollutant that is derived from the NO_x produced by Country C, **explain** how it is produced, and **explain** how that pollutant travels to adjacent countries.

   (c) **Describe** one abiotic and one botic impact that the NO_x pollution will have on any countries adjacent to Country C.

   (d) **Propose** and **describe** one method that Country C could employ to reduce the amount of emitted NO_x.

STOP

END OF EXAM

Practice Test 3:
Answers and
Explanations

PRACTICE TEST 3 ANSWER KEY

1.	D	21.	B	41.	C	61.	C
2.	B	22.	A	42.	B	62.	C
3.	B	23.	B	43.	C	63.	B
4.	A	24.	C	44.	B	64.	B
5.	D	25.	C	45.	B	65.	C
6.	D	26.	D	46.	D	66.	A
7.	A	27.	B	47.	B	67.	D
8.	D	28.	D	48.	B	68.	A
9.	C	29.	D	49.	C	69.	B
10.	A	30.	D	50.	A	70.	C
11.	A	31.	C	51.	A	71.	D
12.	D	32.	D	52.	C	72.	A
13.	A	33.	A	53.	A	73.	D
14.	D	34.	D	54.	B	74.	B
15.	A	35.	B	55.	B	75.	D
16.	D	36.	C	56.	A	76.	D
17.	C	37.	B	57.	B	77.	D
18.	C	38.	D	58.	C	78.	A
19.	B	39.	D	59.	D	79.	C
20.	A	40.	C	60.	D	80.	C

PRACTICE TEST 3: ANSWERS AND EXPLANATIONS

Section I—Multiple-Choice Questions

1. **D** Type III curves tend to describe *r*-selected species, which have high growth rates and produce many offspring, relatively few of which survive to adulthood. Oysters are a good example of an *r*-selected species, while moose are *K*-selected. Songbirds and sea turtles are likely to have Type II survivorship curves, showing them to be somewhere on the continuum between *r*- and *K*-selection.

2. **B** Type II curves are more or less diagonal lines, like that of the chickens in the graph. This shape represents a constant slope, or roughly equal mortality rates along the curve. Choice (A) describes Type I curves, (C) describes Type III curves, and (D) is not a pattern that's commonly seen among most species.

3. **B** On the bar chart, the middle column represents 2010. The portion of the bar that represents the United States' contribution is the second portion from the bottom. Use your answer sheet as a ruler to get approximate values for the top and bottom of this box on the *y*-axis: about 425 and about 550. Do the same for the column total: it's about 750. The units are not important since this is a percentage question. Subtract the top and bottom values for the United States portion to find the U.S. contribution: 550 − 425 = 125. To find the percentage of the total, divide this value by the total and multiply the result by 100: 125 ÷ 750 ≈ 0.167, or about 16.7%. Since the question asks for an approximation and the answer choices are whole numbers, round up to the nearest whole number. Choice (B), 17%, is correct.

4. **A** The sections of each column representing Africa, the United States, and the rest of the world each increase between 2000 and the 2020 predicted value. However, the section representing China remains about the same size.

5. **D** The dead zone was created because of the fertilizer and sewage that the Mississippi River carries into the Gulf. Since the water containing these pollutants is very nutrient-rich and warm, it creates a food source for phytoplankton and zooplankton in the surface waters there, creating a population explosion among these organisms. Since there is then more detritus from dead phytoplankton and zooplankton feeds, the bacteria at the sea floor have abundant food for decomposition, but they metabolize and use up the available oxygen while doing so, creating the hypoxic ("dead") zone. Choice (D) correctly captures this chain of events.

6. **D** The hypoxic zone stays in place from May to September, when cooler and wetter weather helps break it up: during this time, so little oxygen is available that most organisms can't grow there, (B). Another effect is reproductive problems for fish who develop in low-oxygen areas near the dead zone, (C). Both of these contribute to the decline of the fishing industries that depend on ocean life there, (A). While it is easy to imagine that fish would simply migrate, many are simply rendered unconscious and die quickly, and habitat loss kills others, (D).

7. **A** A regulating service is one that provides benefits from the regulation of some ecosystem process. Water purification, (A), is a good example. Pollination, (B), is a supporting service; energy generation, (C), is a provisioning service; ecotherapy, (D), is a therapeutic, and therefore cultural, service.

8. **D** All the choices are consequences of disruption to ecosystem services, but (D) is a positive, not a negative consequence: habitat disruption in this case is providing a cultural service (that of science and education). Lower crop yields and higher costs of mining are examples of negative consequences to provisioning services, while outbreaks of disease are an example of a negative consequence to regulating services.

9. **C** While water vapor, (C), is a greenhouse gas, it doesn't contribute significantly to global climate change because it has a short residence time in the atmosphere. All the other choices are important contributors to global climate change.

10. **A** The depletion of stratospheric ozone, while an important environmental issue that also has anthropogenic causes, is not directly linked to the greenhouse effect. All the other choices are true about human involvement in the greenhouse effect.

11. **A** Chlorofluorocarbons have the highest GWP, followed by nitrous oxide, and then methane. Carbon dioxide has a GWP value of 1, since it is used as a baseline.

12. **D** Noise is measured in decibels (dB). A vacuum cleaner has a loudness of about 70 dB, a chain saw has a loudness of about 100 dB, the average factory has a loudness of about 80 dB, and a rifle has a loudness of about 160 dB. The rifle would most likely cause the greatest damage to a person's hearing because it is the loudest.

13. **A** Population density is a measure of individuals per unit area. On land, it is measured as the number of individuals per square area (hectare, meter, etc.), while in aquatic environments, the measure is based on the number of individuals per unit of volume (milliliters, liters, etc.).

14. **D** This graph illustrates what can happen to supplies of a nonrenewable resource, such as oil. Curve A shows a resource which is being rapidly consumed, which is not being recycled, and for which no new reserves are found. Curve B shows a resource which can be partially recycled, for which there are more reserves, and which is being conserved. Curve C is seen when the use of the resource is reduced and there are relatively ample reserves.

15. **A** The Aral Sea, located in Uzbekistan and Kazakhstan (both countries were part of the former Soviet Union), is a saline lake. It is in the center of a large, flat desert basin. In the past few decades, the Aral Sea's volume has decreased by 75 percent because of both natural and human causes. Choice (B) would be a result, not a cause; (C) is not correct because earthquakes do not cause water loss; (D) is not correct because pesticides would cause water pollution, not water loss.

16. **D** All three of these events can occur when fish populations are overharvested. The U.S. National Fish and Wildlife Foundation found that 14 major commercial fish species are almost depleted. Also, one-fourth of the annual fish catch is by-catch or nontarget organisms.

17. **C** In flood irrigation, water flows via gravity to all parts of the field. In this process, about 20% of the water is lost to evaporation and runoff. Drip irrigation uses above-ground or below-ground pipes to deliver water directly to the plant's roots; this method is highly effective. Furrow irrigation involves cutting furrows between crop rows and filling them with water: about 1/3, or 33%, of the water is lost to evaporation and runoff. Finally, spray irrigation is more efficient than furrow irrigation but still loses about 25% of water.

18. **C** Watch your letters! Option B is (C). Choice (C) represents the exponential growth phase of this population. In exponential growth, birth rates are high and environmental resistance is low, so the population rapidly increases over a short period. Also note that the line is at its steepest slope.

19. **B** In population graphs such as this logistical growth curve, the label (K) represents the carrying capacity. The carrying capacity represents the maximum number of individuals that the habitat can sustain for a long time. The biotic potential (or how rapidly the organisms can reproduce) is balanced by the environmental resistance (factors that lower a population). Note how the population size modulates above and below the carrying capacity.

20. **A** Population growth is studied over time, so the independent axis (x) is time. Choice (B) is the dependent variable, population number, which is on the vertical axis (y). Choice (C) is the name of the growth pattern; it is not a variable. Choice (D) refers to environmental effects, factors that would slow down population growth.

21. **B** "Environmental resistance" refers to those factors that slow the growth of a population. Unfavorable abiotic factors (e.g., the climate is too hot or too dry) or biotic factors (e.g., predators or disease) will limit the number of survivors. At point D, the population is about to increase, so environmental resistance is not high—it's low.

22. **A** Phosphates are most commonly found as phosphate salts containing phosphate ions (PO_4^{-3}). Phosphorous does not dissolve easily in water and doesn't form a gas at normal temperatures and pressures. Most of the Earth's sulfur is stored in rocks as sulfate salts (SO_4^{-2}). Both phosphorus and sulfur are only held in living organisms for short periods.

23. **B** The movement of tectonic plates in the Earth's lithosphere occurs because these plates are floating on the semi-liquid magma underneath them. Choices (A) and (C) both deal with the loss of species, and (D) deals with the movement of populations.

24. **C** The Green Revolution started in the 1950s; farmers used new methods of farming, such as breeding high-yield crops, better irrigation processes, and large amounts of fertilizer and pesticides.

25. **C** Mortality figures are calculated as the number of deaths per 1,000 people aged zero to one. For example, in 1987, there were 9,889 babies born in State X. In that same year, 116 infants (aged zero to one) died in the state. The infant mortality rate would be calculated as

$(116 \div 9,889 = .0117) \times 1,000 = 11.7$

So, the mortality rate is 11.7.

26. **D** Underground storage tanks can leak chemicals into the surrounding ground; these chemicals might then move into the water that lies underground, also known as the aquifer. Choice (A) may pollute aquifers as well, so it is not a good answer. Choice (B) is not associated with underground events. Choice (C) is the boundary between two habitats; it is too broad an answer.

27. **B** Choice (A), CITES, is an agreement that concerns the international trade of endangered species. Choice (B), the Kyoto Protocol, is the correct answer. Choice (C), the Montreal Protocol, concerns the emissions of CFCs. Choice (D), the Clean Air Act, deals with air pollution in the United States; it is not an international treaty.

28. **D** Choice (D) is the best answer. Choices (A) and (C) both require the use of more mineral (and therefore nonrenewable) resources. Choice (B) consumes mineral resources and is also costly. Sustainability is the ability of people to utilize a resource over the long term (for many generations). This can be achieved by reducing demand and finding ways to reduce waste as much as possible. Replacing incandescent light bulbs with florescent light bulbs is one example of a way to decrease waste.

29. **D** The "Energy Policy Conservation Act," enacted into law by Congress in 1975, added Title V, "Improving Automotive Efficiency," to the Motor Vehicle Information and Cost Savings Act and established CAFE standards for passenger cars and light trucks. The act was passed in response to the 1973–1974 Arab Oil Embargo. The near-term goal was to double new car fuel economy by model year 1985. Choice (D) clearly states the goal of the Corporate Average Fuel Economy policy.

30. **D** DDT is an organic insecticide first produced in 1873. It is a neurotoxin and causes nerve cells to fire continuously, leading to spasms and death. It is dangerous because it can accumulate in the fatty tissues of all animals, and humans that are exposed to it can become sick. Birds exposed to DDT lay eggs with shells that are too thin for the embryo to survive incubation.

31. **C** Resistance cannot be acquired or developed during exposure to the pesticide, so eliminate (B) and (D). In order for pests to survive the application of the pesticide, there must have already existed pests with resistance to the pesticide. The pests with resistance to the pesticide survive and reproduce so that the next generation of pests is pesticide resistant. This explains why pesticides become ineffective over time. Therefore, (A) can be eliminated, and (C) is the correct answer.

32. **D** The amount of plastic that is conserved each year by recycling is equal to $\dfrac{0.3 \text{ tons of plastic}}{1 \text{ ton recycled plastic}} \times$ 50 million tons of recycled plastic = 15 tons of plastic from recycling. As 150 million tons of plastic are produced each year, the amount of plastic that needs to be newly manufactured is equal to 150 million tons − 15 tons of plastic from recycling = 135 million tons of newly manufactured plastic each year. Choice (D) is correct.

33. **A** Although climate change does increase the amount of flooding in coastal areas, flooding leads to soil salinization, not desalinization. Choice (A) is correct. Choices (B), (C), and (D) are all true statements.

34. **D** Domestic water use includes all uses of water around the home, and the chart indicates that it is responsible for 1 percent of all water use.

35. **B** Sea levels are rising because water is entering the oceans from melting ice caps and glacial melting. Another contributor to rising sea levels is the thermal expansion of water. Choice (A) is a result of decreased ozone in the stratosphere. Choice (C) is not correct: radon seepage does not correlate with atmospheric temperatures. Choice (D) would occur if there were lower temperatures, not higher temperatures.

36. **C** Arsenic is a groundwater pollutant released by mining activities and is not an air pollutant. The major components of air pollution are sulfur dioxide, particulates, lead, ozone, nitrogen dioxide, and carbon monoxide.

37. **B** Poverty is a force that drives many people to exploit the land they live on. Farming, trade in animals, and building roads are all methods that people use to make money.

38. **D** Increased oil prices would make it less affordable for people to fuel their cars to commute to and from their jobs. This would lead to less urban sprawl, so (D) is correct. The other choices are all reasons why people migrate out of the city and into the suburbs.

39. **D** When two of these massive rock plates collide, often one slips under the other; this process is called subduction. The collision of two plates causes the rock layer of the lithosphere to crack. At these sites, magma may rise from the molten exterior of the Earth, and a volcano is formed. Choice (A) is a human activity; (B) can be caused by many geological processes; and (C) is an effect of the relationship between the Earth and the sun.

40. **C** "*r*" and "*K*" describe the different ways that populations reproduce. *r*-strategists reproduce at their biological potential, the maximum reproductive rate. These populations produce as many offspring as possible, and many of the offspring will not live to reproductive age. These populations are characterized by rapid increases and rapid declines, as seen, for example, in insect populations and weeds. *K*-strategists produce few offspring and nurture them carefully; humans are examples of *K*-strategists.

41. **C** Convectional cooling is an important process that moves heat and moisture around the planet. As warm, moist air rises, it cools. This cooling causes the water vapor to condense and fall as rain. The now cool, dry air becomes denser as it sinks to the Earth's surface, where more moisture and heat are picked up and convection starts again.

42. **B** Sedimentary rock is formed by the compression of eroded rock, silt, and the remains of dead organisms. Under heat, pressure, and stress within the mantle, sedimentary rock can form metamorphic rock. Deeper in the mantle, the metamorphic rock melts and cools, forming igneous rock.

43. **C** Since sedimentary rock is made of the remains of dead organisms, it contains the greatest number of fossils. Choice (D) is not correct because the melting and cooling process that forms igneous rock destroys fossils, which are usually quite fragile.

44. **B** The North Atlantic Current circulates water that was warmed by the sun (at the equator) into the northern latitudes, thus warming the land masses. When it cools, the water becomes more dense, and it picks up large amounts of CO_2 and salt. This cold, salty water circulates into the equatorial areas, cooling them while releasing CO_2 and salt.

45. **B** Choice (A) is a gas formed by combustion. Choice (C) is a gas produced by the radioactive decay of uranium. Choice (D) is a gas that comes from furniture stuffing and foam insulation. Asbestos is a mineral used in insulating pipes and as a fire retardant. If not completely confined it can crumble, and the fibers can be inhaled into the lungs, which can cause certain types of cancer.

46. **D** In commensalism, one organism benefits while the other is neither helped nor hurt. Barnacles grow on whales, where they can "hitch a ride" to nutrient-rich areas. The whale is unaffected by the barnacles, so this is an example of commensalism. Choice (D) is correct. Choice (B) is an example of mutualism, (C) is an example of parasitism, and (A) is an example of predator and prey.

47. **B** The United States and China together emit 40% of all global greenhouse gases. In 2016, both countries ratified the Paris Agreement to lower greenhouse gas emissions. Choice (B) is correct.

48. **B** Choice (A) is used to turn steam into water so that it can be turned back into steam and spin the turbines. Choice (C) is the site where the heat from the core heats water, turning it into steam that spins the turbines. Choice (D) contains uranium, which is the fuel. Control rods contain compounds, such as cadmium, that absorb neutrons; this reduces their ability to perpetuate a chain reaction. Choice (B) is correct.

49. **C** Riparian zones are the areas of vegetation that abut a river or stream, so (C) is correct. They form a corridor of vegetation along the banks of the river or stream. They increase habitat diversity for organisms living in or near the water and can absorb excessive nitrogen, phosphorous, and pesticides, preventing them from entering the water.

50. **A** Trawling involves dragging a net across the ocean floor. This disrupts the habitat and catches a wide variety of bottom-dwelling species. Drift nets float in the water; long lines include baited hooks, and they are towed behind boats; purse seine involves catching surface fish as a net is drawn up from below.

51. **A** In 1987, 36 nations met in Montreal and signed the Montreal Protocol, which cut the emissions of CFCs by about 35 percent between 1989 and 2000. In 1992, the protocol was updated in a meeting in Denmark to accelerate the phasing out of ozone-depleting chemicals. Choice (B), Kyoto Protocol, dealt with CO_2 levels. Choice (C), CAA, affected the United States only. Choice (D), Rio Earth Summit, dealt with many global environmental problems, but Montreal was specific to ozone depletion.

52. **C** Bioaccumulation is the process by which chemicals remain and accumulate in the bodies of animals. These chemicals come from food and cannot be metabolically removed from the tissues. Generally, they are fat-soluble. Biomagnification occurs when a predator eats several organisms and each individual prey has a bit of the chemical in its tissues. Bioaccumulation and biomagnification work together to increase the levels of toxic materials in the bodies of animals. Bioremediation is the removal of toxic compounds by living organisms.

53. **A** A reduction in the creation of pollution is always the least expensive method. Choice (D) is impractical because air pollution can still spread to other areas. Choice (C) does not help acid deposition on land or in the atmosphere. Choice (B), taller stacks, would just spread pollution to other locations.

54. **B** Gamma radiation can penetrate most materials and, of the listed forms of radiation, it is the most harmful to humans. When gamma radiation gets into a cell and damages the DNA, cancer can result. Choice (A), alpha rays, can be stopped by a piece of paper. Choice (C), beta rays, can be stopped by wood or clothing. Choice (D), infrared radiation, is heat, such as that which comes from a stove.

55. **B** The producers in the deep-sea hydrothermal vent ecosystem do not capture sunlight and use it to perform photosynthesis. Instead, they use chemical energy from the hot, mineral-rich water that comes out of the vents. The producers in these vents are bacteria that fall into the domain Archaea. The other animals that live near the vents—tube worms, crabs, and many others—all depend on the bacteria for food. Choices (A), (C), and (D) all refer to ecosystems in which producers use radiant energy to carry out photosynthesis.

56. **A** All of the answer choices represent correct processes except (A), adding it to beverages. In this process, CO_2 would reenter the atmosphere, which is the opposite of the long-term storage goal of sequestering.

57. **B** Choice (B) is not necessarily true. Invasive species are r-selected organisms; they are usually small adults, have many offspring, sexually mature very quickly, and are generalists. Choices (A), (C), and (D) all correctly describe invasive species.

58. **C** Choice (A) is an example of the First Law of Thermodynamics. Choice (B) exemplifies the law of mass conservation. Choice (D), the usefulness of electricity, is the same regardless of how it is produced. The Second Law of Thermodynamics states that there is a loss of energy at each energy transformation.

59. **D** Drip irrigation is a form of irrigation designed to prevent salinization of soil from excessive irrigation. Choice (D) is correct. The other choices are all causes of salinization of soil.

60. **D** Many of the young die in species that are considered early loss; one example of this is seen in fish. In some fish species, most of the individuals die in the first weeks after hatching. Choice (A) describes species such as birds, and (B) and (C) both describe late-loss species, like humans and elephants.

61. **C** Secondary succession is defined as succession that begins on land that was disturbed by an (often human) activity. Primary succession begins on land that was exposed to abiotic factors, e.g., when lava cools or when a glacier leaves an area.

62. **C** By definition, early-stage species are small and exhibit little diversity. At the early stages of succession, an area is populated mostly by r-selected organisms. Middle stage and late-stage communities generally have larger adult species and many more K-selected organisms.

63. **B** Steady winds create polar vortices, which can trap large amounts of CFCs from other parts of the world. In winter, ice crystals form in the upper atmosphere. The CFCs accumulate on the crystals, and when the polar spring returns, the crystals melt and release the CFCs. Then, the free Cl breaks down the ozone.

64. **B** The crown is the top, or most rapidly growing part, of the tree. In very large forest fires, the crowns of the trees burn, weakening or killing the trees. Leaves or needles burning in smaller fires take away fuel for future fires and break down matter to ash, which provides nutrients for the trees. Fires are also responsible for allowing the seeds of many plant species to germinate.

65. **C** Dams can have detrimental impacts on local environments and populations. These detrimental effects include high construction costs, high CO_2 emissions from decaying biomass, flooding of natural areas, conversion of terrestrial ecosystems to aquatic ecosystems, danger of collapse, and blocking migratory fish. Choice II, the generation of low pollution electricity, is a positive effect, so eliminate any answer choices that include II.

66. **A** Evolution is the change in the genetic makeup of a population over time. The key term is *genetic makeup*—also referred to as the gene pool. In this type of question, look for a phrase or word that implies the passage of time; in this question, it's *becomes*. That should give you a hint that, at one time, the population was not resistant and now it is resistant. All the options except (A) are temporary changes. Choice (D) is also incorrect because the change occurs in one individual, not a whole population.

67. **D** Hydrogen cells, (D), are a renewable energy source. The fuel, hydrogen, can be produced by electrolysis of water. The waste product of hydrogen cells is water, which can be reused to produce more hydrogen fuel. Choices (A), (B), and (C) are all nonrenewable energy sources and are, therefore, incorrect.

68. **A** Ozone formation occurs when sunlight (UV) binds an atom of oxygen to a diatomic oxygen molecule (O_2). Neither radioactive decay nor magma's heat influences events in the stratosphere. Wind energy does not cause the chemical reaction between O_2 and oxygen.

69. **B** Ozone is a major component of air pollution, especially pollution that results from the combustion of fossil fuels. In the stratosphere, however, ozone blocks large amounts of UV light from the sun. Choice (A), all animals breathe O_2, so it is vital to survival. Choice (C), chlorofluorocarbons destroy O_3, so it is harmful in the stratosphere. Choice (D), H_2SO_4, or sulfuric acid, is harmful anywhere it's found in the atmosphere.

70. **C** Primary pollutants are created via combustion (coal burning, wood burning, etc.) or from volcanic activity. Examples of pollutants produced this way include oxides of sulfur and nitrogen. Secondary pollutants are created when primary pollutants combine in the atmosphere. These reactions are very complex and involve energy from the sun, water vapor, and the mixing effects of air currents.

71. **D** In order for the activity level of the radioactive iodine-131 sample to drop to 2.5×10^{-5} curies, four half-lives must pass. As the half-life of iodine-131 is 8 days, a total of 32 days must pass.

Days Elapsed	Activity Level (curies)
0	4×10^{-6}
8	2×10^{-6}
16	1×10^{-6}
24	$0.5 \times 10^{-6} = 5 \times 10^{-5}$
32	2.5×10^{-5}

72. **A**   Choices (B) and (C) are both geological processes. For (D), radioactive materials do not form fossil fuels. Coal and other fossil fuels all formed from plants and other organic material that lived some 300 to 400 million years ago. These plants and animals decayed, and the remains were exposed to tremendous pressures and temperatures.

73. **D**   Countries that have more food will have healthier children and adult populations, and thus they will have lower death rates. Increased food production results in increased erosion because after crops are harvested, the bare ground is more readily eroded. Approximately 90 percent of applied pesticides do not reach their targets and can end up in water and soil. The heavy use of large machinery means that more plants can be harvested, but these machines are powered by fossil fuels.

74. **B**   Mid-latitude grasslands alternate between rapid growth and drought. Drought and cold winter temperatures permit only a small amount of decomposition each year, which leads to the accumulation of organic matter. In North America, this layer can be more than 30 m thick. Choice (A) describes coniferous forest soils, (C) is arctic tundra, and (D) describes desert soil.

75. **D**   Choice (D) might reduce the cost of heating the water, but it will not slow down water consumption. All the other answer choices describe ways to use less water.

76. **D**   Evolution results in the development of populations that are fit to live in certain habitats. Sexual reproduction allows for genetic diversity in populations. Organisms with slightly different adaptations can survive in different niches within a habitat. So, as different types of species evolve in different niches, the biodiversity in a habitat increases.

77. **D**   Because fecal coliform bacteria come from animal (and human) waste, the answer is (D). Choices (A) and (C) are not correct, as fecal bacteria can cause a number of diseases. Chlorine is used to kill fecal bacteria, so (B) is also incorrect.

78. **A**   In an El Niño, the winds that normally blow from the eastern Pacific to the western Pacific weaken or stop. This shifts the precipitation from the western Pacific to the central Pacific and can alter weather events over as much as two-thirds of the globe.

79. **C**   Gray water is water from sources such as washing machines, showers, dishwashers—almost any water-using appliances, except toilets. In areas of limited rainfall, gray water is stored in underground tanks and used as irrigation water for gardens, lawns, etc. It may constitute 50 percent to 80 percent of domestic wastewater. Generally, it is not used to irrigate crop plants.

80. **C**   Choices (A), (B), and (D) are all provisions of the CWA. Choice (C) is a provision of the National Environmental Policy Act, so it is the exception and the correct answer.

Section II—Free-Response Questions

Question 1

A class wished to determine the LD_{50} of a particular herbicide, Chemical X, on seedlings of the most common type of pest plant at a local tree farm. Using standard laboratory apparatus and glassware, they accurately made the following dilutions: 1.0M, 10^{-1}M, 10^{-2}M, 10^{-3}M, 10^{-4}M, and 10^{-5}M. They grew the seedlings under standard conditions, varying only the concentrations of Chemical X. Finally, they determined the percentage of seedlings that germinated at each concentration.

   (a) Use the description of the experiment above to answer the following questions.

       i. **Identify** a reasonable hypothesis that this experiment could test.

You should have provided a clearly written statement that describes how the independent variable (Chemical X concentration) will affect the dependent variable (germination of seeds). Some examples might include: "As the concentration of Chemical X increases, the amount (or percentage) of seed germination decreases"; "As the amount of Chemical X decreases, the amount of seed germination increases"; or "The concentration of Chemical X does not affect the rate of seed germination."

       ii. **Describe** the experimental control group and give ONE method for performing repeated trials.

The control group would include seeds grown under exactly the same conditions as the experimental group, except that they would have no Chemical X added to their environment.

You should indicate that the seeds were germinated under the various conditions. Trials might include many seeds in one container or many containers each with a single seed, all watered with the same amount of Chemical X.

(3 points maximum—1 point each for the hypothesis, control, and repeated trials)

   (b) **Identify** a set of hypothetical results.

       i. Using the axes, graph your results. Properly **identify** and label the axes and provide a title for the graph.

       ii. **Calculate** and identify the LD_{50} concentration on your graph. **Show** your work.

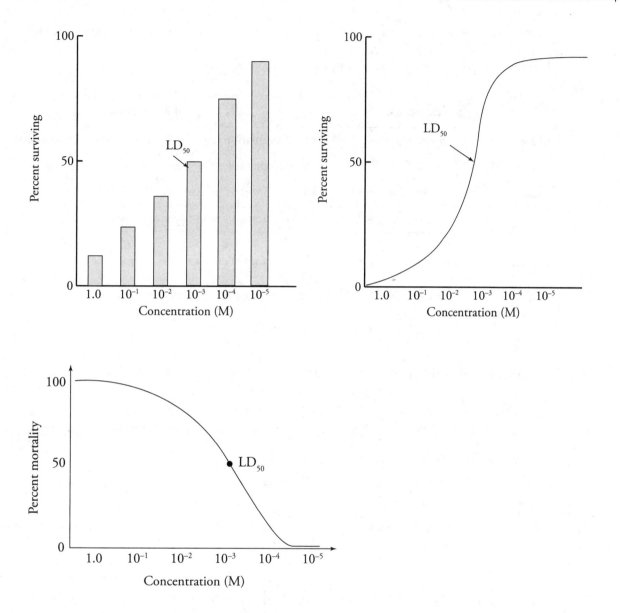

The independent variable (*x*-axis) is concentration (1.0M, 10^{-1}M, 10^{-2}M, 10^{-3}M, 10^{-4}M, and 10^{-5}M). The dependent variable (*y*-axis) is percentage surviving or mortality percentage. The LD_{50} value is indicated by the point at which less than 50 percent of the seeds germinated. For the example data here, since the surviving/mortality percentage of 50 percent corresponds to a concentration of 10^{-3}M, the LD_{50} point on the graph is labeled there. Students can use either a histogram (left) or line graph (right) for a correct graph type.

(3 points maximum—1 point for a correct x-axis and y-axis, 1 point for reasonable line or histogram graph data, and 1 point for illustrating LD_{50})

(c) **Describe** one positive outcome and TWO negative outcomes of using herbicides in the environment.

Possible answers to this question are listed in the following table.

Positive Outcomes	Negative Outcomes
Lower costs for crop production	Bioaccumulation and biomagnification of deadly poisons
Higher yields	Development of resistance to herbicide
Rapidly treated infestations	Possible harm to humans
	Possible death of non-target organisms

(3 points maximum—1 point for a positive effect, 1 point for each negative effect with a maximum of two)

(d) Given the possible negative effects of herbicide use, propose one alternative strategy the tree farm could employ to control this pest plant.

Possible answers:

- Introduce an herbivore to the area that is known to eat the plant in question

- Manual weeding

- Install blockers that will prevent plants from growing in places not designated for the tree crop

(1 point maximum for an acceptable alternative strategy)

Question 2

The diagram below illustrates the demographic transition model of the relationship between economic status and population.

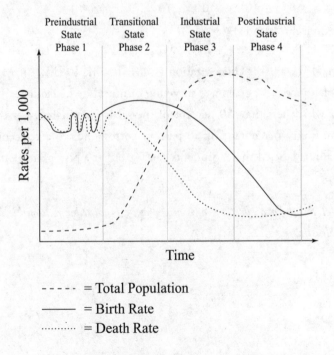

(a) In Phases 2 and 3, there is a large difference between the birth rate and the death rate. **Describe** the effects on the overall population as a result of this difference. **Explain** why the population doubling time during these phases is short.

With a large number of children being born and fewer dying, the population would grow very rapidly. Doubling times would be very short because the percent growth would be very large. The Rule of 70 states that when the percent change is large, the doubling time is rapid (e.g., 70/4 percent = 17.5 years doubling time, while 70/1 percent = 70 years doubling time). When the large number of children mature and procreate, there will be even more children born. This is known as population inertia.

(2 points maximum—1 for effects and 1 for doubling time)

(b) Choose one of the four phases and **identify** an economic factor that would account for the difference between birth rate and death rate. **Explain** how it accounts for the difference.

- Phase 1—Birth rates (BR) and death rates (DR) are approximately equal. While there are a large number of births, they are offset by high death rates due to starvation, disease, unsanitary water, poverty, or any factor that would increase the death rate.

- Phase 2—The death rate falls due to better economic conditions (more food, better health care, more jobs mean there is more money to spend). The birth rate is still high.

- Phase 3—The birth rate begins to drop due to more education opportunities for women, an increase in the cost of a child's education, an increase in the cost of raising children, more available and effective forms of birth control, fewer children needed for child labor, more children surviving past infancy, and any other factor that would lower birth rate.

- Phase 4—The birth rate and death rate are now about equal. Any factor that lowers the birth rate and maintains a low death rate is acceptable.

(2 points maximum—1 point for the factor and 1 point for the explanation)

(c) **Describe** one biological method of birth control.

There are three basic methods.

- Surgical—In females, examples include tubal ligation or cutting of oviducts (Fallopian tubes), hysterectomy (removal of the uterus), and removal of the ovaries. In males, examples include removal of the testicles and vasectomy (cutting of the vas deferens).

- Hormonal—Birth control hormones for women include "the pill" and variations, such as hormonal implants. In men, hormonal methods and chemical control are still under development.

- Obstructive/Non-hormonal Chemical—Methods include cervical cap, condoms, abstinence, vaginal ring, intrauterine device, vaginal pouch, diaphragm, spermicide, rhythm method, withdrawal, and douche.

(2 points maximum—1 for the method and 1 for the explanation)

(d) Population experts have reported that in some developing countries the population is experiencing a reverse transition from Phase 2 to Phase 1. **Describe** what would happen to a country's population during such a reverse transition and **make a claim** about one event that could cause this reverse transition.

A backward transition would mean that the death rate is increasing while a high birth rate is maintained. This means that the overall population would grow very slowly, or even decrease, depending on how close the birth rate and death rate are to each other. Events that might cause a rapid rise in death rate are war, civil unrest, abject poverty, a rapid rise in infectious diseases (such as AIDS), or natural disaster.

(4 points maximum—2 for the description of the effect and 2 for naming the event and explanation)

Question 3

Under certain conditions, an internal combustion car engine produces approximately 3 grams of NO_x per kilometer driven. In Country C, there are 300 million cars and each car is driven only 20,000 km per year.

(a) **Calculate** the number of metric tons of NO_x produced by the cars in Country C under these conditions in one year. 1 metric ton = 1,000,000 g.

3×10^8 cars $\times 2 \times 10^4$ km per year $\times 3$ g NO_x per km $= 18 \times 10^{12}$ or 1.8×10^{13} g NO_x. 1 metric ton is 1×10^6 g, so the final answer is 1.8×10^7 tons of NO_x.

(2 points maximum—1 point for the setup and 1 point for the correct answer)

(b) **Identify** a secondary pollutant that is derived from the NO_x produced by Country C, **explain** how it is produced, and **explain** how that pollutant travels to adjacent countries.

Nitric acid is a secondary pollutant produced by the addition of NO_x to atmospheric water. This acidified water is carried by wind and rain to the countries downwind and released as acid deposition (in the form of acid rain, snow, fog, sleet, or hail). PANs (peroxyacetyl nitrates) result from NO reacting with hydrocarbons.

(2 points maximum—1 point for the explanation of nitric acid production and 1 point for the explanation of acid deposition)

(c) **Describe** one abiotic and one biotic impact that the NO_x pollution will have on any countries adjacent to Country C.

See the table on the next page for possible answers to this question.

Abiotic Effects	Biotic Effects
Water acidification	Death of young fish; loss of primary productivity in bodies of water
Soil acidification	Death of decomposers in soil (bacteria, fungi)
Rain acidification: Acid may burn plant leaves or bark, making plants more susceptible to disease.	Death of mycorrhizal fungi living with plant roots and loss of nutrients to plants
Increased transpiration might lead to lower soil water levels.	Loss of cutin layer on leaves means more transpiration out of plants

(4 points maximum—1 point each for the abiotic factor and description and for the biotic factor and description)

(d) **Propose** and **describe** one method that Country C could employ to reduce the amount of emitted NO_x.

Some methods Country C could use include the following:

- Using cleaner-burning fuels

- Adding or increasing efficiency of catalytic converters

- Increasing public transportation in order to lower the number of cars on roadways

- Taxing or employing other incentives to get people to use alternate-fuel vehicles

(2 points maximum—1 point for the method and 1 point for the description)

HOW TO SCORE PRACTICE TEST 3

Section I: Multiple-Choice

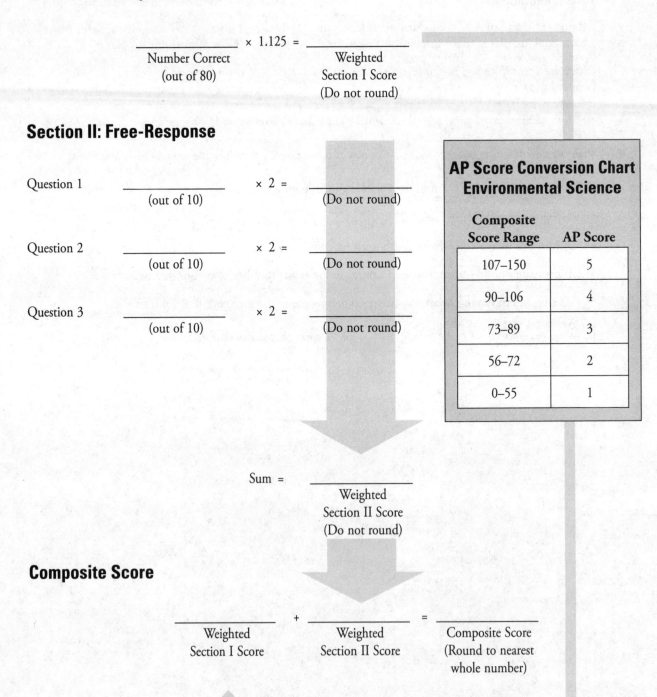

_____ × 1.125 = _____
Number Correct   Weighted
(out of 80)   Section I Score
                              (Do not round)

Section II: Free-Response

Question 1   _____   × 2 =   _____
              (out of 10)   (Do not round)

Question 2   _____   × 2 =   _____
              (out of 10)   (Do not round)

Question 3   _____   × 2 =   _____
              (out of 10)   (Do not round)

**AP Score Conversion Chart
Environmental Science**

Composite Score Range	AP Score
107–150	5
90–106	4
73–89	3
56–72	2
0–55	1

Sum =   _____
         Weighted
         Section II Score
         (Do not round)

Composite Score

_____   +   _____   =   _____
Weighted   Weighted   Composite Score
Section I Score   Section II Score   (Round to nearest
                                                 whole number)

Note: This score sheet is to help you estimate your approximate score for the official exam, not your actual score.

The **Princeton** review®

YOUR NAME: _____
(Print)   Last   First   M.I.

SIGNATURE: _____   DATE: ___ / ___ / ___

HOME ADDRESS: _____
(Print)   Number and Street

                City   State   Zip Code

PHONE NO. : _____
(Print)

IMPORTANT: Please fill in these boxes exactly as shown on the back cover of your test book.

2. TEST FORM

3. TEST CODE

4. REGISTRATION NUMBER

5. YOUR NAME

First 4 letters of last name				FIRST INIT	MID INIT
Ⓐ	Ⓐ	Ⓐ	Ⓐ	Ⓐ	Ⓐ
Ⓑ	Ⓑ	Ⓑ	Ⓑ	Ⓑ	Ⓑ
Ⓒ	Ⓒ	Ⓒ	Ⓒ	Ⓒ	Ⓒ
Ⓓ	Ⓓ	Ⓓ	Ⓓ	Ⓓ	Ⓓ
Ⓔ	Ⓔ	Ⓔ	Ⓔ	Ⓔ	Ⓔ
Ⓕ	Ⓕ	Ⓕ	Ⓕ	Ⓕ	Ⓕ
Ⓖ	Ⓖ	Ⓖ	Ⓖ	Ⓖ	Ⓖ
Ⓗ	Ⓗ	Ⓗ	Ⓗ	Ⓗ	Ⓗ
Ⓘ	Ⓘ	Ⓘ	Ⓘ	Ⓘ	Ⓘ
Ⓙ	Ⓙ	Ⓙ	Ⓙ	Ⓙ	Ⓙ
Ⓚ	Ⓚ	Ⓚ	Ⓚ	Ⓚ	Ⓚ
Ⓛ	Ⓛ	Ⓛ	Ⓛ	Ⓛ	Ⓛ
Ⓜ	Ⓜ	Ⓜ	Ⓜ	Ⓜ	Ⓜ
Ⓝ	Ⓝ	Ⓝ	Ⓝ	Ⓝ	Ⓝ
Ⓞ	Ⓞ	Ⓞ	Ⓞ	Ⓞ	Ⓞ
Ⓟ	Ⓟ	Ⓟ	Ⓟ	Ⓟ	Ⓟ
Ⓠ	Ⓠ	Ⓠ	Ⓠ	Ⓠ	Ⓠ
Ⓡ	Ⓡ	Ⓡ	Ⓡ	Ⓡ	Ⓡ
Ⓢ	Ⓢ	Ⓢ	Ⓢ	Ⓢ	Ⓢ
Ⓣ	Ⓣ	Ⓣ	Ⓣ	Ⓣ	Ⓣ
Ⓤ	Ⓤ	Ⓤ	Ⓤ	Ⓤ	Ⓤ
Ⓥ	Ⓥ	Ⓥ	Ⓥ	Ⓥ	Ⓥ
Ⓦ	Ⓦ	Ⓦ	Ⓦ	Ⓦ	Ⓦ
Ⓧ	Ⓧ	Ⓧ	Ⓧ	Ⓧ	Ⓧ
Ⓨ	Ⓨ	Ⓨ	Ⓨ	Ⓨ	Ⓨ
Ⓩ	Ⓩ	Ⓩ	Ⓩ	Ⓩ	Ⓩ

6. DATE OF BIRTH

Month	Day		Year	
○ JAN				
○ FEB				
○ MAR	Ⓞ Ⓞ		Ⓞ Ⓞ	
○ APR	① ①		① ①	
○ MAY	② ②		② ②	
○ JUN	③ ③		③ ③	
○ JUL	④		④ ④	
○ AUG	⑤		⑤ ⑤	
○ SEP	⑥		⑥ ⑥	
○ OCT	⑦		⑦ ⑦	
○ NOV	⑧		⑧ ⑧	
○ DEC	⑨		⑨ ⑨	

TEST CODE bubbles: ⓪①②③④⑤⑥⑦⑧⑨ and Ⓐ Ⓑ Ⓒ Ⓓ Ⓔ Ⓕ Ⓖ

7. SEX
○ MALE
○ FEMALE

The **Princeton** Review®

Section I

Start with number 1 for each new section.
If a section has fewer questions than answer spaces, leave the extra answer spaces blank.

1. Ⓐ Ⓑ Ⓒ Ⓓ   21. Ⓐ Ⓑ Ⓒ Ⓓ   41. Ⓐ Ⓑ Ⓒ Ⓓ   61. Ⓐ Ⓑ Ⓒ Ⓓ
2. Ⓐ Ⓑ Ⓒ Ⓓ   22. Ⓐ Ⓑ Ⓒ Ⓓ   42. Ⓐ Ⓑ Ⓒ Ⓓ   62. Ⓐ Ⓑ Ⓒ Ⓓ
3. Ⓐ Ⓑ Ⓒ Ⓓ   23. Ⓐ Ⓑ Ⓒ Ⓓ   43. Ⓐ Ⓑ Ⓒ Ⓓ   63. Ⓐ Ⓑ Ⓒ Ⓓ
4. Ⓐ Ⓑ Ⓒ Ⓓ   24. Ⓐ Ⓑ Ⓒ Ⓓ   44. Ⓐ Ⓑ Ⓒ Ⓓ   64. Ⓐ Ⓑ Ⓒ Ⓓ
5. Ⓐ Ⓑ Ⓒ Ⓓ   25. Ⓐ Ⓑ Ⓒ Ⓓ   45. Ⓐ Ⓑ Ⓒ Ⓓ   65. Ⓐ Ⓑ Ⓒ Ⓓ
6. Ⓐ Ⓑ Ⓒ Ⓓ   26. Ⓐ Ⓑ Ⓒ Ⓓ   46. Ⓐ Ⓑ Ⓒ Ⓓ   66. Ⓐ Ⓑ Ⓒ Ⓓ
7. Ⓐ Ⓑ Ⓒ Ⓓ   27. Ⓐ Ⓑ Ⓒ Ⓓ   47. Ⓐ Ⓑ Ⓒ Ⓓ   67. Ⓐ Ⓑ Ⓒ Ⓓ
8. Ⓐ Ⓑ Ⓒ Ⓓ   28. Ⓐ Ⓑ Ⓒ Ⓓ   48. Ⓐ Ⓑ Ⓒ Ⓓ   68. Ⓐ Ⓑ Ⓒ Ⓓ
9. Ⓐ Ⓑ Ⓒ Ⓓ   29. Ⓐ Ⓑ Ⓒ Ⓓ   49. Ⓐ Ⓑ Ⓒ Ⓓ   69. Ⓐ Ⓑ Ⓒ Ⓓ
10. Ⓐ Ⓑ Ⓒ Ⓓ   30. Ⓐ Ⓑ Ⓒ Ⓓ   50. Ⓐ Ⓑ Ⓒ Ⓓ   70. Ⓐ Ⓑ Ⓒ Ⓓ
11. Ⓐ Ⓑ Ⓒ Ⓓ   31. Ⓐ Ⓑ Ⓒ Ⓓ   51. Ⓐ Ⓑ Ⓒ Ⓓ   71. Ⓐ Ⓑ Ⓒ Ⓓ
12. Ⓐ Ⓑ Ⓒ Ⓓ   32. Ⓐ Ⓑ Ⓒ Ⓓ   52. Ⓐ Ⓑ Ⓒ Ⓓ   72. Ⓐ Ⓑ Ⓒ Ⓓ
13. Ⓐ Ⓑ Ⓒ Ⓓ   33. Ⓐ Ⓑ Ⓒ Ⓓ   53. Ⓐ Ⓑ Ⓒ Ⓓ   73. Ⓐ Ⓑ Ⓒ Ⓓ
14. Ⓐ Ⓑ Ⓒ Ⓓ   34. Ⓐ Ⓑ Ⓒ Ⓓ   54. Ⓐ Ⓑ Ⓒ Ⓓ   74. Ⓐ Ⓑ Ⓒ Ⓓ
15. Ⓐ Ⓑ Ⓒ Ⓓ   35. Ⓐ Ⓑ Ⓒ Ⓓ   55. Ⓐ Ⓑ Ⓒ Ⓓ   75. Ⓐ Ⓑ Ⓒ Ⓓ
16. Ⓐ Ⓑ Ⓒ Ⓓ   36. Ⓐ Ⓑ Ⓒ Ⓓ   56. Ⓐ Ⓑ Ⓒ Ⓓ   76. Ⓐ Ⓑ Ⓒ Ⓓ
17. Ⓐ Ⓑ Ⓒ Ⓓ   37. Ⓐ Ⓑ Ⓒ Ⓓ   57. Ⓐ Ⓑ Ⓒ Ⓓ   77. Ⓐ Ⓑ Ⓒ Ⓓ
18. Ⓐ Ⓑ Ⓒ Ⓓ   38. Ⓐ Ⓑ Ⓒ Ⓓ   58. Ⓐ Ⓑ Ⓒ Ⓓ   78. Ⓐ Ⓑ Ⓒ Ⓓ
19. Ⓐ Ⓑ Ⓒ Ⓓ   39. Ⓐ Ⓑ Ⓒ Ⓓ   59. Ⓐ Ⓑ Ⓒ Ⓓ   79. Ⓐ Ⓑ Ⓒ Ⓓ
20. Ⓐ Ⓑ Ⓒ Ⓓ   40. Ⓐ Ⓑ Ⓒ Ⓓ   60. Ⓐ Ⓑ Ⓒ Ⓓ   80. Ⓐ Ⓑ Ⓒ Ⓓ

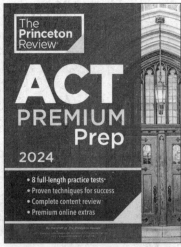

DESIGNER'S

TWO-STORY

HOME PLANS

MW00862033

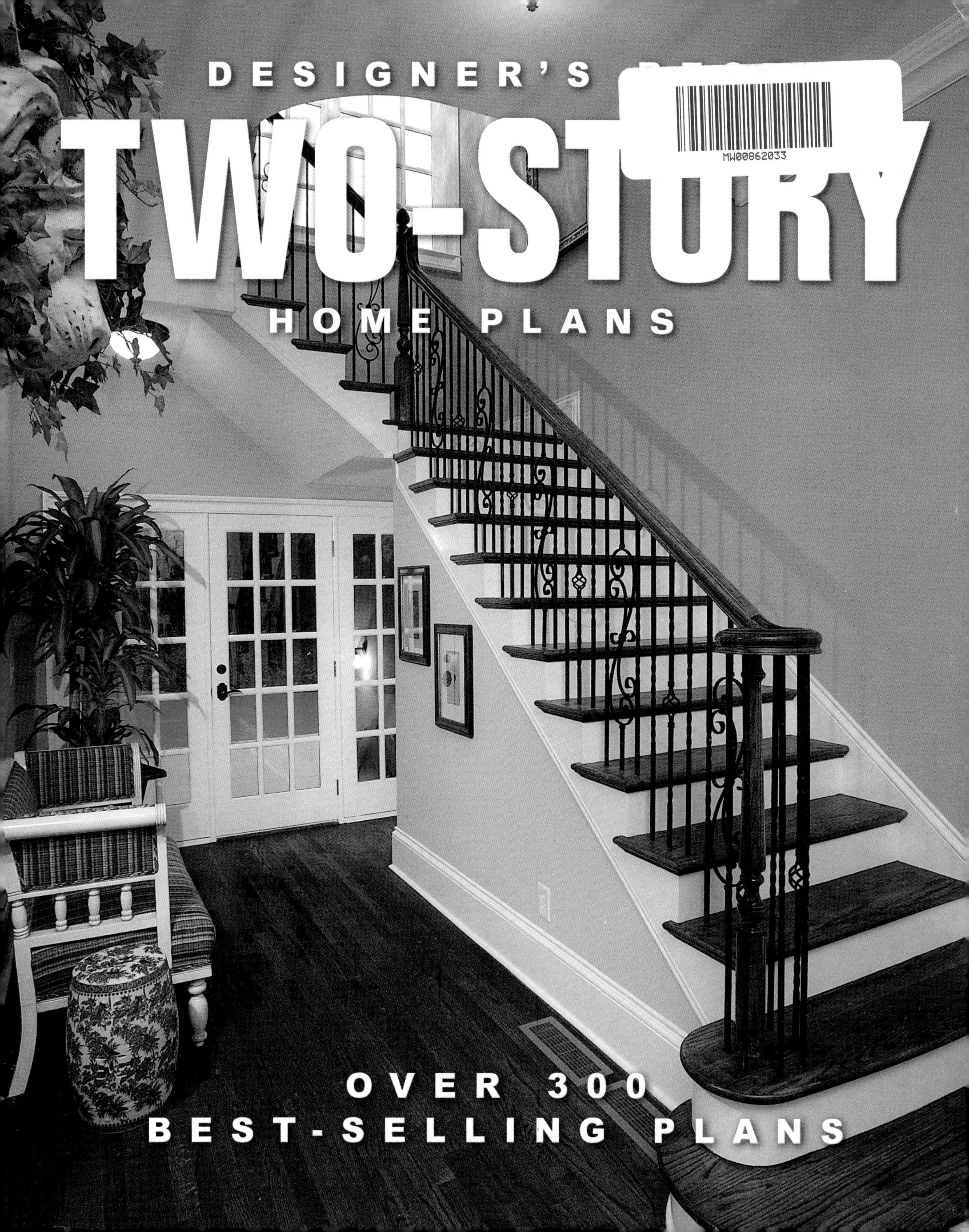

OVER 300
BEST-SELLING PLANS

TABLE OF
CONTENTS

FEATURE PLANS

PRESENTED BY

DESIGNSDIRECT
PUBLISHING ®

PRESIDENT	Angela Santerini
PUBLISHER	Dominic Foley
EDITOR	Kathleen Nalley
WRITER	Jennifer Bacon
GRAPHIC ARTIST	Bishana Shipp
CONTRIBUTING EDITORS	Jennifer Emmons
	Laura Segers
CONTRIBUTING GRAPHIC ARTISTS	Kim Campeau
	Emily Sessa
	Joshua Thomas
CONTRIBUTING WRITERS	Paula Powers
	Claire Ulik
ILLUSTRATORS	Architectural Art
	Allen Bennetts
	Greg Havens
	Holzhauer, Inc.
	Dave Jenkins
	Kurt Kauss
	Barry Nathan
PHOTOGRAPHERS	Walter Kirk
	Matthew Scott
	Laurence Taylor
	Happy Terrebone
	Doug Thompson
	Bryan Willy

A DESIGNS DIRECT PUBLISHING® BOOK

**Printed by Toppan Printing Co., Hong Kong
First Printing, February 2006
10 9 8 7 6 5 4 3 2 1**

Stunning
Combination

A low-maintenance exterior coupled with a friendly wrapping porch creates immediate curb appeal.

Jametowne

- TSPDG01-828
- 1-866-525-9374
- www.twostoryplans.com

Welcome on any streetscape, the *Jamestowne* is inviting from the minute you pull up the front drive. Columns outline the charming front porch, while a decorative transom above the double front doors adds instant warmth and beauty. A prominent center gable and a low-maintenance exterior create excellent curb appeal for this appealing two-story family home.

The interior is equally as impressive, as a lofty two-story foyer immediately greets guests into the home. Adjacent to the foyer, the dining room is defined by dramatic columns. The great room features a two-story ceiling, fireplace and built-in bookshelves as custom features. The generously proportioned kitchen features a nearby pantry and is open to the breakfast room and great room for easy entertaining and family togetherness.

Completing the first floor is the master suite. Topped with a tray ceiling, the master bedroom accesses a second rear porch through French doors that bathe the room with sunlight. Two closets simplify morning rituals and an elegant master bath completes the suite. Dual vanities, private privy and a separate shower and tub provide convenience and luxury.

Above: A stone fireplace and adjacent balcony serve as stunning focal points in the great room.

Above: Built-in shelves in the foyer add flair, while a two-story great room and foyer draw the eye upwards.

Left: Overlooking the breakfast room and great room, the kitchen features an abundance of counter space and convenience.

Upstairs, a balcony overlooks both the two-story great room and two-story foyer, granting striking architectural interest the minute you walk in the door. The remainder of the bedrooms are also on the second floor, giving the master suite complete autonomy downstairs. Complete with two closets each, the secondary bedrooms are spread apart from one another, granting privacy to each.

With a garage to the rear of the home, the *Jamestowne* enjoys massive curb appeal. An easy-to-maintain exterior blended with a practical and modern floor plan becomes a stunning combination.

Center: A formal staircase descends into the two-story foyer, and creates a dramatic impression upon entering the home.

Right: Choosing to add a bonus room over the garage, this homeowner designed the perfect media room.

Donald A. Gardner Architects, Inc.

Jamestowne

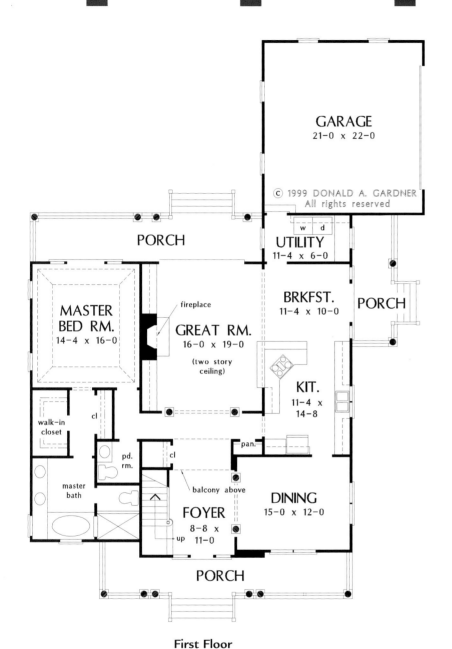

GARAGE
21-0 x 22-0

PORCH

UTILITY
11-4 x 6-0

w   d

MASTER BED RM.
14-4 x 16-0

fireplace

GREAT RM.
16-0 x 19-0

(two story ceiling)

BRKFST.
11-4 x 10-0

PORCH

KIT.
11-4 x 14-8

walk-in closet

cl

pan.

pd. rm.

cl

master bath

balcony above

DINING
15-0 x 12-0

FOYER
8-8 x 11-0

up

PORCH

First Floor

BED RM.
11-4 x 12-8

great room below

attic storage

cl

railing

cl   cl

bath

BED RM.
12-0 x 12-0

down

lin.

cl

foyer below

cl

attic storage

cl

BED RM.
12-8 x 11-0

cl

Second Floor

Rear Elevation

TSPDG01-828 — 1-866-525-9374

Total Living	First Floor	Second Floor	Bonus	Bed	Bath	Width	Depth	Foundation	Price Category
2500 sq ft	1685 sq ft	815 sq ft	N/A	4	2-1/2	52' 8"	72' 4"	Crawl Space*	F

*Other options available. See page 347.

Donald A. Gardner Architects, Inc.

www.twostoryplans.com   **9**

Traditional Twist

Brookhaven

- TSPFB01-963
- 1-866-525-9374
- www.twostoryplans.com

Far Left: Stone, cedar shake and siding create a Craftsman exterior on a Traditional-styled home.

Left: Multiple activities take place in this bonus room. Whether bumper pool or video games, this is sure to be a family favorite.

Below: Rich stained cabinetry sets the tone for this stylish kitchen.

Combining Traditional style with Craftsman materials, the *Brookhaven* proves that beautiful and low-maintenance can describe the same home's exterior. Cedar shakes accent gables, while stone anchors the house to its surroundings. Siding is used to create architectural interest by providing visual rests. Square columns add a touch of Colonial flair to the covered porch, while flagstone flooring showcases a perfect spot for twin rocking chairs. A French door, topped by a radius window, connects the family room to the backyard.

Open and family-efficient, the floor plan incorporates a step-saving design, reducing the number of steps required to perform routine tasks and chores. A circular traffic pattern flows through the gathering areas, and a balcony overlook visually links both floors.

An arched walkway leads from the foyer to the family room, and an elegant tray ceiling and heavy crown moldings enhance the master bedroom. A secondary first-floor bedroom makes a great office or in-law suite.

Upstairs, the foyer's two-story ceiling and Palladian window help brighten the second floor. Each bedroom, and even the bonus room, features a walk-in closet, while the bathrooms include both double and single vanities.

At home in the city or country, the *Brookhaven* complements any streetscape or setting. For those desiring relaxed living and architectural character, this home will not disappoint.

Frank Betz Associates, Inc.

A decorative fireplace makes a dramatic focal point in the family room.

Brookhaven

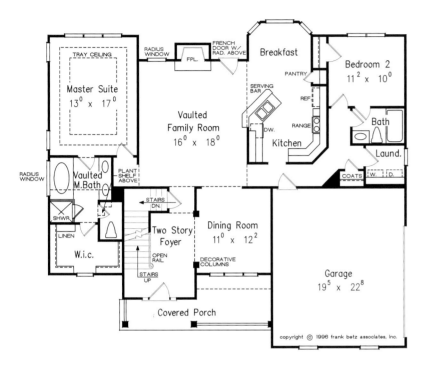

First Floor

Second Floor

Rear Elevation

TSPFB01-963 — 1-866-525-9374

Total Living	First Floor	Second Floor	Opt. Bonus	Bed	Bath	Width	Depth	Foundation	Price Category
2126 sq ft	1583 sq ft	543 sq ft	251 sq ft	4	3	53' 0"	47' 0"	Basement, Crawl Space or Slab	H

Southern Charm

Savannah
- TSPDS01-6688
- 1-866-525-9374
- www.twostoryplans.com

Cream-white stucco sets a contemporary tone for this sensational seaside house. The façade harbors a doorway that leads to an extensive courtyard with a sundeck, lap pool and spa. Two steps up, a wide covered porch heightens the processional experience to the formal entry — a paneled door that opens to the foyer and central stairs. To the right, French doors provide access to a secluded study. With an adjoining powder bath, this space easily converts to guest quarters.

A barrel-vaulted gallery hall leads to a spacious great room through an open arrangement of the formal dining room and gourmet kitchen. Wrapping counters and a food-preparation island permit the cook to enjoy conversations with guests and view the courtyard through the dining room's bay window. A grand fireplace framed by an entertainment center and built-in cabinetry anchors the great room, which offers wide views of the scenery. Upstairs, the master suite boasts a luxe bath harbored within a bay, with separate vanities, a spa-style tub and a step-in shower designed for two owners. Two secondary bedrooms share a bath on this level, and a guest suite above the garage provides a morning kitchen.

Graceful colonnades define the double porticos of the rear elevation. Rows of windows and a series of French doors allow breezes to mingle in an elegant, open interior.

Strikingly contemporary, the interior is sensitive to the vintage aspects of its architecture, with paneled windows, perfect sightlines and restrained symmetry. Walls toned in sienna and earthen hues complement the colors of nature brought in by wide glimpses of scenery.

Above: In the great room, lovely French doors and a row of tall windows under classic trimwork bring the outside in. This stunning room also features built-in cabinets and a fireplace perfect for special evenings spent indoors.

Left: Rows of French doors invite gentle breezes inside and lend an easygoing spirit to the home. The Spanish tile roof and stucco exterior evoke memories of vacations past and make this charming villa a relaxing everyday retreat.

The master retreat features a curved wall of glass and a series of French doors that grant natural light to the sitting area.

Relaxed in feel, exquisite in detail, columns outline the hall and frame the view of the gourmet kitchen. The bar connects the two spaces, allowing for casual dining for two and terrific vistas through a span of glass facing the veranda. Doorway columns have been transformed into cabinetry, providing a unique display space for barware and easy-to-reach storage for favorite vintages.

Upstairs, an extended covered porch wraps the rear and one side of the home, providing unending views and a breezy setting for morning coffee for both the master suite and a secondary bedroom.

Above: A double shower and a step-up tub highlight the master bath, enriched by marble surrounds.

The Sater Design Collection, Inc.

Savannah

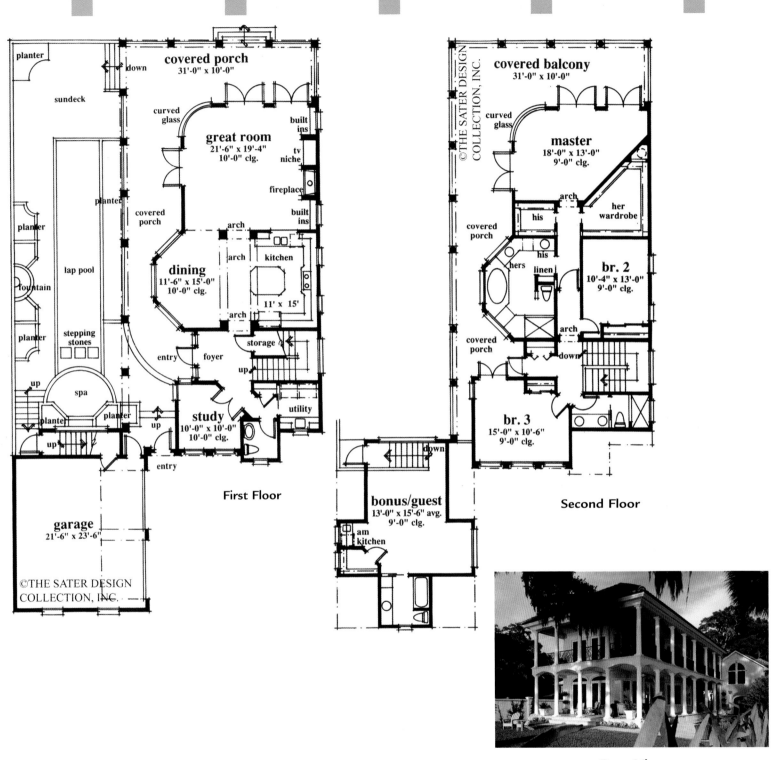

First Floor

- planter
- sundeck
- covered porch 31'-0" x 10'-0"
- down
- curved glass
- great room 21'-6" x 19'-4" 10'-0" clg.
- built ins
- tv niche
- fireplace
- built ins
- covered porch
- planter
- planter
- fountain
- planter
- lap pool
- stepping stones
- arch
- arch
- kitchen
- dining 11'-6" x 15'-0" 10'-0" clg.
- 11' x 15'
- arch
- entry
- foyer
- up
- storage
- up
- spa
- planter
- planter
- study 10'-0" x 10'-0" 10'-0" clg.
- utility
- up
- entry
- garage 21'-6" x 23'-6"
- ©THE SATER DESIGN COLLECTION, INC.

Second Floor

- ©THE SATER DESIGN COLLECTION, INC.
- covered balcony 31'-0" x 10'-0"
- curved glass
- master 18'-0" x 13'-0" 9'-0" clg.
- arch
- his
- her wardrobe
- covered porch
- his
- hers
- linen
- br. 2 10'-4" x 13'-0" 9'-0" clg.
- covered porch
- arch
- down
- br. 3 15'-0" x 10'-6" 9'-0" clg.
- down
- bonus/guest 13'-0" x 15'-6" avg. 9'-0" clg.
- am kitchen

Rear View

Home photographed may differ from actual construction documents.

TSPDS01-6688 — 1-866-525-9374

Total Living	First Floor	Second Floor	Bonus	Bed	Bath	Width	Depth	Foundation	Price Category
2447 sq ft	1293 sq ft	1154 sq ft	426 sq ft	4	3-1/2	50' 0"	90' 0"	Slab	G

INVITING
spaces

Rear Elevation

© 1996 Frank Betz Associates, Inc.

Chestnut Hill
- TSPFB01-3435
- 1-866-525-9374
- www.twostoryplans.com

Chestnut Hill

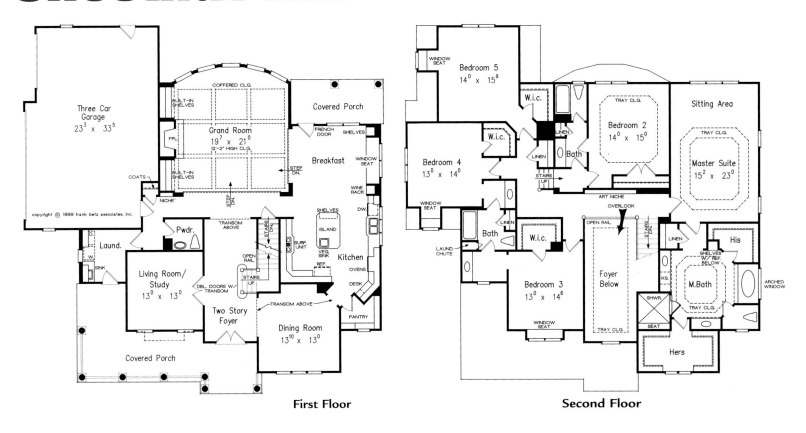

First Floor

Second Floor

TSPFB01-3435 — 1-866-525-9374

Home photographed may differ from actual construction documents.

Total Living	First Floor	Second Floor	Bonus	Bed	Bath	Width	Depth	Foundation	Price Category
4159 sq ft	1839 sq ft	2320 sq ft	N/A	5	3-1/2	61' 6"	61' 0"	Basement, Crawl Space or Slab	I

This amazing floor plan is the perfect retreat for the sophisticated homeowner. The wraparound porch features enough room for rocking chairs and a side entrance to the house, through a laundry room complete with a utility sink.

The coffered grand room offers built-in shelves flanking the fireplace and an exceptional wall of windows, making this the perfect home for a lake-front or golf course community. Off the foyer is a room with two French doors that can double as either a living room or a study for the stay-at-home mom or the telecommuter.

The kitchen is an entertainer's dream and boasts a window seat, wine racks, shelves and a spacious walk-in pantry. The center island includes a vegetable sink, a cookbook library and plenty of counter space for food preparation.

Upstairs, more surprises await; the niche at the top of the stairs is the perfect place for a treasured piece of art. Window seats are in three secondary bedrooms, while the master suite is something to behold. Besides an oversized room with a tray ceiling, there is also a sitting area where the lucky homeowners can relax after a busy day. The luxurious master bath features a tray ceiling with dual vanities. There is also a separate tub and shower with seat. Large his-and-her closets and a built-in refrigerator finish out this truly spectacular suite.

Above: This homeowner chose to convert the flexible living room/study into an elegant home office.

Country Tradition

This brick farmhouse features curved transoms on the second level that echo the barrel-vaulted entry for an impressive exterior.

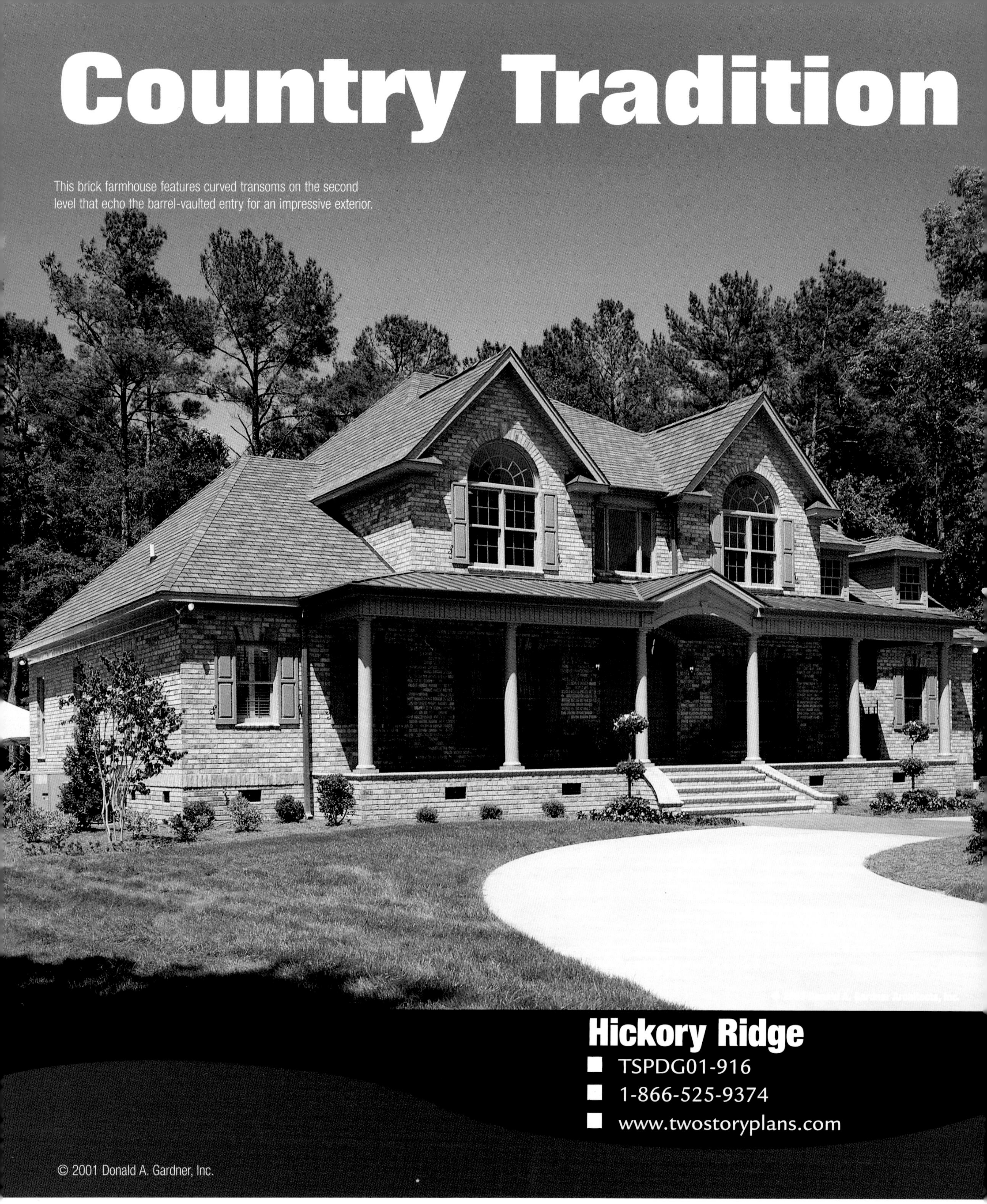

Hickory Ridge

- TSPDG01-916
- 1-866-525-9374
- www.twostoryplans.com

Above: Honey-colored cabinetry and stainless-steel appliances elegantly blend for a modern look in the kitchen of this updated farmhouse.

The *Hickory Ridge* has been given a generous amount of square footage and a plethora of custom-styled features. The exterior consists of a welcoming façade of twin gables and sidelights. Columns punctuate the friendly front porch and arched keystones mimic the barrel-vaulted entryway. Inside, elegance permeates throughout the entire home. Ceiling treatments crown several rooms, adding drama to each. Other exciting details include bay windows, French doors and a balcony.

The kitchen is any chef's dream. With an abundance of counter space, there's no such thing as too many people in the kitchen. Overlooking the great room, the kitchen has two pantries and a completely open feel. The large walk-in pantry creates additional storage space and the adjacent utility room makes it easy to separate food from cleaning supplies. Overflowing from the breakfast room is the spacious, two-story great room. Impressive columns support a curved balcony that overlooks the great room with built-in fireplace and bookshelves. Looking out onto a rear porch, the great room becomes flooded with natural light through several windows.

Above: Sophisticated furnishings combine with a bay window to grant a formal appearance to the breakfast room.

Donald A. Gardner Architects, Inc.

Above: Modified from the plan, the living room/study's closet was removed to allow a wet bar.
Below: The large garden tub, dual vanities and separate shower provide convenience and luxury in the master bath.

Angled corners soften the dining room, which also features columns to create a stunning entrance from the foyer. The stunning master suite is crowned with a decorative tray ceiling, which is echoed in the master bath. Two walk-in closets, separate shower and tub, and dual vanities complete the suite.

Above: Expanding volume, a tray ceiling creates the right environment for a four-post bed.

Three additional bedrooms are located on the second floor, two with walk-in closets and adjacent full baths. The bonus room completes the second level. For added flexibility, a versatile living room/study with adjacent powder room rests just off the foyer, and a large storage closet creates extra space in the garage.

For the nights when enjoying Mother Nature is a must, the *Hickory Ridge* features both a front and back porch. Stretching almost the entire width of the home, these two porches provide an abundance of space for entertaining or place for solitude. Clearly refined, this home is filled with luxury.

Donald A. Gardner Architects, Inc.

Hickory Ridge

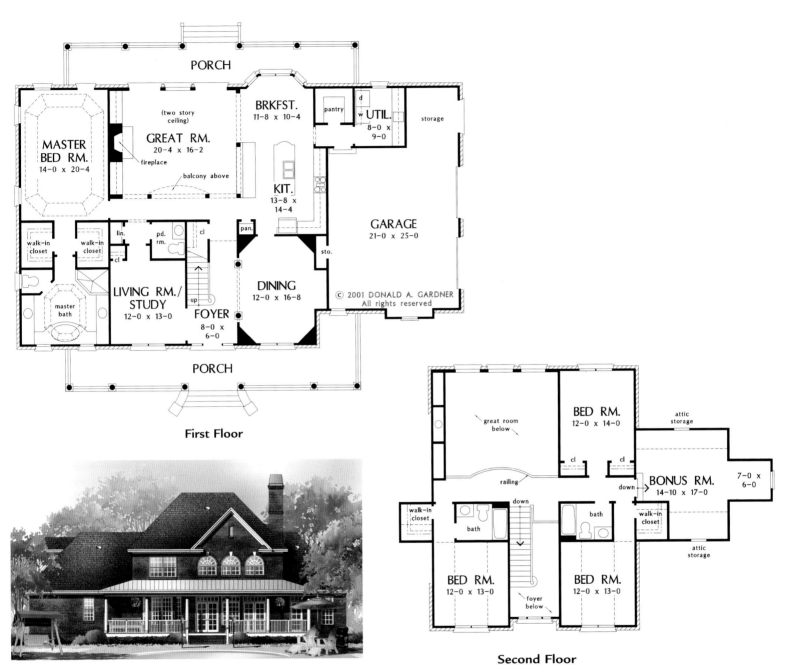

PORCH

MASTER BED RM.
14-0 x 20-4

(two story ceiling)

GREAT RM.
20-4 x 16-2

fireplace

balcony above

BRKFST.
11-8 x 10-4

pantry

UTIL.
8-0 x 9-0

d
w

storage

walk-in closet

walk-in closet

lin.

pd. rm.

cl

KIT.
13-8 x 14-4

pan.

GARAGE
21-0 x 25-0

master bath

LIVING RM./ STUDY
12-0 x 13-0

FOYER
8-0 x 6-0

up

DINING
12-0 x 16-8

sto.

© 2001 DONALD A. GARDNER
All rights reserved

PORCH

First Floor

Rear Elevation

great room below

railing

BED RM.
12-0 x 14-0

cl

cl

attic storage

BONUS RM.
14-10 x 17-0

7-0 x 6-0

down

walk-in closet

bath

down

bath

walk-in closet

attic storage

BED RM.
12-0 x 13-0

foyer below

BED RM.
12-0 x 13-0

Second Floor

TSPDG01-916 — 1-866-525-9374

Total Living	First Floor	Second Floor	Bonus	Bed	Bath	Width	Depth	Foundation	Price Category
3167 sq ft	2194 sq ft	973 sq ft	281 sq ft	4	3-1/2	71' 11"	54' 4"	Crawl Space*	G

*Other options available. See page 347.

Donald A. Gardner Architects, Inc.

Floor-to-ceiling woodwork gives this kitchen visual country warmth.

Right: A red metal roof caps this Craftsman classic.

Donald A. Gardner Architects, Inc.

Rustic Retreat

Blue Ridge

- TSPDG01-1130-D
- 1-866-525-9374
- www.twostoryplans.com

A stunning center dormer with arched window and decorative wood brackets cap the entry to this extraordinary hillside estate. Floor-to-ceiling windows invite the natural light inside, while stable-style garage doors and a decorative clerestory window add extra curb appeal to this open floor plan. The three-car garage with a triplet of dormers overhead decoratively contribute to the impressive exterior.

Exposed wood beams embrace the rustic feel and enhance the magnificent cathedral ceilings of the foyer, great room, dining room, master bedroom and screened porch, while ten-foot ceilings top the remainder of the first floor. The great room takes in scenic rear views through a wall of windows. Relaxing spaces are abundant as fireplaces add warmth and ambience to the great room, basement and screened porch.

The kitchen is complete with a center cook-top island, pantry and ample room for two or more chefs. Overlooking the great room, the kitchen's close proximity makes entertaining guests a cinch. When cool weather calls for alfresco meals, the screen porch is adjacent to the kitchen and accesses two porches, providing the perfect place for outdoor dining. The three-and-a-half car garage allows space for a golf cart, while the basement level leaves plenty of space to design a luxury game room and bar for those who love entertaining.

The luxurious master suite is nothing short of impressive. Off on its own wing and including a huge walk-in closet and double vanities in the bath, this sprawling space makes retiring to bed an easy task. Two secondary bedrooms complete the first floor.

Ideal for a game room, the expansive basement level's media/rec room also features a built-in wet bar. The adjacent bedrooms include private baths that become great guest suites for overnight guests or rooms for larger families. Two covered patios off the rec room are excellent places to convene outdoors when entertaining larger groups. Able to accommodate several people comfortably, the rooms in this plan contain the flexibility to change as your family grows.

Above: A wall of windows and hipped cathedral ceiling highlight the master bedroom.

Right: The rec room in the lower level is complete with a full service bar.

Donald A. Gardner Architects, Inc.

Blue Ridge

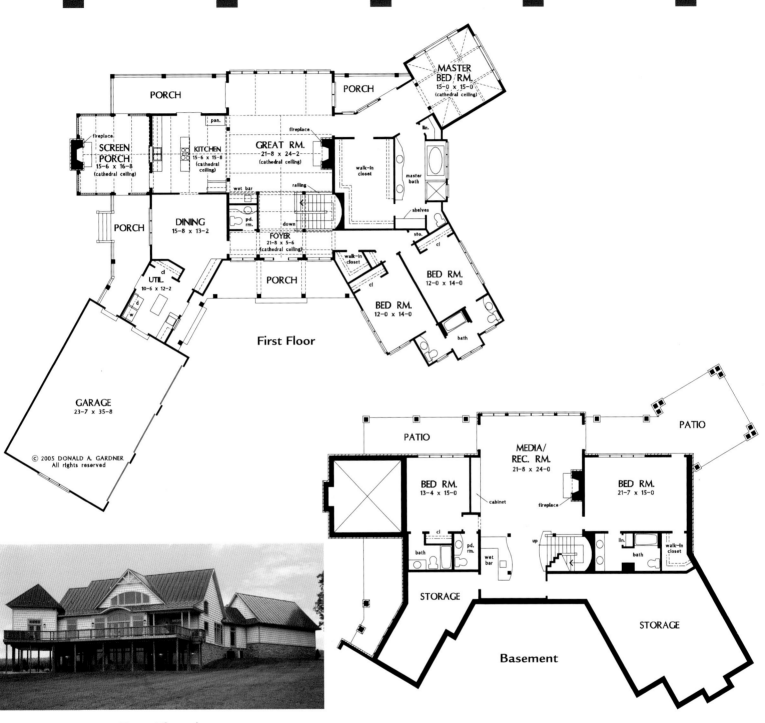

First Floor

- PORCH
- PORCH
- MASTER BED RM. 15-0 x 15-0 (cathedral ceiling)
- SCREEN PORCH 15-6 x 16-8 (cathedral ceiling)
- fireplace
- KITCHEN 15-6 x 15-8 (cathedral ceiling)
- pan.
- GREAT RM. 21-8 x 24-2 (cathedral ceiling)
- fireplace
- walk-in closet
- master bath
- lin.
- wet bar
- railing
- shelves
- sto.
- cl
- PORCH
- DINING 15-8 x 13-2
- pd. rm.
- FOYER 21-8 x 5-6 (cathedral ceiling)
- down
- walk-in closet
- cl
- BED RM. 12-0 x 14-0
- UTIL. 10-6 x 12-2
- d
- d w
- PORCH
- BED RM. 12-0 x 14-0
- bath
- GARAGE 23-7 x 35-8

Basement

- PATIO
- PATIO
- MEDIA/ REC. RM. 21-8 x 24-0
- BED RM. 13-4 x 15-0
- cabinet
- fireplace
- BED RM. 21-7 x 15-0
- cl
- bath
- pd. rm.
- wet bar
- up
- lin.
- bath
- walk-in closet
- STORAGE
- STORAGE
- STORAGE

Rear Elevation

TSPDG01-1130-D — 1-866-525-9374

Home photographed may differ from actual construction documents.

Total Living	First Floor	Basement	Bonus	Bed	Bath	Width	Depth	Foundation	Price Category
4853 sq ft	3098 sq ft	1755 sq ft	N/A	5	4F/2H	106' 1"	99' 3"	Hillside Walkout	J

Donald A. Gardner Architects, Inc.

Entertainer's Dream

Summerlyn

- TSPFB01-3675
- 1-866-525-9374
- www.twostoryplans.com

© 1995 Frank Betz Associates, Inc.

Above: Stainless-steel appliances add a modern touch to this French-country kitchen.

Left: Soft lighting highlights the Old-World details of this brick-and-stone-accented elevation.

At a glance it's clear to see that the *Summerlyn* is a unique and extraordinary design.

A warm blend of brick and cedar shake with stone accents graces the exterior of the home. The courtyard entry adds the finishing touch to this façade, while an attention-grabbing, two-story turret is perhaps the most unique feature.

More pleasant surprises await you inside. The foyer opens into the two-story family room, and the impressive bowed wall of windows allows light to stream in, making this room bright and sunny. A sunken sunroom shares a see-through fireplace with the family room. The kitchen boasts plenty of cabinet space, double ovens and a center island, making the space an entertainer's dream. The uniquely shaped dining room offers a niche to showcase a special piece of art or floral arrangement. Finishing out the first floor is a study and a guest bedroom, perfect for the aging parent or the visitor needing additional privacy.

The second floor offers three additional bedrooms, each with private access to a bathroom. The master bedroom and bath must be seen to be believed. The master suite offers a spacious bedroom with an equally impressive sitting room. His-and-her closets ensure that the homeowners have plenty of space for their belongings. The master bath features decorative niches on each end of the tub, allowing the homeowner to add personal touches. Additionally, the dual vanities, separate tub and shower make this the perfect retreat at the end of the day.

Above Left: The master bath features a tray ceiling finished with painted wood and elaborate molding.

Above: The rear wall of windows creates a panorama in the two-story family room.

Frank Betz Associates, Inc.

Summerlyn

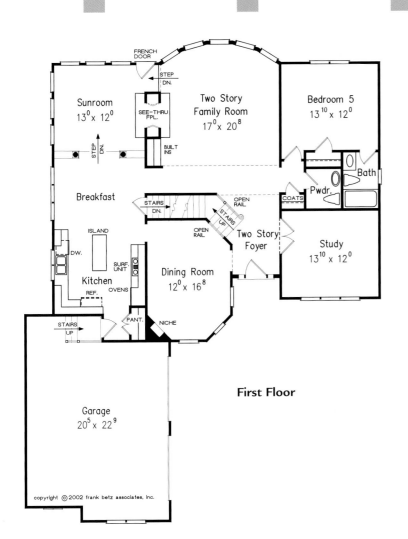

First Floor

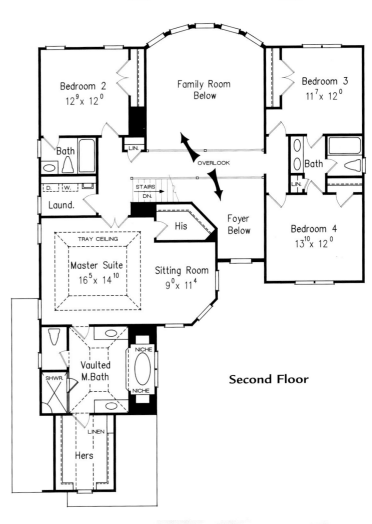

Second Floor

Rear Elevation

TSPFB01-3675 — 1-866-525-9374

Home photographed may differ from actual construction documents.

Total Living	First Floor	Second Floor	Bonus	Bed	Bath	Width	Depth	Foundation	Price Category
3281 sq ft	1685 sq ft	1596 sq ft	N/A	5	4-1/2	51' 0"	66' 10"	Basement or Crawl Space	I

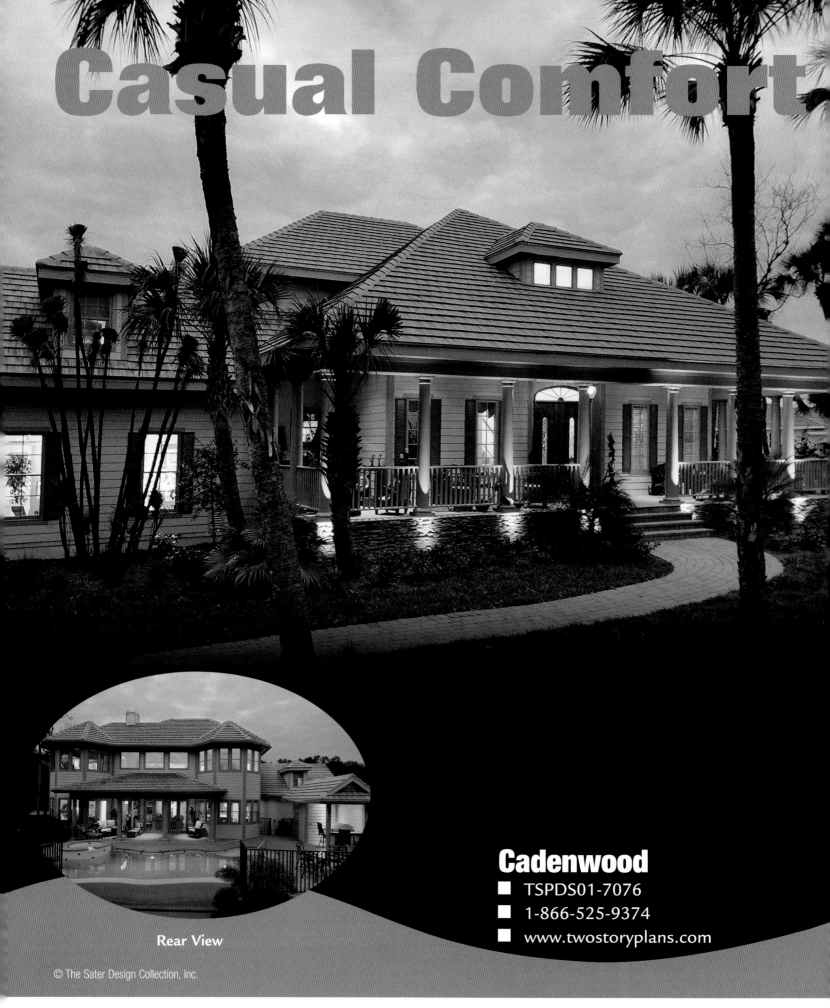

Casual Comfort

Rear View

© The Sater Design Collection, Inc.

Cadenwood

- TSPDS01-7076
- 1-866-525-9374
- www.twostoryplans.com

Cadenwood

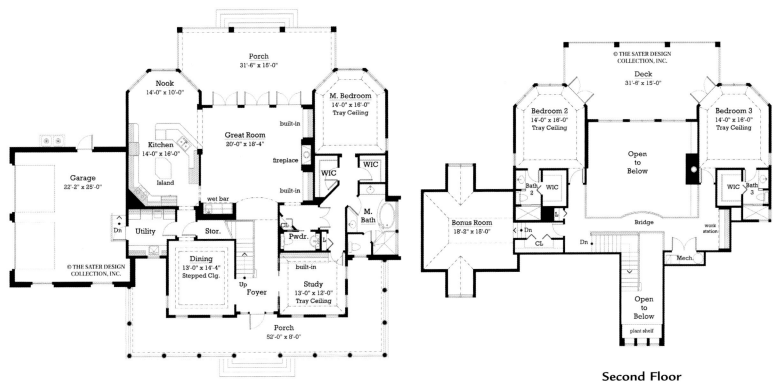

First Floor

Second Floor

TSPDS01-7076 — 1-866-525-9374

Home photographed may differ from actual construction documents.

Total Living	First Floor	Second Floor	Bonus	Bed	Bath	Width	Depth	Foundation	Price Category
3082 sq ft	2138 sq ft	944 sq ft	427 sq ft	3	3-1/2	77' 8"	64' 0"	Basement or Crawl Space	H

An abundance of outdoor entertaining areas makes this luxury resort villa the perfect place for enjoying the great outdoors with the company of close friends.

One of the defining characteristics of *Cadenwood* is its front porch. Like the "good old days" when folks sat on their front porches sipping lemonade or ice tea and watching the world go by, this home's porch brings us back to a time when life was simpler. A few rocking chairs, a table or two, and this outdoor oasis serves as a summer retreat where evenings can be spent enjoying a cool drink and good friends.

The theme of casual comfort continues once you enter the home. As you pass through the foyer and under the second-floor bridge, you're greeted by the spectacularly inviting great room. Earth tones abound, as do places for overstuffed chairs, giving the room warmth and making the hearth almost unnecessary except on the coldest of evenings.

Tucked away in the rear corner of the first floor is a spacious master suite. With its limited access, it is extremely private. On the opposite side of the plan, the kitchen and nook form a generous family zone while above, two additional bedrooms and a spacious bonus room provide house guests or younger family members their own private retreats.

Right: The soaring stepped ceiling, carved coffers, hardwood floors and sky-high windows of the great room epitomize the casual, comfortable elegance that makes Cadenwood the perfect home for entertaining.

Brewster — TSPFB01-1175 — 1-866-525-9374

© 1998 Frank Betz Associates, Inc.

Total Living	First Floor	Second Floor	Bonus	Bed	Bath	Width	Depth	Foundation	Price Category
1467 sq ft	1001 sq ft	466 sq ft	292 sq ft	4	2-1/2	42' 0"	42' 0"	Basement, Crawl Space or Slab	F

Design Features

- An optional bonus room upstairs makes the perfect playroom or exercise area.
- The impressive dining room has a rare two-story ceiling, however an optional loft or fourth bedroom can be easily incorporated.
- Vaulted ceilings and a decorative plant shelf make the main floor of this design interesting and dimensional.
- The laundry room and a handy coat closet are strategically placed just off the garage.

First Floor

Second Floor

Opt. Loft or Bedroom

Rear Elevation

© 1995 Frank Betz Associates, Inc.

Wynterhall — TSPFB01-835 — 1-866-525-9374

Total Living	First Floor	Second Floor	Bonus	Bed	Bath	Width	Depth	Foundation	Price Category
1531 sq ft	1067 sq ft	464 sq ft	207 sq ft	3	2-1/2	41' 0"	44' 4"	Basement, Crawl Space or Slab	F

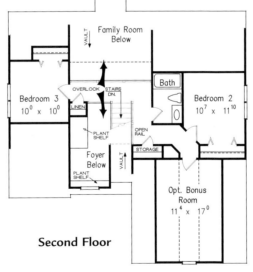

Second Floor

First Floor

copyright © 1995 frank betz associates, inc.

Design Features

- This house was designed to combine spaciousness and affordability.

- An overlook and decorative plant shelves add appealing detail to the home.

- A main-level master suite is a highly sought-after feature these days and was accomplished in a 1531 square-foot home.

- The open floor plan allows easy movement, which is perfect for entertaining.

Rear Elevation

Aberdeen Place — TSPFB01-3809 — 1-866-525-9374

© 2003 Frank Betz Associates, Inc.

Total Living	First Floor	Second Floor	Opt. Bonus	Bed	Bath	Width	Depth	Foundation	Price Category
1593 sq ft	1118 sq ft	475 sq ft	223 sq ft	3	2-1/2	41' 0"	45' 0"	Basement, Crawl Space or Slab	G

Design Features

- From the *Southern Living® Design Collection*

- The *Aberdeen Place* has that cozy cottage appeal with its board and batten exterior and stone accents. Its conservative square footage is rich in style and functional design.

- The breakfast room is bordered by windows that allow natural light to pour in.

- A generously sized master bedroom includes a tray ceiling and views to the back yard.

- An optional bonus area provides the extra space that each homeowner can personalize.

Rear Elevation

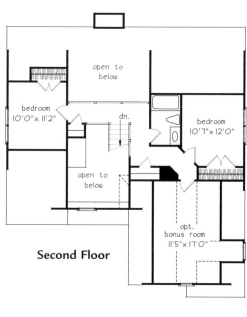

Second Floor

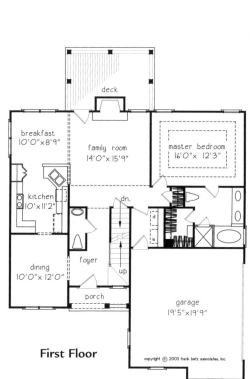

First Floor

© 1997 Frank Betz Associates, Inc.

Willowbrook — TSPFB01-1086 — 1-866-525-9374

Total Living	First Floor	Second Floor	Opt. Bonus	Bed	Bath	Width	Depth	Foundation	Price Category
1597 sq ft	1205 sq ft	392 sq ft	190	3	2-1/2	50' 6"	44' 4"	Basement, Crawl Space or Slab	D

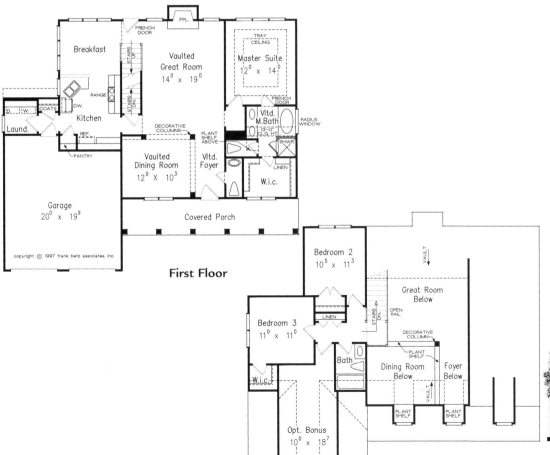

copyright © 1997 frank betz associates, inc.

First Floor

Second Floor

Design Features

- The rocking-chair front porch gives guests a glimpse of what's inside.
- The main-level master suite offers homeowners the privacy so many desire.
- Decorative columns frame the dining room, which leads to the vaulted great room.
- Two bedrooms and an optional bonus room are located on the second floor.

Rear Elevation

Seymour — TSPDG01-287 — 1-866-525-9374

© 1992 Donald A. Gardner Architects, Inc.

Total Living	First Floor	Second Floor	Bonus	Bed	Bath	Width	Depth	Foundation	Price Category
1622 sq ft	1039 sq ft	583 sq ft	N/A	3	2	37' 9"	44' 8"	Crawl Space*	D

Design Features

*Other options available. See page 347.

- This rustic three-bedroom plan takes its relaxed manner from spacious, covered front and back porches.

- The openness from the great room to the kitchen/dining area provides a spacious feeling.

- The master suite pampers with separate shower, whirlpool tub and generous walk-in closet.

- One of the two upstairs bedrooms overlooks the great room for added drama.

Rear Elevation

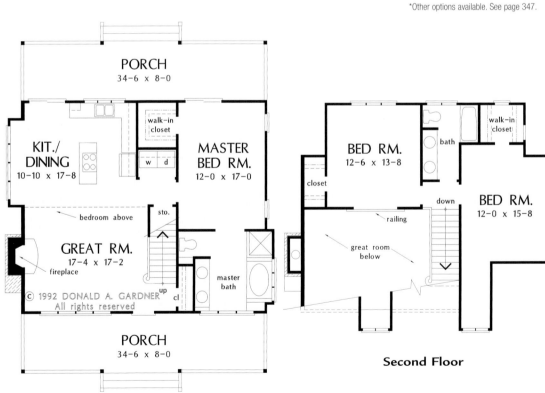

First Floor

Second Floor

PORCH 34-6 x 8-0

KIT./DINING 10-10 x 17-8

walk-in closet

MASTER BED RM. 12-0 x 17-0

w   d

bedroom above

sto.

GREAT RM. 17-4 x 17-2

fireplace

up

cl

master bath

PORCH 34-6 x 8-0

BED RM. 12-6 x 13-8

bath

walk-in closet

closet

down

BED RM. 12-0 x 15-8

railing

great room below

© 2002 Frank Betz Associates, Inc.

Brentwood — TSPFB01-3711 — 1-866-525-9374

Total Living	First Floor	Second Floor	Opt. Bonus	Bed	Bath	Width	Depth	Foundation	Price Category
1634 sq ft	1177 sq ft	457 sq ft	249 sq ft	3	2-1/2	41' 0"	48' 4"	Basement, Crawl Space or Slab	F

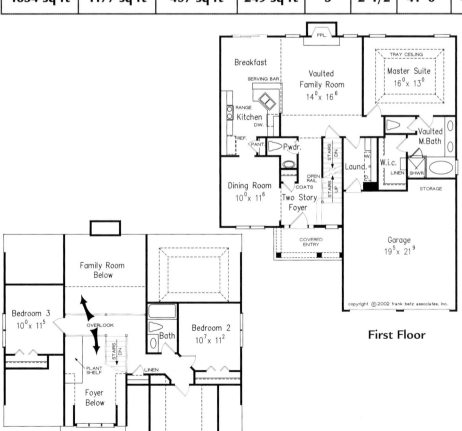

First Floor

Second Floor

Design Features

- Brick and stone, set off by multi-pane windows, highlight the street presence of this classic home.

- A sheltered entry leads to a two-story foyer and wide interior vistas that extend to the back property. Rooms in the public zone are open, allowing the spaces to flex for planned events as well as family gatherings.

- At the heart of the home, the vaulted family room frames a fireplace with tall windows that bring in natural light.

- The main-level master suite boasts a tray ceiling, while two upper-level bedrooms are connected by a balcony bridge that overlooks the foyer and family room.

Rear Elevation

Broadmoor — TSPFB01-1088 — 1-866-525-9374

© 1997 Frank Betz Associates, Inc.

Total Living	First Floor	Second Floor	Bonus	Bed	Bath	Width	Depth	Foundation	Price Category
1658 sq ft	1179 sq ft	479 sq ft	338 sq ft	3	2-1/2	41' 6"	54' 4"	Basement, Crawl Space or Slab	D

Design Features

- Quaint and charming outside and in, the *Broadmoor* combines function with beauty to make a wonderful home. The courtyard entry leads to a stunning two-story foyer.

- The great room, breakfast area and kitchen are all connected, making conversation and traffic flow effortless.

- The laundry room and a coat closet are thoughtfully placed just off the garage, keeping coats and shoes in their place.

- An optional bonus area upstairs has endless possibilities. This could be the ideal playroom, fitness area or home office.

Rear Elevation

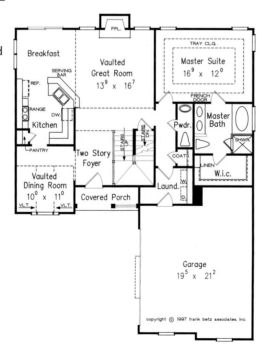

First Floor

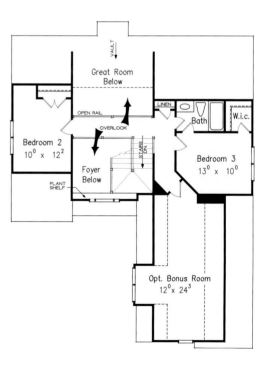

Second Floor

© 1992 Donald A. Gardner Architects, Inc.

Thackery — TSPDG01-283 — 1-866-525-9374

Total Living	First Floor	Second Floor	Bonus	Bed	Bath	Width	Depth	Foundation	Price Category
1663 sq ft	1145 sq ft	518 sq ft	380 sq ft	3	2-1/2	59' 4"	56' 6"	Crawl Space*	D

*Other options available. See page 347.

Design Features

- A covered porch wraps around this classic farmhouse with multiple dormers.
- A clerestory window bathes the two-story foyer in natural light.
- The large great room with fireplace opens to the dining/breakfast/kitchen space.
- A first-level master bedroom offers privacy and luxury with a separate shower.

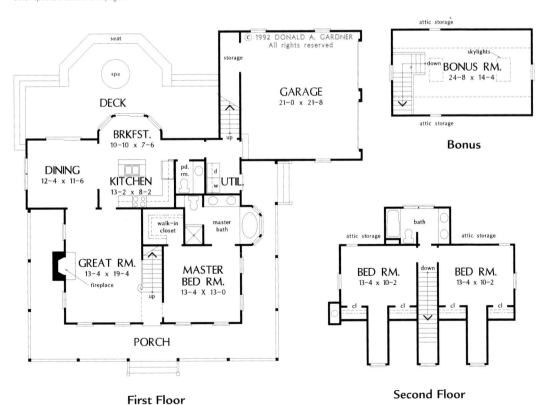

First Floor

Bonus

Second Floor

Rear Elevation

Hudson — TSPDG01-512 — 1-866-525-9374

Total Living	First Floor	Second Floor	Bonus	Bed	Bath	Width	Depth	Foundation	Price Category
1669 sq ft	1219 sq ft	450 sq ft	406 sq ft	3	2-1/2	50' 4"	49' 2"	Crawl Space*	D

Design Features

- This narrow-lot home offers some of the extras usually reserved for wider lots.

- Columns and a bay window add distinction to the dining room.

- The kitchen is designed for efficiency and offers access to the side porch.

- The master suite is located on the first floor, while two secondary bedrooms are upstairs.

- For growing families, the bonus room can be turned into a fourth bedroom with bath for growing families.

*Other options available. See page 347.

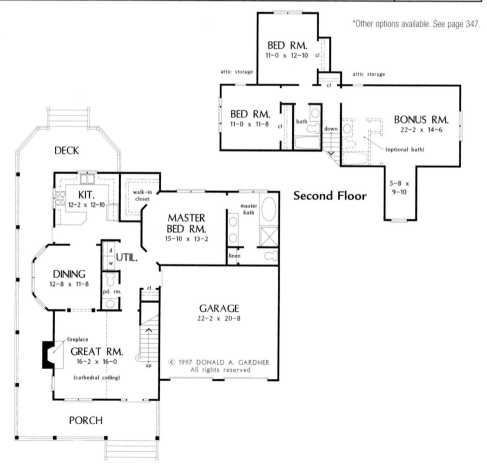

Second Floor

BED RM.
11-0 x 12-10

attic storage

BED RM.
11-0 x 11-8

bath

down

BONUS RM.
22-2 x 14-6

(optional bath)

5-8 x
9-10

DECK

KIT.
12-2 x 12-10

walk-in closet

MASTER BED RM.
15-10 x 13-2

master bath

linen

DINING
12-8 x 11-8

UTIL.

pd. rm.

cl

GARAGE
22-2 x 20-8

fireplace

GREAT RM.
16-2 x 16-0

(cathedral ceiling)

up

PORCH

First Floor

Rear Elevation

© 1999 Frank Betz Associates, Inc.

Jessamine Park — TSPFB01-1233 — 1-866-525-9374

Total Living	First Floor	Second Floor	Bonus	Bed	Bath	Width	Depth	Foundation	Price Category
1675 sq ft	882 sq ft	793 sq ft	458 sq ft	4	2-1/2	49' 6"	35' 4"	Basement, Crawl Space or Slab	G

First Floor

breakfast 9'11"× 8'6"

family room 17'5"× 12'0"

garage 19'8"× 24'11"

kitchen 11'3"×9'3"

up

dining 11'3"×10'0"

foyer

dn.

copyright © 1999 frank betz associates, inc.

living 12'5"× 11'3"

porch

master bedroom 17'0"×12'0"

D. W.

bedroom 11'3"×10'0"

dn.

foyer below

bedroom 10'2"×11'3"

Second Floor

bedroom 15'5"×19'8"

D. W.

bedroom 11'3"×10'0"

Opt. Bedroom

Design Features

■ From the *Southern Living® Design Collection*

■ A bay window in the breakfast room invites in natural light, creating the perfect spot for a quick morning coffee.

■ Upstairs, the master suite offers a decorative ceiling treatment and a vaulted master bath.

■ An optional second floor lends flexibility to this home, with a large bonus room that can serve as the perfect game room or upstairs den.

Rear Elevation

Bellwoode — TSPFB01-821 — 1-866-525-9374

© 1995 Frank Betz Associates, Inc.

Total Living	First Floor	Second Floor	Opt. Bonus	Bed	Bath	Width	Depth	Foundation	Price Category
1675 sq ft	882 sq ft	793 sq ft	416 sq ft	3	2-1/2	49' 6"	35' 4"	Basement, Crawl Space or Slab	F

Design Features

- Simple in design and conservative in square footage, the *Bellwoode* makes an excellent starter home with plenty of character.

- A bay window keeps the breakfast room bright and sunny.

- Laundry facilities are designed into the upper level of this home, convenient to all the bedrooms.

- An optional second-floor bonus room would make a great playroom or work area.

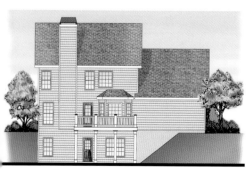

Rear Elevation

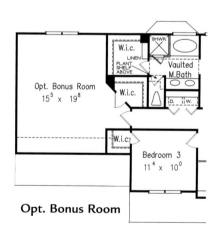

Opt. Bonus Room

First Floor

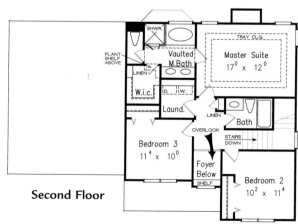

Second Floor

© The Sater Design Collection, Inc.

Shadow Lane — TSPDS01-6686 — 1-866-525-9374

Total Living	First Floor	Second Floor	Bonus	Bed	Bath	Width	Depth	Foundation	Price Category
1684 sq ft	1046 sq ft	638 sq ft	N/A	3	3	25' 0"	72' 2"	Post/Pier	E

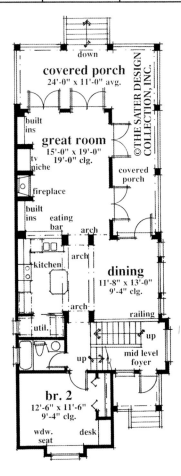

First Floor

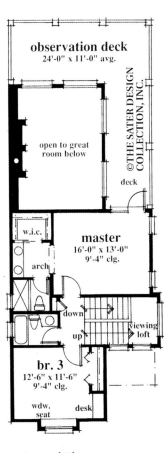

Second Floor

Design Features

- Built-ins and a media niche frame the fireplace in the great room.

- The formal dining room opens to the wraparound covered porch.

- The gourmet kitchen shares an eating bar with the great room.

- The second-level master suite includes a private observation deck.

- A vestibule leads to a viewing-loft stair, where a built-in window seat offers quiet relaxation.

Rear Elevation

Runaway Bay — TSPDS01-6616 — 1-866-525-9374

© The Sater Design Collection, Inc.

Total Living	First Floor	Second Floor	Bonus	Bed	Bath	Width	Depth	Foundation	Price Category
1772 sq ft	1136 sq ft	636 sq ft	N/A	3	2	39' 9"	45' 0"	Post/Pier	E

Design Features

- The entry leads through a well-lit foyer to an expansive great room warmed by a fireplace.

- The great room, gourmet kitchen and dining area all open onto a screened verandah.

- Split sleeping quarters offer a secondary bedroom and study on the main level.

- The upper level is dedicated to a luxurious master with a spacious private bath.

Rear Elevation

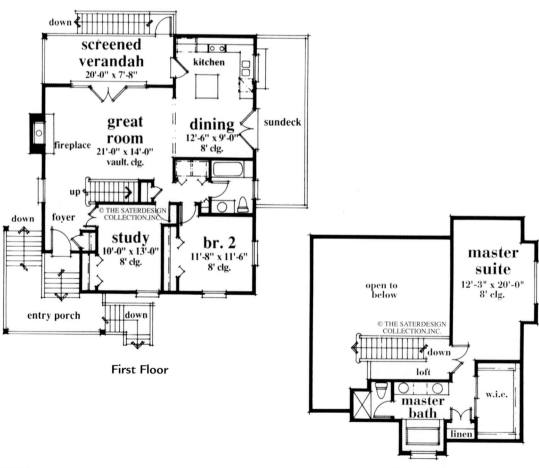

First Floor

Second Floor

© 1994 Donald A. Gardner Architects, Inc.

Meredith — TSPDG01-355 — 1-866-525-9374

Total Living	First Floor	Second Floor	Bonus	Bed	Bath	Width	Depth	Foundation	Price Category
1694 sq ft	1100 sq ft	594 sq ft	N/A	3	2	36' 8"	45' 0"	Crawl Space*	D

*Other options available. See page 347.

Design Features

- This rustic home invites casual, comfortable living in impressive indoor spaces.

- The clever kitchen features an island cooktop with counter, which is ideal for entertaining.

- The second-floor master suite pampers owners with a whirlpool tub and a walk-in closet.

- Extra storage space is available in the attic.

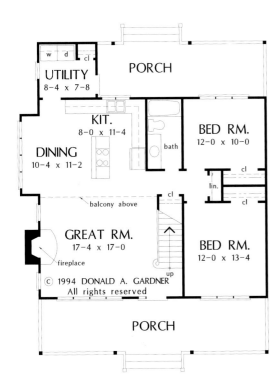

First Floor

Second Floor

Rear Elevation

Foxcrofte — TSPFB01-3735 — 1-866-525-9374

© 2002 Frank Betz Associates, Inc.

Total Living	First Floor	Second Floor	Bonus	Bed	Bath	Width	Depth	Foundation	Price Category
1698 sq ft	1245 sq ft	453 sq ft	246 sq ft	3	2-1/2	50' 6"	44' 4"	Basement or Crawl Space	F

Design Features

- The *Foxcrofte* is quaint and charming with its gabled roofline and covered front porch.

- Vaulted ceilings give a roomy feeling inside in the foyer and great room. The staircase to the upper floor is tucked away near the back of the home, adding to this design's spaciousness.

- The dining room is bordered by decorative columns, leaving this space open and accessible.

- Optional bonus space has been incorporated upstairs and presents the opportunity for adding a playroom, home office of craft room.

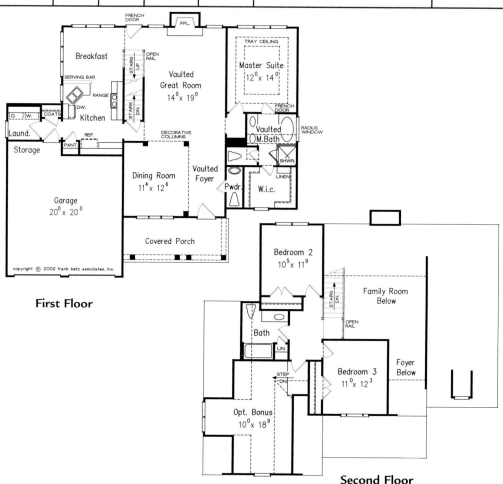

First Floor

Second Floor

Rear Elevation

Sunderland — TSPDG01-545 — 1-866-525-9374

Total Living	First Floor	Second Floor	Bonus	Bed	Bath	Width	Depth	Foundation	Price Category
1761 sq ft	1271 sq ft	490 sq ft	N/A	3	2-1/2	77' 8"	50' 0"	Crawl Space*	D

*Other options available. See page 347.

Design Features

- This country farmhouse looks and lives bigger than its square footage indicates.

- The large center dormer directs light through clerestory windows into the two-story foyer.

- Interior columns mark entrance to the formal dining room.

- The heart of the home is the central great room with fireplace and built-ins.

- Upstairs bedrooms sport dormer alcoves and are separated by a balcony.

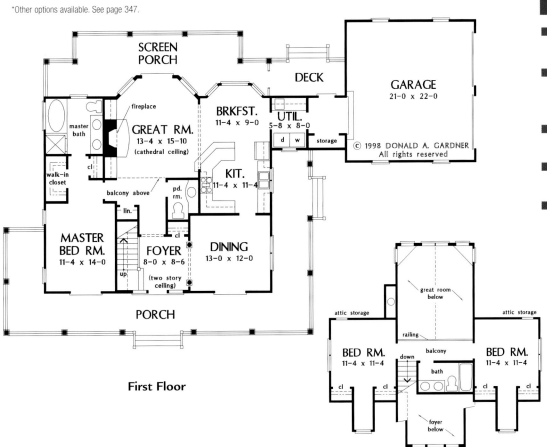

First Floor

Second Floor

Rear Elevation

Morninglory — TSPDG01-236 — 1-866-525-9374

© 1991 Donald A. Gardner Architects, Inc.

Total Living	First Floor	Second Floor	Bonus	Bed	Bath	Width	Depth	Foundation	Price Category
1778 sq ft	1325 sq ft	453 sq ft	N/A	3	2-1/2	48' 4"	40' 4"	Crawl Space*	D

Design Features

- Custom-style windows dress up the exterior.
- The U-shaped kitchen has a pass-thru to the vaulted great room.
- The private, first-level master suite opens to a rear deck.
- The master bath pampers with double lavs, garden tub and shower.

*Other options available. See page 347.

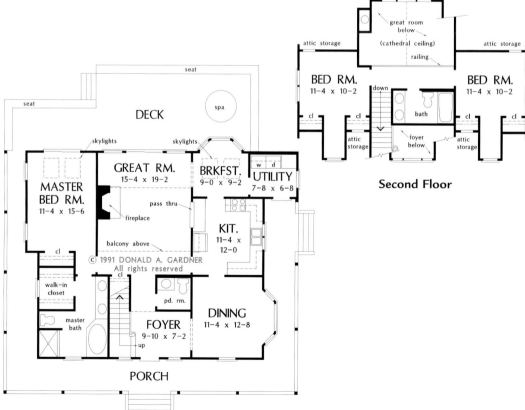

First Floor

clerestory with palladian window

great room below

(cathedral ceiling)

attic storage

railing

attic storage

BED RM.
11-4 x 10-2

down

BED RM.
11-4 x 10-2

cl   cl

bath

cl   cl

attic storage

foyer below

attic storage

Second Floor

DECK

spa

seat

seat

skylights   skylights

MASTER BED RM.
11-4 x 15-6

GREAT RM.
15-4 x 19-2

BRKFST.
9-0 x 9-2

UTILITY
7-8 x 6-8

w   d

pass thru

fireplace

cl

balcony above

KIT.
11-4 x 12-0

walk-in closet

cl

pd. rm.

master bath

FOYER
9-10 x 7-2

up

DINING
11-4 x 12-8

PORCH

Rear Elevation

© 2000 Donald A. Gardner, Inc.

Swansboro — TSPDG01-853 — 1-866-525-9374

Total Living	First Floor	Second Floor	Bonus	Bed	Bath	Width	Depth	Foundation	Price Category
1787 sq ft	964 sq ft	823 sq ft	331 sq ft	3	2-1/2	57' 2"	42' 8"	Crawl Space*	D

*Other options available. See page 347.

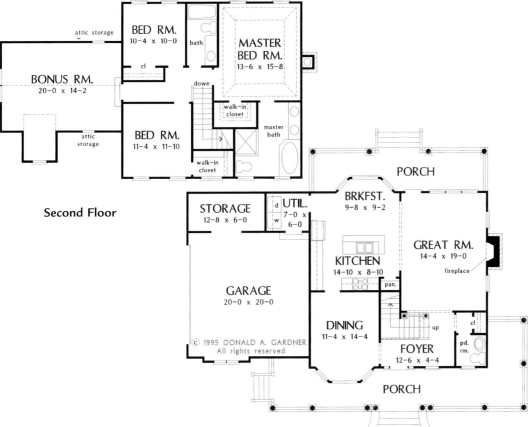

Second Floor

First Floor

Design Features

- A trio of dormers and a wrapping front porch with gable entry create a genteel façade.

- The great room is generously proportioned and open to the kitchen and breakfast area.

- Another bay window expands the dining room.

- Front and back porches provide ample outdoor living space.

- A tray ceiling tops the master bedroom, which features a walk-in closet and private bath.

Rear Elevation

Reidville — TSPDG01-424 — 1-866-525-9374

Total Living	First Floor	Second Floor	Bonus	Bed	Bath	Width	Depth	Foundation	Price Category
1792 sq ft	959 sq ft	833 sq ft	344 sq ft	3	2-1/2	52' 6"	42' 8"	Crawl Space*	D

*Other options available. See page 347.

Design Features

- This comfortable farmhouse appeals to everyone with an easy-to-build floor plan.
- Active families will enjoy the great room, open to the kitchen and breakfast bay.
- For narrower lot restrictions, the garage can be modified to be a front-load style.
- Bedrooms on the second floor share a full bath.

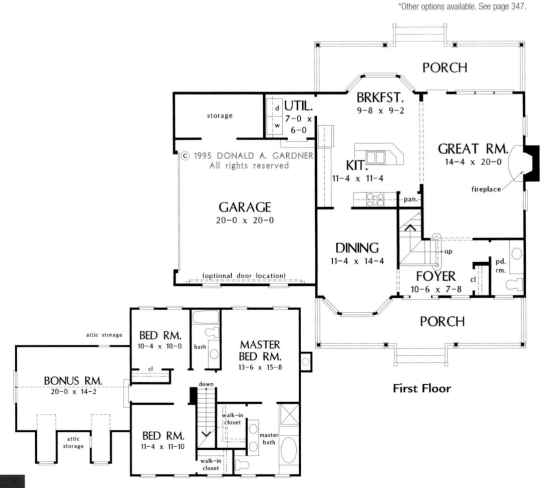

First Floor

Second Floor

Rear Elevation

Luxembourg — TSPDG01-979 — 1-866-525-9374

Total Living	First Floor	Second Floor	Bonus	Bed	Bath	Width	Depth	Foundation	Price Category
1797 sq ft	1345 sq ft	452 sq ft	349 sq ft	3	2-1/2	63' 0"	40' 0"	Crawl Space*	D

*Other options available. See page 347.

Second Floor

First Floor

Design Features

- Incorporating Old-World style, this house combines stone and stucco in its façade.

- The grand portico leads to an open floor plan, which is equally impressive.

- Built-in cabinetry, French doors and a fireplace enhance the great room.

- The first-floor master suite is located in the quiet zone with no rooms above it.

- Upstairs, the bonus room features convenient second-floor access.

Rear Elevation

Gershwin — TSPFB01-1122 — 1-866-525-9374

© 1998 Frank Betz Associates, Inc.

Total Living	First Floor	Second Floor	Bonus	Bed	Bath	Width	Depth	Foundation	Price Category
1811 sq ft	916 sq ft	895 sq ft	262 sq ft	3	2-1/2	44' 0"	38' 0"	Basement, Crawl Space or Slab	G

Design Features

- Front porches are once again popular among today's homeowners, so we situated a cozy wraparound porch on this home.

- The plan maintains an open feeling inside as the family room, kitchen and breakfast area enjoy common space.

- The master suite features a tray ceiling, double vanities and his-and-her closets.

First Floor

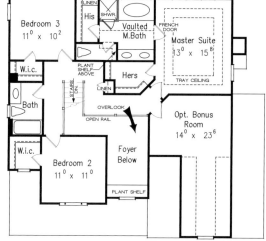

Second Floor

Rear Elevation

© 1997 Frank Betz Associates, Inc.

Willow — TSPFB01-1085 — 1-866-525-9374

Total Living	First Floor	Second Floor	Bonus	Bed	Bath	Width	Depth	Foundation	Price Category
1818 sq ft	1382 sq ft	436 sq ft	298 sq ft	3	2-1/2	52' 4"	45' 10"	Basement, Crawl Space or Slab	G

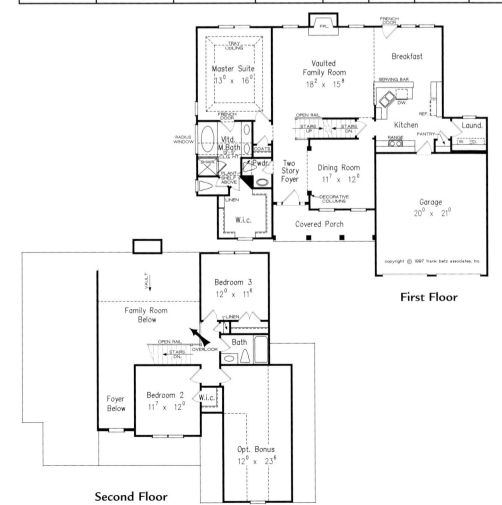

First Floor

Second Floor

Design Features

■ Classic and tasteful, the combination of brick and siding is long-appreciated and always a welcome addition to any neighborhood. This home uses its space wisely, with function in every corner.

■ The master suite encompasses an entire wing of the home, giving homeowners the privacy of being the only residents on the main floor.

■ The kitchen, breakfast area and family room blend together to create one common living space.

■ Decorative columns accent the formal dining room, making a bold statement from the foyer.

Rear Elevation

Gibson — TSPDG01-821 — 1-866-525-9374

© 1999 Donald A. Gardner, Inc.

Total Living	First Floor	Second Floor	Bonus	Bed	Bath	Width	Depth	Foundation	Price Category
1821 sq ft	1293 sq ft	528 sq ft	355 sq ft	3	2-1/2	48' 8"	50' 0"	Crawl Space*	D

*Other options available. See page 347.

Design Features

- A charming gable and a wrapping front porch enhance this bungalow.

- The kitchen features a practical design and includes a handy pantry.

- A nearby utility room boasts a sink, additional cabinets and countertop space.

- The downstairs master suite enjoys a private bath and walk-in closet.

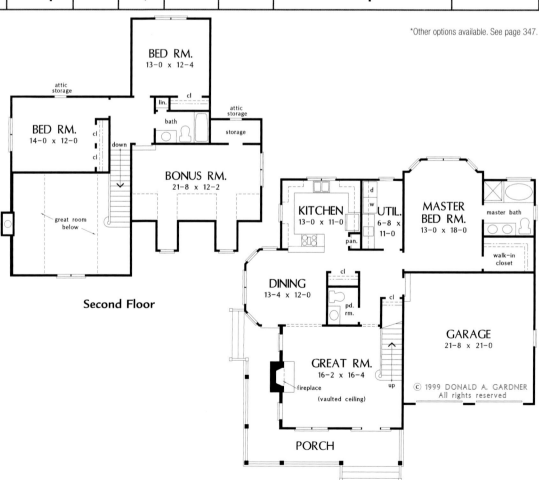

Second Floor

First Floor

Rear Elevation

© 1990 Donald A. Gardner Architects, Inc.

Glenwood — TSPDG01-224 — 1-866-525-9374

Total Living	First Floor	Second Floor	Bonus	Bed	Bath	Width	Depth	Foundation	Price Category
1831 sq ft	1289 sq ft	542 sq ft	393 sq ft	3	2-1/2	66' 4"	40' 4"	Crawl Space*	D

*Other options available. See page 347.

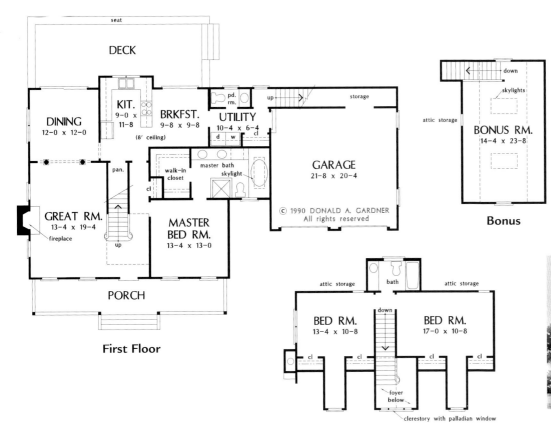

First Floor

Second Floor

Design Features

- An unfinished bonus room and an optional basement offer flexibility.

- An elegant Palladian window in a clerestory dormer washes the two-story foyer in natural light.

- Columns between the great and dining rooms add drama and accent nine-foot ceilings.

- A luxurious first-level master suite makes a great parent get-away.

- The master bath features a cheery skylight above the whirlpool tub.

Rear Elevation

Photographed home may have been modified from original construction documents.

Periwinkle Way — TSPDS01-6683 — 1-866-525-9374

© The Sater Design Collection, Inc.

Total Living	First Floor	Second Floor	Bonus	Bed	Bath	Width	Depth	Foundation	Price Category
1838 sq ft	1290 sq ft	548 sq ft	N/A	3	2-1/2	38' 0"	51' 0"	Basement/Crawl Space	E

Design Features

- The two-story great room features a corner fireplace.

- Columns and sweeping archways define the formal dining room.

- The main-level master suite has two lavatories.

- The second level includes a large deck.

- Two upper-level bedrooms enjoy a balcony overlook.

Rear Elevation

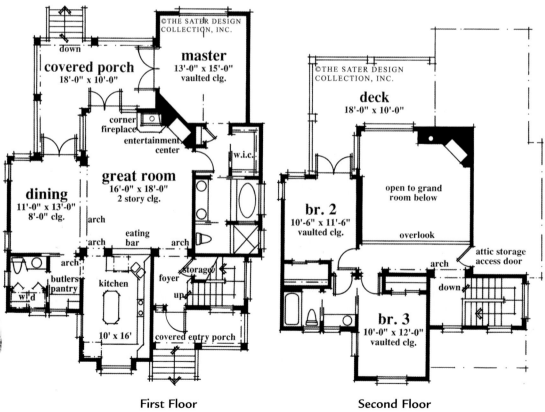

First Floor

Second Floor

© The Sater Design Collection, Inc.

Nassau Cove — TSPDS01-6654 — 1-866-525-9374

Total Living	First Floor	Second Floor	Bonus	Bed	Bath	Width	Depth	Foundation	Price Category
1853 sq ft	1342 sq ft	511 sq ft	N/A	3	2	44' 0"	40' 0"	Post/Pier	E

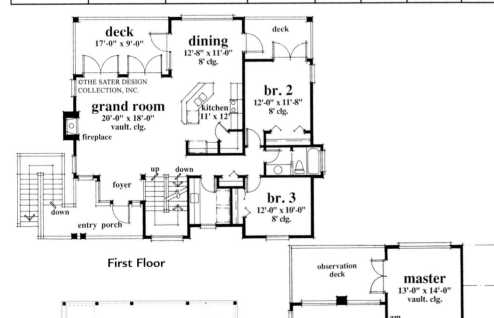

First Floor

deck
17'-0" x 9'-0"

dining
12'-8" x 11'-0"
8' clg.

deck

©THE SATER DESIGN COLLECTION, INC.

grand room
20'-0" x 18'-0"
vault. clg.

fireplace

kitchen
11' x 12'

br. 2
12'-0" x 11'-8"
8' clg.

up   down

foyer

down

entry porch

br. 3
12'-0" x 10'-0"
8' clg.

Lower Level

lattice panels

Garage
42'-0" x 22'-6"

Storage

Storage

UP

Second Floor

observation deck

master
13'-0" x 14'-0"
vault. clg.

am kitchen

open to grand room below

©THE SATER DESIGN COLLECTION, INC.

down

Design Features

- Double French doors lead to the deck from the grand room.

- Both sides of the dining room open to decks.

- The well-appointed kitchen overlooks the living area.

- Upstairs, a hall with balcony overlook leads to the master's retreat.

- The master bath boasts a windowed whirlpool tub.

Rear Elevation

Courtney — TSPDG01-706 — 1-866-525-9374

© 1998 Donald A. Gardner, Inc.

Total Living	First Floor	Second Floor	Bonus	Bed	Bath	Width	Depth	Foundation	Price Category
1859 sq ft	1336 sq ft	523 sq ft	225 sq ft	3	2-1/2	45' 0"	53' 0"	Crawl Space*	D

*Other options available. See page 347.

Design Features

- This classic cottage offers maximum comfort for its economic design and narrow-lot width.
- The foyer features a generous coat closet and a niche for displaying collectibles.
- Two clerestory dormers and a balcony overlooking the second floor grant drama to the great room.
- The kitchen services both dining room and great room with ease.

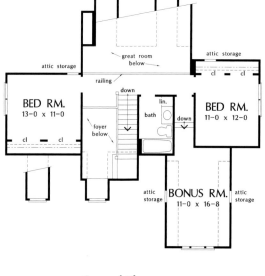

Second Floor

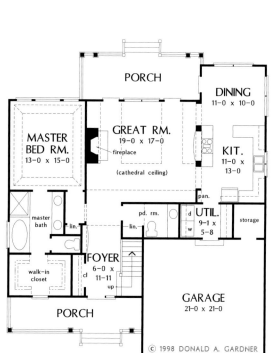

First Floor

Rear Elevation

Barclay — TSPDG01-248 — 1-866-525-9374

Total Living	First Floor	Second Floor	Bonus	Bed	Bath	Width	Depth	Foundation	Price Category
1861 sq ft	1416 sq ft	445 sq ft	284 sq ft	3	2-1/2	58' 3"	53' 5"	Crawl Space*	D

*Other options available. See page 347.

Design Features

- Interior columns add elegance while visually dividing the foyer from the dining room.
- A box-bay window adds space to the formal dining room.
- The master suite boasts his-and-her walk-in closets and garden tub with skylight.
- Two bedrooms upstairs share another skylit bath.

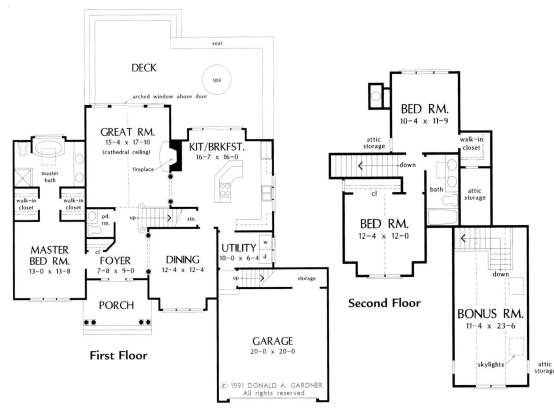

First Floor

Second Floor

Rear Elevation

Jasmine Lane — TSPDS01-6680 — 1-866-525-9374

© The Sater Design Collection, Inc.

Total Living	First Floor	Second Floor	Bonus	Bed	Bath	Width	Depth	Foundation	Price Category
1876 sq ft	1007 sq ft	869 sq ft	N/A	3	3	43' 8"	53' 6"	Pier/Crawl Space	E

Design Features

- A columned porch and romantic fretwork enhance the façade.

- The kitchen features a corner walk-in pantry.

- A box bay fills the dining room with light.

- A secondary bedroom has French doors to a balcony.

- The plan includes pier and crawl-space foundation options.

Second Floor

First Floor

Rear Elevation

© 2002 Frank Betz Associates, Inc.

Bouldercrest — TSPFB01-3674 — 1-866-525-9374

Total Living	First Floor	Second Floor	Opt. Bonus	Bed	Bath	Width	Depth	Foundation	Price Category
1879 sq ft	1407 sq ft	472 sq ft	321 sq ft	3	2-1/2	48' 0"	53' 10"	Basement or Crawl Space	F

First Floor

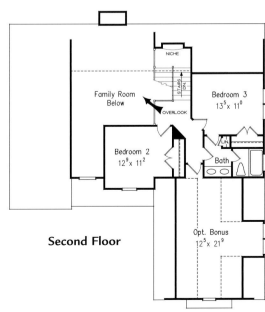

Second Floor

Design Features

- Stacked stone and siding give this Old-World elevation an up-to-date look.
- The spacious his-and-her closets are just one special feature in this main-level master home.
- The kitchen boasts a serving bar, perfect for doing homework or a quick meal on the run.
- Two bedrooms, a bath and an optional bonus room complete the second floor.

Rear Elevation

Rivermeade — TSPFB01-3668 — 1-866-525-9374

© 2002 Frank Betz Associates, Inc.

Total Living	First Floor	Second Floor	Opt. Bonus	Bed	Bath	Width	Depth	Foundation	Price Category
1879 sq ft	1359 sq ft	520 sq ft	320 sq ft	3	2-1/2	45' 0"	52' 4"	Basement or Crawl Space	G

Design Features

- Volume makes this home feel larger than it is, with vaulted, tray and two-story ceilings throughout the main level.

- The laundry area can also be a mudroom with direct access off the garage.

- A handy pass-thru from the kitchen to the great room makes entertaining easier.

- A large covered porch adds interest to the front façade.

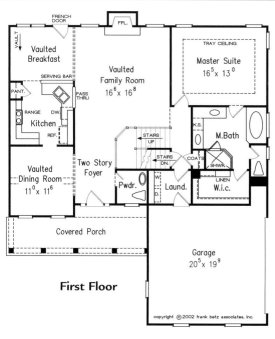

First Floor

copyright © 2002 frank betz associates, inc.

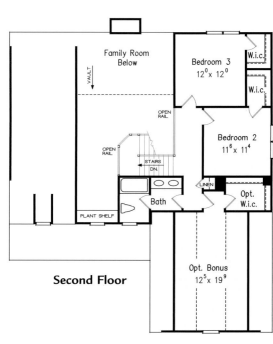

Second Floor

Rear Elevation

Liberty Hill — TSPDG01-414 — 1-866-525-9374

Total Living	First Floor	Second Floor	Bonus	Bed	Bath	Width	Depth	Foundation	Price Category
1883 sq ft	1803 sq ft	80 sq ft	918 sq ft	3	2	63' 8"	57' 4"	Crawl Space*	D

*Other options available. See page 347.

Design Features

- Growing families will adore this farm-house with plenty of versatile unfinished bonus space.

- Living and entertaining space expands to the deck.

- A tray ceiling adds interest and volume to the master bedroom.

- The master suite includes a walk-in closet and skylit bath with a garden tub.

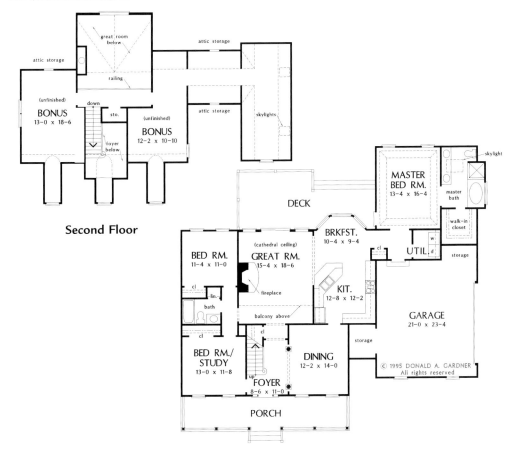

Second Floor

First Floor

Rear Elevation

Merrifield — TSPDG01-235 — 1-866-525-9374

Total Living	First Floor	Second Floor	Bonus	Bed	Bath	Width	Depth	Foundation	Price Category
1898 sq ft	1356 sq ft	542 sq ft	378 sq ft	3	2-1/2	59' 0"	64' 4"	Crawl Space*	D

Design Features

*Other options available. See page 347.

- A Palladian window in the clerestory dormer bathes the two-story foyer in natural light.

- Nine-foot ceilings throughout the first level add drama.

- The first-floor master suite has a whirlpool tub, shower, double lavs and walk-in closet.

- Upstairs, two bedrooms with dormers and attic-storage access share a full bath.

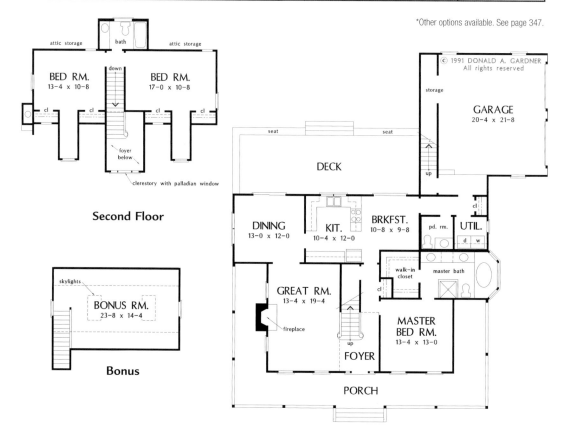

Second Floor

Bonus

First Floor

Rear Elevation

Photographed home may have been modified from original construction documents.

© 1999 Donald A. Gardner, Inc.

Saluda — TSPDG01-795 — 1-866-525-9374

Total Living	First Floor	Second Floor	Bonus	Bed	Bath	Width	Depth	Foundation	Price Category
1891 sq ft	1309 sq ft	582 sq ft	570 sq ft	3	2-1/2	65' 8"	39' 4"	Crawl Space*	D

*Other options available. See page 347.

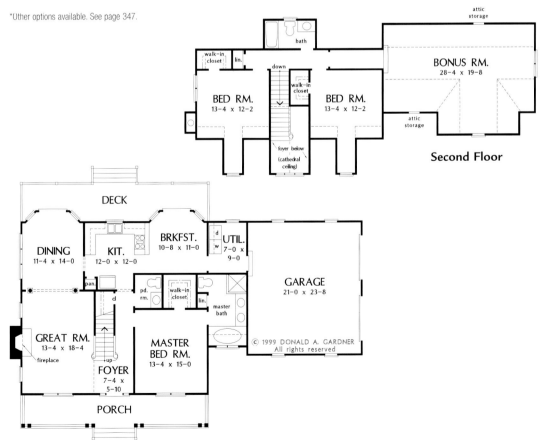

Second Floor

First Floor

Design Features

- This one-and-a-half-story home features a relaxing front porch and a generous great room.

- Like the master suite, both upstairs bedrooms include walk-in closets.

- Dormer alcoves add interest and illumination to the second-floor family bedrooms.

- An oversized bonus room offers an abundance of space for storage or future expansion.

Rear Elevation

Georgetown Cove — TSPDS01-6690 — 1-866-525-9374

© The Sater Design Collection, Inc.

Total Living	First Floor	Second Floor	Bonus	Bed	Bath	Width	Depth	Foundation	Price Category
1910 sq ft	873 sq ft	1037 sq ft	N/A	3	2-1/2	27' 6"	64' 0"	Piling/Garage on Slab	E

Design Features

- The quaint front balcony has a glass-paneled entry to the foyer.

- The great room has French doors to the outside and a fireplace framed by built-in cabinetry.

- The formal dining room opens to a private area of the covered porch.

- Double French doors fill the upper-level master suite with sunlight and open to a sundeck.

- The study has a walk-in closet and views to the side property.

Rear Elevation

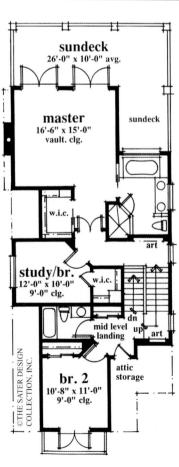

Second Floor

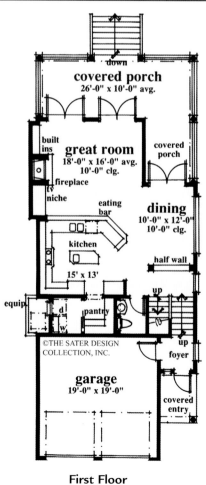

First Floor

Drysdale — TSPDG01-841 — 1-866-525-9374

Total Living	First Floor	Second Floor	Bonus	Bed	Bath	Width	Depth	Foundation	Price Category
1918 sq ft	1412 sq ft	506 sq ft	320 sq ft	3	2-1/2	49' 8"	52' 0"	Crawl Space*	D

*Other options available. See page 347.

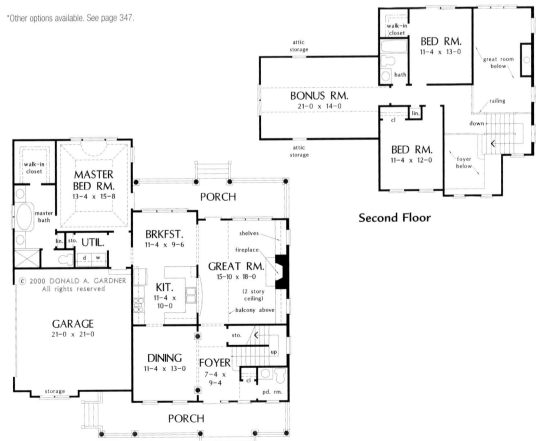

Second Floor

First Floor

Design Features

- A relaxing front porch and plenty of windows create a superb façade.

- The foyer is stunning, with a two-story ceiling, plant shelf and balcony.

- Built-in shelving flanks the fireplace in the great room.

- A back porch encourages outdoor relaxation.

Rear Elevation

Whitley — TSPDG01-863 — 1-866-525-9374

© 2000 Donald A. Gardner, Inc.

Total Living	First Floor	Second Floor	Bonus	Bed	Bath	Width	Depth	Foundation	Price Category
1921 sq ft	1408 sq ft	513 sq ft	444 sq ft	3	2-1/2	40' 4"	55' 2"	Crawl Space*	D

*Other options available. See page 347.

Design Features

- Double gables, a cozy front porch, and arched windows create charm.
- A vaulted ceiling expands the great room, overlooked by the second-floor hallway.
- Built-in cabinets topped by small windows flank the fireplace in the great room.
- Two bedrooms with ample closet space are located on the second floor.

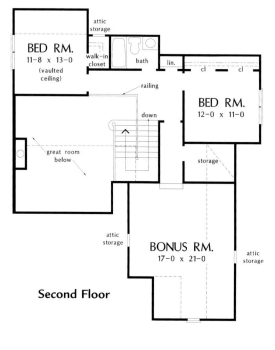

Second Floor

First Floor

Rear Elevation

© 2002 Frank Betz Associates, Inc.

Bentridge — TSPFB01-3666 — 1-866-525-9374

Total Living	First Floor	Second Floor	Bonus	Bed	Bath	Width	Depth	Foundation	Price Category
1928 sq ft	947 sq ft	981 sq ft	N/A	4	2-1/2	41' 0"	39' 4"	Basement, Crawl Space or Slab	G

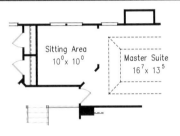

Opt. Sitting Room

Second Floor

First Floor

Design Features

- Simply elegant outside and in, the *Bentridge* is well planned and functional.

- The main floor is dedicated solely to living space, while four bedrooms share the upper level.

- Ornamental columns and knee walls offset the family room, separating it from the breakfast area, yet allowing for easy transition from one spot to the next.

- A handy serving bar provides additional seating or serving space in the kitchen.

- If four bedrooms are not necessary, an alternate design is included with a master sitting room option.

Rear Elevation

Pickney — TSPDG01-260 — 1-866-525-9374

© 1991 Donald A. Gardner Architects, Inc.

Total Living	First Floor	Second Floor	Bonus	Bed	Bath	Width	Depth	Foundation	Price Category
1936 sq ft	1025 sq ft	911 sq ft	410 sq ft	3	2-1/2	53' 8"	58' 8"	Crawl Space*	D

*Other options available. See page 347.

Design Features

- Windows and gables give a romantic feel to this farmhouse with expansive deck area.
- Visual excitement continues on the interior with a two-story foyer and upstairs balcony.
- A center-island kitchen opens to the dining room and large breakfast area.
- The master suite has double walk-in closets plus whirlpool tub and shower.
- Access the skylit bonus room from the second floor through an additional storage area.

Rear Elevation

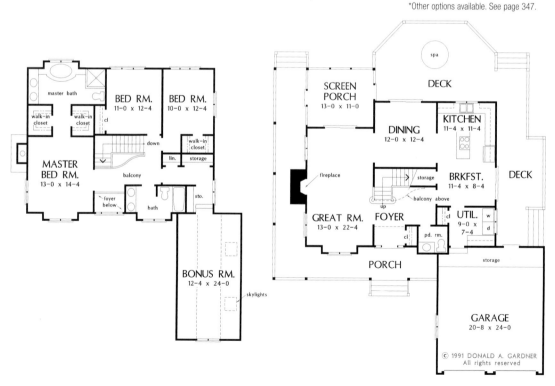

Second Floor

First Floor

Creston — TSPDG01-420 — 1-866-525-9374

Total Living	First Floor	Second Floor	Bonus	Bed	Bath	Width	Depth	Foundation	Price Category
1972 sq ft	1436 sq ft	536 sq ft	296 sq ft	3	2-1/2	67' 10"	47' 4"	Crawl Space*	D

*Other options available. See page 347.

Design Features

- Palladian windows and covered porches give this farmhouse special refinement.

- Interior columns distinguish the inviting two-story foyer from the dining room.

- Nine-foot ceilings add volume and drama throughout the first floor.

- The master suite is secluded downstairs and features a tray ceiling.

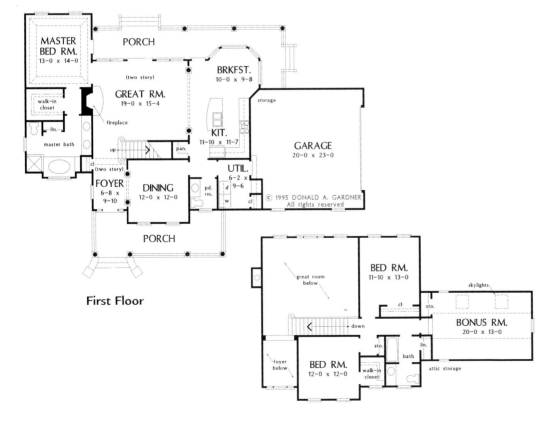

First Floor

Second Floor

Rear Elevation

Stonechase — TSPFB01-3662 — 1-866-525-9374

© 2002 Frank Betz Associates, Inc.

Total Living	First Floor	Second Floor	Bonus	Bed	Bath	Width	Depth	Foundation	Price Category
1974 sq ft	1458 sq ft	516 sq ft	168 sq ft	3	2-1/2	50' 0"	46' 0"	Basement or Crawl Space	G

Design Features

- Stone accents, carriage doors and board and batten shutters come together to create the cottage-like appeal that is so desirable today. The creativity continues inside, where an art niche is positioned as a focal point in the foyer.

- The master suite is secluded from secondary bedrooms, encompassing an entire wing of the main level.

- A powder room and coat closet are strategically placed near the garage entrance, keeping shoes and coats in their place.

- The secondary bedrooms have an overlook to the vaulted family room.

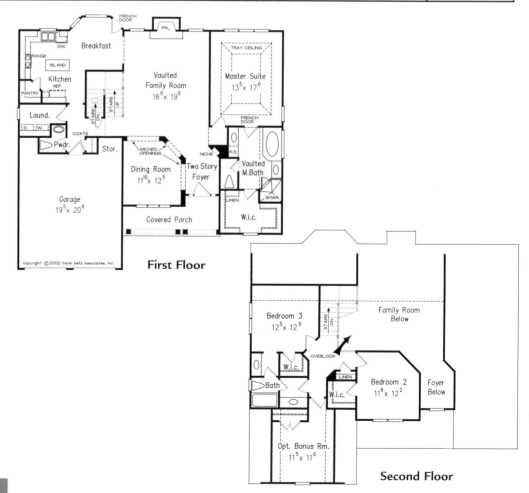

First Floor

Second Floor

Rear Elevation

Willow Creek — TSPFB01-3539 — 1-866-525-9374

Total Living	First Floor	Second Floor	Bonus	Bed	Bath	Width	Depth	Foundation	Price Category
1975 sq ft	1399 sq ft	576 sq ft	221 sq ft	3	2-1/2	52' 4"	46' 10"	Basement, Crawl Space or Slab	H

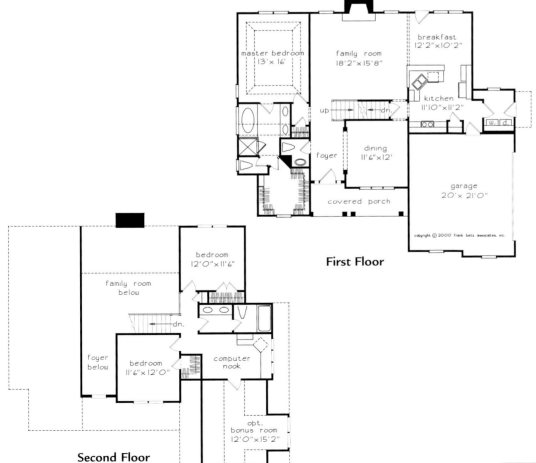

First Floor

Second Floor

Design Features

- From the *Southern Living® Design Collection*
- The *Willow Creek's* inviting exterior combines an attractive combination of stone and siding, graced by a columned front porch.
- The kitchen is conveniently linked to a large laundry room, which is a must-have for growing families.
- The master bath contains all the luxuries including double vanities, a garden tub, separate shower and walk-in closet.

Rear Elevation

Santa Rosa — TSPDS01-6808 — 1-866-525-9374

© The Sater Design Collection, Inc.

Total Living	First Floor	Second Floor	Bonus	Bed	Bath	Width	Depth	Foundation	Price Category
1978 sq ft	1383 sq ft	595 sq ft	N/A	3	2	48' 0"	42' 0"	Island Basement	F

Design Features

- The great room is bright with soaring ceilings and a fireplace.

- A chef-friendly kitchen features an angled island.

- Two generous bedrooms enjoy private porches.

- The upper level features a dramatic master suite.

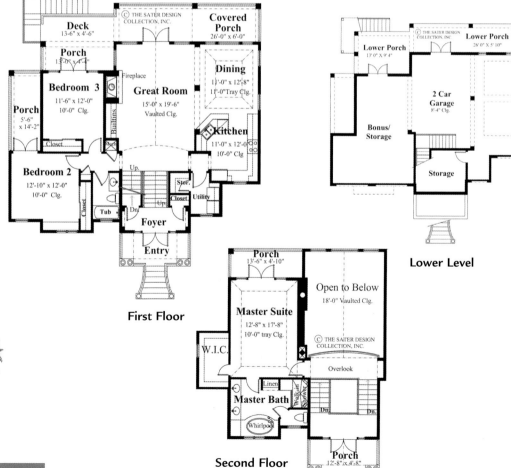

Rear Elevation

First Floor

Second Floor

Lower Level

© 1995 Donald A. Gardner Architects, Inc.

Williamston — TSPDG01-391 — 1-866-525-9374

Total Living	First Floor	Second Floor	Bonus	Bed	Bath	Width	Depth	Foundation	Price Category
1991 sq ft	1480 sq ft	511 sq ft	363 sq ft	3	2-1/2	73' 0"	45' 0"	Crawl Space*	D

*Other options available. See page 347.

Design Features

- The kitchen features an angled island peninsula for effortless entertaining.

- Bay windows accent the formal dining room and breakfast bay.

- The secluded master suite accesses the deck and includes a private bath.

- Upstairs, two bedrooms with dormers share a bath from a balcony that overlooks the great room.

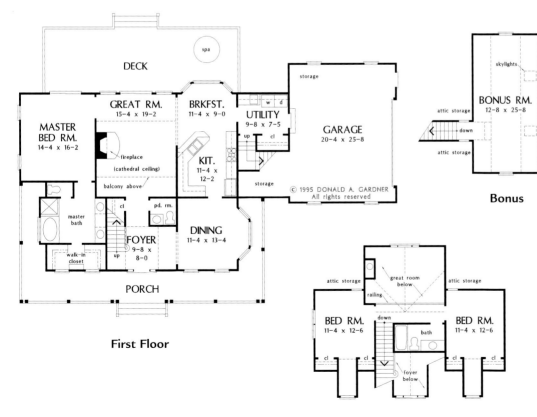

First Floor

Bonus

Second Floor

Rear Elevation

Winchester — TSPDG01-329 — 1-866-525-9374

Total Living	First Floor	Second Floor	Bonus	Bed	Bath	Width	Depth	Foundation	Price Category
2019 sq ft	1506 sq ft	513 sq ft	397 sq ft	3	2-1/2	65' 4"	67' 10"	Crawl Space*	E

Design Features

- This open floor plan with real country appeal makes the most of space.

- A second-level balcony overlooks the great room with cathedral ceiling.

- Privately located downstairs, the master suite opens to the back porch.

- A two-story foyer provides a grand welcome into the home.

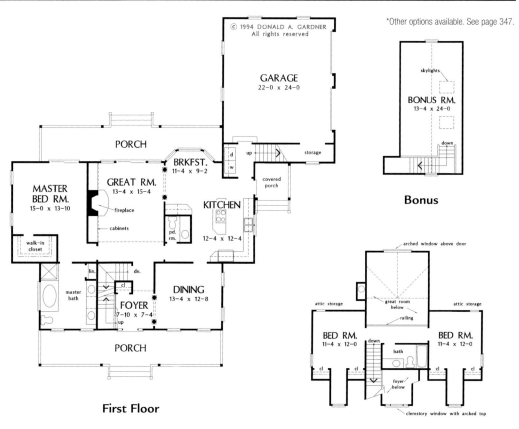

*Other options available. See page 347.

First Floor

Bonus

Second Floor

Rear Elevation

Hanover — TSPDG01-489 — 1-866-525-9374

Total Living	First Floor	Second Floor	Bonus	Bed	Bath	Width	Depth	Foundation	Price Category
2023 sq ft	1489 sq ft	534 sq ft	393 sq ft	3	2-1/2	59' 4"	58' 7"	Crawl Space*	E

*Other options available. See page 347.

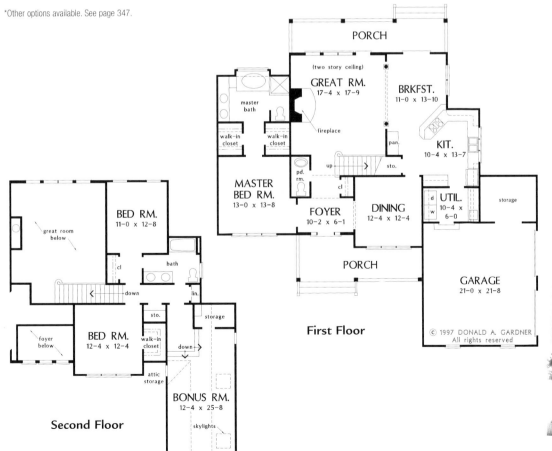

First Floor

Second Floor

Design Features

- A smart exterior and open interior space combine to create this roomy, yet practical, home.
- The two-story foyer leads to a two-story great room and back-porch access.
- Columns divide the great room from breakfast room.
- A handy utility room leads to a two-car garage with ample storage space.
- A split-bedroom plan places the master suite down and additional bedrooms upstairs.

Rear Elevation

McArthur Park — TSPFB01-3837 — 1-866-525-9374

© 2003 Frank Betz Associates, Inc.

Total Living	First Floor	Second Floor	Opt. Bonus	Bed	Bath	Width	Depth	Foundation	Price Category
2024 sq ft	1480 sq ft	544 sq ft	253 sq ft	3	2-1/2	52' 0"	46' 4"	Basement, Crawl Space or Slab	G

Design Features

- It has been said that beauty is often found in simplicity -- and the McArthur Park brings truth to that statement. Its understated exterior and uncomplicated roofline make this home both appealing and cost-effective to build.

- The kitchen, breakfast area and keeping room share common space, making an ideal space for entertaining.

- An optional bonus room is available on the upper floor that has endless possibilities. It can be easily finished as a fourth bedroom, playroom, exercise area or craft room.

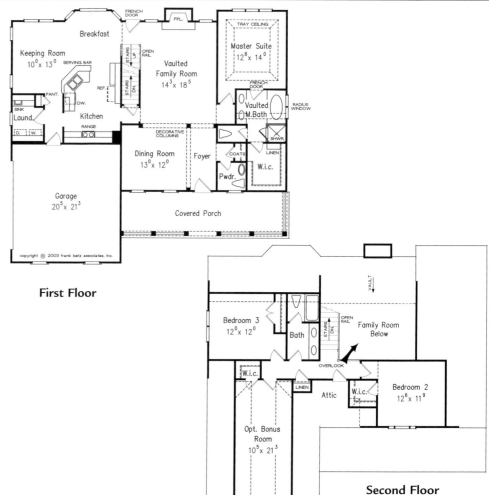

First Floor

Second Floor

Rear Elevation

Brock — TSPDG01-1071 — 1-866-525-9374

Total Living	First Floor	Second Floor	Bonus	Bed	Bath	Width	Depth	Foundation	Price Category
2030 sq ft	1507 sq ft	523 sq ft	322 sq ft	3	2-1/2	42' 5"	63' 2"	Crawl Space*	E

*Other options available. See page 347.

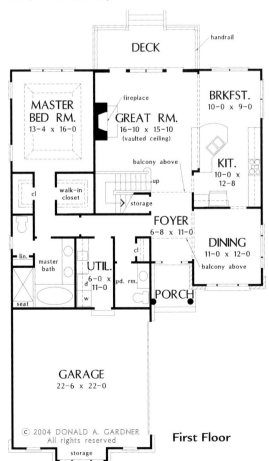

First Floor

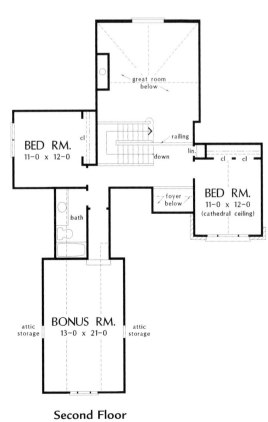

Second Floor

Design Features

- Brick and siding form an exciting elevation for this charming traditional home.

- Gables contrast with arched transoms, and a box-bay window disguises the side of the garage.

- The sidelights and transom in the foyer brighten the entrance.

- Emphasized by a balcony, the two-story ceiling in the great room is showcased upstairs.

Rear Elevation

Seymour — TSPFB01-1210 — 1-866-525-9374

© 1998 Frank Betz Associates, Inc.

Total Living	First Floor	Second Floor	Opt. Bonus	Bed	Bath	Width	Depth	Foundation	Price Category
2034 sq ft	1559 sq ft	475 sq ft	321 sq ft	4	3	50' 0"	56' 4"	Basement, Crawl Space or Slab	G

Design Features

- Style and sensibility meet to create the *Seymour*, a design that is as practical as it is beautiful.

- Fieldstone, copper accents and a courtyard entry distinguish this home from the others on the block.

- The main floor consists of the common living areas, the master bedroom and a secondary bedroom that can easily serve as the home office.

- Optional bonus space is also available on the second floor, leaving homeowners choices on how to finish it.

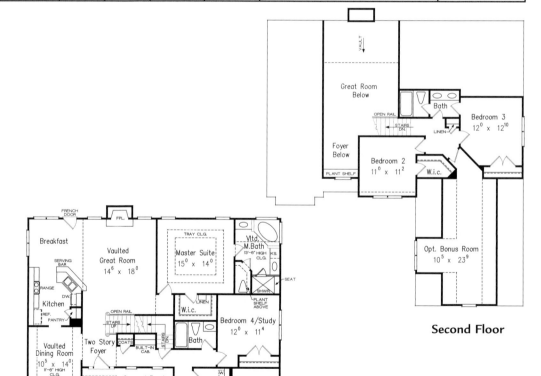

Second Floor

First Floor

Rear Elevation

Prynwood — TSPDG01-818 — 1-866-525-9374

Total Living	First Floor	Second Floor	Bonus	Bed	Bath	Width	Depth	Foundation	Price Category
2037 sq ft	1502 sq ft	535 sq ft	275 sq ft	3	2-1/2	43' 0"	57' 6"	Crawl Space*	E

*Other options available. See page 347.

Design Features

- A stunning center dormer with Palladian-style window complements this home's façade.

- A pass-thru from the kitchen to the vaulted great room makes entertaining easy.

- The first-floor master suite enjoys a tray ceiling, back-porch access and a private bath.

- Two upstairs bedrooms are divided by a balcony that overlooks both the foyer and the great room.

Second Floor

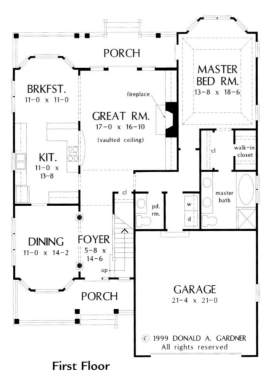

First Floor

Rear Elevation

Gastonia — TSPFB01-3721 — 1-866-525-9374

© 2002 Frank Betz Associates, Inc.

Total Living	First Floor	Second Floor	Bonus	Bed	Bath	Width	Depth	Foundation	Price Category
2040 sq ft	935 sq ft	1105 sq ft	N/A	4	2-1/2	44' 0"	39' 0"	Basement, Crawl Space or Slab	G

Design Features

- Smaller homes don't have to lack upscale amenities! The *Gastonia* includes many special features often hard to find even in larger homes.

- The master suite has a comfortable sitting room, large enough for lounging furniture. If a fourth bedroom is a higher priority, this space can be easily converted to accommodate it.

- Each secondary bedroom features a walk-in closet.

- An island serves as the center point of the kitchen and helps with meal preparation.

First Floor

Rear Elevation

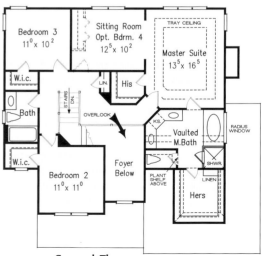

Second Floor

B. NATHAN.

Cottonwood — TSPDG01-523 — 1-866-525-9374

Total Living	First Floor	Second Floor	Bonus	Bed	Bath	Width	Depth	Foundation	Price Category
2048 sq ft	1471 sq ft	577 sq ft	368 sq ft	3	2-1/2	75' 5"	52' 0"	Crawl Space*	E

*Other options available. See page 347.

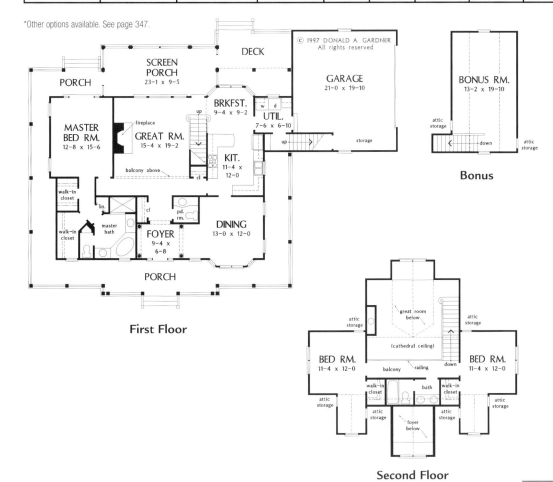

First Floor

Bonus

Second Floor

Design Features

- Surrounded by porches, this farmhouse boasts front and back clerestory dormers for natural light.
- The unconventional rear staircase is located in the great room for convenience.
- The master suite is privately situated downstairs and features a magnificent bath.
- Both second-story bedrooms boast dormer alcoves and walk-in closets.

Rear Elevation

Ashby — TSPDG01-1061 — 1-866-525-9374

Total Living	First Floor	Second Floor	Bonus	Bed	Bath	Width	Depth	Foundation	Price Category
2051 sq ft	1502 sq ft	549 sq ft	285 sq ft	3	2-1/2	43' 0"	57' 6"	Crawl Space*	E

*Other options available. See page 347.

Design Features

- This Craftsman cottage starts with an exterior of stone and siding for incredible curb appeal.

- A balcony divides the two-story foyer from the great room.

- Bay windows extend floor space in the dining room, breakfast nook and master bedroom.

- Above the garage, a bonus room provides space for a family's changing needs.

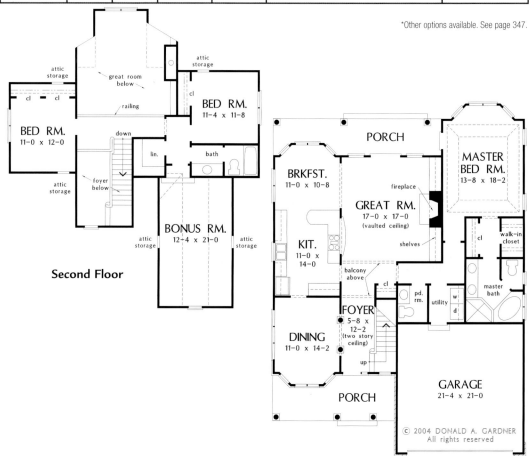

Second Floor

First Floor

Rear Elevation

1800 sq ft—2200 sq ft

Donald A. Gardner Architects, Inc.

Topeka — TSPDG01-1000 — 1-866-525-9374

Total Living	First Floor	Second Floor	Bonus	Bed	Bath	Width	Depth	Foundation	Price Category
2064 sq ft	1562 sq ft	502 sq ft	416 sq ft	3	2-1/2	54' 0"	55' 10"	Crawl Space*	E

*Other options available. See page 347.

Design Features

- Capturing the heartland feel, this farmhouse is designed to make an impression.

- A welcoming front porch guides guests inside to the two-story foyer.

- The great room features numerous windows, French doors and a stunning fireplace.

- A cathedral ceiling visually expands the master suite, and a French door leads to the rear porch.

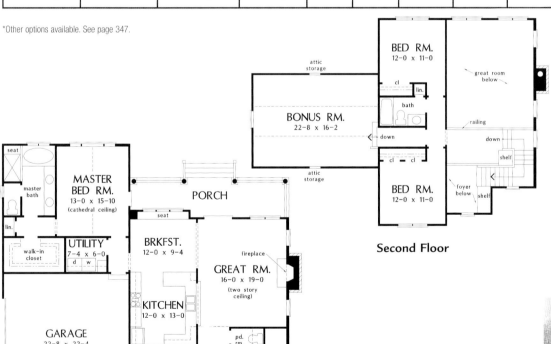

Second Floor

BED RM.
12-0 x 11-0

attic storage

BONUS RM.
22-8 x 16-2

attic storage

great room below

bath

railing

down

shelf

down

foyer below

shelf

BED RM.
12-0 x 11-0

MASTER BED RM.
13-0 x 15-10
(cathedral ceiling)

master bath

walk-in closet

lin.

seat

UTILITY
7-4 x 6-0

d   w

PORCH

seat

BRKFST.
12-0 x 9-4

fireplace

GREAT RM.
16-0 x 19-0
(two story ceiling)

KITCHEN
12-0 x 13-0

GARAGE
22-8 x 22-4

pd. rm.

DINING
12-0 x 13-4

FOYER
5-8 x 8-4

up

cl

PORCH

First Floor

Rear Elevation

Dayton — TSPDG01-1008 — 1-866-525-9374

© 2003 Donald A. Gardner, Inc.

Total Living	First Floor	Second Floor	Bonus	Bed	Bath	Width	Depth	Foundation	Price Category
2073 sq ft	1569 sq ft	504 sq ft	320 sq ft	3	2-1/2	47' 0"	55' 0"	Crawl Space*	E

*Other options available. See page 347.

Design Features

- An abundance of windows floods this home with natural light.

- Custom-styled features include a plant shelf in the foyer, fireplace and two-story great room ceiling.

- The master suite is complete with a walk-in and wardrobe closets, and a luxurious bathroom.

- Two secondary bedrooms share a full bath with the bonus room.

- Note storage closets on both floors.

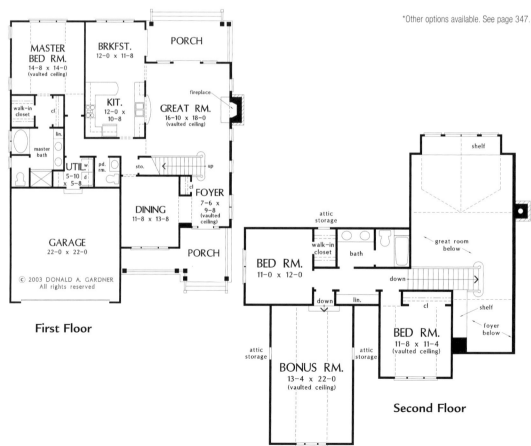

First Floor

Second Floor

Rear Elevation

© 1999 Donald A. Gardner, Inc.

Radcliffe — TSPDG01-799 — 1-866-525-9374

Total Living	First Floor	Second Floor	Bonus	Bed	Bath	Width	Depth	Foundation	Price Category
2075 sq ft	1588 sq ft	487 sq ft	363 sq ft	3	2-1/2	60' 1"	50' 11"	Crawl Space*	E

*Other options available. See page 347.

Design Features

■ With its hip roof, gables and brick and siding exterior, this home possesses traditional elegance.

■ The generous great room with cathedral ceiling and fireplace is centrally located.

■ A patio extends living space beyond the great room.

■ A cozy back porch expands the breakfast area.

■ Two secondary bedrooms and a bonus room share a full bath upstairs.

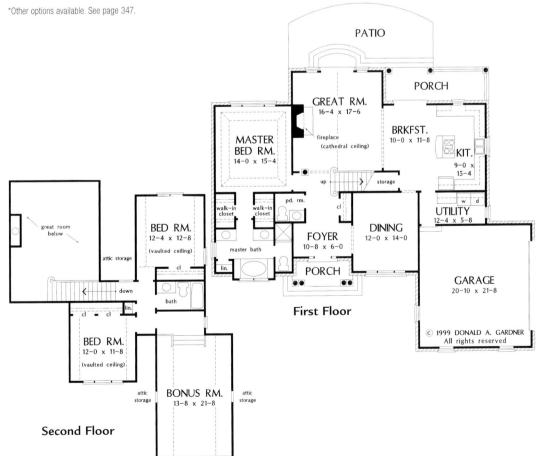

First Floor

Second Floor

Rear Elevation

Madaridge — TSPDG01-974 — 1-866-525-9374

© 2002 Donald A. Gardner, Inc.

Total Living	First Floor	Second Floor	Bonus	Bed	Bath	Width	Depth	Foundation	Price Category
2111 sq ft	1496 sq ft	615 sq ft	277 sq ft	3	2-1/2	40' 4"	70' 0"	Crawl Space*	E

Design Features

- Craftsman materials on this Traditional design create incredible curb appeal.

- The front-entry garage is ideal for narrow lots.

- The open floor plan features a first-floor master suite and screened porch.

- Upstairs, a loft overlooks the open great room and foyer.

- Note the built-in hutch in the dining room and double vanities in both full baths.

*Other options available. See page 347.

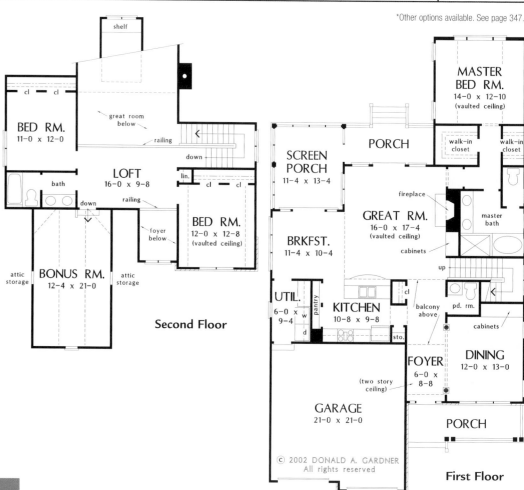

Second Floor

First Floor

Rear Elevation

Dartmouth — TSPDG01-488 — 1-866-525-9374

Total Living	First Floor	Second Floor	Bonus	Bed	Bath	Width	Depth	Foundation	Price Category
2121 sq ft	1572 sq ft	549 sq ft	384 sq ft	3	2-1/2	59' 4"	53' 11"	Crawl Space*	E

*Other options available. See page 347.

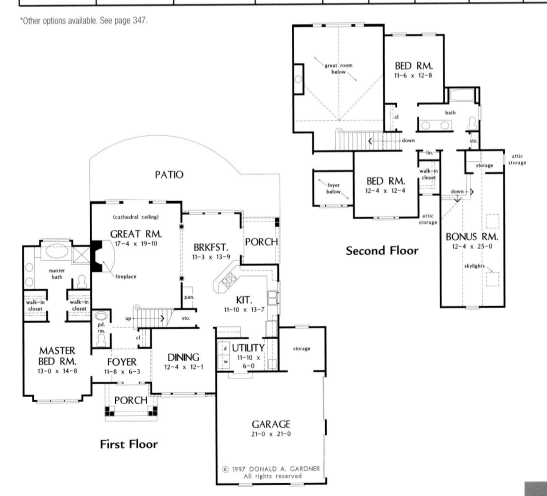

First Floor

Second Floor

Design Features

- This home is dressed to impress with its stone and stucco exterior and dramatic square-columned entry.

- The great room with cathedral ceiling adjoins a breakfast area which opens onto a side porch.

- A separate utility room has built-in cabinets and countertop with laundry sink.

- Double doors lead into the first-floor master suite with walk-in closets and lavish bath.

Rear Elevation

Duvall Street — TSPDS01-6701 — 1-866-525-9374

© The Sater Design Collection, Inc.

Total Living	First Floor	Second Floor	Bonus	Bed	Bath	Width	Depth	Foundation	Price Category
2123 sq ft	878 sq ft	1245 sq ft	N/A	4	2-1/2	27' 6"	64' 0"	Post/Pier	F

Design Features

- Two sets of French doors open the great room to the porch.

- A gourmet kitchen boasts a prep sink and eating bar.

- The mid-level landing leads to two bedrooms and a bath.

- French doors open the master suite to a sundeck.

- The master bath has a windowed soaking tub.

Rear Elevation

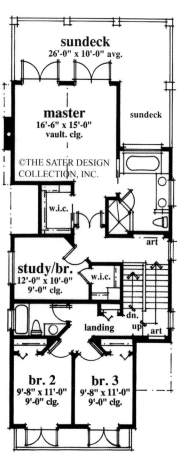

Second Floor

First Floor

Holly Hill — TSPFB01-1237 — 1-866-525-9374

Total Living	First Floor	Second Floor	Bonus	Bed	Bath	Width	Depth	Foundation	Price Category
2126 sq ft	1583 sq ft	543 sq ft	251 sq ft	4	3	53' 0"	47' 0"	Basement, Crawl Space or Slab	I

First Floor

Second Floor

Design Features

- From the *Southern Living® Design Collection*

- The *Holly Hill* is an attentive and articulate design that has quickly become a favorite.

- Set just off the entry is a formal dining room set apart by decorative columns.

- Thoughtful planning has made the dining room easily accessible from the kitchen, ideal for entertaining.

- The kitchen offers ample cabinet space and a serving bar for extra seating.

Rear Elevation

Via Pascoli — TSPDS01-6842 — 1-866-525-9374

© The Sater Design Collection, Inc.

Total Living	First Floor	Lower Level	Bonus	Bed	Bath	Width	Depth	Foundation	Price Category
2137 sq ft	2137 sq ft	N/A	N/A	3	2	44' 0"	61' 0"	Island Basement	F

Design Features

- Lovely architecture and brilliant windows adorn this villa.

- A hip vaulted ceiling highlights the great room.

- French doors open the family space to a sheltered porch.

- Lower-level bonus spaces convert to hobby or storage rooms.

- The master suite is secluded and has many views.

Rear Elevation

First Floor

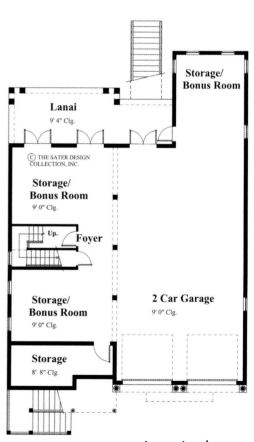

Lower Level

© 2002 Frank Betz Associates, Inc.

Colonnade — TSPFB01-3699 — 1-866-525-9374

Total Living	First Floor	Second Floor	Opt. Bonus	Bed	Bath	Width	Depth	Foundation	Price Category
2138 sq ft	1589 sq ft	549 sq ft	248 sq ft	4	3	53' 0"	47' 6"	Basement or Crawl Space	H

First Floor

Second Floor

Design Features

- Many homeowners today want function with flexibility. The *Colonnade* was created to provide both of these elements.

- The family room, kitchen and breakfast area connect to create the home's center point.

- The size and location of the main-floor bedroom also make it a perfect home office or den.

- Optional bonus space on the upper level of this home gives homeowners the opportunity to add additional living space to their home.

Rear Elevation

Pasadena — TSPFB01-3756 — 1-866-525-9374

© 2002 Frank Betz Associates, Inc.

Total Living	First Floor	Second Floor	Opt. Bonus	Bed	Bath	Width	Depth	Foundation	Price Category
2139 sq ft	1561 sq ft	578 sq ft	274 sq ft	3	2-1/2	50' 0"	57' 0"	Basement, Crawl Space or Slab	H

Design Features

- Charm and character abound from the façade of the *Pasadena*, with its tapered architectural columns and carriage doors.

- Inside, the master suite is tucked away on the rear of the main level, giving the homeowner a peaceful place to unwind.

- An art niche is situated in the breakfast area, providing the perfect spot for a favorite art piece or floral arrangement.

- The kitchen is complete with a large island, making mealtime easier.

Rear Elevation

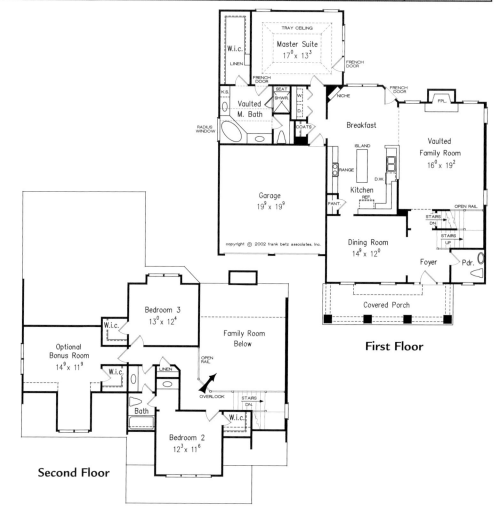

First Floor

Second Floor

© 1996 Frank Betz Associates, Inc.

Mallory — TSPFB01-992 — 1-866-525-9374

Total Living	First Floor	Second Floor	Bonus	Bed	Bath	Width	Depth	Foundation	Price Category
2155 sq ft	1628 sq ft	527 sq ft	207 sq ft	3	2-1/2	54' 0"	46' 10"	Basement, Crawl Space or Slab	I

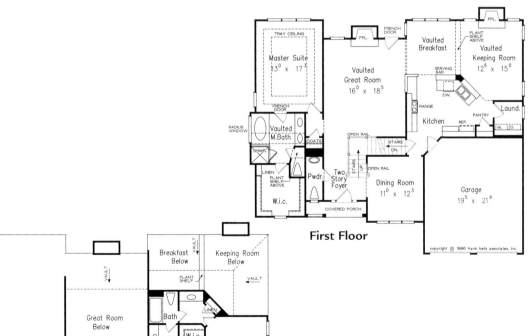

First Floor

Second Floor

Design Features

- Earthy fieldstone and cedar shake accents give the *Mallory* a casual elegance that Old-World style encompasses.
- A vaulted breakfast area and keeping room with fireplace adjoin the kitchen.
- Two secondary bedrooms—each with a walk-in closet—share a divided bathing area on the second floor.
- An optional bonus room is ready to finish into a fourth bedroom, playroom or exercise area.

Rear Elevation

Midland — TSPDG01-371 — 1-866-525-9374

Total Living	First Floor	Second Floor	Bonus	Bed	Bath	Width	Depth	Foundation	Price Category
2164 sq ft	1499 sq ft	665 sq ft	332 sq ft	4	2-1/2	69' 8"	43' 6"	Crawl Space*	E

*Other options available. See page 347.

Design Features

- The great room opens to both the deck and the island kitchen with convenient pantry.

- Nine-foot ceilings on the first level expand volume.

- The master suite welcomes owners with whirlpool tub, dual vanities and deck access.

- The bonus room can be finished at initial construction or at a later time.

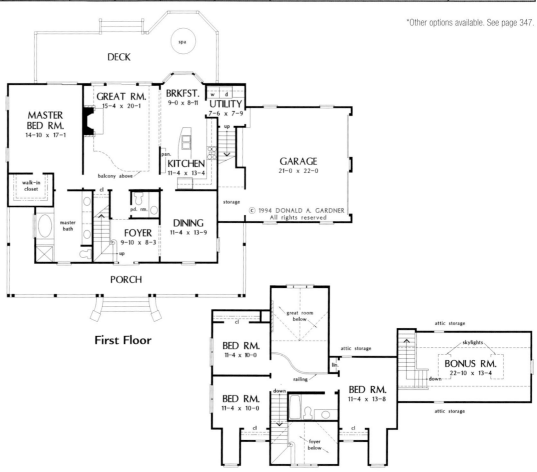

First Floor

Second Floor

Rear Elevation

Wyndham — TSPDG01-793 — 1-866-525-9374

Total Living	First Floor	Second Floor	Bonus	Bed	Bath	Width	Depth	Foundation	Price Category
2163 sq ft	1668 sq ft	495 sq ft	327 sq ft	4	3	52' 7"	50' 11"	Crawl Space*	E

*Other options available. See page 347.

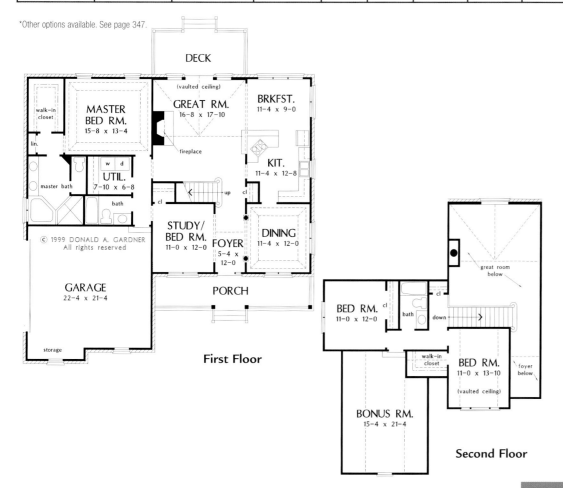

First Floor

Second Floor

Design Features

- A stunning combination of a hip roof and bold, front-facing gables creates curb appeal.
- Tray and vaulted ceilings increase the feeling of spaciousness in key rooms.
- The home's openness and easy flow create a comfortable, casual atmosphere.
- The first-floor study/bedroom is a versatile space with a nearby full bath.
- Upstairs, two bedrooms share a hall bath and access to the large bonus room.

Rear Elevation

Julian — TSPFB01-1262 — 1-866-525-9374

© 1999 Frank Betz Associates, Inc.

Total Living	First Floor	Second Floor	Opt. Bonus	Bed	Bath	Width	Depth	Foundation	Price Category
2167 sq ft	1626 sq ft	541 sq ft	256 sq ft	3	2-1/2	53' 0"	43' 4"	Basement, Crawl Space or Slab	H

Design Features

- Decorative columns separate the kitchen area from the vaulted family room.

- Built-in cabinetry and a fireplace make the family room a comfortable and charming place to spend time with family and friends.

- The master suite has a private sitting area offset from the rest of the room by pedestal columns.

- An optional bonus room is available on the second floor, making space for a playroom, crafting room or exercise area.

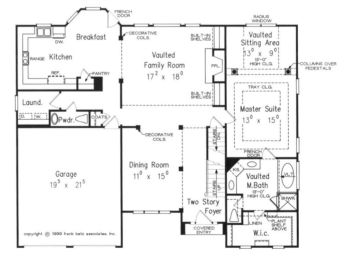

First Floor

Rear Elevation

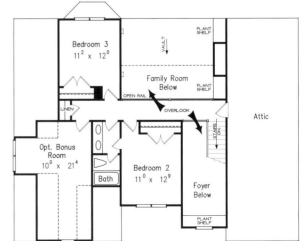

Second Floor

Spaulding — TSPDG01-316 — 1-866-525-9374

Total Living	First Floor	Second Floor	Bonus	Bed	Bath	Width	Depth	Foundation	Price Category
2182 sq ft	1346 sq ft	836 sq ft	N/A	4	3-1/2	49' 8"	45' 4"	Crawl Space*	E

*Other options available. See page 347.

Design Features

- Palladian windows flood the two-level foyer and great room with natural light.
- Both the master bedroom and great room access the covered rear porch.
- The master bath features a walk-in closet, twin vanities, separate shower and whirlpool tub.
- One of three upstairs bedrooms enjoys a private bath and walk-in closet.

First Floor

PORCH

GREAT RM.
15-4 x 14-8

BRKFST.
11-0 x 9-0

UTIL.
6-2 x
cl 5-10

w d

MASTER BED RM.
12-0 x 15-0

fireplace

balcony above

KIT.
11-0 x 12-0

© 1993 DONALD A. GARDNER
All rights reserved

cl

walk-in closet

cl

master bath

pd. rm.

DINING
13-4 x 12-8

FOYER
9-10 x 8-6

up

PORCH

Second Floor

clerestory with palladian window

bath

walk-in closet

great room below

cl

BED RM.
11-0 x 12-0

railing

cl

lin.

walk-in closet

down

bath

BED RM.
11-0 x 17-8

foyer below

BED RM.
11-0 x 12-8

clerestory with palladian window

Rear Elevation

Burgess — TSPDG01-290 — 1-866-525-9374

Total Living	First Floor	Second Floor	Bonus	Bed	Bath	Width	Depth	Foundation	Price Category
2188 sq ft	1618 sq ft	570 sq ft	495 sq ft	3	2-1/2	54' 0"	57' 0"	Crawl Space*	E

*Other options available. See page 347.

Design Features

- A two-story great room and two-story foyer welcome natural light into this country classic.

- The large kitchen features a center-cooking island with counter and large breakfast area.

- Columns punctuate the interior spaces, and a separate dining room provides a formal touch.

- The semi-detached garage features a large bonus room above.

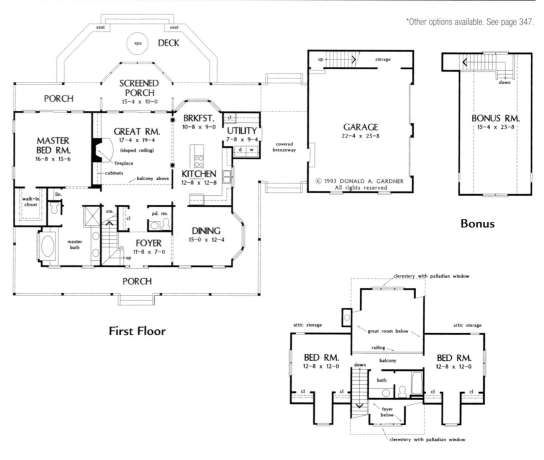

First Floor

Bonus

Second Floor

Rear Elevation

© The Sater Design Collection, Inc.

Tucker Town Way — TSPDS01-6692 — 1-866-525-9374

Total Living	First Floor	Lower Level	Bonus	Bed	Bath	Width	Depth	Foundation	Price Category
2190 sq ft	2190 sq ft	N/A	N/A	3	2	59' 8"	54' 0"	Slab	F

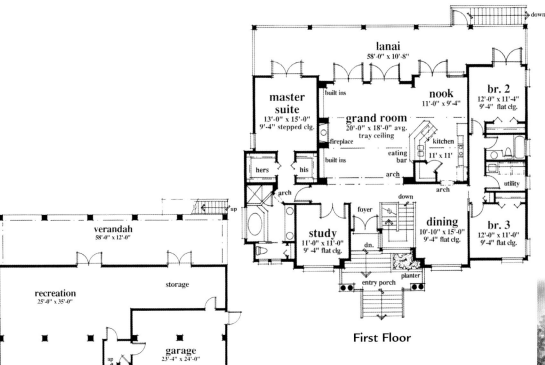

First Floor

Lower Level

Design Features

- The foyer opens to the grand room with fireplace.

- A well-crafted kitchen has wrapping counter space.

- A secluded master suite offers lanai access.

- The opulent master bath has twin lavatories.

- Two secondary bedrooms share a full bath.

Rear Elevation

Walterboro — TSPDG01-436 — 1-866-525-9374

Total Living	First Floor	Second Floor	Bonus	Bed	Bath	Width	Depth	Foundation	Price Category
2190 sq ft	1577 sq ft	613 sq ft	360 sq ft	3	2-1/2	65' 2"	43' 8"	Crawl Space*	E

*Other options available. See page 347.

Design Features

- Brick accents and a hip roof with gables polish the exterior of this traditional home.
- Inside, a curved balcony overlooks the spacious great room with vaulted ceiling.
- A cased opening with round columns introduces the kitchen and breakfast room.
- The private master suite, located off the great room, features a tray ceiling.
- Upstairs, the skylit bonus room makes a great hobby or playroom for the kids.

Rear Elevation

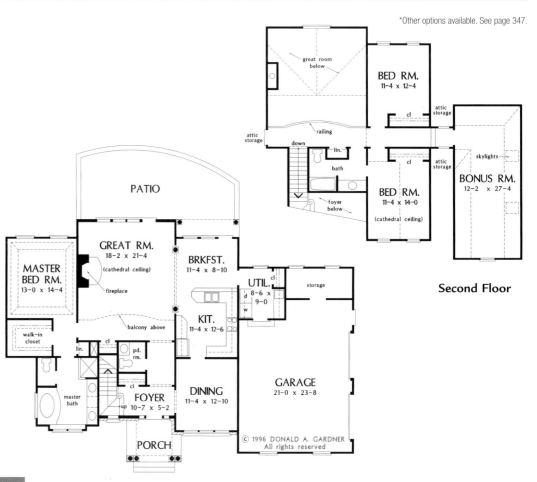

Second Floor

First Floor

© 1999 Frank Betz Associates, Inc.

Crestwood Place — TSPFB01-3485 — 1-866-525-9374

Total Living	First Floor	Second Floor	Opt. Bonus	Bed	Bath	Width	Depth	Foundation	Price Category
2196 sq ft	1652 sq ft	544 sq ft	222 sq ft	3	2-1/2	56' 4"	44' 6"	Basement, Crawl Space or Slab	H

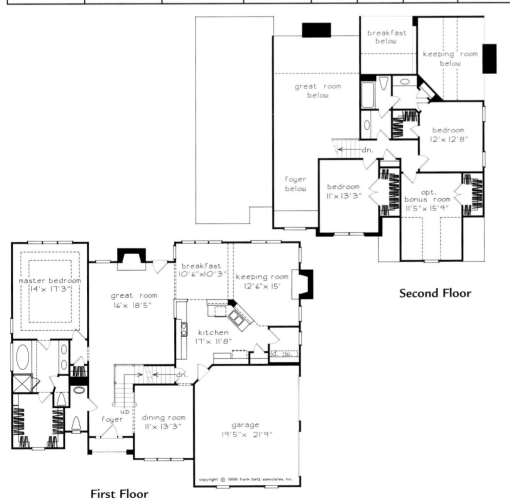

Second Floor

First Floor

Design Features

- From the *Southern Living® Design Collection*

- This charming two-story home uses a subtle mix of brick and cedar shingles to create a cottage-like look and feel.

- Inside, this 2100+ square-foot home offers a two-story family room that overlooks the kitchen and breakfast areas for easy interaction.

- One of the home's most delightful spaces is the keeping room, a casual gathering place with its own fireplace and two-story vaulted ceiling.

Rear Elevation

North Hampton — TSPDG01-542 — 1-866-525-9374

© 1997 Donald A. Gardner, Inc.

Total Living	First Floor	Second Floor	Bonus	Bed	Bath	Width	Depth	Foundation	Price Category
2201 sq ft	1687 sq ft	514 sq ft	336 sq ft	4	3	59' 2"	49' 4"	Crawl Space*	E

Design Features

- The stunning great room makes a statement with a magnificent cathedral ceiling.

- A cozy fireplace with space-saving built-ins shields the great room from kitchen noise.

- With two bedrooms up and two bedrooms down, this home promises privacy.

- The first-floor study/bedroom provides ample flexibility for any family.

Rear Elevation

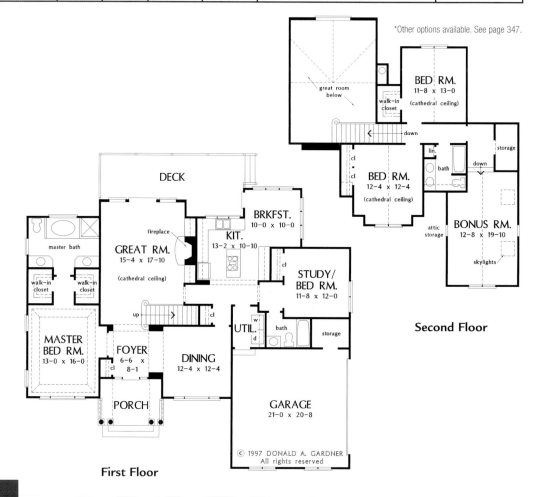

*Other options available. See page 347.

Second Floor

First Floor

© 1997 DONALD A. GARDNER
All rights reserved

Gasden — TSPDG01-431 — 1-866-525-9374

Total Living	First Floor	Second Floor	Bonus	Bed	Bath	Width	Depth	Foundation	Price Category
2202 sq ft	1585 sq ft	617 sq ft	353 sq ft	3	2-1/2	65' 8"	42' 6"	Crawl Space*	E

*Other options available. See page 347.

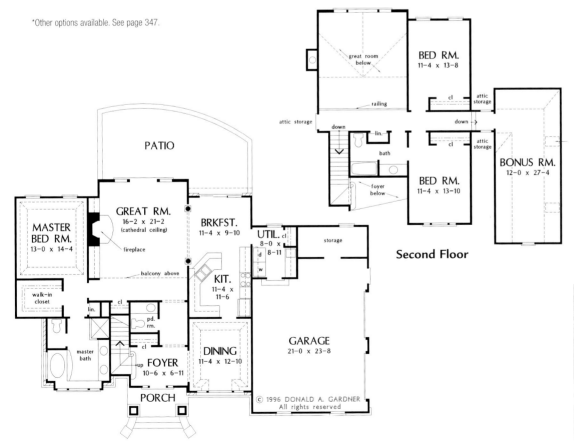

First Floor

Second Floor

Design Features

- Stone, stucco and refined styling add to the Old-World charm of this home.
- An open foyer accentuates a graceful stair and leads to the dining room.
- The great room features built-in cabinets and a wall of windows for extra drama.
- The first-floor master suite enjoys a tray ceiling.
- A skylit bonus room and attic storage are easily accessed from the second floor.

Rear Elevation

Donald A. Gardner Architects, Inc.

Pineville — TSPDG01-405 — 1-866-525-9374

Total Living	First Floor	Second Floor	Bonus	Bed	Bath	Width	Depth	Foundation	Price Category
2203 sq ft	1561 sq ft	642 sq ft	324 sq ft	3	2-1/2	68' 0"	50' 4"	Crawl Space*	E

Design Features

*Other options available. See page 347.

- Brick steps and a covered porch with metal roof give this relaxed farmhouse extra finesse.

- Interior accent columns and a Palladian window distinguish the inviting two-story foyer.

- Transom windows over French doors brighten the dining room.

- A spacious great room is set off by two-story windows.

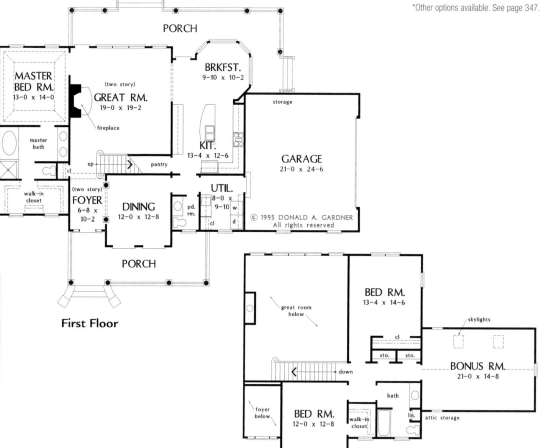

First Floor

Second Floor

Rear Elevation

Meadowsweet — TSPDG01-918 — 1-866-525-9374

Total Living	First Floor	Second Floor	Bonus	Bed	Bath	Width	Depth	Foundation	Price Category
2211 sq ft	1476 sq ft	735 sq ft	374 sq ft	4	2-1/2	48' 4"	51' 4"	Crawl Space*	E

*Other options available. See page 347.

Design Features

■ A turret-styled bay window is showcased on a wall of stone.

■ Dormers and columns combine both country and traditional styles.

■ The common rooms remain open, and the smart angled kitchen is the heart of the home.

■ French doors access a rear porch from the great room and master suite.

■ The U-shaped staircase has a decorative shelf for accessories.

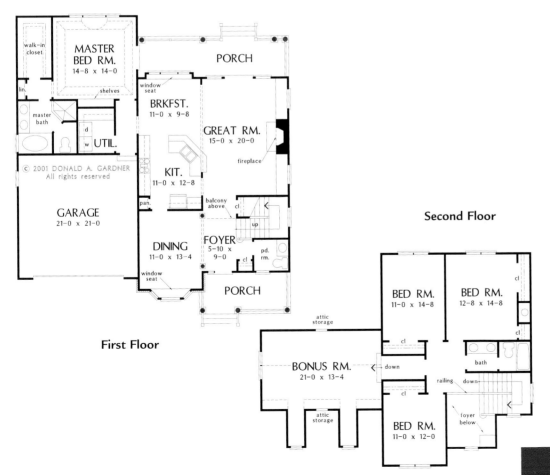

First Floor

MASTER BED RM. 14-8 x 14-0
walk-in closet
lin.
shelves
master bath
UTIL.
GARAGE 21-0 x 21-0
window seat
BRKFST. 11-0 x 9-8
PORCH
GREAT RM. 15-0 x 20-0
fireplace
KIT. 11-0 x 12-8
pan.
balcony above
up
DINING 11-0 x 13-4
FOYER 5-10 x 9-0
pd. rm.
window seat
PORCH

Second Floor

attic storage
BONUS RM. 21-0 x 13-4
attic storage
BED RM. 11-0 x 14-8
BED RM. 12-8 x 14-8
down
bath
railing
down
BED RM. 11-0 x 12-0
foyer below

Rear Elevation

Donald A. Gardner Architects, Inc.

Downing — TSPDG01-1067 — 1-866-525-9374

Total Living	First Floor	Second Floor	Bonus	Bed	Bath	Width	Depth	Foundation	Price Category
2213 sq ft	1726 sq ft	487 sq ft	N/A	4	3	34' 6"	103' 6"	Crawl Space*	E

Design Features

- In charming row house fashion, this home makes the most of long, narrow lots.
- Two spacious porches provide room for outdoor living.
- A linen closet is conveniently tucked alongside the large utility room.
- A built-in desk accentuates the bedroom/study.

*Other options available. See page 347.

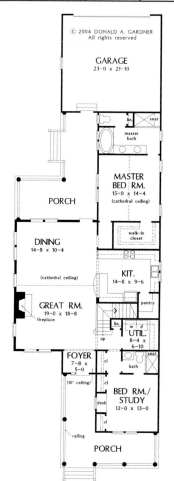

First Floor

Second Floor

Rear Elevation

Greystone — TSPDG01-919 — 1-866-525-9374

Total Living	First Floor	Second Floor	Bonus	Bed	Bath	Width	Depth	Foundation	Price Category
2221 sq ft	1707 sq ft	514 sq ft	211 sq ft	4	2-1/2	50' 0"	71' 8"	Crawl Space*	E

*Other options available. See page 347.

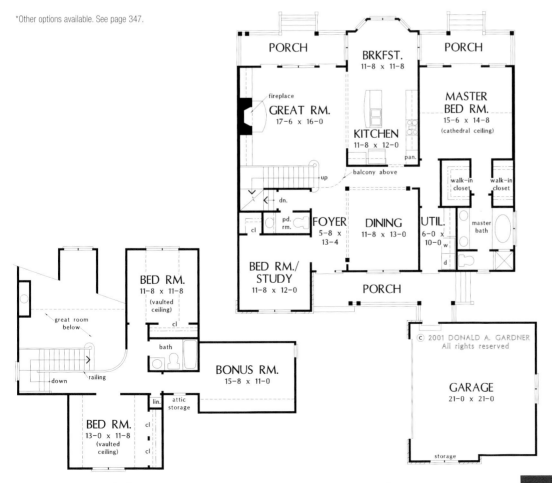

Design Features

- Elegant columns give distinction to the covered breezeway and dining room.

- The great room features a two-story vaulted ceiling, built-ins and French doors.

- The breakfast area is extended by a bay window.

- The master suite also enjoys private access to its own porch.

- French doors flanked by windows allow for spectacular views and morning light.

Rear Elevation

Second Floor

First Floor

Verdigre — TSPDG01-936 — 1-866-525-9374

© 2001 Donald A. Gardner, Inc.

Total Living	First Floor	Second Floor	Bonus	Bed	Bath	Width	Depth	Foundation	Price Category
2231 sq ft	1547 sq ft	684 sq ft	300 sq ft	3	2-1/2	59' 2"	44' 4"	Crawl Space*	E

*Other options available. See page 347.

Design Features

- A metal roof embellishes the garage's box-bay window.

- Arches are seen in and above windows as well as the front entrance.

- The loft includes a study area that receives light from the open, two-story great room.

- The second floor bathroom includes twin vanities.

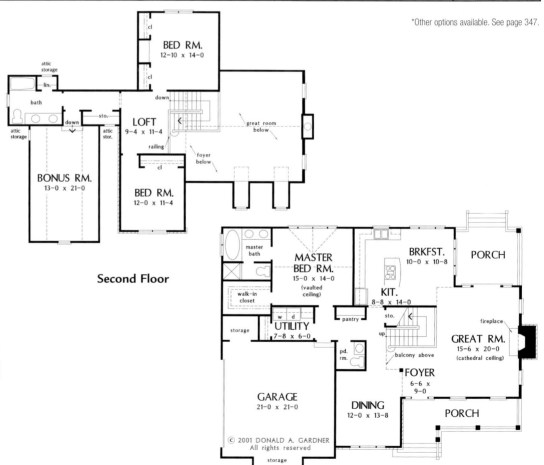

Second Floor

First Floor

Rear Elevation

Chalkrock — TSPDG01-1054 — 1-866-525-9374

Total Living	First Floor	Second Floor	Bonus	Bed	Bath	Width	Depth	Foundation	Price Category
2231 sq ft	1699 sq ft	532 sq ft	549 sq ft	3	2-1/2	48' 10"	73' 2"	Crawl Space*	E

*Other options available. See page 347.

Design Features

- Stone, siding and metal roofing add a touch of Arts-N-Crafts to the exterior.

- A sidelight and transom enhance the front door and usher in natural light.

- Inside, a natural traffic flow follows a circular pattern.

- Built-in cabinetry lies on both sides of the fireplace, and French doors lead to the rear porch.

- The versatile bonus room works perfectly as a guest suite.

First Floor

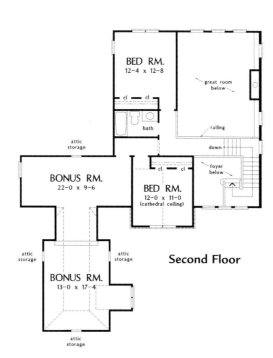

Second Floor

Rear Elevation

Roxbury — TSPDG01-722 — 1-866-525-9374

Total Living	First Floor	Second Floor	Bonus	Bed	Bath	Width	Depth	Foundation	Price Category
2235 sq ft	1701 sq ft	534 sq ft	274 sq ft	3	2-1/2	65' 4"	43' 5"	Crawl Space*	E

Design Features

*Other options available. See page 347.

- Stone and stucco embellish the exterior of this stately traditional.

- Volume ceilings and architectural columns enhance its interior.

- Ceiling treatments cap the living room/study, family room and master bedroom.

- A second-floor loft with curved railing overlooks both the foyer and family room.

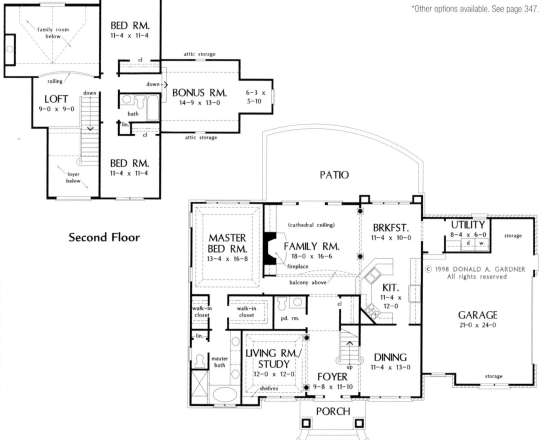

Second Floor

First Floor

Rear Elevation

Sullivan — TSPFB01-1224 — 1-866-525-9374

2200 sq ft—2600 sq ft

Frank Betz Associates, Inc.

Total Living	First Floor	Second Floor	Opt. Bonus	Bed	Bath	Width	Depth	Foundation	Price Category
2246 sq ft	1688 sq ft	558 sq ft	269 sq ft	4	3	54' 0"	48' 0"	Basement, Crawl Space or Slab	H

First Floor

Second Floor

Design Features

- The full brick façade, as well as the classic turret, have stood the test of time.

- A well-planned and thoughtful design accommodates the lifestyle of today's homeowner.

- The home is anchored by a vaulted great room, adjoining the kitchen and breakfast areas, allowing easy interaction between family and guests.

- An additional main-floor bedroom makes the ideal guest room or can also serve as a home office.

Rear Elevation

Manchester — TSPDG01-718 — 1-866-525-9374

© 1998 Donald A. Gardner, Inc.

Total Living	First Floor	Second Floor	Bonus	Bed	Bath	Width	Depth	Foundation	Price Category
2261 sq ft	1699 sq ft	562 sq ft	235 sq ft	3	2-1/2	69' 3"	45' 10"	Crawl Space*	E

*Other options available. See page 347.

Design Features

■ Traditional brick with a touch of stucco creates a formidable presence for this stunning three-bedroom home.

■ The first-floor master suite gains refinement from tray ceilings and his-and-her walk-in closets

■ Sharing a hall bath, two upstairs bedrooms enjoy ample closet space.

■ The bonus room features a dormer alcove and access to attic storage.

Rear Elevation

2200 sq ft—2600 sq ft

Donald A. Gardner Architects, Inc.

© 2005 Donald A. Gardner, Inc.

Stratton — TSPDG01-1095 — 1-866-525-9374

Total Living	First Floor	Second Floor	Bonus	Bed	Bath	Width	Depth	Foundation	Price Category
2264 sq ft	1635 sq ft	629 sq ft	351 sq ft	4	4	44' 8"	52' 4"	Crawl Space*	E

*Other options available. See page 347.

Design Features

- An angled cooktop island creates a serving bar in the kitchen.

- Just off the foyer is a flexible bedroom/study with adjacent full bath.

- The second level consists of two secondary bedrooms and a large bonus room.

- A balcony overlooks the great room and foyer.

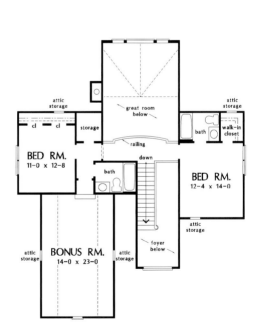

Second Floor

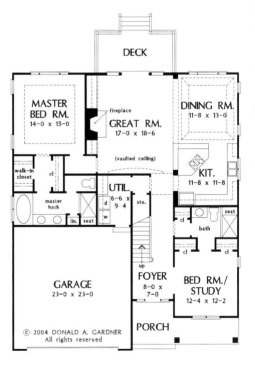

First Floor

Rear Elevation

www.twostoryplans.com **117**

Donald A. Gardner Architects, Inc.

Derbyville — TSPDG01-1032 — 1-866-525-9374

Total Living	First Floor	Second Floor	Bonus	Bed	Bath	Width	Depth	Foundation	Price Category
2276 sq ft	1778 sq ft	498 sq ft	315 sq ft	4	3	54' 8"	53' 2"	Crawl Space*	E

*Other options available. See page 347.

Design Features

- Front and side stairs show the way to the porch, creating a warm, inviting welcome.

- Double doors lead into a versatile bedroom/study.

- The dining room is connected to the kitchen by a butler's pantry.

- While a bay window extends the breakfast nook, French doors access the rear porch.

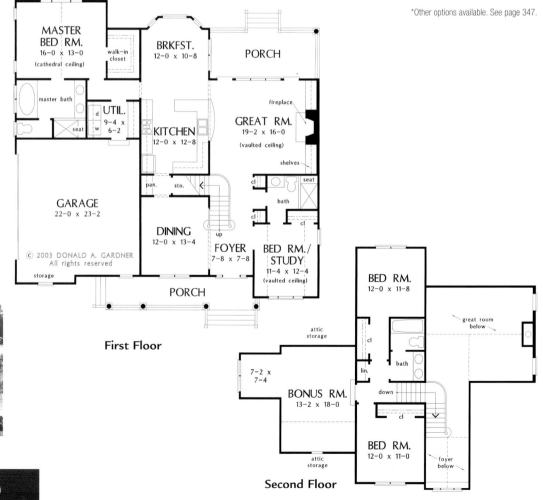

First Floor

Second Floor

Rear Elevation

Frank Betz Associates, Inc.

© 2003 Frank Betz Associates, Inc.

Durham Park — TSPFB01-3810 — 1-866-525-9374

Total Living	First Floor	Second Floor	Opt. Bonus	Bed	Bath	Width	Depth	Foundation	Price Category
2283 sq ft	1704 sq ft	579 sq ft	235 sq ft	3	2-1/2	55' 0"	58' 0"	Basement or Crawl Space	H

First Floor

Second Floor

Design Features

- From the *Southern Living® Design Collection*

- The *Durham Park* exudes charm and character with its inviting stone and siding exterior.

- A bayed breakfast area adjoins the kitchen and accesses a tranquil screened porch and deck area.

- The master bedroom also has access to the deck and features its own private sitting area. A vaulted ceiling and decorative columns make this space distinct.

Rear Elevation

Amelia — TSPFB01-3807 — 1-866-525-9374

© 2003 Frank Betz Associates, Inc.

Total Living	First Floor	Second Floor	Opt. Bonus	Bed	Bath	Width	Depth	Foundation	Price Category
2286 sq ft	1663 sq ft	623 sq ft	211 sq ft	4	3	54' 0"	48' 0"	Basement or Crawl Space	H

Design Features

- The *Amelia's* façade is clean and simple, with a cozy front porch and gabled roofline.

- Louvered shutters and stone accents bring a sense of warmth that is echoed inside with a homey floor plan.

- A bedroom on the main level is also the perfect location for a home office or den.

- Upstairs, flexible areas are incorporated, giving the homeowners choices on how to use their space.

- A loft is situated among the bedrooms, making an ideal homework station or lounging area for kids.

Rear Elevation

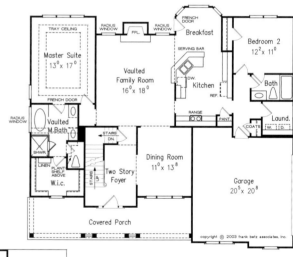

First Floor

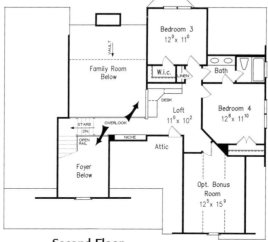

Second Floor

Wicklow — TSPDG01-950 — 1-866-525-9374

Total Living	First Floor	Second Floor	Bonus	Bed	Bath	Width	Depth	Foundation	Price Category
2294 sq ft	1542 sq ft	752 sq ft	370 sq ft	3	2-1/2	44' 4"	54' 0"	Crawl Space*	E

*Other options available. See page 347.

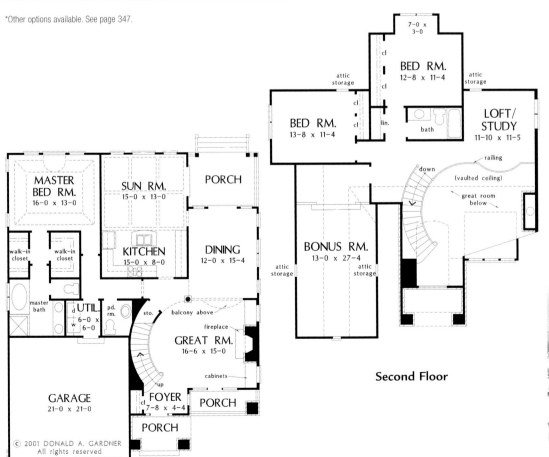

First Floor

Second Floor

Design Features

- A unique mixture of stone, siding and windows create character in this unique design.
- Columns, decorative railing and a metal roof add architectural interest to an intimate front porch.
- A rock entryway frames a French door flanked by sidelights and crowned with a transom.
- An elegant, curved staircase highlights the grand two-story foyer and great room.
- A delightful sunroom can be accessed from the dining room and is open to the kitchen.

Rear Elevation

Donald A. Gardner Architects, Inc.

Peachtree — TSPDG01-524 — 1-866-525-9374

Total Living	First Floor	Second Floor	Bonus	Bed	Bath	Width	Depth	Foundation	Price Category
2298 sq ft	1743 sq ft	555 sq ft	350 sq ft	4	3	78' 0"	53' 2"	Crawl Space*	E

Design Features

*Other options available. See page 347.

- Nine-foot ceilings are standard throughout the first and second floors of this classic country design.

- The foyer, great room and screened porch enjoy vaulted and cathedral ceilings.

- Bay windows perk up the dining room and breakfast area.

- The master suite features back-porch access, walk-in closet and a lavish bath.

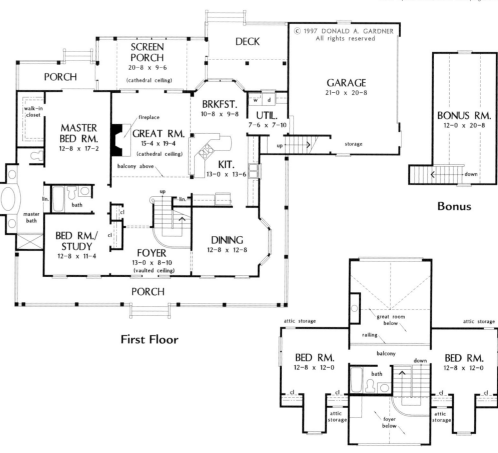

First Floor

Bonus

Second Floor

Rear Elevation

Dunwood — TSPDG01-299 — 1-866-525-9374

Total Living	First Floor	Second Floor	Bonus	Bed	Bath	Width	Depth	Foundation	Price Category
2301 sq ft	1632 sq ft	669 sq ft	528 sq ft	3	2-1/2	72' 6"	46' 10"	Crawl Space*	E

*Other options available. See page 347.

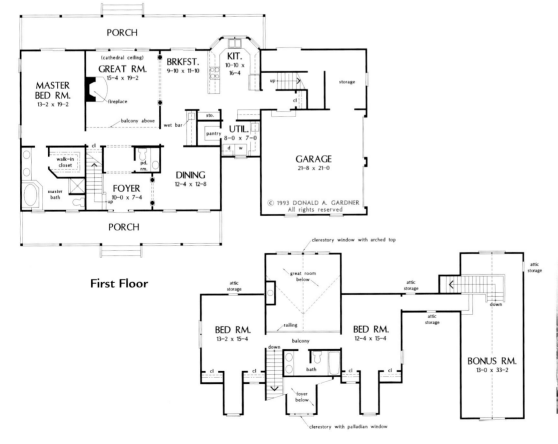

First Floor

Second Floor

Design Features

- An open floor plan plus bonus room make this home great for today's family.

- Columns visually separate the great room from the breakfast area and smart, U-shaped kitchen.

- A large utility room with walk-in pantry provides ultimate convenience.

- The privately located master suite accesses the back porch and features a luxurious bath.

Rear Elevation

Wentworth — TSPDG01-522 — 1-866-525-9374

© 1997 Donald A. Gardner, Inc.

Total Living	First Floor	Second Floor	Bonus	Bed	Bath	Width	Depth	Foundation	Price Category
2308 sq ft	1829 sq ft	479 sq ft	228 sq ft	4	3	63' 7"	50' 5"	Crawl Space*	E

Design Features

- This traditional home's foyer is enhanced by a graceful cathedral ceiling.

- Columns add definition to the casually elegant, open dining room.

- A cathedral ceiling in the great room creates a feeling of spaciousness.

- The kitchen is more than generous, featuring a center work island, pantry and breakfast counter.

Other options available. See page 347.

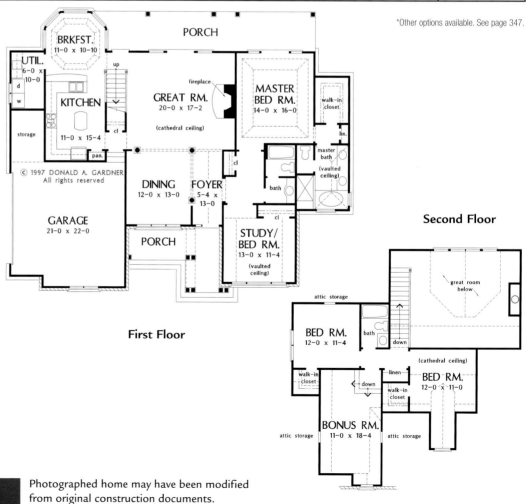

First Floor

Second Floor

Rear Elevation

Photographed home may have been modified from original construction documents.

© The Sater Design Collection, Inc.

Lanchester — TSPDS01-7007 — 1-866-525-9374

Total Living	First Floor	Second Floor	Bonus	Bed	Bath	Width	Depth	Foundation	Price Category
2334 sq ft	1716 sq ft	618 sq ft	N/A	3	3	47' 0"	50' 0"	Crawl Space	F

Design Features

- Double transom fanlights accentuate the front entry.

- The columned front porch is brightened by a planter box.

- The island kitchen has a serving counter, perfect for entertaining.

- French doors to the porch and a fireplace make the great room cozy.

- One of two upstairs bedrooms features a built-in desk.

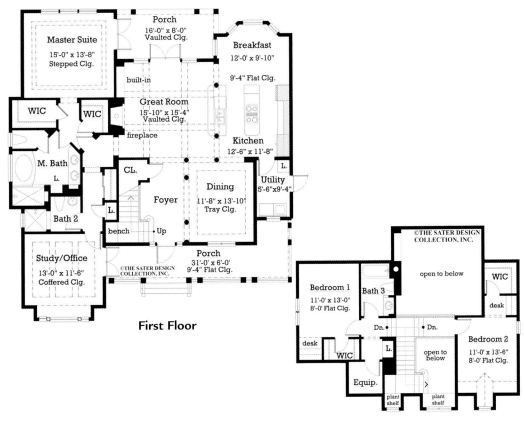

First Floor

Second Floor

Rear Elevation

Donald A. Gardner Architects, Inc.

Longworth — TSPDG01-362 — 1-866-525-9374

© 1994 Donald A. Gardner Architects, Inc.

Total Living	First Floor	Second Floor	Bonus	Bed	Bath	Width	Depth	Foundation	Price Category
2335 sq ft	1715 sq ft	620 sq ft	265 sq ft	3	2-1/2	58' 6"	50' 9"	Crawl Space*	E

Design Features

- Keystone arches, bay windows and a stone and stucco veneer give this home classic appeal.
- A fireplace nestled between built-in bookshelves warms a two-story family room.
- The master suite, secluded on the first floor, is a private haven offering a luxurious bath.
- Upstairs, a loft/study overlooks the family room and separates two additional bedrooms.

*Other options available. See page 347.

First Floor

Second Floor

Rear Elevation

© 2003 Frank Betz Associates, Inc.

Holly Springs — TSPFB01-3821 — 1-866-525-9374

Total Living	First Floor	Second Floor	Opt. Bonus	Bed	Bath	Width	Depth	Foundation	Price Category
2338 sq ft	1761 sq ft	577 sq ft	305 sq ft	4	3	56' 0"	48' 0"	Basement or Crawl Space	G

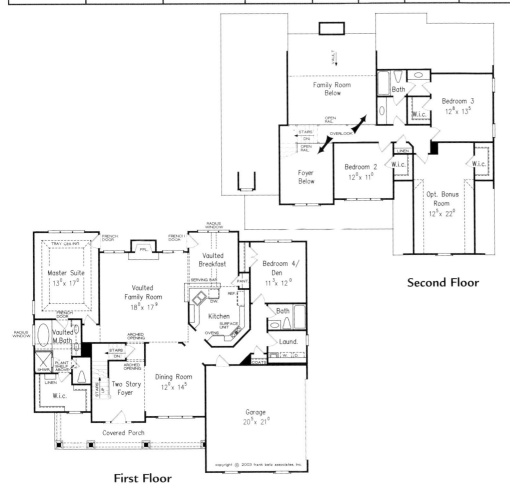

First Floor

Second Floor

Design Features

▪ The façade of the *Holly Springs* is eye-catching and original, with its unique windows, battered columns and varied exterior materials.

▪ Its main floor houses the master suite, as well as an additional bedroom that makes an ideal guest bedroom with a bath in close proximity.

▪ A vaulted breakfast area connects to the kitchen, complete with double ovens and a serving bar.

▪ Additional niceties include an arched opening leading to the family room from the foyer and a linen closet in the master suite.

Rear Elevation

Savoy — TSPFB01-851 — 1-866-525-9374

© 1995 Frank Betz Associates, Inc.

Total Living	First Floor	Second Floor	Opt. Bonus	Bed	Bath	Width	Depth	Foundation	Price Category
2341 sq ft	1761 sq ft	580 sq ft	276 sq ft	4	3	56' 0"	47' 6"	Basement, Crawl Space or Slab	H

Design Features

- The hipped roofline and decorative quoin accents give interesting and eye-catching dimension to the *Savoy*'s façade.

- The master suite encompasses an entire wing of the main level, giving homeowners ample space to unwind.

- An additional main-floor bedroom makes the perfect guest room or can be used as a home office.

- Two bedrooms share a divided bath upstairs.

Rear Elevation

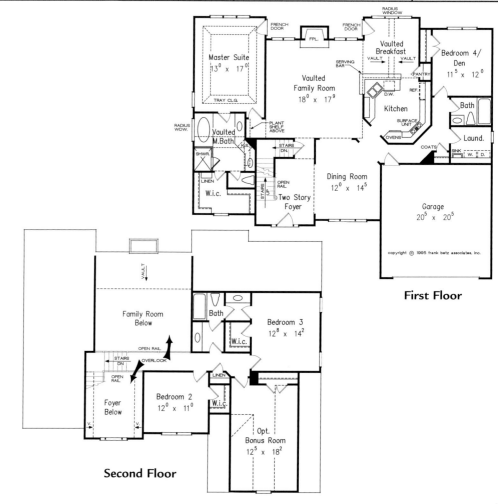

First Floor

Second Floor

© 1999 Frank Betz Associates, Inc.

Hanley Hall — TSPFB01-1256 — 1-866-525-9374

Total Living	First Floor	Second Floor	Opt. Bonus	Bed	Bath	Width	Depth	Foundation	Price Category
2345 sq ft	1761 sq ft	584 sq ft	302 sq ft	4	3	56' 0"	47' 6"	Basement, Crawl Space or Slab	I

First Floor

Design Features

- From the *Southern Living® Design Collection*

- A spacious family room—with a two-story ceiling—is the focal point for activity.

- The kitchen also functions as the nucleus of the home; its openness allows the cook to interact with anyone in adjoining rooms.

- One bedroom with a full bath is located in its own corner of the plan.

Rear Elevation

Second Floor

2200 sq ft—2600 sq ft

Frank Betz Associates, Inc.

Defoors Mill — TSPFB01-3712 — 1-866-525-9374

© 2002 Frank Betz Associates, Inc.

Total Living	First Floor	Second Floor	Opt. Bonus	Bed	Bath	Width	Depth	Foundation	Price Category
2351 sq ft	1803 sq ft	548 sq ft	277 sq ft	4	3	55' 0"	48' 0"	Basement, Crawl Space or Slab	H

Design Features

- A covered porch is situated on the front of the home and leads to a charming and practical floor plan.

- The master suite encompasses an entire wing of the home for comfort and privacy.

- An optional bonus room is available on the second floor

- Special details in this home include a handy island in the kitchen, decorative columns around the dining area and a coat closet just off the garage.

Second Floor

Rear Elevation

First Floor

Vissage — TSPDG01-1063 — 1-866-525-9374

Total Living	First Floor	Second Floor	Bonus	Bed	Bath	Width	Depth	Foundation	Price Category
2351 sq ft	1575 sq ft	776 sq ft	394 sq ft	3	2-1/2	45' 0"	54' 0"	Crawl Space*	E

*Other options available. See page 347.

Design Features

- Open and family-efficient, each common room flows into the next.

- A fireplace creates a stunning focal point in the great room.

- A breakfast bar also doubles as a serving counter to the dining room.

- The loft/study features a curved balcony and overlooks the great room.

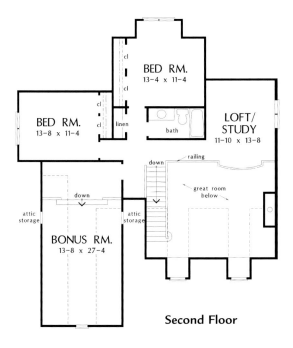

Second Floor

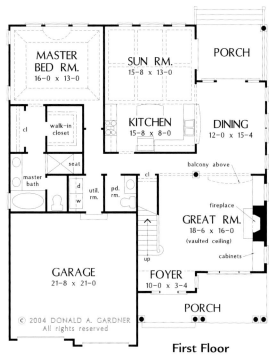

First Floor

Rear Elevation

Donald A. Gardner Architects, Inc.

Saddlebrook — TSPDG01-762 — 1-866-525-9374

© 1998 Donald A. Gardner, Inc.

Total Living	First Floor	Second Floor	Bonus	Bed	Bath	Width	Depth	Foundation	Price Category
2356 sq ft	1718 sq ft	638 sq ft	341 sq ft	4	3	71' 0"	42' 8"	Crawl Space*	E

*Other options available. See page 347.

Design Features

- Twin gables and a generous front porch give this graceful farmhouse stature and appeal.
- Vaulted and cathedral ceilings amplify the great room and master bedroom.
- The great room features a fireplace, built-ins, convenient rear staircase and balcony.
- The first-floor master suite enjoys his-and-her walk-in closets and a private bath with garden tub.
- Upstairs are two family bedrooms, separated by a loft with built-in bookshelves.

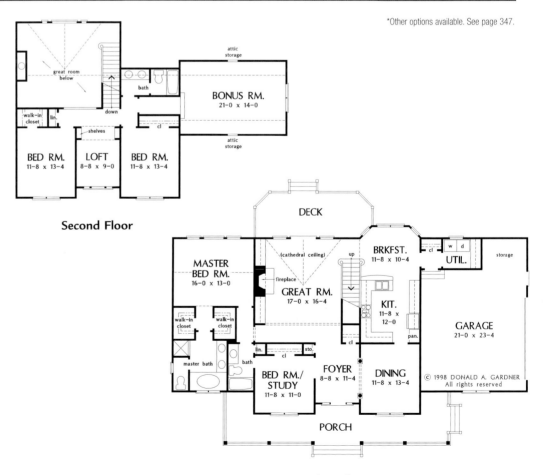

Second Floor

First Floor

Rear Elevation

© 1999 Frank Betz Associates, Inc.

Jackson Springs — TSPFB01-1238 — 1-866-525-9374

Total Living	First Floor	Second Floor	Opt. Bonus	Bed	Bath	Width	Depth	Foundation	Price Category
2358 sq ft	1289 sq ft	1069 sq ft	168 sq ft	4	3	57' 7"	40' 1"	Basement, Crawl Space or Slab	I

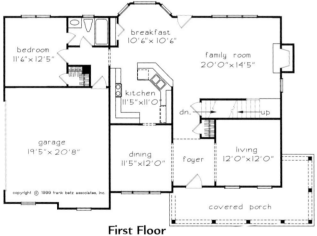

First Floor

Design Features

■ From the *Southern Living®
Design Collection*

■ A gracious family room is anchored by an inviting fireplace and a wall of windows.

■ A convenient serving bar and bay window in the breakfast area create a warm, inviting atmosphere.

■ Upstairs, the roomy master suite offers a decorative tray ceiling, luxurious bath and an expansive walk-in closet.

Second Floor

Rear Elevation

Donald A. Gardner Architects, Inc.

Sutcliffe — TSPDG01-807 — 1-866-525-9374

© 1999 Donald A. Gardner, Inc.

Total Living	First Floor	Second Floor	Bonus	Bed	Bath	Width	Depth	Foundation	Price Category
2360 sq ft	1869 sq ft	491 sq ft	297 sq ft	4	3	65' 0"	48' 0"	Crawl Space*	E

Design Features

- The vaulted great room is open and spacious, illuminated by a rear clerestory dormer.

- The kitchen is open with easy access to the dining room, great room and screened porch.

- The master suite and a secondary bedroom are located on the first floor.

- The master bedroom features a tray ceiling, linen closet, walk-in closet and a private bath.

*Other options available. See page 347.

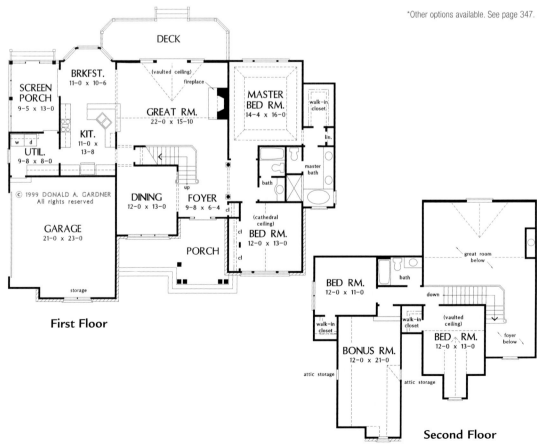

First Floor

Second Floor

Rear Elevation

Juniper — TSPDG01-704 — 1-866-525-9374

Total Living	First Floor	Second Floor	Bonus	Bed	Bath	Width	Depth	Foundation	Price Category
2363 sq ft	1575 sq ft	788 sq ft	251 sq ft	4	3-1/2	65' 0"	49' 0"	Crawl Space*	E

*Other options available. See page 347.

Design Features

- Multiple gables and a gracious front porch add charm to this country farmhouse.
- The practical floor plan allows for easy traffic patterns around the U-shaped staircase.
- A space-enhancing cathedral ceiling tops the great room.
- The master suite on the first floor features a built-in entertainment center.

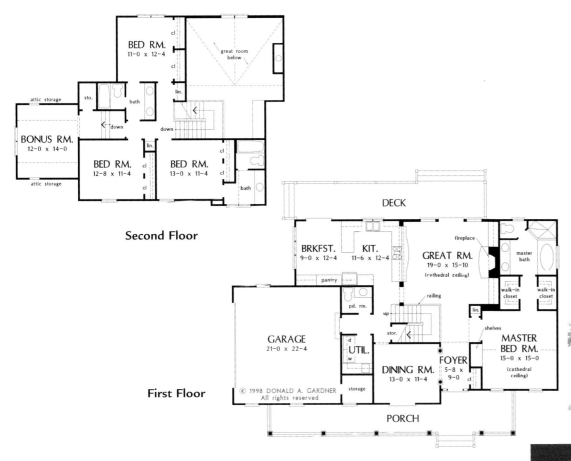

Second Floor

First Floor

Rear Elevation

Donald A. Gardner Architects, Inc.

Brentwood — TSPDG01-998 — 1-866-525-9374

© 2002 Donald A. Gardner, Inc.

Total Living	First Floor	Second Floor	Bonus	Bed	Bath	Width	Depth	Foundation	Price Category
2384 sq ft	1633 sq ft	751 sq ft	359 sq ft	3	3-1/2	69' 8"	44' 0"	Crawl Space*	E

*Other options available. See page 347.

Design Features

- This farmhouse captures Old-Time ambiance with a large wrapping porch and columns.

- A trio of gables and a side-entry garage add to the curb appeal.

- Filled with light by clerestory windows and French doors, this floor plan is efficient.

- A two-story ceiling extends from the foyer into the great room, while a balcony overlooks both areas.

- Upstairs, a loft separates two additional bedrooms and full baths.

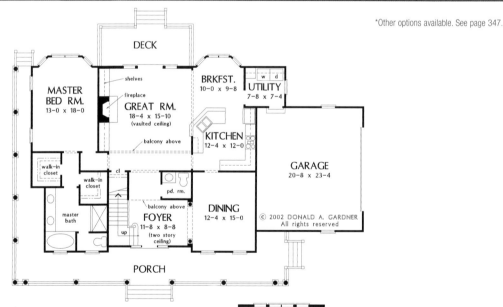

First Floor

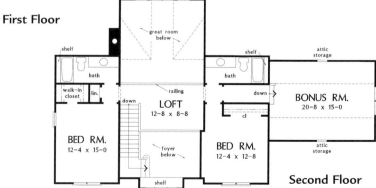

Second Floor

Rear Elevation

© 1996 Frank Betz Associates, Inc.

Macgregor — TSPFB01-773 — 1-866-525-9374

Total Living	First Floor	Second Floor	Opt. Bonus	Bed	Bath	Width	Depth	Foundation	Price Category
2386 sq ft	1223 sq ft	1163 sq ft	204 sq ft	4	2-1/2	50' 0"	48' 0"	Basement, Crawl Space or Slab	H

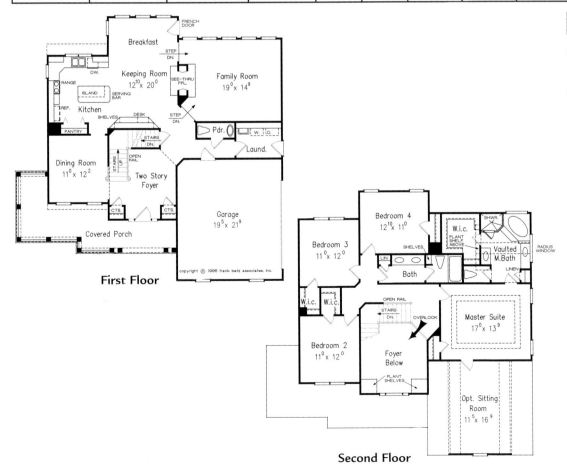

First Floor

Second Floor

Design Features

- The family room flows into the breakfast and kitchen area.

- A convenient serving bar and bay window in the breakfast area create a warm, inviting atmosphere.

- The roomy master suite offers a decorative tray ceiling, luxurious bath and an expansive walk-in closet.

- Both upstairs bedrooms feature a walk-in closet.

Rear Elevation

Hyde Park — TSPDG01-816 — 1-866-525-9374

Total Living	First Floor	Second Floor	Bonus	Bed	Bath	Width	Depth	Foundation	Price Category
2387 sq ft	1918 sq ft	469 sq ft	374 sq ft	4	3	73' 3"	43' 6"	Crawl Space*	E

*Other options available. See page 347.

Design Features

- A two-story ceiling adds height and drama to the formal foyer.

- The adjacent dining room is enriched by a box-bay window with lovely window seat.

- The vaulted great room features a stunning rear clerestory dormer and a fireplace.

- A second-floor balcony overlooks both the great room and foyer.

Rear Elevation

First Floor

Second Floor

© 2003 Frank Betz Associates, Inc.

Catawba Ridge — TSPFB01-3823 — 1-866-525-9374

Total Living	First Floor	Second Floor	Opt. Bonus	Bed	Bath	Width	Depth	Foundation	Price Category
2389 sq ft	1593 sq ft	796 sq ft	238 sq ft	3	3-1/2	59' 8"	50' 6"	Basement, Crawl Space or Slab	I

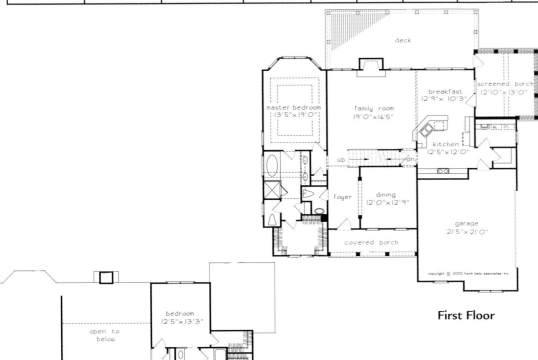

First Floor

Second Floor

Design Features

- From the *Southern Living®* *Design Collection*
- A cozy front porch graces the front of the home.
- The kitchen, breakfast area and family room are conveniently grouped together for easy family interaction.
- The master suite encompasses an entire wing of the home, giving the homeowner added privacy.

Rear Elevation

Tullamore Square — TSPFB01-3801 — 1-866-525-9374

© 2003 Frank Betz Associates, Inc.

Total Living	First Floor	Second Floor	Opt. Bonus	Bed	Bath	Width	Depth	Foundation	Price Category
2398 sq ft	1805 sq ft	593 sq ft	255 sq ft	4	3	55' 0"	48' 0"	Basement or Crawl Space	H

Design Features

- This home is every bit as quaint as its name, with a cedar shake exterior accented with a rooftop cupola.

- A seated shower, soaking tub and his-and-her closets make the master bath feel like five-star luxury.

- A coat closet and a laundry room are placed just off the garage, keeping coats and shoes in their place.

- A second main-floor bedroom makes a great guest room or home office.

Second Floor

First Floor

Rear Elevation

Ledgestone — TSPDG01-1006 — 1-866-525-9374

Total Living	First Floor	Second Floor	Bonus	Bed	Bath	Width	Depth	Foundation	Price Category
2425 sq ft	1648 sq ft	777 sq ft	353 sq ft	4	2-1/2	48' 8"	57' 8"	Crawl Space*	E

*Other options available. See page 347.

Second Floor

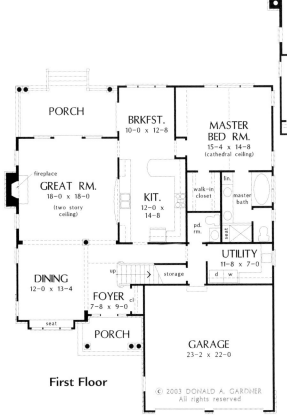

First Floor

Design Features

- With traditional appeal, this builder-friendly design starts with a stone and siding exterior.
- Columns make a dramatic entry, while metal tops the front porch and a box-bay window.
- A front-entry garage adds convenience.
- Custom-styled features include a fireplace, French doors and cathedral ceiling.
- Note the large utility room and closet storage under the staircase.

Rear Elevation

Clarkston — TSPFB01-3751 — 1-866-525-9374

© 2002 Frank Betz Associates, Inc.

Total Living	First Floor	Second Floor	Bonus	Bed	Bath	Width	Depth	Foundation	Price Category
2426 sq ft	1327 sq ft	1099 sq ft	290 sq ft	4	3	58' 0"	42' 6"	Basement, Crawl Space or Slab	H

Design Features

- A two-story foyer is encompassed by a dining room and study.

- Three bedrooms share the upper level of this design—the master suite and two secondary bedrooms that share a bath.

- A fourth bedroom with private bath is on the main level and makes perfect secluded guest accommodations.

- An optional bonus room on the upper level presents the opportunity to expand—ideal for a playroom, craft room or exercise area.

First Floor

Rear Elevation

Second Floor

Dobbins — TSPDG01-370 — 1-866-525-9374

Total Living	First Floor	Second Floor	Bonus	Bed	Bath	Width	Depth	Foundation	Price Category
2435 sq ft	1841 sq ft	594 sq ft	391 sq ft	4	3	82' 2"	48' 10"	Crawl Space*	E

*Other options available. See page 347.

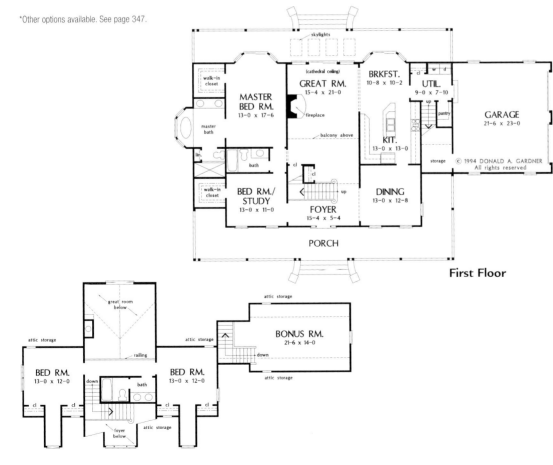

First Floor

Second Floor

Design Features

- This four-bedroom country home defines casual elegance with its front and back wrapping porches.

- An open two-level foyer with Palladian window leads to the expansive great room.

- Windows all around and bays in the master suite and breakfast area add space and light.

- The master suite features back-porch access, a walk-in closet and a luxurious bath.

Rear Elevation

Riverside — TSPDG01-269 — 1-866-525-9374

Total Living	First Floor	Second Floor	Bonus	Bed	Bath	Width	Depth	Foundation	Price Category
2436 sq ft	1766 sq ft	670 sq ft	N/A	4	3-1/2	59' 10"	53' 4"	Crawl Space*	E

Design Features

*Other options available. See page 347.

- This four-bedroom farmhouse celebrates sunlight with a Palladian window and screened porch.

- The center island kitchen and breakfast area flow into the great room through an elegant colonnade.

- The first-level master suite opens to the screened porch through a bay area.

- A garden tub, double lavs and separate shower provide added luxury.

- Upstairs, two more bedrooms each have a private bath.

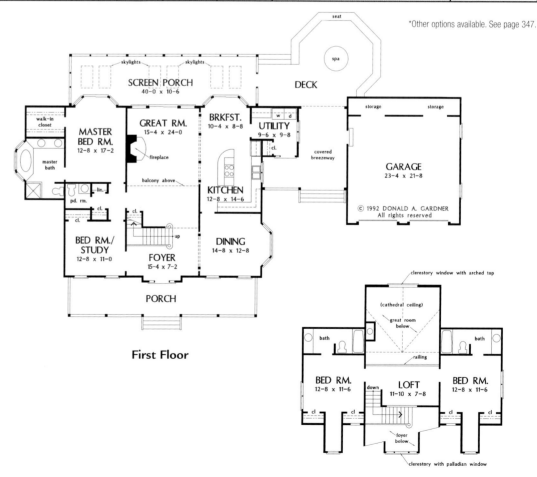

First Floor

Second Floor

Rear Elevation

© 2000 Frank Betz Associates, Inc.

Lancaster Place — TSPFB01-3512 — 1-866-525-9374

Total Living	First Floor	Second Floor	Opt. Bonus	Bed	Bath	Width	Depth	Foundation	Price Category
2437 sq ft	1879 sq ft	558 sq ft	340 sq ft	3	2-1/2	56' 0"	54' 4"	Basement, Crawl Space or Slab	I

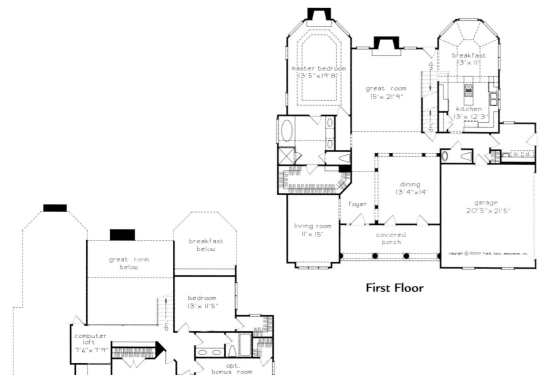

First Floor

Second Floor

Design Features

- From the *Southern Living®* *Design Collection*
- Traditional columns and an inviting front porch recall the essence of home.
- The first floor is open and spacious, with a formal living room and a generous dining room accented by columns.
- A large kitchen island gives the cook ample space to prepare the family's favorite meal.
- A luxurious master suite features a fireplace and a tray ceiling.

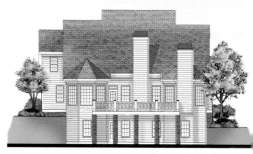

Rear Elevation

Culbertson — TSPFB01-3860 — 1-866-525-9374

© 2004 Frank Betz Associates, Inc.

Total Living	First Floor	Second Floor	Bonus	Bed	Bath	Width	Depth	Foundation	Price Category
2443 sq ft	1214 sq ft	1229 sq ft	N/A	4	2-1/2	52' 4"	55' 10"	Basement or Crawl Space	H

Design Features

- The kitchen overlooks a keeping room with a fireplace, giving families a cozy place to gather for a board game.

- Double ovens, serving bar and an ample pantry accommodate the family.

- The family room is open to the breakfast area, allowing for easy traffic flow from one space to the next.

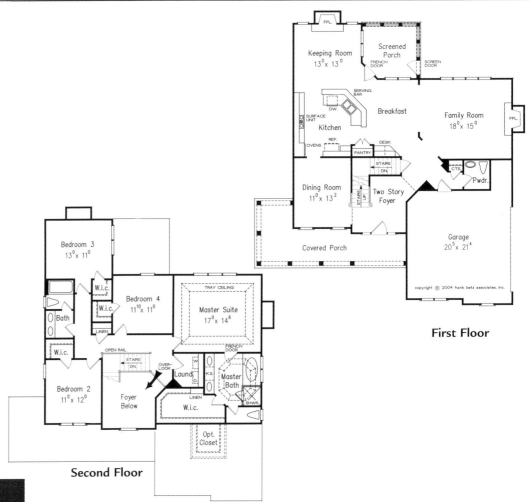

First Floor

Second Floor

Rear Elevation

© 2000 Frank Betz Associates, Inc.

Musgrave — TSPFB01-3526 — 1-866-525-9374

Total Living	First Floor	Second Floor	Opt. Bonus	Bed	Bath	Width	Depth	Foundation	Price Category
2444 sq ft	1720 sq ft	724 sq ft	212 sq ft	4	3	58' 0"	47' 0"	Basement, Crawl Space or Slab	H

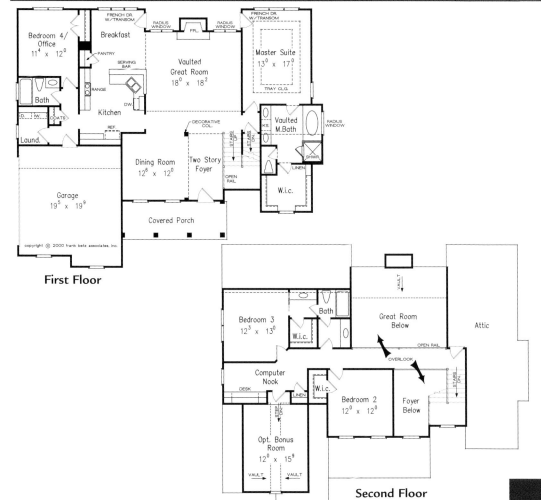

First Floor

Second Floor

Design Features

- A main-floor bedroom is the ideal location for a guest suite with direct access to a full bath.

- Telecommuters may opt to use this space as a home office.

- The second-floor bedrooms have walk-in closets and share a divided bath.

- A computer nook has been thoughtfully incorporated into the upper level, giving children a private and convenient homework station.

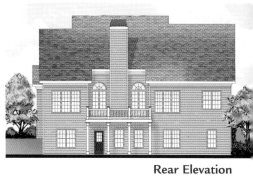

Rear Elevation

Southhampton Bay — TSPDS01-6684 — 1-866-525-9374

© The Sater Design Collection, Inc.

Total Living	First Floor	Lower Level	Bonus	Bed	Bath	Width	Depth	Foundation	Price Category
2465 sq ft	2385 sq ft	80 sq ft	N/A	3	2-1/2	60' 4"	59' 4"	Slab	F

Design Features

- A cupola tops a classic pediment and low-pitched roof.
- A sunburst and sidelights set off the entry.
- The great room has a wall of built-ins for high-tech media.
- The master suite features a luxurious bath and dressing area.
- The lower level includes a game room and bonus room.

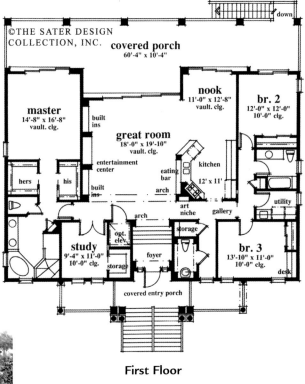

First Floor

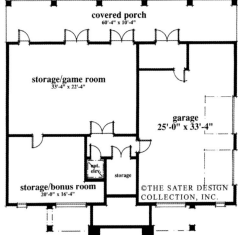

Lower Level

Rear Elevation

Abbott — TSPDG01-1109 — 1-866-525-9374

Total Living	First Floor	Second Floor	Bonus	Bed	Bath	Width	Depth	Foundation	Price Category
2465 sq ft	1820 sq ft	645 sq ft	456 sq ft	4	3	42' 4"	79' 10"	Crawl Space*	E

*Other options available. See page 347.

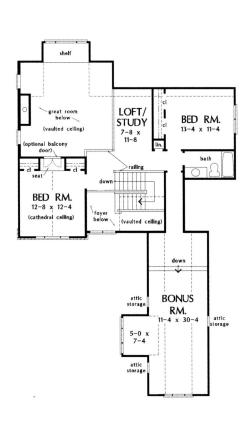

Second Floor

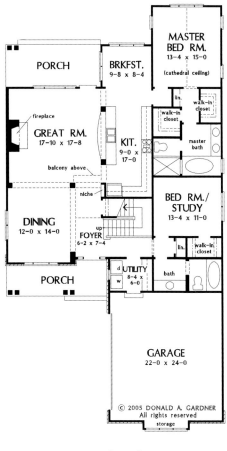

First Floor

Design Features

- A grand staircase immediately greets guests in the formal foyer.

- An adjacent art niche adds flair.

- The open dining and great rooms permit a natural traffic flow.

- The breakfast room overlooks a rear porch.

- The bonus room has a perfect alcove for a computer station.

Rear Elevation

Coltraine — TSPDG01-966 — 1-866-525-9374

© 2002 Donald A. Gardner, Inc.

Total Living	First Floor	Second Floor	Bonus	Bed	Bath	Width	Depth	Foundation	Price Category
2466 sq ft	1856 sq ft	610 sq ft	322 sq ft	3	2-1/2	59' 0"	47' 8"	Crawl Space*	E

*Other options available. See page 347.

Design Features

- Full of charm, this home combines stone, cedar shake and siding for curb appeal.
- Triple dormers usher light into an upstairs bedroom and the two-story foyer.
- Columns in the dining room and a tray ceiling in the master suite add architectural interest.
- Built-in cabinetry, a kitchen island and a butler's pantry provide convenience.
- French doors lead outside from the great room and master suite.

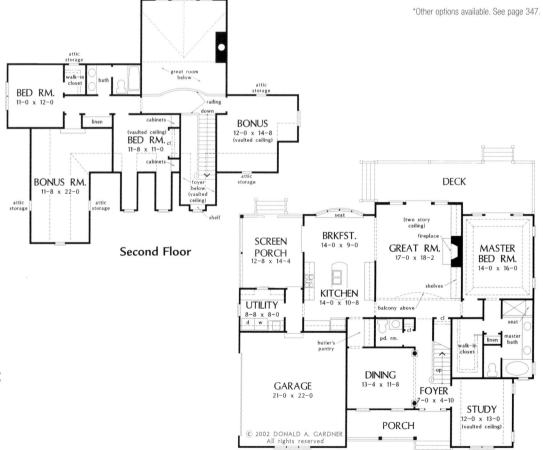

Second Floor

First Floor

Rear Elevation

2200 sq ft—2600 sq ft

Donald A. Gardner Architects, Inc.

Holloway — TSPDG01-1064 — 1-866-525-9374

Total Living	First Floor	Second Floor	Bonus	Bed	Bath	Width	Depth	Foundation	Price Category
2469 sq ft	1977 sq ft	492 sq ft	374 sq ft	4	3	75' 1"	44' 5"	Crawl Space*	E

*Other options available. See page 347.

Design Features

▪ The hip roof and strong use of brick reflect a popular traditional style.

▪ French doors open to a rear patio for outdoor entertaining and relaxation.

▪ Built-in cabinetry flanks the fireplace, and columns punctuate the dining room.

▪ A large bonus room provides space for a rec/media room, home gym or guest suite.

Second Floor

First Floor

Rear Elevation

Carlton Square — TSPFB01-3588 — 1-866-525-9374

© 2001 Frank Betz Associates, Inc.

Total Living	First Floor	Second Floor	Opt. Bonus	Bed	Bath	Width	Depth	Foundation	Price Category
2481 sq ft	1961 sq ft	520 sq ft	265 sq ft	4	3	60' 0"	53' 0"	Basement or Crawl Space	G

Design Features

- A clean-lined combination of brick and siding creates a façade of understated simplicity.

- The interior spaces connect with ease, generating effortless traffic flow from one room to the next.

- Also on the main level, a bedroom can be easily converted into a home office—perfect for teleworkers or retirees.

- Optional bonus space is designed into the second floor.

Rear Elevation

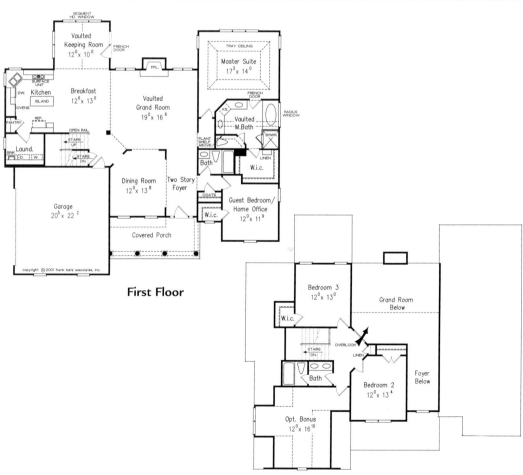

First Floor

Second Floor

Azalea Crossing — TSPDG01-849 — 1-866-525-9374

Total Living	First Floor	Second Floor	Bonus	Bed	Bath	Width	Depth	Foundation	Price Category
2482 sq ft	1706 sq ft	776 sq ft	414 sq ft	4	2-1/2	54' 8"	43' 0"	Crawl Space*	E

*Other options available. See page 347.

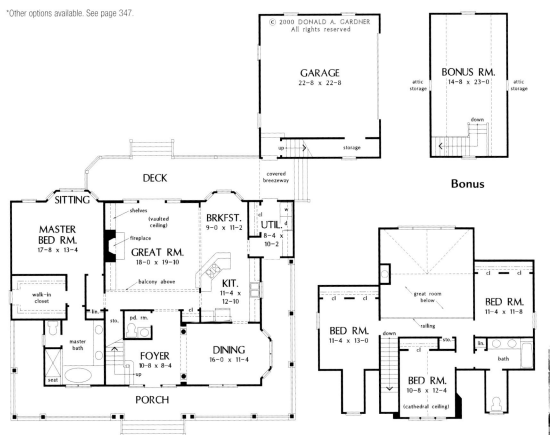

First Floor

Second Floor

Design Features

- A relaxing front porch wraps this three-dormered, four-bedroom home in unbeatable curb appeal.

- An open, vaulted great room is the home's main attraction.

- Built-in bookshelves flank the fireplace, and French doors open to the rear deck.

- Bay windows expand the dining room, breakfast area and master bedroom.

- A lovely sitting area, generous walk-in closet and lavish bath enhance the master suite.

Rear Elevation

Heywood — TSPDG01-991 — 1-866-525-9374

Total Living	First Floor	Second Floor	Bonus	Bed	Bath	Width	Depth	Foundation	Price Category
2485 sq ft	1420 sq ft	1065 sq ft	411 sq ft	4	3	57' 8"	49' 0"	Crawl Space*	E

*Other options available. See page 347.

Design Features

- Twin dormers flank a prominent gable, and a Palladian-style window mimics the arched entryway.

- Numerous windows flood the home with natural light and allow wonderful views.

- French doors accent the bedroom/study as well as the great room.

- A striking fireplace highlights the great room, and the kitchen has plenty of workspace.

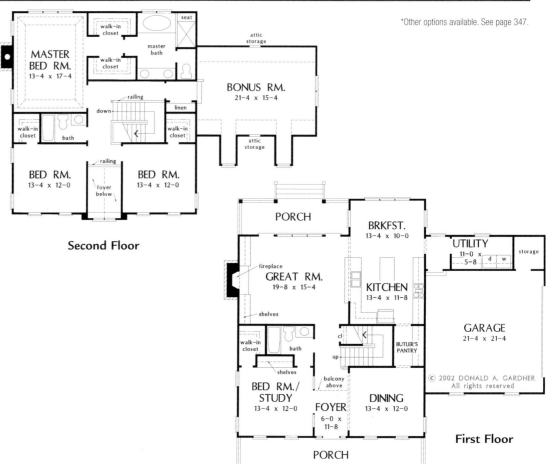

Second Floor

First Floor

Rear Elevation

© 2002 Donald A. Gardner, Inc.

Trotterville — TSPDG01-984 — 1-866-525-9374

Total Living	First Floor	Second Floor	Bonus	Bed	Bath	Width	Depth	Foundation	Price Category
2490 sq ft	1687 sq ft	803 sq ft	N/A	4	2-1/2	52' 8"	67' 0"	Crawl Space*	E

*Other options available. See page 347.

Design Features

- A shed dormer, metal porch roof and gable with bracket display Craftsman charm.

- Two spacious porches welcome outdoor relaxation and entertaining.

- Columns accentuate the great room and dining room.

- Built-in cabinetry, French doors and a fireplace highlight the great room.

- Located for convenience, the utility/mud room connects the garage to the house.

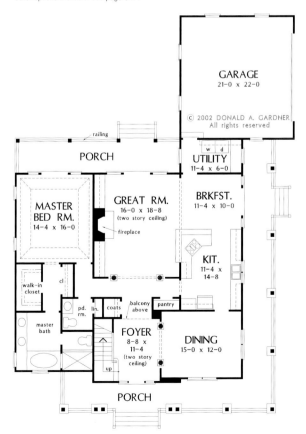

First Floor

Second Floor

Rear Elevation

Mission Hills — TSPDS01-6845 — 1-866-525-9374

© The Sater Design Collection, Inc.

Total Living	First Floor	Lower Level	Bonus	Bed	Bath	Width	Depth	Foundation	Price Category
2494 sq ft	2385 sq ft	109 sq ft	N/A	3	3	60' 0"	52' 0"	Island Basement	F

Design Features

- The study has double doors to the front balcony.
- Vaulted ceilings create openness throughout the living areas.
- Disappearing glass doors connect several rooms to a full-length veranda.
- A large utility room enjoys French doors to the balcony.
- A U-shaped kitchen has a center island and corner pantry.

Rear Elevation

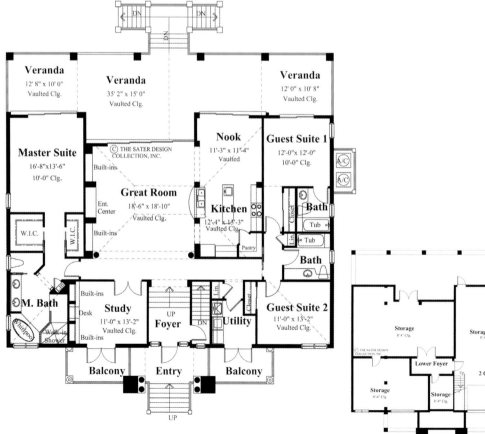

First Floor

Lower Level

Donald A. Gardner Architects, Inc.

© 1998 Donald A. Gardner, Inc.

Mulberry — TSPDG01-547 — 1-866-525-9374

Total Living	First Floor	Second Floor	Bonus	Bed	Bath	Width	Depth	Foundation	Price Category
2506 sq ft	1614 sq ft	892 sq ft	341 sq ft	4	2-1/2	71' 10"	50' 0"	Crawl Space*	F

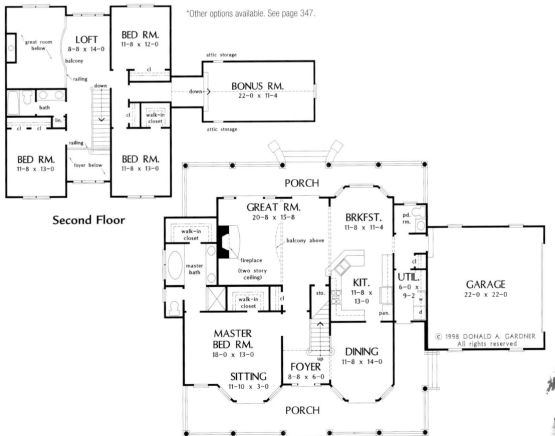

*Other options available. See page 347.

Second Floor

First Floor

Design Features

- The foyer and great room are each brightened by clerestory windows.
- Bay windows grace the dining room, breakfast area and master suite with sitting area.
- Generous front and back porches afford options for outdoor relaxation.
- Upstairs, three roomy bedrooms and a bonus room utilize a hall bath with linen closet.

Rear Elevation

Lakeshore — TSPFB01-3752 — 1-866-525-9374

© 2002 Frank Betz Associates, Inc.

Total Living	First Floor	Second Floor	Opt. Bonus	Bed	Bath	Width	Depth	Foundation	Price Category
2507 sq ft	1483 sq ft	1024 sq ft	252 sq ft	4	3	55' 0"	47' 10"	Basement or Crawl Space	H

Design Features

- Thoughtful design details are apparent on the inside with functional amenities that make everyday living easier.

- A large island, double ovens and a butler's pantry take the work out of entertaining.

- The main-floor bedroom has access to a bath, making it ideal for accommodating guests.

- Radius windows keep the family room bright and sunny.

Rear Elevation

Second Floor

First Floor

Edisto — TSPDG01-764 — 1-866-525-9374

Total Living	First Floor	Second Floor	Bonus	Bed	Bath	Width	Depth	Foundation	Price Category
2509 sq ft	1830 sq ft	679 sq ft	346 sq ft	4	4	81' 2"	48' 0"	Crawl Space*	F

*Other options available. See page 347.

Second Floor

Design Features

- A wrapping front porch creates a charming façade for this four-bedroom home.
- The vaulted foyer receives light from the center clerestory dormer.
- The master bedroom features a tray ceiling for added elegance and back porch access.
- Another bedroom with walk-in closet and adjacent bath are nearby.
- Both upstairs bedrooms enjoy dormer windows, walk-in closets and private baths.

Rear Elevation

First Floor

Cherryvale — TSPDG01-533 — 1-866-525-9374

Total Living	First Floor	Second Floor	Bonus	Bed	Bath	Width	Depth	Foundation	Price Category
2511 sq ft	1914 sq ft	597 sq ft	487 sq ft	3	2-1/2	79' 2"	51' 6"	Crawl Space*	F

*Other options available. See page 347.

Design Features

- The main staircase is positioned in the family room for convenience.

- Both the family room and breakfast bay area are optimized by exciting two story ceilings.

- The second floor includes a loft that overlooks both the two-story family room and foyer.

- The bonus room is accessed by a second staircase as well as the upstairs hallway.

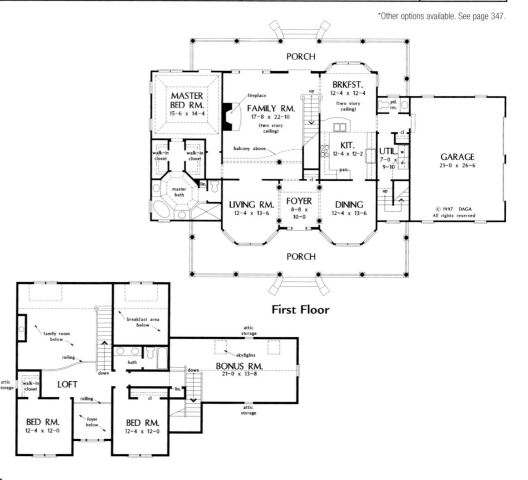

First Floor

Second Floor

Rear Elevation

Newcastle — TSPDG01-994 — 1-866-525-9374

Total Living	First Floor	Second Floor	Bonus	Bed	Bath	Width	Depth	Foundation	Price Category
2515 sq ft	1834 sq ft	681 sq ft	365 sq ft	3	3-1/2	50' 8"	66' 8"	Crawl Space*	F

*Other options available. See page 347.

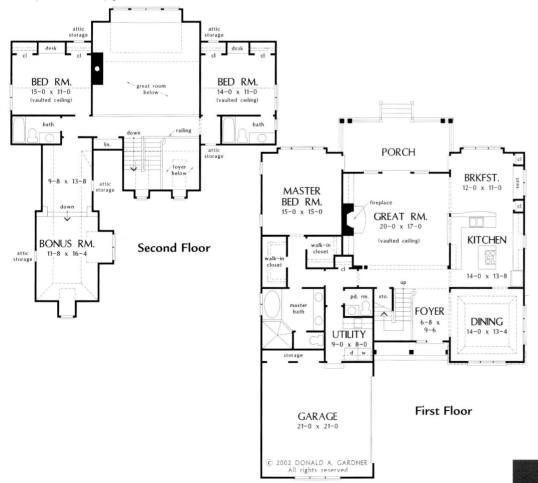

Second Floor

First Floor

Design Features

- With Old-World charm, this cottage's impressive exterior is made of stone and cedar shake.
- Inside, gathering rooms are open to each other and distinguished by columns.
- Both the foyer and great room have two-story ceilings that are brightened by dormers.
- The breakfast nook includes two pantries, while the kitchen features a cooktop island.
- Upstairs, a flexible bonus room can accommodate a family's growing needs or wishes.

Rear Elevation

Southerland — TSPDG01-971 — 1-866-525-9374

Total Living	First Floor	Second Floor	Bonus	Bed	Bath	Width	Depth	Foundation	Price Category
2521 sq ft	1798 sq ft	723 sq ft	349 sq ft	4	3-1/2	66' 8"	49' 8"	Crawl Space*	F

*Other options available. See page 347.

Design Features

- Columns make a grand impression both inside and outside.

- Transoms above French doors brighten both the front and rear of the floor plan.

- The mud room/utility area is complete with a sink.

- An upstairs bedroom has its own bath and can be used as a guest suite.

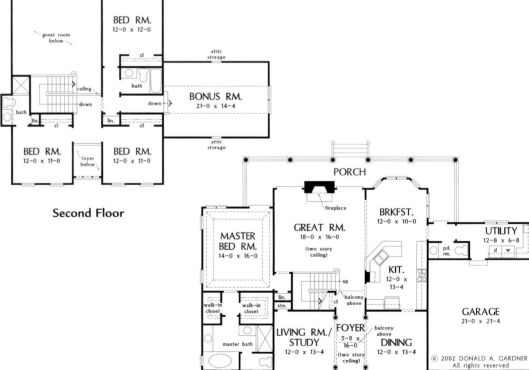

Second Floor

First Floor

Rear Elevation

Rousseau — TSPDG01-451 — 1-866-525-9374

Total Living	First Floor	Second Floor	Bonus	Bed	Bath	Width	Depth	Foundation	Price Category
2549 sq ft	1904 sq ft	645 sq ft	434 sq ft	3	2-1/2	71' 2"	45' 8"	Crawl Space*	F

*Other options available. See page 347.

Design Features

■ This home is packed with visual excitement created by a stone and stucco exterior.

■ The two-story great room has a fireplace, vaulted ceiling and overlooking balcony.

■ The first-floor master suite features dual walk-in closets and a private bath with linen closet.

■ Upstairs includes two bedrooms with full bath, bonus room, loft and attic storage.

First Floor

Second Floor

Rear Elevation

Photographed home may have been modified from original construction documents.

Coventry — TSPFB01-1103 — 1-866-525-9374

© 1997 Frank Betz Associates, Inc.

Total Living	First Floor	Second Floor	Bonus	Bed	Bath	Width	Depth	Foundation	Price Category
2551 sq ft	1972 sq ft	579 sq ft	256 sq ft	3	2-1/2	57' 4"	51' 2"	Basement, Crawl Space or Slab	H

Design Features

- A turret sitting room graces the master suite, which has its own wing on the main level.

- The vaulted ceilings add volume to the great room, keeping room and master bath.

- Windows across the rear of the home showcase backyard views.

- An informal keeping room is open to the breakfast nook and kitchen.

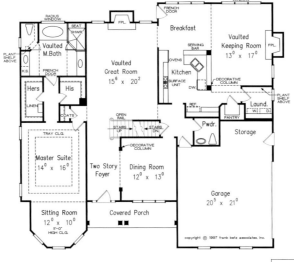

First Floor

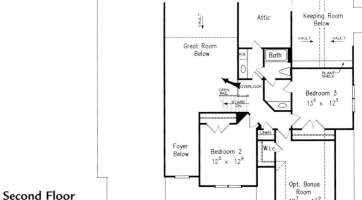

Second Floor

Rear Elevation

Warrenton — TSPDG01-261 — 1-866-525-9374

Total Living	First Floor	Second Floor	Bonus	Bed	Bath	Width	Depth	Foundation	Price Category
2561 sq ft	1357 sq ft	1204 sq ft	N/A	4	2-1/2	80' 0"	45' 9"	Crawl Space*	F

*Other options available. See page 347.

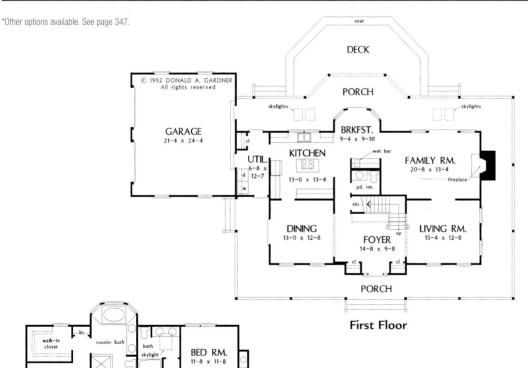

First Floor

Second Floor

Design Features

- A wraparound porch, double gables and custom window details create an impressive exterior.

- Formal living and dining rooms provide plenty of space for entertaining.

- An expansive family room opens to the lavish kitchen/breakfast area.

- Bay windows and a screened porch make this four-bedroom home a haven for outdoor enthusiasts.

- The spacious master suite features a walk-in closet and luxurious bath.

Rear Elevation

Orlina — TSPDS01-7030 — 1-866-525-9374

© The Sater Design Collection, Inc.

Total Living	First Floor	Second Floor	Bonus	Bed	Bath	Width	Depth	Foundation	Price Category
2562 sq ft	1387 sq ft	1175 sq ft	362 sq ft	3	2-1/2	54' 6"	78' 6"	Crawl Space	G

Design Features

- This charming home sports a full-width front porch.
- The rear of the home features a veranda and second-story deck.
- A bonus room with full bath is part of the second floor.
- A curved glass bay provides panoramic views to the nook.
- An oversized island with built-in hutch completes the kitchen.

Rear Elevation

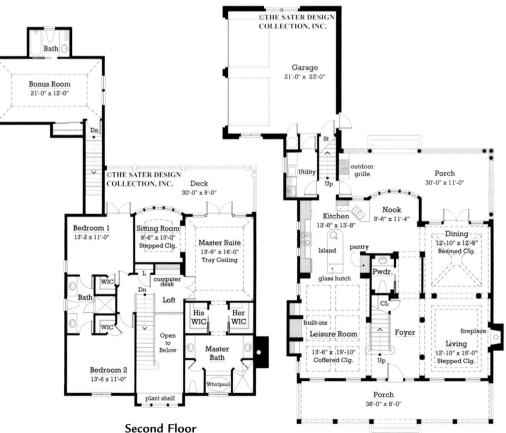

Second Floor

First Floor

© 2003 Frank Betz Associates, Inc.

Bainbridge Court — TSPFB01-3815 — 1-866-525-9374

Total Living	First Floor	Second Floor	Bonus	Bed	Bath	Width	Depth	Foundation	Price Category
2563 sq ft	1795 sq ft	768 sq ft	N/A	5	3	55' 0"	61' 0"	Basement, Crawl Space or Slab	I

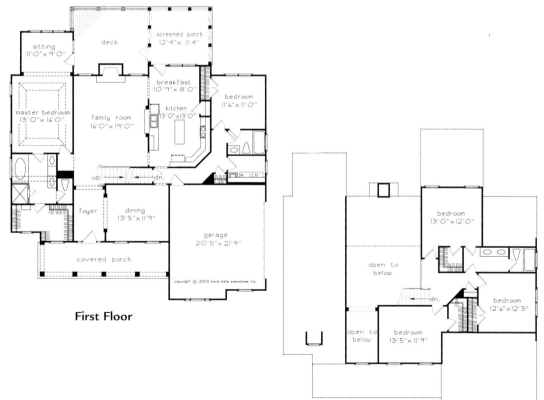

First Floor

Second Floor

Design Features

- From the *Southern Living® Design Collection*

- The classic brick-and-siding exterior of the *Bainbridge Court* gives this home a time-honored elegant appeal.

- The staircase is strategically tucked away to keep the foyer open and roomy.

- A cozy screened porch adjoins the kitchen area and has access to an expansive deck, making a great space for grilling and entertaining.

Rear Elevation

Beckett — TSPDG01-357 — 1-866-525-9374

Total Living	First Floor	Second Floor	Bonus	Bed	Bath	Width	Depth	Foundation	Price Category
2563 sq ft	1907 sq ft	656 sq ft	467 sq ft	4	2-1/2	89' 10"	53' 4"	Crawl Space*	F

Design Features

- Bay windows and a screened porch make this four-bedroom home a haven for outdoor enthusiasts.

- The open foyer takes advantage of light from the central dormer with Palladian window.

- A grand gathering space emerges by opening the contemporary kitchen to the breakfast and great rooms.

- The versatile front bedroom can also be a study.

- A low maintenance exterior adds to the home's appeal.

*Other options available. See page 347.

Rear Elevation

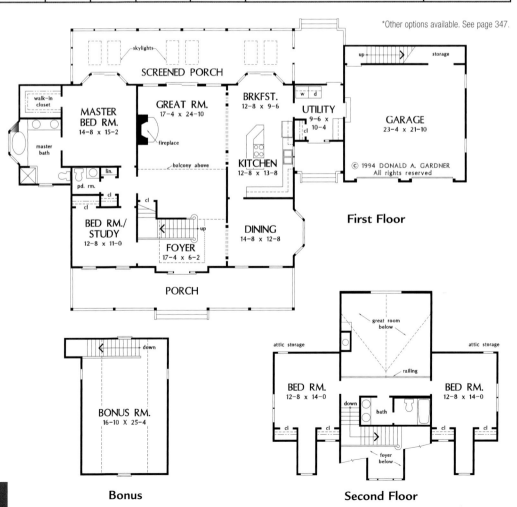

First Floor

Bonus

Second Floor

Oak Knoll — TSPFB01-3734 — 1-866-525-9374

Total Living	First Floor	Second Floor	Opt. Bonus	Bed	Bath	Width	Depth	Foundation	Price Category
2577 sq ft	1894 sq ft	683 sq ft	210 sq ft	4	3	57' 0"	53' 6"	Basement/Crawl Space	H

First Floor

Second Floor

Design Features

- This design was created to give homeowners living spaces that are supplementary to the traditional rooms found in most plans.

- The master bedroom is enhanced by a private sitting area with views overlooking the backyard.

- Just off the breakfast area is a bright and sunny keeping room with a radius window.

- Kids will appreciate having a private spot to study in the loft designed into the upper level of this home.

Rear Elevation

Ambrose — TSPFB01-3551 — 1-866-525-9374

© 2000 Frank Betz Associates, Inc.

Total Living	First Floor	Second Floor	Bonus	Bed	Bath	Width	Depth	Foundation	Price Category
2582 sq ft	2003 sq ft	579 sq ft	262 sq ft	4	3	54' 0"	60' 0"	Basement, Crawl Space or Slab	H

Design Features

- A distinctive turret with arched windows is the focal point of the façade on the *Ambrose*.

- A covered entry leads to an interesting and thoughtful layout inside.

- Family time is well spent in the large sunroom, situated just beyond the breakfast area.

- With its fireplace and built-in cabinetry, this room adds a comfortable and casual element to this home.

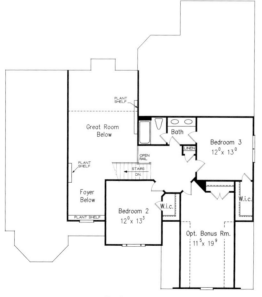

Second Floor

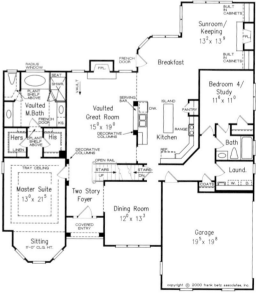

First Floor

Rear Elevation

© 2003 Frank Betz Associates, Inc.

Bakersfield — TSPFB01-3783 — 1-866-525-9374

Total Living	First Floor	Second Floor	Bonus	Bed	Bath	Width	Depth	Foundation	Price Category
2584 sq ft	1322 sq ft	1262 sq ft	N/A	4	3	48' 0"	50' 0"	Basement, Crawl Space or Slab	H

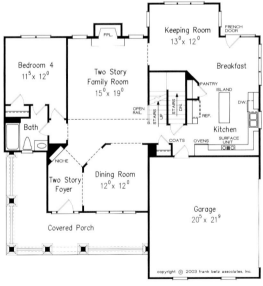

First Floor

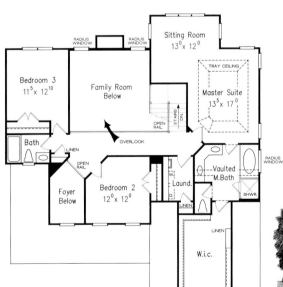

Second Floor

Design Features

- Tapered columns and Mission-styled windows, combined with earthen stone and cedar shakes, make the *Bakersfield* reminiscent of the Craftsman era.

- A two-story foyer with an art niche creates an exciting first impression.

- The keeping room is strategically placed just off the kitchen for easy interaction from room to room.

- A bedroom connected to a full bath is incorporated into the main floor, making an ideal guest suite.

Rear Elevation

Peppermill — TSPDG01-1034 — 1-866-525-9374

Total Living	First Floor	Second Floor	Bonus	Bed	Bath	Width	Depth	Foundation	Price Category
2586 sq ft	1809 sq ft	777 sq ft	264 sq ft	4	3-1/2	70' 7"	48' 4"	Crawl Space*	F

Design Features

- A traditional brick exterior with two country porches creates a modern exterior.
- Bold columns and a metal roof welcome guests inside an equally impressive interior.
- Both the foyer and great room have two-story ceilings, and a tray ceiling tops the master bedroom.
- A bay window expands the breakfast nook, while French doors lead to the rear porch.
- The living room/study and bonus room add flexibility for changing needs.

Rear Elevation

*Other options available. See page 347.

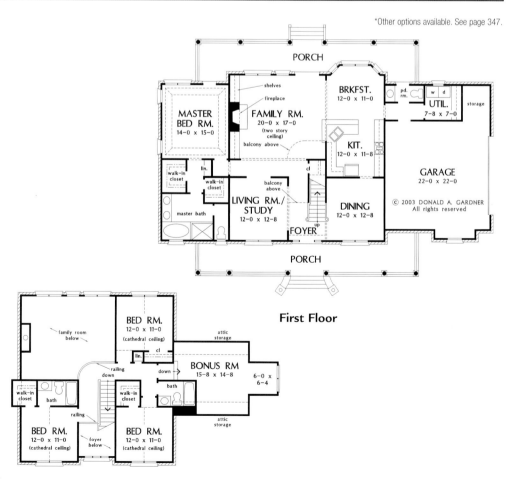

First Floor

Second Floor

© 2002 Frank Betz Associates, Inc.

Greythorne — TSPFB01-3764 — 1-866-525-9374

Total Living	First Floor	Second Floor	Bonus	Bed	Bath	Width	Depth	Foundation	Price Category
2587 sq ft	2047 sq ft	540 sq ft	278 sq ft	4	3	60' 0"	56' 0"	Basement or Crawl Space	I

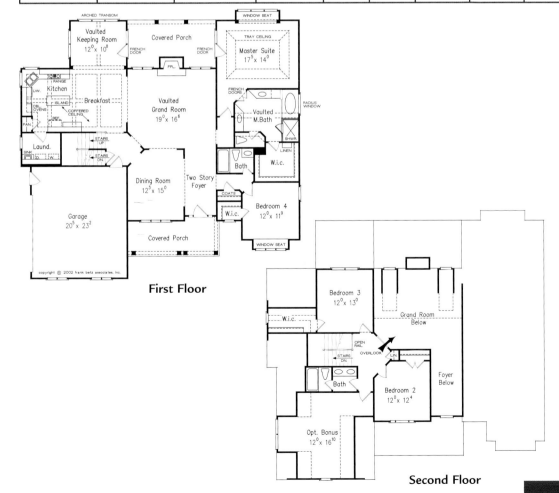

First Floor

Second Floor

Design Features

■ Thoughtful and creative design makes the *Greythorne* unique in both the layout and the details.

■ The full-service kitchen has an attention-grabbing coffered ceiling.

■ Accessible from both the master suite and the keeping room is a covered back porch.

■ Window seats are incorporated into the master suite and the fourth bedroom.

Rear Elevation

Donald A. Gardner Architects, Inc.

Avalon — TSPDG01-726 — 1-866-525-9374

Total Living	First Floor	Second Floor	Bonus	Bed	Bath	Width	Depth	Foundation	Price Category
2588 sq ft	1896 sq ft	692 sq ft	N/A	3	2-1/2	84' 10"	60' 0"	Crawl Space*	F

Design Features

*Other options available. See page 347.

- Visually, the great room, dining room, kitchen, breakfast area and loft flow effortlessly together.

- A curved staircase, as well as cathedral ceilings in the great and dining rooms, enhance volume.

- His-and-her vanities make mornings easy.

- An abundance of storage space is accessible upstairs as well as in the garage.

Rear Elevation

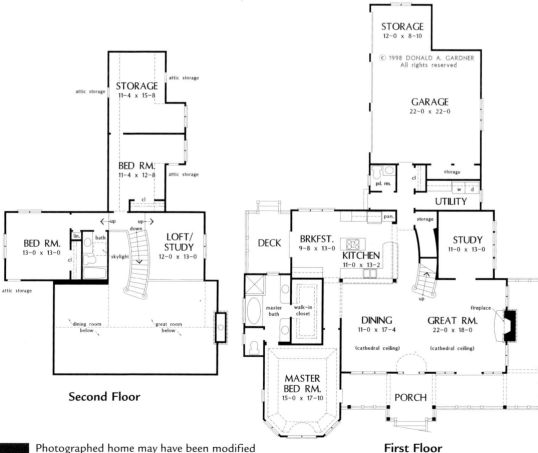

Second Floor

First Floor

Photographed home may have been modified from original construction documents.

© 1997 Donald A. Gardner Architects, Inc.

Sunnybrook — TSPDG01-484 — 1-866-525-9374

Total Living	First Floor	Second Floor	Bonus	Bed	Bath	Width	Depth	Foundation	Price Category
2596 sq ft	1939 sq ft	657 sq ft	386 sq ft	4	3	80' 10"	55' 8"	Crawl Space*	F

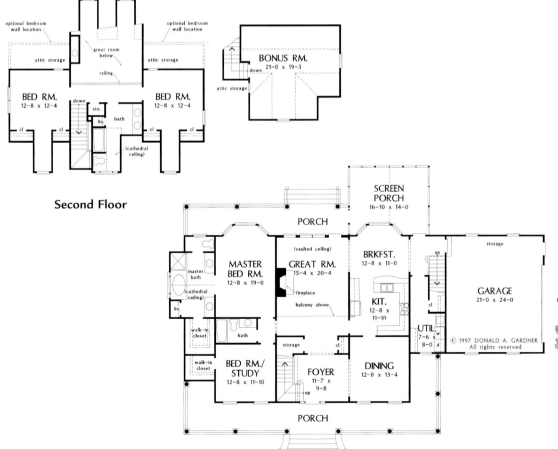

Second Floor

First Floor

Design Features

- This country farmhouse charms with its wrapping front porch and gabled dormers.

- The foyer gives way to a generous great room with fireplace, cathedral ceiling and balcony above.

- Nearby, a flexible bedroom/study has a walk-in closet and adjacent full bath.

- The first-floor master suite with sunny bay window features a private bath with cathedral ceiling.

- A bonus room over the garage completes the plan.

Rear Elevation

Donald A. Gardner Architects, Inc.

Fayette — TSPDG01-954 — 1-866-525-9374

Total Living	First Floor	Second Floor	Bonus	Bed	Bath	Width	Depth	Foundation	Price Category
2606 sq ft	1500 sq ft	1106 sq ft	366 sq ft	4	2-1/2	63' 3"	48' 1"	Crawl Space*	F

Design Features

- The floor plan features open spaces with more room definition for those seeking a truly traditional design.
- Columns separate the formal living and dining rooms, while a bay window extends the breakfast area.
- French doors access the rear porch as well as connect the family room to the outdoors.
- An angled counter allows the kitchen to take part in casual entertaining.

*Other options available. See page 347.

First Floor

Second Floor

Rear Elevation

Magnolia — TSPDG01-544 — 1-866-525-9374

Total Living	First Floor	Second Floor	Bonus	Bed	Bath	Width	Depth	Foundation	Price Category
2617 sq ft	1878 sq ft	739 sq ft	383 sq ft	4	3	79' 8"	73' 4"	Crawl Space*	F

*Other options available. See page 347.

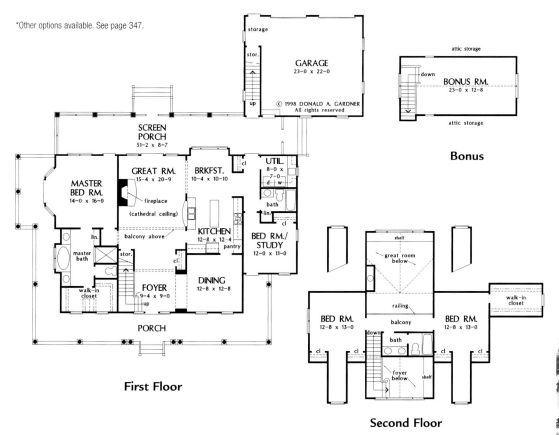

First Floor

Bonus

Second Floor

Design Features

- This Southern farmhouse features a wraparound porch and dormer windows for instant curb appeal.
- Illuminating the vaulted foyer is an arched clerestory window within the large, center dormer.
- The kitchen offers easy service to the great room by way of a convenient pass-thru.
- The master suite includes screened-porch access, a bay window and a splendid bath.
- The first floor bedroom/study, with its own entrance, makes an ideal home office or guest room.

Rear Elevation

Merchan — TSPDS01-7032 — 1-866-525-9374

© The Sater Design Collection, Inc.

Total Living	First Floor	Second Floor	Bonus	Bed	Bath	Width	Depth	Foundation	Price Category
2626 sq ft	1627 sq ft	999 sq ft	N/A	3	3-1/2	78' 6"	80' 6"	Crawl Space/Opt. Basement	G

Design Features

- A loft overlooks the octagonal-shaped great room.

- Grand columns and archways infuse elegance into the dining room.

- The U-shaped kitchen features an island with cooktop.

- The expansive master suite has a private-entry art niche.

- A dual wraparound porch, fireplace and built-in media center are wonderful.

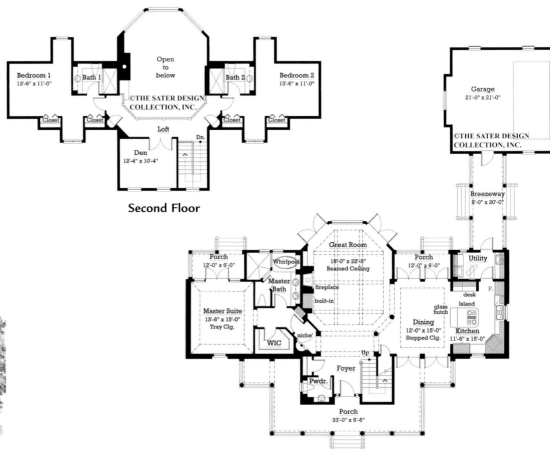

Second Floor

First Floor

Rear Elevation

Battery Park — TSPDG01-1093 — 1-866-525-9374

Total Living	First Floor	Second Floor	Bonus	Bed	Bath	Width	Depth	Foundation	Price Category
2629 sq ft	1308 sq ft	1321 sq ft	N/A	3	2-1/2	31' 11"	73' 3"	Crawl Space*	F

*Other options available. See page 347.

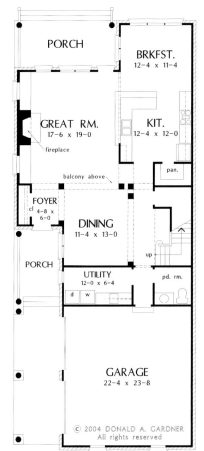

First Floor

PORCH

BRKFST.
12-4 x 11-4

GREAT RM.
17-6 x 19-0
fireplace

KIT.
12-4 x 12-0

pan.

balcony above

FOYER
cl 4-8 x 6-0

DINING
11-4 x 13-0

PORCH

up

UTILITY
12-0 x 6-4
d | w

pd. rm.

GARAGE
22-4 x 23-8

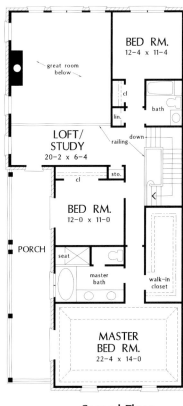

Second Floor

BED RM.
12-4 x 11-4

great room below

cl

bath

lin.

LOFT/STUDY
20-2 x 6-4

down

railing

sto.

cl

BED RM.
12-0 x 11-0

PORCH

seat

master bath

walk-in closet

MASTER BED RM.
22-4 x 14-0

Design Features

- Dual porches and decorative keystone detailing create an appealing exterior.

- Columns define the dining and great rooms.

- A rear porch provides an additional area to enjoy the outdoors.

- The second floor boasts an extended front porch and large loft overlooking the great room.

Rear Elevation

Michelle — TSPFB01-1013 — 1-866-525-9374

© 1996 Frank Betz Associates, Inc.

Total Living	First Floor	Second Floor	Bonus	Bed	Bath	Width	Depth	Foundation	Price Category
2643 sq ft	2015 sq ft	628 sq ft	315 sq ft	4	3	56' 0"	52' 6"	Basement, Crawl Space or Slab	H

Design Features

- Arched and bay windows complement vaulted and tray ceilings to make this home bright and cheery.

- Built-in bookshelves flanking the fireplace make wonderful areas for storage or display.

- The master suite has a tray ceiling and an arched window.

- A main-level secondary bedroom makes an ideal guest suite or home office.

Second Floor

First Floor

Rear Elevation

Briarcliff — TSPDG01-259 — 1-866-525-9374

Total Living	First Floor	Second Floor	Bonus	Bed	Bath	Width	Depth	Foundation	Price Category
2647 sq ft	1759 sq ft	888 sq ft	324 sq ft	4	3-1/2	85' 0"	53' 0"	Crawl Space*	F

*Other options available. See page 347.

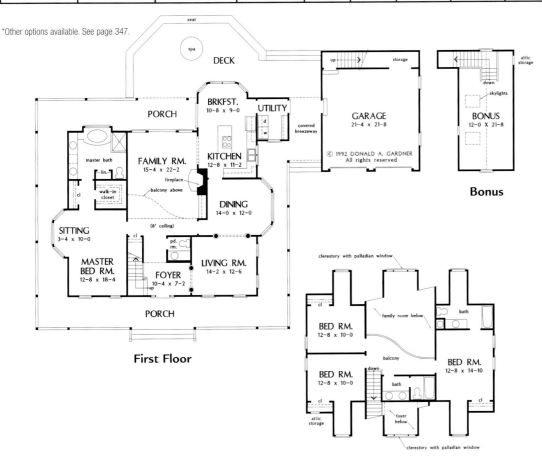

First Floor

Bonus

Second Floor

Design Features

- This spectacular country dream home is wrapped in porches and a massive deck.
- Front and rear Palladian windows let light flood this open plan.
- The family room with sloped ceiling receives added drama from an open, curved balcony.
- The first-floor master suite contains a sitting area, ample closet space and a luxurious bath.
- The bonus room provides extra elbow room to growing families.

Rear Elevation

Sommerset — TSPDS01-6827 — 1-866-525-9374

© The Sater Design Collection, Inc.

Total Living	First Floor	Second Floor	Bonus	Bed	Bath	Width	Depth	Foundation	Price Category
2650 sq ft	1296 sq ft	1354 sq ft	N/A	3	2-1/2	34' 0"	63' 2"	Slab	G

Design Features

- A gallery-style foyer leads to a powder room and a walk-in pantry.

- Wrapping counter space provides an overlook to a breakfast nook.

- French doors bring the outdoors into the dining and great rooms.

- Upstairs, the luxurious master suite has an octagonal sitting area and deck.

- Also opening to the upper deck are two more bedrooms and a study.

Rear Elevation

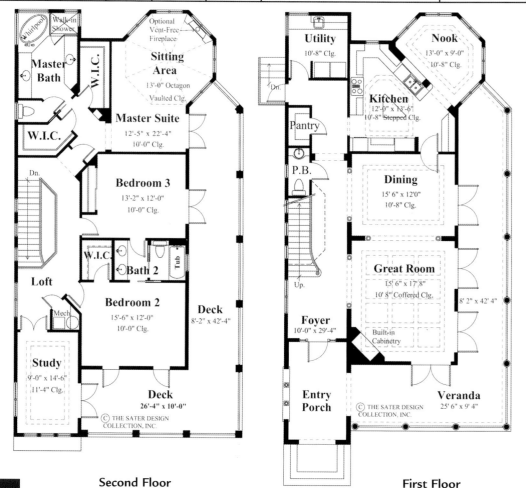

Second Floor

First Floor

© 2003 Donald A. Gardner, Inc.

Colridge — TSPDG01-1012-D — 1-866-525-9374

Total Living	First Floor	Basement	Bonus	Bed	Bath	Width	Depth	Walls/Foundation	Price Category
2652 sq ft	1732 sq ft	920 sq ft	N/A	3	3	70' 6"	59' 6"	Hillside Walkout	F

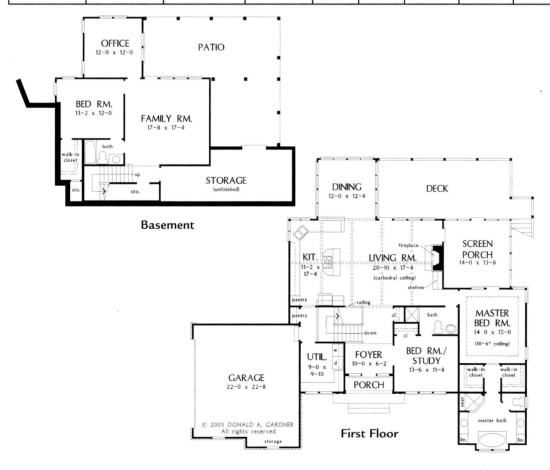

Basement

First Floor

Design Features

- A mixture of exterior materials enhances the curb appeal of this hillside walkout.

- A large deck, screened porch and patio promote outdoor living.

- A cathedral ceiling with exposed beams crowns the kitchen and great room, creating volume.

- The master suite is complete with a tray ceiling in the bedroom and screened-porch access.

Rear Elevation

Photographed home may have been modified from original construction documents.

www.twostoryplans.com **183**

Betonville — TSPDG01-308 — 1-866-525-9374

Total Living	First Floor	Second Floor	Bonus	Bed	Bath	Width	Depth	Walls/Foundation	Price Category
2658 sq ft	2064 sq ft	594 sq ft	483 sq ft	4	3-1/2	92' 0"	57' 8"	Crawl Space*	F

Design Features

*Other options available. See page 347.

- Light floods the interior through a front Palladian window over the two-level foyer.

- The great room, breakfast room and master bedroom access the porch for open circulation.

- The first floor enjoys nine-foot ceilings throughout.

- Special features include built-in bookshelves, curved balcony, wet bar and bedroom/study combo.

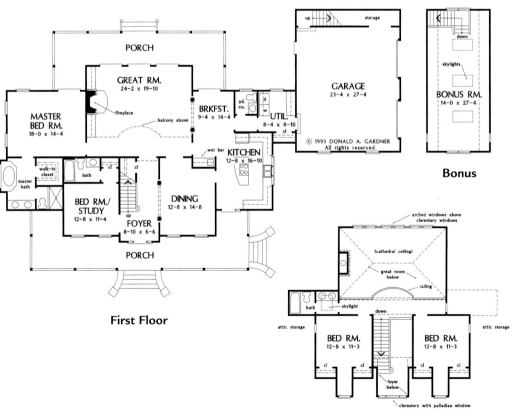

First Floor

Bonus

Second Floor

Rear Elevation

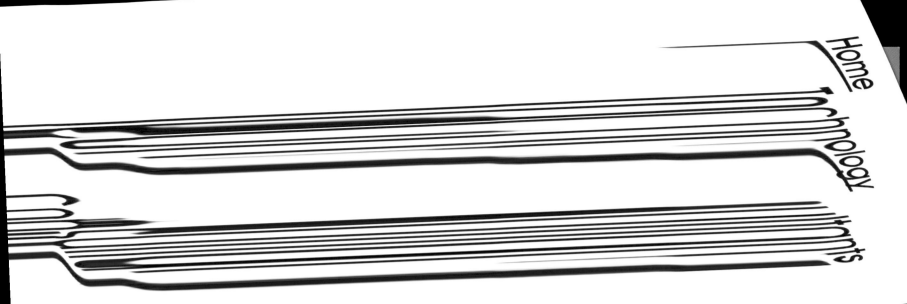

Hollyhock — TSPDG01-864 — 1-866-525-9374

Total Living	First Floor	Second Floor	Bonus	Bed	Bath	Width	Depth	Walls/Foundation	Price Category
2661 sq ft	2065 sq ft	596 sq ft	483 sq ft	4	3-1/2	92' 0"	57' 8"	Crawl Space*	F

*Other options available. See page 347.

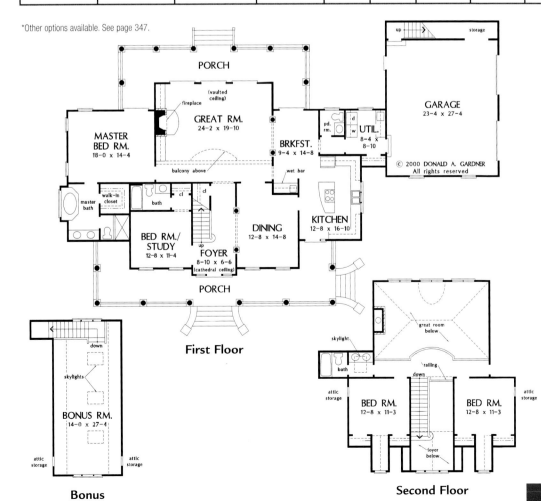

First Floor

Bonus

Second Floor

Design Features

- Sophistication abounds in this exquisite four-bedroom home with wrapping porches.

- A cathedral ceiling in the home's foyer makes a notable first impression.

- A large dormer with Palladian-style window adds to the drama.

- The expansive great room is enhanced by a vaulted ceiling and multiple doors that open onto the back porch.

- Note the nearby wet bar.

Rear Elevation

Waycross — TSPDG01-1039 — 1-866-525-9374

Total Living	First Floor	Second Floor	Bonus	Bed	Bath	Width	Depth	Walls/Foundation	Price Category
2665 sq ft	2006 sq ft	659 sq ft	527 sq ft	4	4	86'	49' 4"	Crawl Space*	F

Design Features

- Two massive porches, a prominent gable with a Palladian window and twin dormers showcase charm.

- An art niche, plant shelf, walk-in pantry and built-in cabinetry merge beauty with convenience.

- A striking staircase creates a grand focal point upon entry.

- The bonus room—with private staircase—provides space for a home office, gym or studio.

*Other options available. See page 347.

First Floor

Second Floor

Rear Elevation

Highlands — TSPDG01-852-D — 1-866-525-9374

Total Living	First Floor	Basement	Bonus	Bed	Bath	Width	Depth	Walls/Foundation	Price Category
2665 sq ft	1694 sq ft	971 sq ft	N/A	3	2-1/2	60' 6"	61' 2"	Hillside Walkout	F

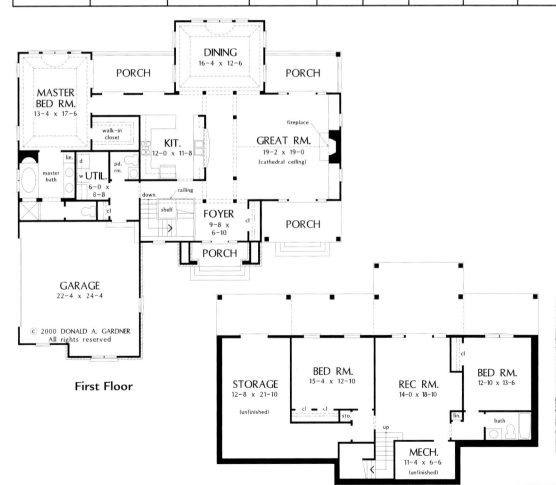

First Floor

Basement

Design Features

- An arched and gabled entry, Craftsman details and a mixture of stone and stucco create an appealing façade.

- Flanked by front and back porches, the great room's cathedral ceiling adds flair.

- Two back porches border the dining room with tray ceiling and arch-topped picture window.

- Downstairs, the walkout basement includes two bedrooms, storage and a recreation room.

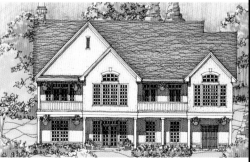

Rear Elevation

Donald A. Gardner Architects, Inc.

Riverbend — TSPDG01-225 — 1-866-525-9374

Total Living	First Floor	Second Floor	Bonus	Bed	Bath	Width	Depth	Walls/Foundation	Price Category
2677 sq ft	1734 sq ft	943 sq ft	N/A	4	3-1/2	55' 0"	44' 0"	Crawl Space*	F

Design Features

*Other options available. See page 347.

- A double-gabled roof, with front and rear Palladian windows, give this plan a stately elegance.

- Vaulted ceilings in the foyer and great room reinforce the visual drama of the Palladian windows.

- The covered front porch and out-standing rear deck expand living space to the outdoors.

- The spacious first-floor master suite accesses the large sunroom from a luxurious master bath.

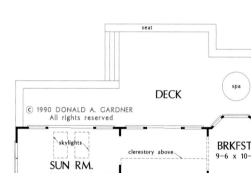

Second Floor

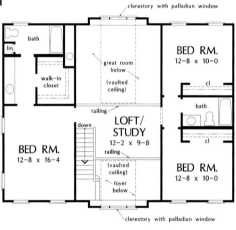

Rear Elevation

First Floor

Summerfield — TSPFB01-3550 — 1-866-525-9374

Total Living	First Floor	Second Floor	Opt. Bonus	Bed	Bath	Width	Depth	Foundation	Price Category
2680 sq ft	2087 sq ft	593 sq ft	249 sq ft	3	2-1/2	58' 4"	55' 2"	Basement or Crawl Space	G

First Floor

Second Floor

Design Features

- The stone front, porch with columns and multiple rooflines give this home the appearance of a storybook cottage.

- A vaulted keeping room with a fireplace also boasts radius windows along the rear of the home, allowing views to the backyard.

- The main-level master suite has a large sitting area that is flanked by decorative columns.

- Two secondary bedrooms and an optional bonus room reside on the second floor.

Rear Elevation

Sable Ridge — TSPDG01-710-D — 1-866-525-9374

Total Living	First Floor	Basement	Bonus	Bed	Bath	Width	Depth	Foundation	Price Category
2683 sq ft	1472 sq ft	1211 sq ft	N/A	3	2-1/2	53' 8"	40' 4"	Hillside Walkout	F

Design Features

- Designed for sloping lots, this stone and stucco home takes advantage of rear views.

- The floor plan features an open design, with cathedral ceilings in the great room and master bedroom.

- A wet bar is conveniently situated between the kitchen and great room.

- Downstairs, two bedrooms share an adjoining bath, and the enormous rec room offers flexibility.

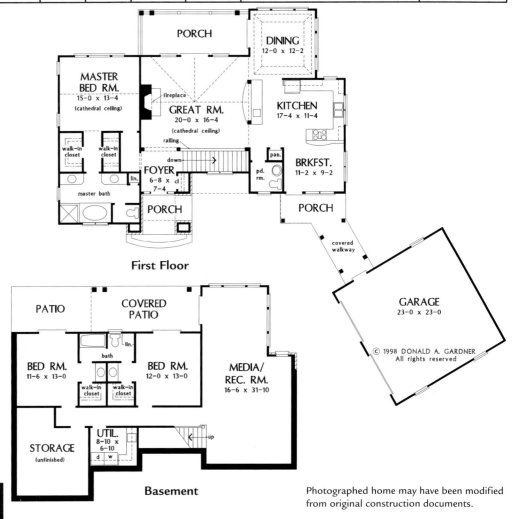

First Floor

Basement

Rear Elevation

Photographed home may have been modified from original construction documents.

© 2003 Frank Betz Associates, Inc.

Stewarts Landing — TSPFB01-3811 — 1-866-525-9374

Total Living	First Floor	Second Floor	Bonus	Bed	Bath	Width	Depth	Foundation	Price Category
2691 sq ft	2140 sq ft	551 sq ft	265 sq ft	4	3	57' 4"	62' 0"	Basement or Crawl Space	I

First Floor

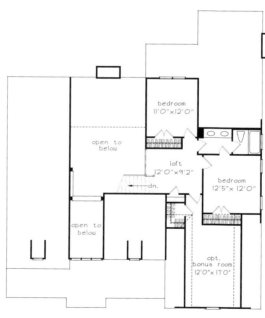

Second Floor

Design Features

- From the *Southern Living® Design Collection*

- The pleasant use of stone and friendly front porch of the *Stewarts Landing* beckons you in to see more.

- The master suite includes a gracious sitting area, providing a quiet place to enjoy views off the front porch.

- An optional bonus room can be easily finished into a fifth bedroom, playroom or exercise area. Situated among the uptsairs rooms is a loft, suitable for a computer desk or additional lounging space.

Rear Elevation

Santee — TSPDG01-1011 — 1-866-525-9374

Total Living	First Floor	Second Floor	Bonus	Bed	Bath	Width	Depth	Walls/Foundation	Price Category
2693 sq ft	1972 sq ft	721 sq ft	377 sq ft	4	3	77' 4"	50' 8"	Crawl Space*	F

Design Features

*Other options available. See page 347.

- Columns and dormers accent a deep porch, while gables and transoms add angles and soft curves.

- Floor space is expanded by bay windows, which also usher light into the home.

- Convenient features include a kitchen island, large utility room with sink and two built-in desks.

- A two-story cathedral ceiling and balcony highlight the foyer and great room.

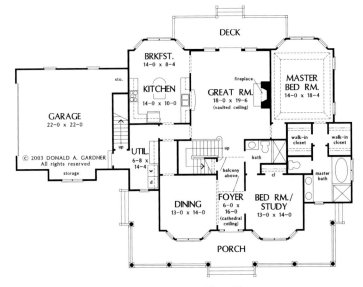

First Floor

Rear Elevation

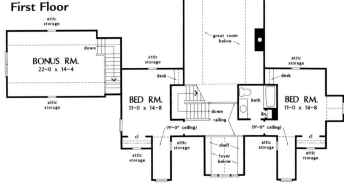

Second Floor

© 2003 Frank Betz Associates, Inc.

Rosemore Place — TSPFB01-3787 — 1-866-525-9374

Total Living	First Floor	Second Floor	Opt. Bonus	Bed	Bath	Width	Depth	Foundation	Price Category
2696 sq ft	2113 sq ft	583 sq ft	341 sq ft	4	3	58' 4"	61' 0"	Basement or Crawl Space	I

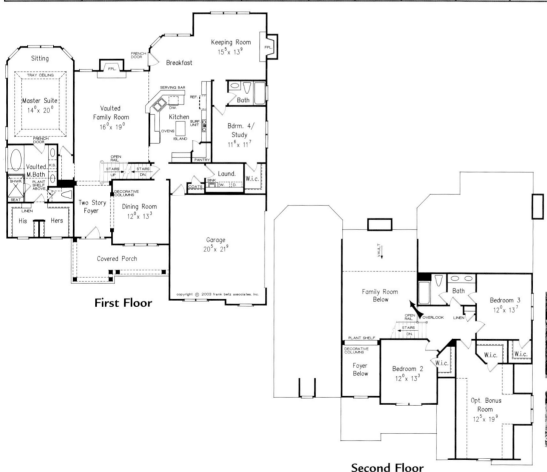

First Floor

Second Floor

Design Features

- Distinctive fieldstone and cedar shake give the façade of the *Rosemore Place* warm texture and dimension.

- This warmth radiates inside as well, with a fire-lit keeping room just off the kitchen.

- The master suite is private and secluded on the main level of the home.

- The main-floor bedroom can act as the home office, perfect for the telecommuter or entrepreneur.

Rear Elevation

Dominion — TSPFB01-3754 — 1-866-525-9374

Total Living	First Floor	Second Floor	Opt. Bonus	Bed	Bath	Width	Depth	Foundation	Price Category
2702 sq ft	1355 sq ft	1347 sq ft	285 sq ft	5	4	41' 0"	66' 0"	Basement or Crawl Space	I

Design Features

- Unique angles and special details inside make a floor plan that is not easily forgotten.

- The kitchen overlooks a keeping room that is nestled on the back of the home with great views to the backyard.

- A coffered ceiling and transom windows in the family room make this space bright and comfortable.

- The main-floor bedroom adjoins a full bath, creating an ideal guest suite.

Rear Elevation

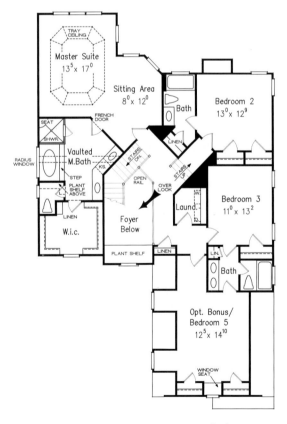

Second Floor

First Floor

© 2004 Frank Betz Associates, Inc.

North Easton — TSPFB01-3901 — 1-866-525-9374

Total Living	First Floor	Second Floor	Opt. Third	Bed	Bath	Width	Depth	Foundation	Price Category
2717 sq ft	1725 sq ft	992 sq ft	351 sq ft	3	2-1/2	46' 0"	85' 0"	Crawl Space or Slab	H

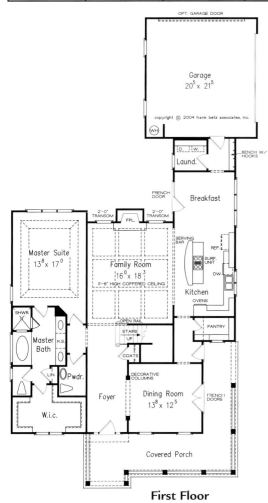

First Floor

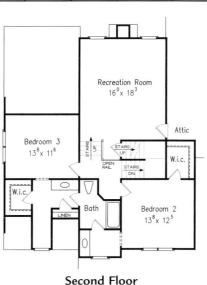

Second Floor

Opt. Third Floor

Design Features

- The *North Easton* is packed with pleasant surprises in every direction.

- A coffered ceiling in the family room makes a dramatic focal point for the design.

- Raised-bar seating on the kitchen island provides a casual dining spot or additional seating for entertaining.

- Upstairs, a children's recreation room provides the ideal place for playing and lounging.

Rear Elevation

Eton — TSPDG01-1051 — 1-866-525-9374

© 2004 Donald A. Gardner, Inc.

Total Living	First Floor	Second Floor	Bonus	Bed	Bath	Width	Depth	Foundation	Price Category
2717 sq ft	2065 sq ft	652 sq ft	636 sq ft	4	4	65' 2"	52' 8"	Crawl Space*	F

Design Features

- Starting with a sophisticated mix of stone and siding, this façade hints of another time.

- Columns frame a towering entrance, while curved transoms usher natural light inside the home.

- The large kitchen island includes a breakfast bar and a built-in desk.

- Blurring the line between indoor and outdoor living, a screened porch creates a special haven.

*Other options available. See page 347.

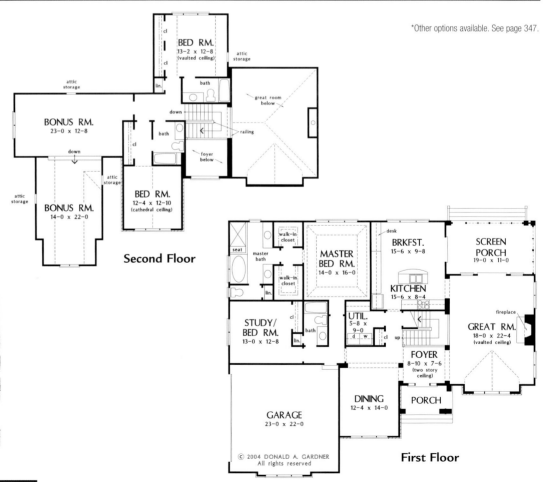

Second Floor

First Floor

© 2004 DONALD A. GARDNER
All rights reserved

Rear Elevation

NEW DORMS

60 rooms for each - students

Terry and Lander

88 students -
New Hall

48 students -
Hagget

76 students -
McMahon

780 students -
McCarty

Breyerton — TSPFB01-3766 — 1-866-525-9374

Total Living	First Floor	Second Floor	Bonus	Bed	Bath	Width	Depth	Foundation	Price Category
2723 sq ft	1847 sq ft	876 sq ft	N/A	4	3-1/2	56' 4"	50' 6"	Basement or Crawl Space	I

First Floor

Design Features

- The *Breyerton* incorporates features that make everyday living more convenient.
- A keeping room is connected to the kitchen area, giving families that cozy, casual spot to reconnect at day's end.
- The coat closet and laundry room are located just inside the garage, making a handy drop spot for coats and shoes.
- The master suite is its own slice of heaven with a private sitting area, step-up jetted tub and seated shower.

Second Floor

Rear Elevation

Fallston — TSPDG01-1107 — 1-866-525-9374

Total Living	First Floor	Second Floor	Bonus	Bed	Bath	Width	Depth	Walls/Foundation	Price Category
2728 sq ft	2090 sq ft	638 sq ft	N/A	4	4	55' 0"	67' 4"	Crawl Space*	F

*Other options available. See page 347.

Design Features

- With symmetrical gables and multiple bay windows, this exterior emits a classic look.

- A sweeping staircase accents the foyer, which opens into the great room.

- Vaulted ceilings enhance volume, while the kitchen and breakfast room open to one another.

- Additional bedrooms are upstairs, separated by an open loft that overlooks the great room.

- A versatile fourth bedroom/study can double as an office or guestroom.

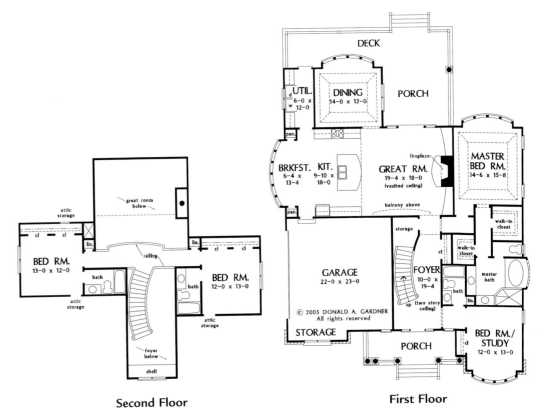

Second Floor

First Floor

Rear Elevation

© The Sater Design Collection, Inc.

Bartolini — TSPDS01-8022 — 1-866-525-9374

Total Living	First Floor	Second Floor	Bonus	Bed	Bath	Width	Depth	Foundation	Price Category
2736 sq ft	2084 sq ft	652 sq ft	375 sq ft	3	2-1/2	60' 6"	94' 0"	Opt. Basement/Slab	G

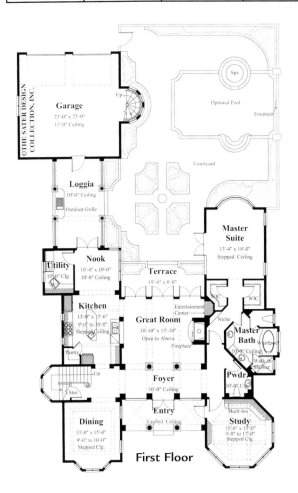

First Floor

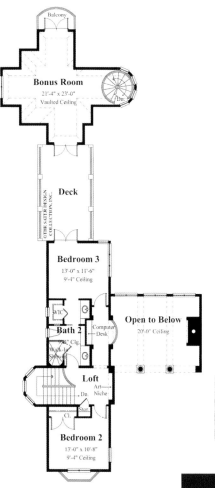

Second Floor

Design Features

- A graceful entry arcade leads to a grand foyer.

- Three sets of French doors are all that seperate the great room from a veranda and courtyard.

- The kitchen and morning nook open to a covered loggia.

- The secluded master suite has private courtyard access.

- The upper level harbors two bedroom suites and a bonus room.

Rear Elevation

Ballard — TSPFB01-3633 — 1-866-525-9374

© 2001 Frank Betz Associates, Inc.

Total Living	First Floor	Second Floor	Bonus	Bed	Bath	Width	Depth	Foundation	Price Category
2759 sq ft	1565 sq ft	1194 sq ft	280 sq ft	4	4	53' 0"	48' 6"	Basement, Crawl Space or Slab	H

Design Features

- This smart design uses every inch of space to create a practical and charming home.

- Every family spends a great deal of time in the kitchen, so a keeping room with a fireplace adjoins this area.

- A built-in desk has been designed into this area as well, keeping household clutter in its place.

- Each bedroom has direct access to a bath, and most feature walk-in closets.

Rear Elevation

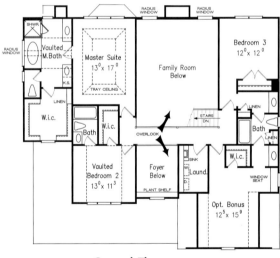

Second Floor

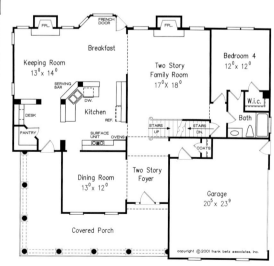

First Floor

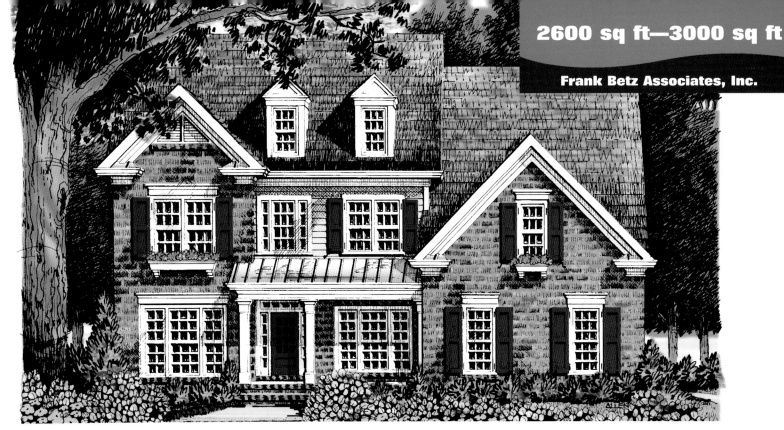

© 2003 Frank Betz Associates, Inc.

Neyland — TSPFB01-3789 — 1-866-525-9374

Total Living	First Floor	Second Floor	Bonus	Bed	Bath	Width	Depth	Foundation	Price Category
2762 sq ft	1364 sq ft	1398 sq ft	N/A	5	4	51' 0"	45' 4"	Basement or Crawl Space	I

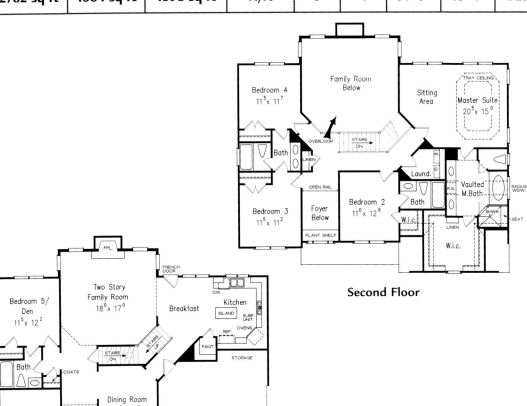

Second Floor

First Floor

Design Features

- Warm brick and friendly dormers create a well-established, time-tested façade.

- The design inside is fresh and innovative, featuring some of today's most popular trends.

- The master suite houses its own private sitting area, giving the homeowner a place to retreat to after a busy day.

- A fifth bedroom on the main level makes a great guest room or can be easily changed to a den.

Rear Elevation

Burnside — TSPFB01-1018 — 1-866-525-9374

© 1996 Frank Betz Associates, Inc.

Total Living	First Floor	Second Floor	Opt. Bonus	Bed	Bath	Width	Depth	Foundation	Price Category
2764 sq ft	1904 sq ft	860 sq ft	388 sq ft	4	3-1/2	56' 0"	61' 6"	Basement or Crawl Space	H

Design Features

- The kitchen features a double oven, large pantry, center island and a serving bar adjoining the family room.

- Each of the four bedrooms has a walk-in closet.

- Other main-level rooms use nine-foot-high ceilings for an added sense of spaciousness.

- An optional bonus room above the garage adds 338 square feet of living space.

- Master suite luxuries include a tray ceiling in the bedroom, a private toilet compartment and a walk-in closet.

Rear Elevation

First Floor

Second Floor

© The Sater Design Collection, Inc.

Cloverdale — TSPDS01-7058 — 1-866-525-9374

Total Living	First Floor	Second Floor	Bonus	Bed	Bath	Width	Depth	Foundation	Price Category
2775 sq ft	1874 sq ft	901 sq ft	382 sq ft	3	3-1/2	90' 0"	58' 6"	Basement/Crawl Space	G

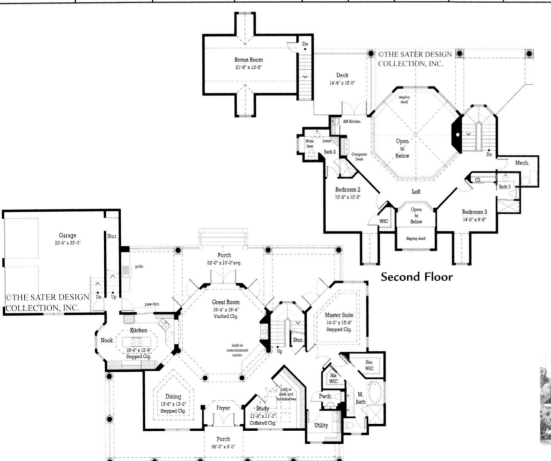

Second Floor

First Floor

Design Features

- Classic columns line the welcoming front porch.
- The octagonal-shaped great room boasts a row of French doors to the rear porch.
- A bay-window kitchen has serving counters to the great room and porch.
- Two upstairs bedrooms and a bonus room share a deck.
- The second-floor loft offers spectacular interior views.

Rear Elevation

Lake Placid — TSPFB01-1131 — 1-866-525-9374

© 1998 Frank Betz Associates, Inc.

Total Living	First Floor	Second Floor	Opt. Bonus	Bed	Bath	Width	Depth	Foundation	Price Category
2792 sq ft	1980 sq ft	812 sq ft	255 sq ft	4	3-1/2	60' 6"	51' 2"	Basement, Crawl Space or Slab	I

Design Features

- Broad bay windows brighten the living room, breakfast room, master suite and a secondary bedroom.

- The two-story family room is enhanced with French doors to the backyard and a fireplace flanked by built-in shelves.

- The bonus room on the upper level can be a fifth bedroom.

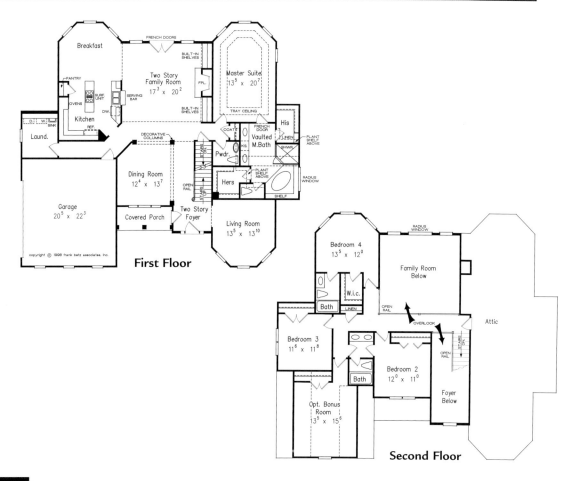

First Floor

Second Floor

Rear Elevation

© 2004 Frank Betz Associates, Inc.

Arramore — TSPFB01-3869 — 1-866-525-9374

Total Living	First Floor	Second Floor	Bonus	Bed	Bath	Width	Depth	Foundation	Price Category
2792 sq ft	1365 sq ft	1427 sq ft	N/A	4	3	44' 0"	58' 0"	Basement, Crawl Space or Slab	H

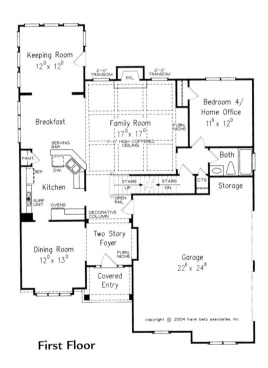

First Floor

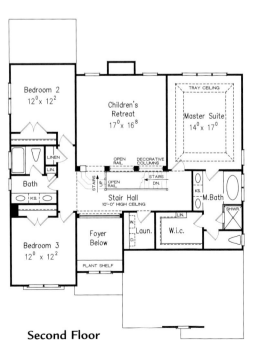

Second Floor

Design Features

- Board-and-batten siding and shutters blend distinctively with brick and cedar shake.

- A furniture niche is incorporated into the foyer, creating the necessary space to decorate.

- A bright and sunny keeping room is connected to the kitchen and breakfast areas.

- A children's retreat is included in the upper level of this design, giving kids plenty of space to play.

Rear Elevation

Torrey Pines Way — TSPDS01-6608 — 1-866-525-9374

© The Sater Design Collection, Inc.

Total Living	First Floor	Second Floor	Bonus	Bed	Bath	Width	Depth	Foundation	Price Category
2796 sq ft	2368 sq ft	428 sq ft	N/A	3	3	72' 8"	72' 0"	Slab	G

Design Features

- Two pairs of French doors open the grand room to the lanai.

- The family kitchen has an island counter and eating bar.

- The owner's wing includes a study and bayed sitting area.

- An inviting master bath has dual vanities and views of a private garden.

- The second floor houses a guest room, loft and full bath.

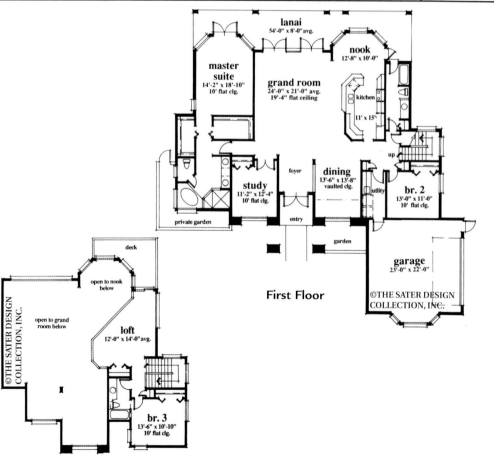

First Floor

©THE SATER DESIGN COLLECTION, INC.

Second Floor

Rear Elevation

© 1995 Frank Betz Associates, Inc.

Balmoral — TSPFB01-925 — 1-866-525-9374

Total Living	First Floor	Second Floor	Bonus	Bed	Bath	Width	Depth	Foundation	Price Category
2806 sq ft	1952 sq ft	854 sq ft	N/A	4	2-1/2	56' 6"	50' 6"	Basement or Crawl Space	I

First Floor

Second Floor

Design Features

- A traditional design style is recognized in the *Balmoral*, with a combination of brick and stucco on the front elevation.

- The kitchen is generously sized and highly functional, incorporating a built-in desk, pantry, island and double ovens.

- Decorative columns create a soft border around the dining area, defining its space but keeping the floor plan open and roomy.

- The master suite resides on the main level, a popular design feature.

Rear Elevation

Biscayne Bay — TSPDS01-6830 — 1-866-525-9374

© The Sater Design Collection, Inc.

Total Living	First Floor	Second Floor	Bonus	Bed	Bath	Width	Depth	Foundation	Price Category
2815 sq ft	1810 sq ft	1005 sq ft	N/A	3	3	54' 0"	57' 0"	Crawl Space	G

Design Features

- The stately great room hosts French doors and cabinetry.

- A two-sided fireplace warms the great room and dining room.

- The spacious kitchen is open to the main living areas.

- A gentlemen's study with stepped ceiling precedes the master suite.

- The second floor has two large bedrooms with private baths.

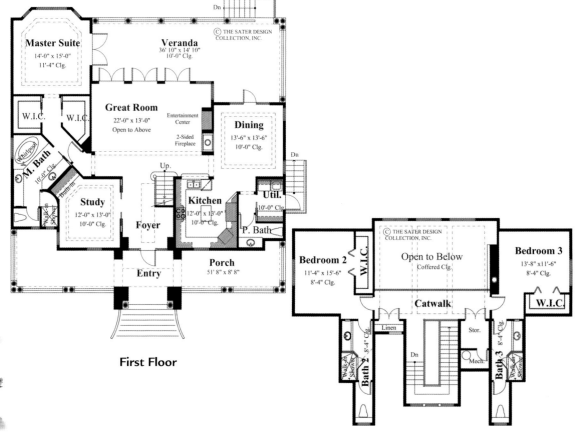

Rear Elevation

First Floor

Second Floor

Ryecroft — TSPDG01-824-D — 1-866-525-9374

Total Living	First Floor	Basement	Bonus	Bed	Bath	Width	Depth	Walls/Foundation	Price Category
2815 sq ft	1725 sq ft	1090 sq ft	N/A	3	3-1/2	59' 0"	59' 4"	Hillside Walkout	F

First Floor

Basement

Design Features

- Designed for sloping lots, this home positions its common living areas and master suite on the first floor.

- A generous rec room and two family bedrooms reside on the lower level.

- With a bay window and back-porch access, the master suite boasts dual walk-ins and a luxurious bath.

- Downstairs, two bedrooms and baths flank the rec room with fireplace and wet bar.

Rear Elevation

Springdale — TSPDG01-419 — 1-866-525-9374

© 1995 Donald A. Gardner Architects, Inc.

Total Living	First Floor	Second Floor	Bonus	Bed	Bath	Width	Depth	Walls/Foundation	Price Category
2832 sq ft	1483 sq ft	1349 sq ft	486 sq ft	4	2-1/2	66' 10"	47' 8"	Crawl Space*	F

*Other options available. See page 347.

Design Features

- Columns between the foyer and living room/study hint at the extras in this four-bedroom estate.

- Transom windows over French doors open up the living room/study to the front porch.

- A generous family room accesses the covered back porch.

- Nine-foot ceilings amplify the first floor.

- All bedrooms are upstairs, including a deluxe master suite with a tray ceiling.

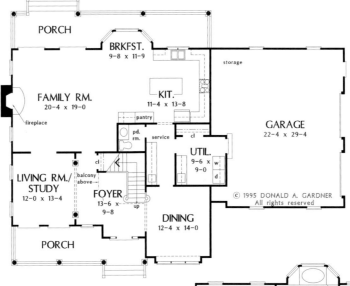

First Floor

Second Floor

Rear Elevation

Mercer — TSPDG01-372 — 1-866-525-9374

Total Living	First Floor	Second Floor	Bonus	Bed	Bath	Width	Depth	Walls/Foundation	Price Category
2833 sq ft	2162 sq ft	671 sq ft	345 sq ft	4	3	61' 11"	54' 8"	Crawl Space*	F

*Other options available. See page 347.

Design Features

- Although impressive and airy, this elegant executive home stays warm from dual fireplaces.

- Rear bays expand space and maximize natural light.

- The master suite has an angled entrance for privacy and includes a sitting bay and lavish bath.

- Extra room is added by a skylit bonus room and ample attic storage.

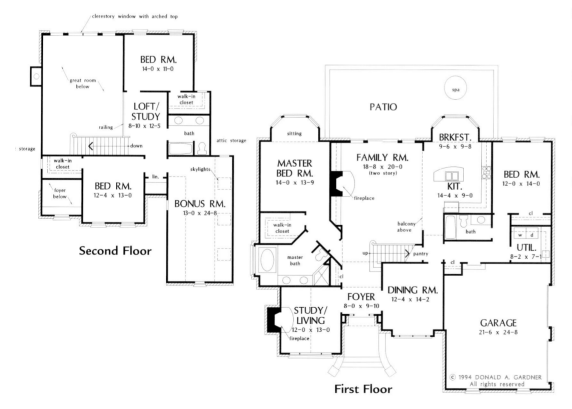

Second Floor

First Floor

Rear Elevation

Photographed home may have been modified from original construction documents.

Gentry — TSPFB01-913 — 1-866-525-9374

© 1995 Frank Betz Associates, Inc.

Total Living	First Floor	Second Floor	Opt. Bonus	Bed	Bath	Width	Depth	Foundation	Price Category
2840 sq ft	1347 sq ft	1493 sq ft	243 sq ft	5	4-1/2	58' 4"	46' 6"	Basement, Crawl Space or Slab	I

Design Features

- French Country warmth is often achieved by blending European stucco and stacked stone, like on the *Gentry*.
- Formal living and dining rooms border the two-story foyer, giving it a time-honored flair.
- The two-story family room is anything but traditional, surrounded by arched openings and decorative columns.
- Family gatherings just got easier in the oversized kitchen, complete with double ovens and a large prep island.

Rear Elevation

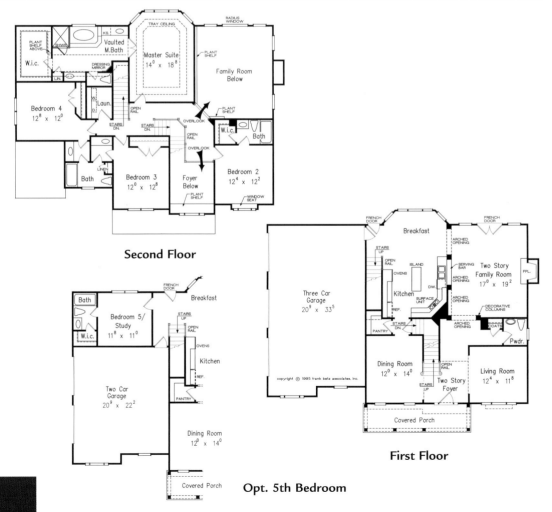

Second Floor

Opt. 5th Bedroom

First Floor

McPherson — TSPDG01-1102 — 1-866-525-9374

Total Living	First Floor	Second Floor	Bonus	Bed	Bath	Width	Depth	Foundation	Price Category
2844 sq ft	1552 sq ft	1292 sq ft	348 sq ft	4	3	51' 0"	46' 0"	Crawl Space*	F

*Other options available. See page 347.

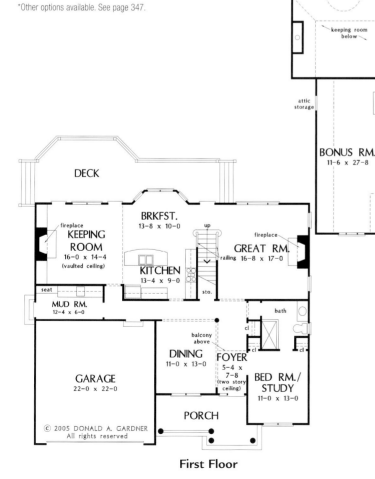

Second Floor

First Floor

Design Features

- Three high-pitched gables and dramatic window design create a lavish, lofty exterior.

- Stately columns define the entryway and lead into a spacious two-story foyer.

- Complete with a built-in seat and sink, the mudroom works well for families that desire a highly functional utility room.

- The generous upstairs includes all secondary bedrooms in addition to the master suite.

Rear Elevation

Braddock — TSPFB01-3725 — 1-866-525-9374

© 2002 Frank Betz Associates, Inc.

Total Living	First Floor	Second Floor	Bonus	Bed	Bath	Width	Depth	Foundation	Price Category
2858 sq ft	1967 sq ft	891 sq ft	N/A	5	4	60' 10"	55' 0"	Basement or Crawl Space	I

Design Features

- Brick and siding give the *Braddock* a traditional flair on the outside.

- A vaulted keeping room acts as a cozy extension of the kitchen and breakfast areas.

- French doors lead to a lush master bath, equipped with all the creature comforts—dual sinks, linen closet and a soaking tub.

- Flexible space is incorporated into this design with a fifth bedroom that can remain as such, or easily convert into a study.

Rear Elevation

© 2001 Frank Betz Associates, Inc.

Windward — TSPFB01-3652 — 1-866-525-9374

Total Living	First Floor	Second Floor	Opt. Bonus	Bed	Bath	Width	Depth	Foundation	Price Category
2863 sq ft	1969 sq ft	894 sq ft	213 sq ft	4	2-1/2	55' 0"	54' 0"	Basement or Crawl Space	I

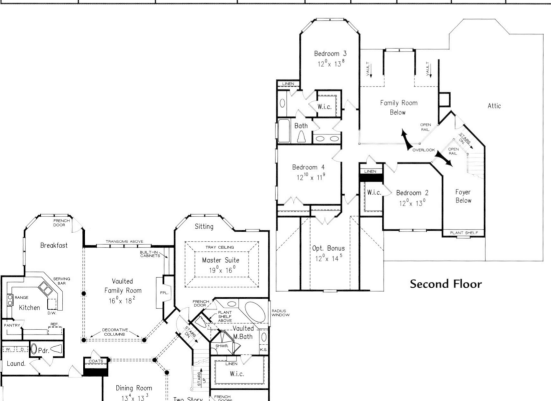

Second Floor

First Floor

Design Features

- Fieldstone and cedar shake, accented by board-and-batten shutters, create the ideal cottage façade.

- A study is located just off the foyer, making the perfect home office.

- The master suite earns its name with a sitting area, luxurious bath and corner soaking tub.

- Optional bonus space is available on the upper level that can be finished into a playroom, exercise area or craft room.

Rear Elevation

Montpelier — TSPDG01-483 — 1-866-525-9374

Total Living	First Floor	Second Floor	Bonus	Bed	Bath	Width	Depth	Foundation	Price Category
2869 sq ft	2249 sq ft	620 sq ft	308 sq ft	4	3-1/2	69' 6"	52' 0"	Crawl Space*	F

*Other options available. See page 347.

Design Features

- A clerestory window and detailed square columns lend drama to the arched and gabled entryway.

- Inside, luxury abounds with a formal living room with fireplace and box-bay window.

- The open, two-story family room boasts a fireplace with built-in cabinets on either side.

- Additionally, this plan offers a convenient second master suite on the first floor.

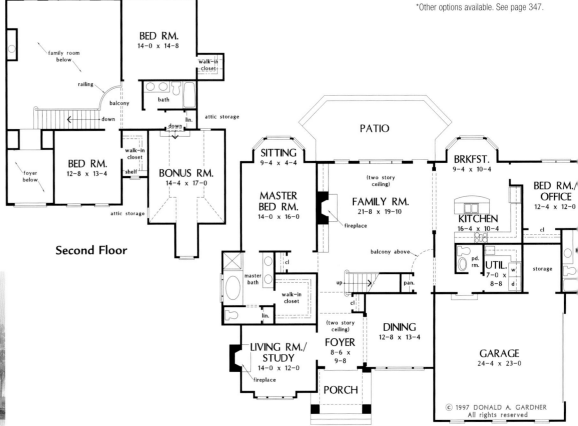

Second Floor

First Floor

Rear Elevation

© 2000 Frank Betz Associates, Inc.

Greenlaw — TSPFB01-3559 — 1-866-525-9374

Total Living	First Floor	Second Floor	Opt. Bonus	Bed	Bath	Width	Depth	Foundation	Price Category
2884 sq ft	2247 sq ft	637 sq ft	235 sq ft	4	4	64' 0"	55' 2"	Basement, Crawl Space or Slab	I

Second Floor

First Floor

Design Features

- Stately brick and a turret embellished with radius windows create a distinguished façade.

- A vaulted keeping room, breakfast area, kitchen and covered porch come together to create a comfortable core of the home.

- The home office on the main level can be effortlessly converted into a nursery with easy access from the master suite.

- A loft upstairs makes a great lounging spot for the kids.

Rear Elevation

Prescott Ridge — TSPFB01-3746 — 1-866-525-9374

© 2002 Frank Betz Associates, Inc.

Total Living	First Floor	Second Floor	Bonus	Bed	Bath	Width	Depth	Foundation	Price Category
2885 sq ft	2052 sq ft	833 sq ft	N/A	5	4	59' 0"	55' 6"	Basement or Crawl Space	H

Design Features

- Original and thoughtful design talent went into every detail of the *Prescott Ridge*.

- The kitchen is complete with a cooktop island and double ovens.

- A window seat creates the back wall of the keeping room, adding charm and comfort.

- The keeping room shares a two-sided fireplace with the neighboring two-story great room.

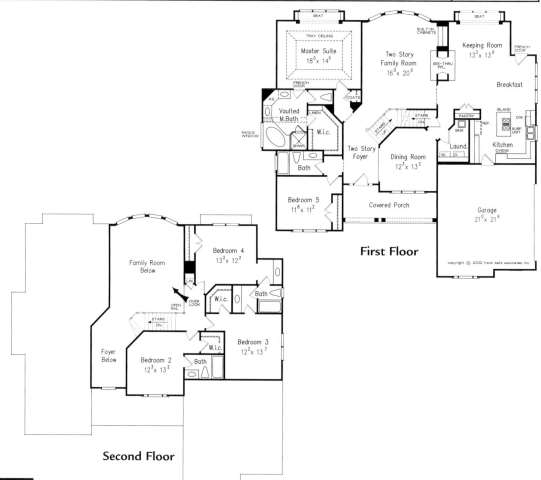

First Floor

Second Floor

Rear Elevation

© The Sater Design Collection, Inc.

Ansel Arbor — TSPDS01-7023 — 1-866-525-9374

Total Living	First Floor	Second Floor	Bonus	Bed	Bath	Width	Depth	Foundation	Price Category
2889 sq ft	2151 sq ft	738 sq ft	534 sq ft	3	2-1/2	99' 0"	56' 0"	Basement/Crawl Space	G

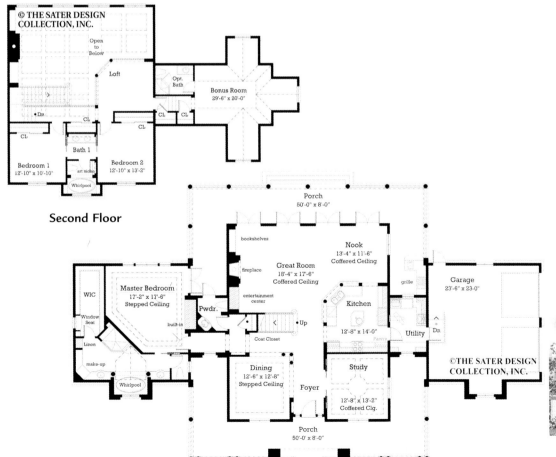

Second Floor

First Floor

Design Features

- Wraparound front and rear porches join the home's wings.
- Speciality ceilings and interior columns provide light and elegance.
- The study opens through pocket doors off the foyer.
- The master suite features a projected whirlpool tub.
- Two bedrooms with an adjoining bath are near the great room.

Rear Elevation

Elk River Lane— TSPDS01-6652 — 1-866-525-9374

© The Sater Design Collection, Inc.

Total Living	First Floor	Second Floor	Bonus	Bed	Bath	Width	Depth	Foundation	Price Category
2891 sq ft	2181 sq ft	710 sq ft	N/A	3	3	66' 4"	79' 0"	Slab	G

Design Features

- The grand room has French doors, a fireplace and built-ins.

- A study can easily be a home office or exercise area.

- The master suite has a step ceiling and sitting space.

- A step-up tub and a make-up space highlight the master bath.

- Upstairs, two bedrooms—one with a deck—share a bath.

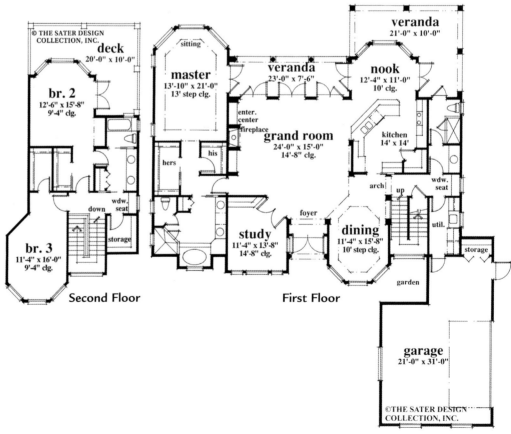

Second Floor

First Floor

Rear Elevation

Copperleaf — TSPDG01-1070 — 1-866-525-9374

Total Living	First Floor	Second Floor	Bonus	Bed	Bath	Width	Depth	Foundation	Price Category
2896 sq ft	2218 sq ft	678 sq ft	462 sq ft	4	3	85' 4"	43' 4"	Crawl Space*	F

*Other options available. See page 347.

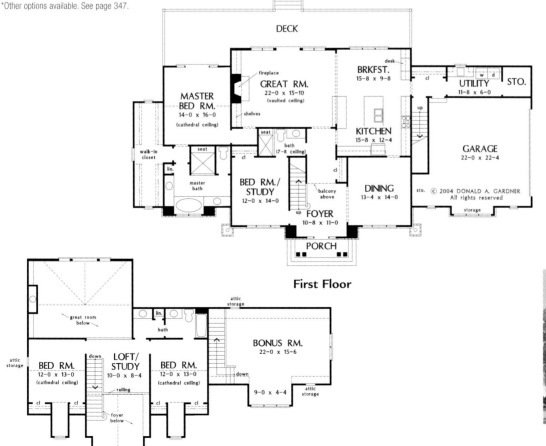

First Floor

Second Floor

Design Features

- A metal-topped portico joins an arched clerestory, creating a towering entrance.

- Twin dormers flank a stone gable, while sidelights and a transom frame the front door.

- The foyer is capped by a cathedral ceiling and features a plant shelf.

- French doors lead to outdoor entertaining, and built-in cabinetry provides architectural interest.

- A private staircase ascends to the large bonus room that's perfect for a home office.

Rear Elevation

Candler Park — TSPFB01-3777 — 1-866-525-9374

© 2003 Frank Betz Associates, Inc.

Total Living	First Floor	Second Floor	Bonus	Bed	Bath	Width	Depth	Foundation	Price Category
2900 sq ft	2262 sq ft	638 sq ft	252 sq ft	4	4	64' 0"	56' 4"	Basement, Crawl Space or Slab	I

Design Features

■ A covered front porch gives a warm welcome to guests as they enter the *Candler Park*.

■ A two-story foyer extends the friendly greeting.

■ Accessorizing the dining room will be fun and easy with a furniture niche and decorative columns.

■ A main-floor bedroom makes the perfect nursery, with easy access to the master suite.

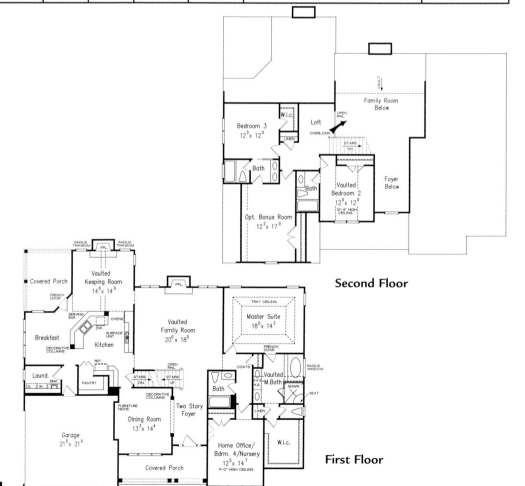

Second Floor

First Floor

Rear Elevation

© The Sater Design Collection, Inc.

Raphaello — TSPDS01-8037 — 1-866-525-9374

Total Living	First Floor	Second Floor	Bonus	Bed	Bath	Width	Depth	Foundation	Price Category
2913 sq ft	2250 sq ft	663 sq ft	N/A	3	3-1/2	72' 0"	68' 3"	Slab	G

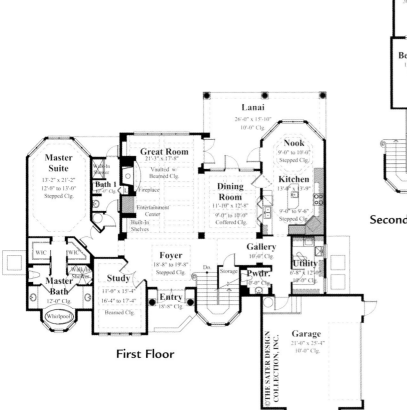

First Floor

Second Floor

Design Features

- Glass towers and Palladian windows lend an Italian Country theme.

- Living spaces focus to the rear through great walls of glass.

- Varying ceiling treatments define open arrangements of space.

- On the upper level, secondary bedrooms share a deck.

- A computer loft leads to a spacious bonus room.

Rear Elevation

Donald A. Gardner Architects, Inc.

Dakota — TSPDG01-789-D — 1-866-525-9374

Total Living	First Floor	Basement	Bonus	Bed	Bath	Width	Depth	Foundation	Price Category
2916 sq ft	2105 sq ft	811 sq ft	453 sq ft	4	3	61' 8"	67' 4"	Hillside Walkout	F

Design Features

- Stone, siding and cedar shake enhance the home's façade.

- A cathedral ceiling caps the great room with central fireplace and flanking built-in cabinets.

- With center island stovetop, the kitchen is spacious and efficiently designed.

- A fourth bedroom can be found on the lower level along with a sizable recreation room.

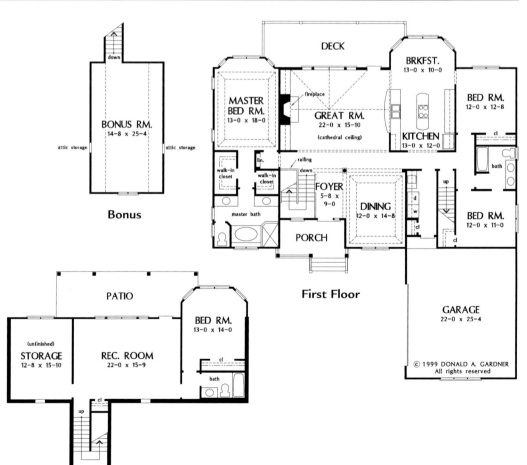

Rear Elevation

Nadine — TSPDS01-7047 — 1-866-525-9374

Total Living	First Floor	Second Floor	Bonus	Bed	Bath	Width	Depth	Foundation	Price Category
2923 sq ft	2215 sq ft	708 sq ft	420 sq ft	3	3	75' 4"	69' 10"	Basement/Crawl Space	G

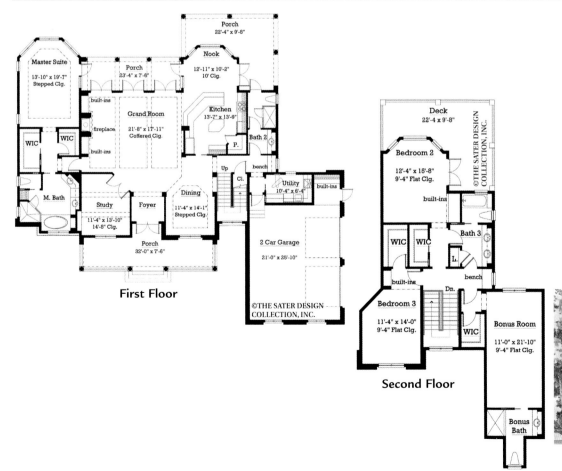

First Floor

Second Floor

Design Features

- Columns and coffered ceilings enlarge the interior spaces.

- The grand room features triple French doors and a fireplace.

- Built-ins and walk-in closets complete two upstairs bedrooms.

- An over-the-garage bonus room houses a bath.

- A bay-window breakfast nook and central island create an ideal kitchen.

Rear Elevation

Shelby — TSPFB01-866 — 1-866-525-9374

© 1995 Frank Betz Associates, Inc.

Total Living	First Floor	Second Floor	Bonus	Bed	Bath	Width	Depth	Foundation	Price Category
2940 sq ft	2044 sq ft	896 sq ft	197 sq ft	4	3-1/2	63' 0"	54' 0"	Basement, Crawl Space or Slab	H

Design Features

- The larger rooms and added amenities that many homeowners want are found in the *Shelby*.

- For an elegant effect when entering this home, a flared staircase was designed in the foyer.

- The kitchen features an island, walk-in pantry, double ovens and a serving bar adjoining the two-story family room.

- Decorative columns and arches grace a gallery hall.

Second Floor

First Floor

Rear Elevation

Elliot — TSPDG01-421 — 1-866-525-9374

Total Living	First Floor	Second Floor	Bonus	Bed	Bath	Width	Depth	Foundation	Price Category
2943 sq ft	1943 sq ft	1000 sq ft	403 sq ft	4	2-1/2	79' 10"	51' 8"	Crawl Space*	F

*Other options available. See page 347.

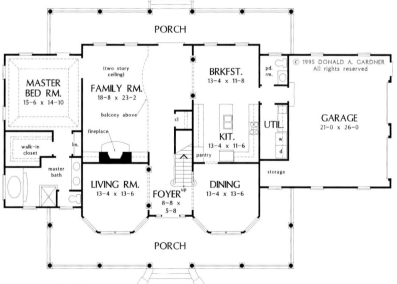

First Floor

Design Features

- Young families grow in style in this attractive and versatile home.

- The open great room and kitchen are enlarged by a cathedral ceiling.

- The master suite includes a walk-in closet and a bath with garden tub and dual vanity.

- The second-floor bonus space could be finished to allow two additional bedrooms and a bath.

Second Floor

Rear Elevation

2600 sq ft—3000 sq ft

Donald A. Gardner Architects, Inc.

Peekskill — TSPDG01-780-D — 1-866-525-9374

Total Living	First Floor	Second Floor	Basement	Bonus	Bed	Bath	Width	Depth	Foundation	Price Category
2953 sq ft	1662 sq ft	585 sq ft	706 sq ft	575 sq ft	4	3-1/2	81' 4"	68' 8"	Hillside Walkout	F

Design Features

- A stunning center dormer with arched window embellishes the exterior.

- The dormer's arched window floods light into the foyer with built-in niche.

- A generous back porch extends the great room, which features a vaulted ceiling.

- The master bedroom, which has a tray ceiling, enjoys back-porch access.

- Note the huge bonus room over the three-car garage.

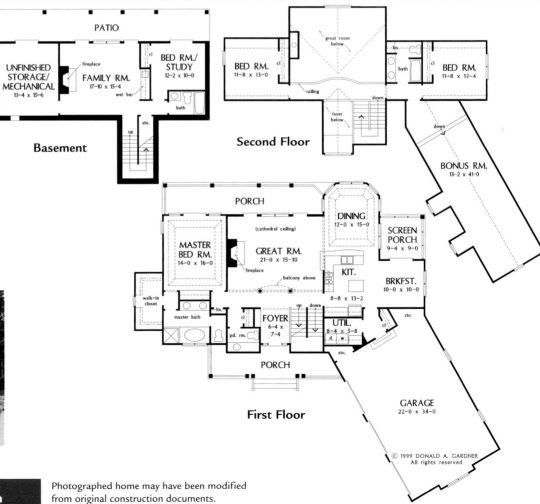

Rear Elevation

Photographed home may have been modified from original construction documents.

Donald A. Gardner Architects, Inc.

Santerini — TSPDG01-868 — 1-866-525-9374

Total Living	First Floor	Second Floor	Bonus	Bed	Bath	Width	Depth	Foundation	Price Category
2955 sq ft	2270 sq ft	685 sq ft	563 sq ft	3	2-1/2	75' 1"	53' 6"	Crawl Space*	F

*Other options available. See page 347.

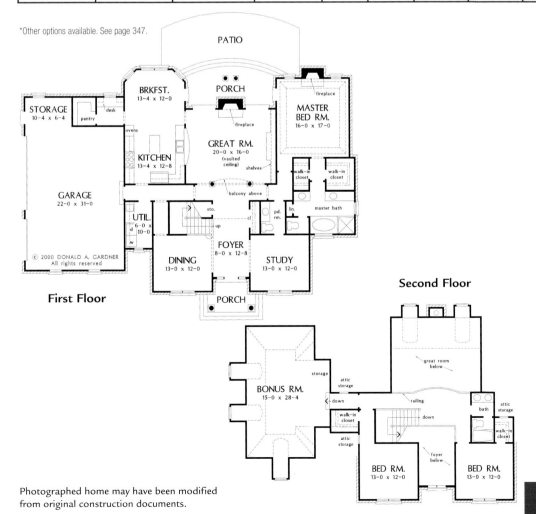

First Floor

Second Floor

Design Features

- This home has a commanding brick exterior with several arch-topped windows.

- An exciting second-floor balcony overlooks the vaulted foyer and great room.

- The adjacent kitchen features a sizable work island and nearby built-in desk and walk-in pantry.

- A short hall provides extra privacy for the first-floor master suite.

Rear Elevation

Donald A. Gardner Architects, Inc.

Vandenberg — TSPDG01-746-D — 1-866-525-9374

Total Living	First Floor	Basement	Bonus	Bed	Bath	Width	Depth	Foundation	Price Category
2956 sq ft	1810 sq ft	1146 sq ft	N/A	4	3	68' 4"	60' 10"	Hillside Walkout	F

Design Features

- This hillside home combines stucco, stone and cedar shake for exceptional Craftsman character.

- The breakfast and dining rooms enjoy screened-porch access.

- The master bedroom includes a tray ceiling, a lovely private bath and walk-in closet.

- A versatile bedroom/study and full bath are nearby.

- Downstairs are two more bedrooms, each with an adjacent covered patio.

Rear Elevation

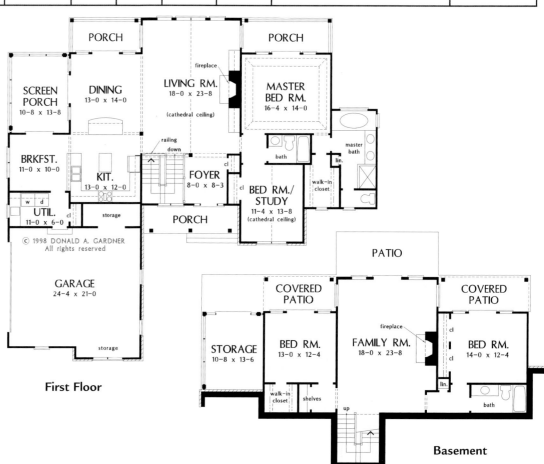

First Floor

Basement

© 2001 Frank Betz Associates, Inc.

Westbury — TSPFB01-3638 — 1-866-525-9374

Total Living	First Floor	Second Floor	Opt. Bonus	Bed	Bath	Width	Depth	Foundation	Price Category
2970 sq ft	2104 sq ft	866 sq ft	245 sq ft	4	3-1/2	65' 10"	53' 0"	Basement or Crawl Space	H

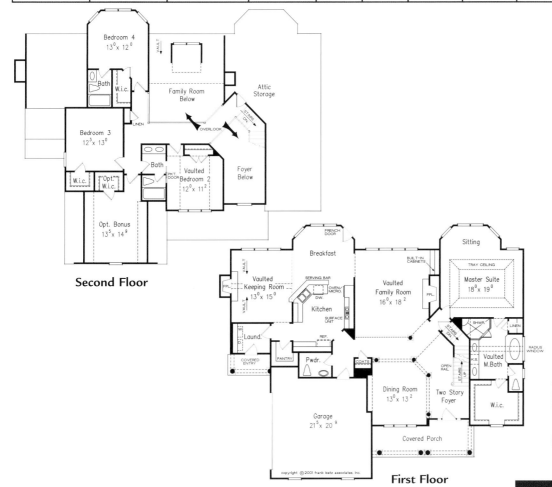

Second Floor

First Floor

Design Features

- A covered porch not only enhances the façade of the *Westbury*, but also adds to its character.

- A crackling fire will add warmth to any gathering in the cozy keeping room that adjoins the kitchen area.

- Decorative columns create gentle boundaries in the dining and family rooms.

- A bayed sitting area with views to the backyard has been incorporated into the master suite.

Rear Elevation

Morningside — TSPFB01-1257 — 1-866-525-9374

© 1999 Frank Betz Associates, Inc.

Total Living	First Floor	Second Floor	Bonus	Bed	Bath	Width	Depth	Foundation	Price Category
2970 sq ft	2074 sq ft	896 sq ft	209 sq ft	4	3-1/2	63' 0"	56' 0"	Basement, Crawl Space or Slab	I

Design Features

- From the *Southern Living® Design Collection*

- Architectural features include arched openings and columns framing the dining and family rooms.

- A bay with large windows brings light into the living room.

- The family room, kitchen and breakfast room are grouped together for convenience.

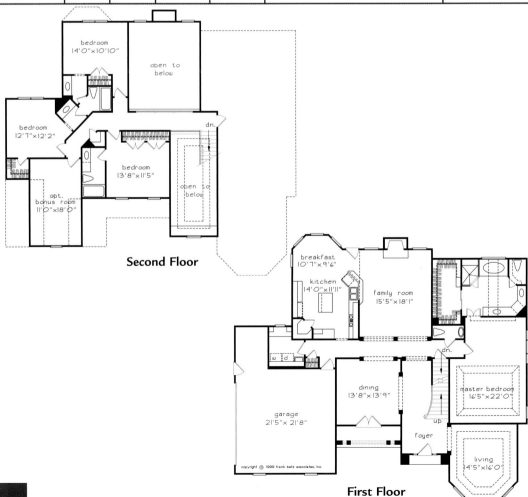

Second Floor

First Floor

Rear Elevation

Questling — TSPDG01-1004-D — 1-866-525-9374

Total Living	First Floor	Basement	Bonus	Bed	Bath	Width	Depth	Foundation	Price Category
2971 sq ft	1938 sq ft	1033 sq ft	N/A	4	4	77' 6"	42' 7"	Hillside Walkout	F

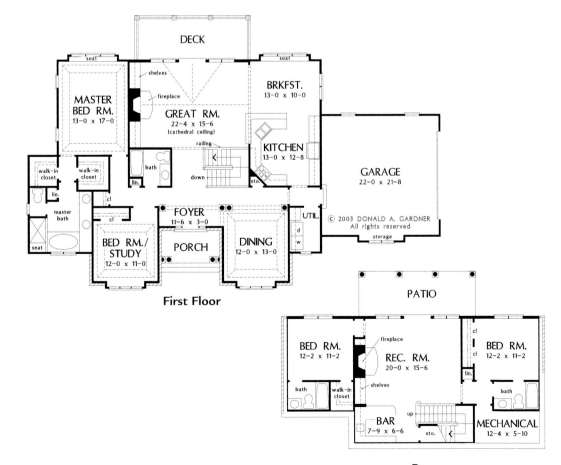

First Floor

Basement

Design Features

■ This hillside walkout features traditional style and a more defined floor plan.

■ Box-bay windows with metal roofs, decorative windows and transoms create curb appeal.

■ A U-shaped stair leads to the lower level with rec room, bar and secondary bedrooms with baths.

■ Decorative ceilings, window seats and built-in cabinetry add custom features.

Rear Elevation

MacLachlan — TSPDG01-825-D — 1-866-525-9374

© 1999 Donald A. Gardner, Inc.

Total Living	First Floor	Basement	Bonus	Bed	Bath	Width	Depth	Foundation	Price Category
2976 sq ft	1901 sq ft	1075 sq ft	N/A	4	3	64' 0"	62' 4"	Hillside Walkout	F

Design Features

- This stylish stone and stucco home features a partially finished walkout basement for sloping lots.

- The foyer is vaulted, receiving light from two clerestory dormer windows, and includes a niche.

- A recreation room is located on the basement level.

- Two bedrooms can be found on the first floor, while two more flank the downstairs recreation room.

First Floor

Basement

Rear Elevation

© 2002 Frank Betz Associates, Inc.

Montaigne — TSPFB01-3731 — 1-866-525-9374

Total Living	First Floor	Second Floor	Bonus	Bed	Bath	Width	Depth	Foundation	Price Category
2983 sq ft	1897 sq ft	1086 sq ft	N/A	4	3-1/2	62' 4"	50' 0"	Basement or Crawl Space	H

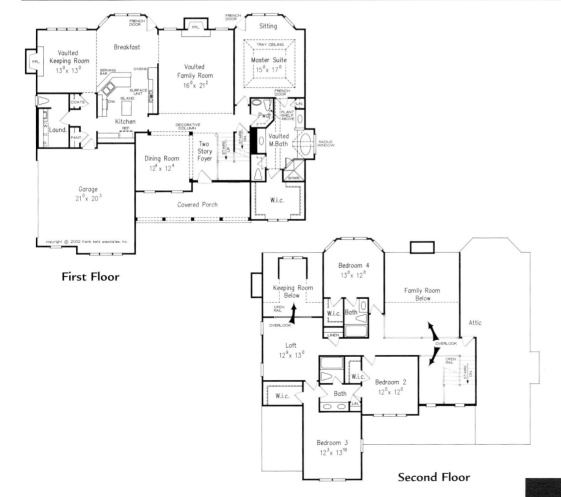

First Floor

Second Floor

Design Features

- Shutters, native stone and shingles create the right mix of rugged and refined elements.
- A gallery foyer defined by arches and columns grants vistas that extend from the entry to the back property.
- Twin windows flank a centered fireplace in the family room
- One wing of the home is dedicated to the owners' retreat, with a sitting area in the bedroom.

Rear Elevation

Montego Bay — TSPDS01-6800 — 1-866-525-9374

© The Sater Design Collection, Inc.

Total Living	First Floor	Second Floor	Bonus	Bed	Bath	Width	Depth	Foundation	Price Category
2988 sq ft	2096 sq ft	892 sq ft	N/A	3	3-1/2	58' 0"	54' 0"	Island Basement	G

Design Features

- An arched foyer leads into a great room with coffered ceiling and fireplace.

- The large kitchen has a center island with cooktop and a bayed dining nook.

- The main-floor master suite has a pampering bath and a walk-in shower.

- Upstairs, each of the two bedrooms has a walk-in closet.

- A loft provides bonus space overlooking the great room below.

Rear Elevation

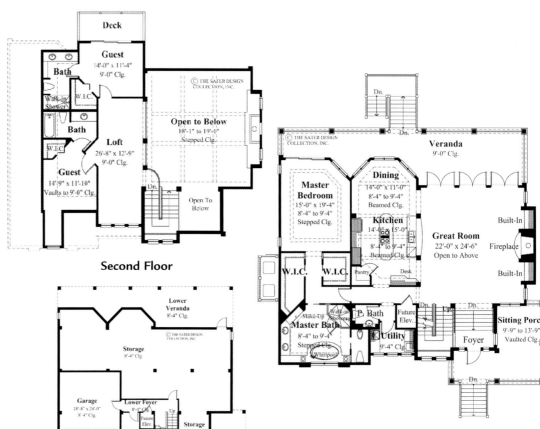

Second Floor

Lower Level

First Floor

Heathridge — TSPDG01-763-D — 1-866-525-9374

Total Living	First Floor	Basement	Bonus	Bed	Bath	Width	Depth	Foundation	Price Category
2998 sq ft	2068 sq ft	930 sq ft	N/A	3	3-1/2	72' 4"	66' 0"	Hillside Walkout	F

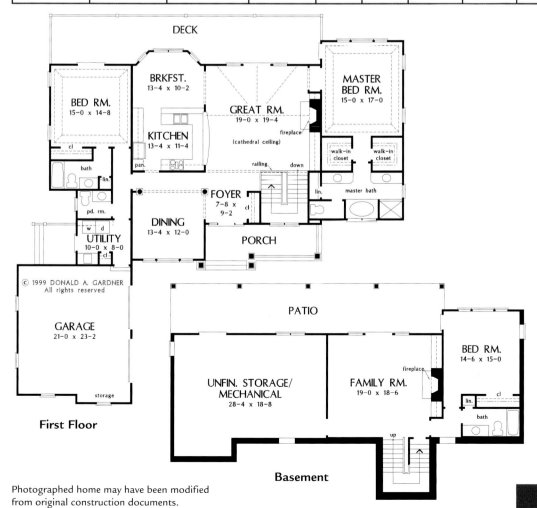

First Floor

Basement

Design Features

- This Craftsman-style home takes advantage of views with its deck, patio and rear windows.

- The great room features a cathedral ceiling and a fireplace with built-in cabinets and shelves.

- The efficient kitchen serves the great and dining rooms and breakfast area with ease.

- Downstairs is a spacious family room with fireplace, a third bedroom and full bath.

Rear Elevation

Photographed home may have been modified from original construction documents.

Westhampton — TSPFB01-3767 — 1-866-525-9374

© 2002 Frank Betz Associates, Inc.

Total Living	First Floor	Second Floor	Bonus	Bed	Bath	Width	Depth	Foundation	Price Category
3012 sq ft	1974 sq ft	1038 sq ft	N/A	4	3-1/2	72' 0"	57' 0"	Basement or Crawl Space	I

Design Features

- A cozy keeping room is situated adjacent to the kitchen, creating uncommon angles rarely found in stock home plans.

- The master suite is secluded on the main level and features a bayed sitting area—a great spot to unwind.

- Three additional bedrooms have private access to bathing areas and walk-in closets.

- A built-in desk is incorporated to the upper floor, giving children a perfect place to do homework or crafts.

First Floor

Second Floor

Rear Elevation

© 2001 Frank Betz Associates, Inc.

Ashton — TSPFB01-3598 — 1-866-525-9374

Total Living	First Floor	Second Floor	Bonus	Bed	Bath	Width	Depth	Foundation	Price Category
3024 sq ft	2146 sq ft	878 sq ft	341 sq ft	4	3-1/2	61' 0"	60' 4"	Basement, Crawl Space or Slab	I

First Floor

Second Floor

Design Features

- The brick and siding exterior gives the ageless appeal that every neighborhood welcomes.

- Inside, a keeping room connects to the kitchen area, providing the perfect place for relaxing family time.

- The master suite is truly luxurious, with a bayed sitting area, his-and-her closets and a decorative art niche.

- An optional bonus room is available on the upper floor that can be used as the homeowner wishes.

Rear Elevation

Feretti — TSPDS01-6786 — 1-866-525-9374

© The Sater Design Collection, Inc.

Total Living	First Floor	Second Floor	Bonus	Bed	Bath	Width	Depth	Foundation	Price Category
3031 sq ft	2254 sq ft	777 sq ft	N/A	4	5	52' 2"	95' 8"	Slab	H

Design Features

- Arched iron gates in the portico offer entry to this courtyard home.

- Glass-walled rooms are situated around a fountain pool.

- The master suite includes a private garden.

- The kitchen has a large island and corner pantry.

- A second floor features guest suites, loft, balcony and deck.

Interior Courtyard

First Floor

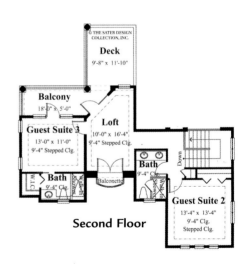

Second Floor

Forrester — TSPDG01-306 — 1-866-525-9374

Total Living	First Floor	Second Floor	Bonus	Bed	Bath	Width	Depth	Foundation	Price Category
3037 sq ft	2316 sq ft	721 sq ft	545 sq ft	4	3-1/2	95' 4"	54' 10"	Crawl Space*	G

*Other options available. See page 347.

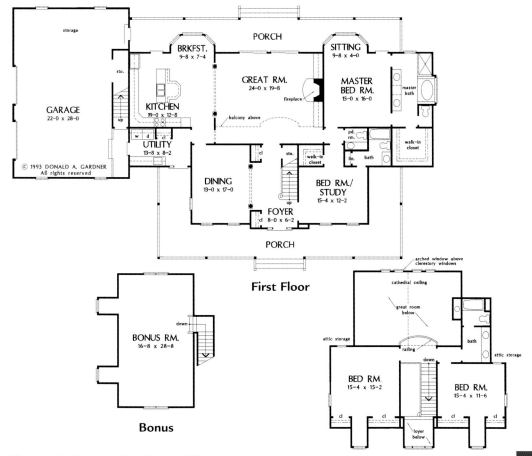

First Floor

Bonus

Second Floor

Design Features

- This large, graceful farmhouse with wraparound porch features an island kitchen.

- Light splashes through four clerestory windows above the vaulted great room and foyer.

- Columns punctuate the open interior, while nine-foot ceilings expand the entire first floor.

- A second first-floor bedroom/study with bath can serve many purposes.

- The master bath has a whirlpool tub with separate shower, double vanity and large walk-in closet.

Rear Elevation

Jessica — TSPFB01-772 — 1-866-525-9374

© 1994 Frank Betz Associates, Inc.

Total Living	First Floor	Second Floor	Bonus	Bed	Bath	Width	Depth	Foundation	Price Category
3039 sq ft	1488 sq ft	1551 sq ft	N/A	5	4	55' 0"	57' 4"	Basement, Crawl Space or Slab	I

Design Features

- Traditional style defines this home, both outside and inside.

- The formal living and dining rooms flank the two-story foyer.

- A bedroom located on the main level can be used for guests or can be converted to a study or home office.

- The master suite boasts a bayed sitting area.

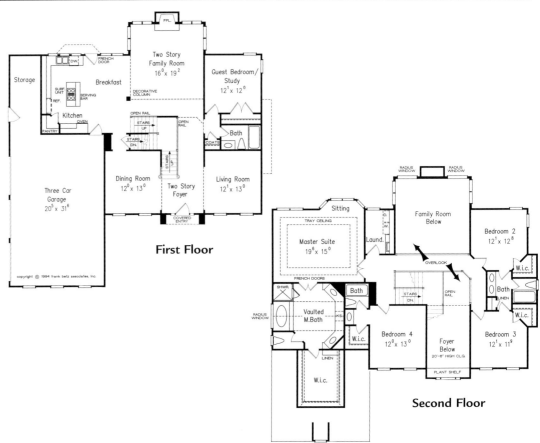

First Floor

Second Floor

Rear Elevation

© The Sater Design Collection, Inc.

Mimosa — TSPDS01-6861 — 1-866-525-9374

Total Living	First Floor	Lower Level	Bonus	Bed	Bath	Width	Depth	Foundation	Price Category
3074 sq ft	3074 sq ft	N/A	N/A	3	3-1/2	77' 0"	66' 8"	Island Basement	H

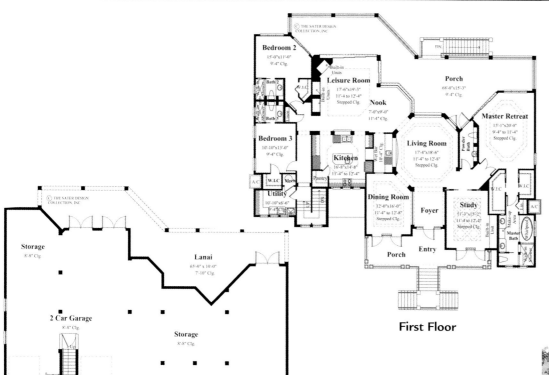

First Floor

Lower Level

Design Features

- The enchanting portico leads to a mid-level foyer.
- A wide-open leisure room hosts a corner fireplace.
- Lower-level space offers storage and a walkout lanai.
- Glass walls and sliding doors open the leisure and living rooms to the outdoors.
- Two guest suites are secluded to one side.

Rear Elevation

Vernay — TSPDS01-7063 — 1-866-525-9374

© The Sater Design Collection, Inc.

Total Living	First Floor	Second Floor	Bonus	Bed	Bath	Width	Depth	Foundation	Price Category
3082 sq ft	2138 sq ft	944 sq ft	427 sq ft	3	3-1/2	77' 2"	64' 0"	Basement/Crawl Space	H

Design Features

- A wraparound porch is a great place for rocking chairs.

- A gourmet kitchen is packed with amenities.

- A host of French doors open the leisure room to inspiring views.

- A lavish master bath includes separate vanities and garden tub.

- The upper level boasts two secondary suites and a bonus room.

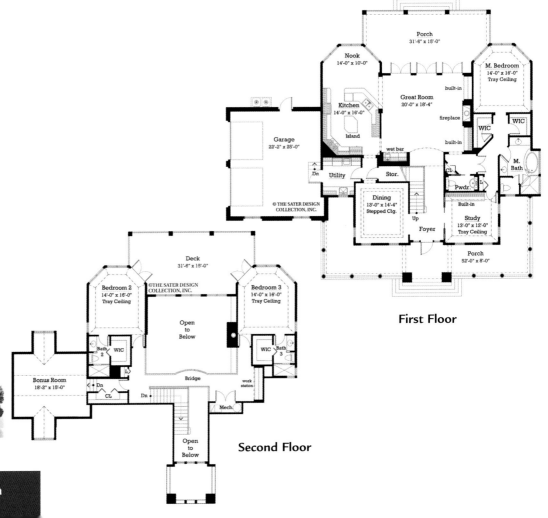

Rear Elevation

© 1996 Frank Betz Associates, Inc.

Northampton — TSPFB01-1005 — 1-866-525-9374

Total Living	First Floor	Second Floor	Bonus	Bed	Bath	Width	Depth	Foundation	Price Category
3083 sq ft	2429 sq ft	654 sq ft	420 sq ft	3	3-1/2	63' 6"	71' 4"	Basement, Crawl Space or Slab	I

First Floor

Second Floor

Design Features

■ This home's façade features a turret and a terrace area, giving it phenomenal curb appeal.

■ The dining room is defined by architectural columns.

■ The turreted room makes a beautiful study, but can be easily altered to make a stunning sitting area for the master bedroom.

■ The kitchen overlooks a vaulted keeping room.

Rear Elevation

Sunset Beach — TSPDS01-6848 — 1-866-525-9374

© The Sater Design Collection, Inc.

Total Living	First Floor	Second Floor	Bonus	Bed	Bath	Width	Depth	Foundation	Price Category
3096 sq ft	2083 sq ft	1013 sq ft	N/A	4	3-1/2	74' 0"	88' 0"	Crawl Space	H

Design Features

- A gazebo-style front porch accents this country villa.

- The entry leads to the grand foyer and sweeping radius staircase.

- The main level master suite boasts a private lanai.

- A study has a window seat and built-in cabinetry.

- A bayed breakfast nook and butler's pantry complete the kitchen.

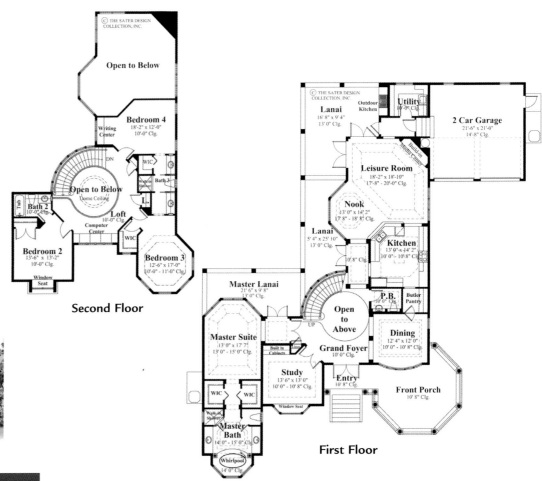

Second Floor

First Floor

Rear Elevation

© The Sater Design Collection, Inc.

Alexandre — TSPDS01-6849 — 1-866-525-9374

Total Living	First Floor	Second Floor	Bonus	Bed	Bath	Width	Depth	Foundation	Price Category
3096 sq ft	2083 sq ft	1013 sq ft	N/A	4	3-1/2	74' 0"	88' 0"	Crawl Space	H

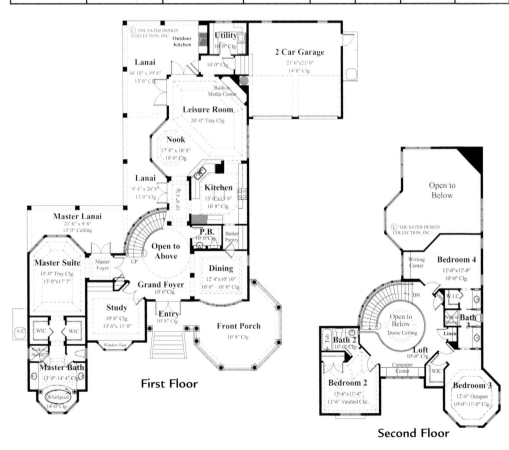

First Floor

Second Floor

Design Features

- A radius staircase leads up to secondary sleeping quarters.

- Tray, stepped or vaulted ceilings add spaciousness and elegance.

- The first-floor master suite boasts a private lanai.

- A study features a window seat and built-in cabinetry.

- The upstairs has three bedrooms, a loft and computer center.

Rear Elevation

Chelsea — TSPDS01-7057 — 1-866-525-9374

© The Sater Design Collection, Inc.

Total Living	First Floor	Second Floor	Bonus	Bed	Bath	Width	Depth	Foundation	Price Category
3096 sq ft	2083 sq ft	1013 sq ft	N/A	4	3-1/2	74' 0"	88' 6"	Crawl Space	H

Design Features

- A circular front porch creates a friendly façade.

- The first-floor master suite includes a private porch.

- The study boasts a cozy window seat and built-ins.

- The relaxing leisure room sports doors to the wraparound porch.

- The upstairs houses three bedrooms, two baths, a loft and computer center.

Rear Elevation

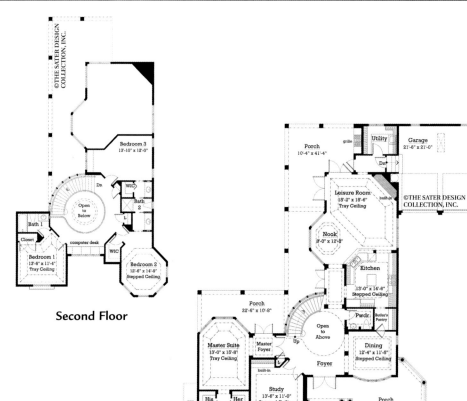

Second Floor

First Floor

© 1998 Frank Betz Associates, Inc.

Hopewell — TSPFB01-1230 — 1-866-525-9374

Total Living	First Floor	Second Floor	Bonus	Bed	Bath	Width	Depth	Foundation	Price Category
3097 sq ft	1510 sq ft	1587 sq ft	N/A	5	4	55' 0"	54' 10"	Basement, Crawl Space or Slab	I

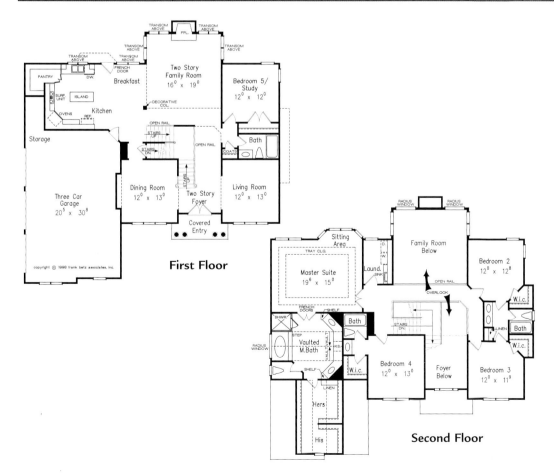

First Floor

Second Floor

Design Features

- This plan combines spacious living spaces with traditional charm.

- The two-story great room and kitchen are divided only by a decorative column for open flow of traffic.

- A generous master suite includes a private sitting area and his-and-her closets.

- The laundry room is situated upstairs for convenient access to the bedrooms.

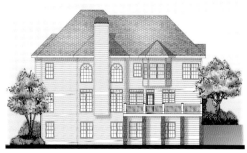

Rear Elevation

Gilchrist — TSPDG01-734-D — 1-866-525-9374

© 1998 Donald A. Gardner, Inc.

Total Living	First Floor	Basement	Bonus	Bed	Bath	Width	Depth	Foundation	Price Category
3132 sq ft	2094 sq ft	1038 sq ft	494 sq ft	4	3-1/2	62' 3"	76' 7"	Hillside Walkout	G

Design Features

- A partial basement foundation makes this home perfect for hillside lots.

- The great room features a cathedral ceiling and fireplace with flanking built-in shelves.

- A convenient pass-thru in the step-saving kitchen keeps everyone connected.

- A generous recreation room and guest suite comprise the lower level.

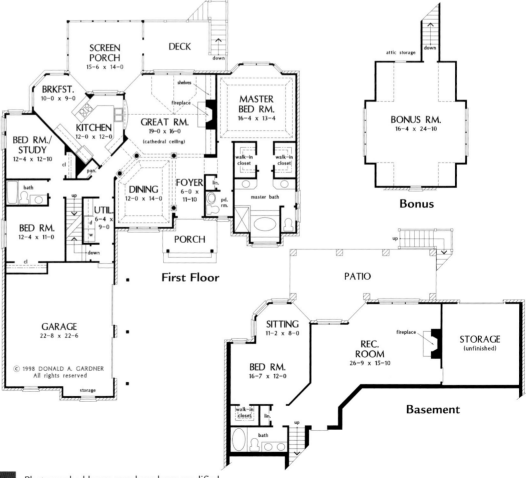

Rear Elevation

Photographed home may have been modified from original construction documents.

Donald A. Gardner Architects, Inc.

Rochester — TSPDG01-1085 — 1-866-525-9374

Total Living	First Floor	Second Floor	Bonus	Bed	Bath	Width	Depth	Foundation	Price Category
3134 sq ft	2172 sq ft	962 sq ft	337 sq ft	4	3	50' 0"	67' 6"	Crawl Space*	G

*Other options available. See page 347.

Design Features

- French doors, windows and an open floor plan allow this home to be airy and bright.

- Strategically placed columns help define rooms without enclosing space.

- A built-in desk provides a computer hub in a centralized location.

- A loft, which overlooks the great room, can be enclosed to form a bedroom.

- The bonus room can be a guest suite, home theatre or play room.

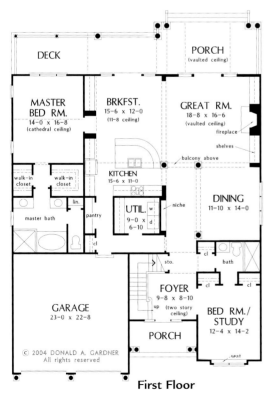

First Floor

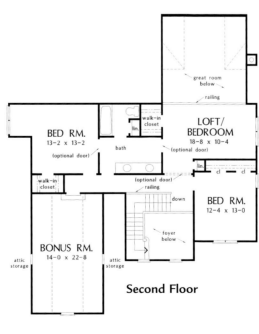

Second Floor

Rear Elevation

Sorrell Grove — TSPDS01-7020 — 1-866-525-9374

© The Sater Design Collection, Inc.

Total Living	First Floor	Second Floor	Bonus	Bed	Bath	Width	Depth	Foundation	Price Category
3136 sq ft	1673 sq ft	1463 sq ft	N/A	3	2-1/2	60' 10"	62' 0"	Opt. Basement/Crawl Space	H

Design Features

- A stone chimney and wraparound porch adorn this French Country home.

- The living and dining areas have two-story ceilings.

- Built-in cabinetry frames a fireplace in the leisure room/nook.

- A space-saving kitchen provides a food-preparation island.

- The master suite has a morning kitchen and sitting room.

Rear Elevation

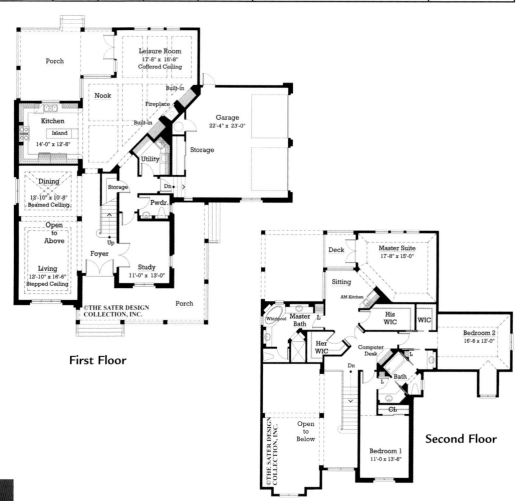

First Floor

Second Floor

Hartford — TSPDG01-1048 — 1-866-525-9374

Total Living	First Floor	Second Floor	Bonus	Bed	Bath	Width	Depth	Foundation	Price Category
3155 sq ft	2395 sq ft	760 sq ft	454 sq ft	4	4	63' 9"	63' 11"	Crawl Space*	G

*Other options available. See page 347.

First Floor

PORCH

BRKFST.
13-0 x 11-8

BED RM./
STUDY
12-0 x 13-0

KITCHEN
14-8 x 14-8

GREAT RM.
20-0 x 18-0
(two story ceiling)

fireplace

MASTER
BED RM.
14-0 x 17-0

balcony above   shelves

UTIL.
8-4 x
9-2

bath   cl

pantry

FOYER
6-8 x
11-8
(two story
ceiling)

DINING
12-0 x 14-0

balcony
above

walk-in
closet

shelves
walk-in
closet

sto.

up

GARAGE
24-0 x 24-0

PORCH

cl

lin.

master bath
(11' ceiling)

seat

storage

First Floor

Second Floor

BED RM.
13-0 x 13-2
(vaulted ceiling)

cl   cl

bath

great room
below

railing

attic
storage

bath

cl   cl

down

attic
storage

BED RM.
12-8 x 12-4
(vaulted ceiling)

foyer
below

shelf

attic
storage

BONUS RM.
16-8 x 24-0

attic
storage

Second Floor

Design Features

- A metal roof caps the front porch, and arched transoms add architectural interest.

- Inside, a curved balcony and columns divide the foyer from the great room.

- Built-in cabinetry, a walk-in pantry and garage-entry closet add convenience.

- A central island and service counter complete the kitchen.

Rear Elevation

Arbordale — TSPDG01-452 — 1-866-525-9374

Total Living	First Floor	Second Floor	Bonus	Bed	Bath	Width	Depth	Foundation	Price Category
3163 sq ft	2086 sq ft	1077 sq ft	403 sq ft	4	3-1/2	81' 10"	51' 8"	Crawl Space*	G

*Other options available. See page 347.

Design Features

- The master suite is quietly tucked away downstairs with no bedrooms directly above.

- The cook of the family will love the spacious U-shaped kitchen.

- The bonus room is easily accessible from the back stairs or second floor.

- Storage space abounds with walk-ins, hall shelves and a linen closet upstairs.

- A curved balcony borders a versatile loft/study that overlooks the two-story great room.

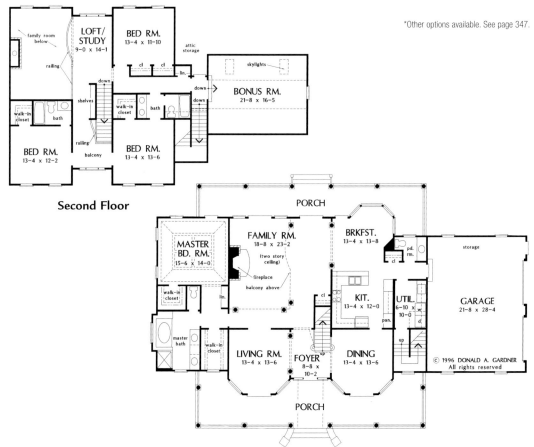

Second Floor

First Floor

Rear Elevation

Photographed home may have been modified from original construction documents.

3000 sq ft—4000 sq ft

Donald A. Gardner Architects, Inc.

Ballenger — TSPDG01-1007 — 1-866-525-9374

Total Living	First Floor	Second Floor	Bonus	Bed	Bath	Width	Depth	Foundation	Price Category
3152 sq ft	2194 sq ft	958 sq ft	462 sq ft	4	3-1/2	82' 7"	51' 1"	Crawl Space*	G

*Other options available. See page 347.

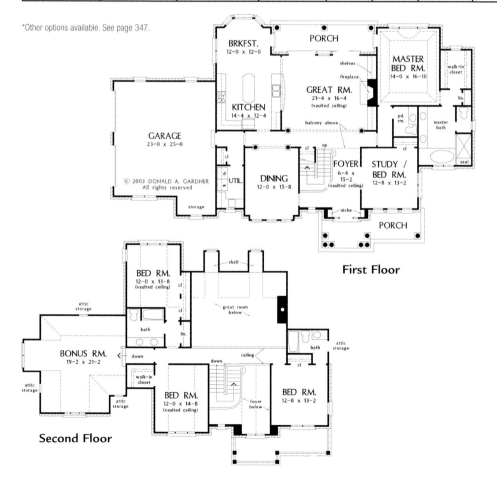

First Floor

Second Floor

Design Features

- Columns and an elegant balustrade create a regal entryway.

- A box-bay window is crowned with a metal roof and accentuated by a Palladian window.

- A hip roof, decorative brickwork and multiple gables create striking curb appeal.

- The foyer features a two-story cathedral ceiling and grand staircase.

Rear Elevation

Frank Betz Associates, Inc.

Muirfield — TSPFB01-3769 — 1-866-525-9374

© 2002 Frank Betz Associates, Inc.

Total Living	First Floor	Second Floor	Opt. Bonus	Bed	Bath	Width	Depth	Foundation	Price Category
3189 sq ft	2153 sq ft	1036 sq ft	N/A	4	3-1/2	72' 0"	60' 6"	Basement or Crawl Space	I

Design Features

- The kitchen, breakfast area and vaulted keeping room come together to create a unified space for casual family time.

- His-and-her closets and a lavish master bath create a "suite."

- The second floor bedrooms all feature walk-in closets and bath access.

- A built-in desk in the second-floor loft creates the perfect homework station for kids.

Rear Elevation

Second Floor

First Floor

© The Sater Design Collection, Inc.

Edgewood Trail — TSPDS01-6667 — 1-866-525-9374

Total Living	First Floor	Second Floor	Bonus	Bed	Bath	Width	Depth	Foundation	Price Category
3190 sq ft	2241 sq ft	949 sq ft	N/A	4	2-1/2	69' 8"	61' 10"	Basement/Slab	H

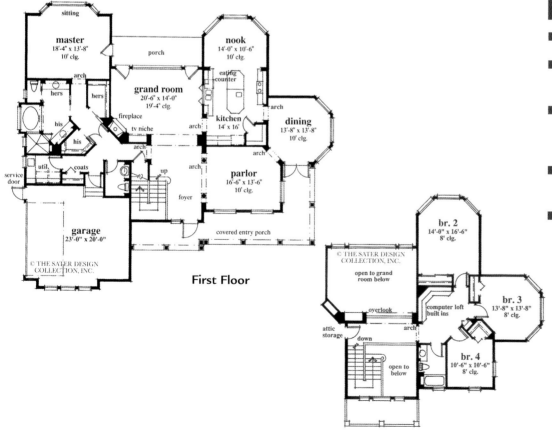

First Floor

Second Floor

Design Features

- Formal rooms cluster near the foyer.

- A wraparound porch is perfect for stargazing.

- Dominated by a bay window, the master suite offers a spacious sitting area.

- An upper-level overlook leads to a computer loft.

- The kitchen has a food prep island and snack counter.

Rear View

Photographed home may have been modified from original construction documents.

Forrest Hills — TSPFB01-3802 — 1-866-525-9374

© 2003 Frank Betz Associates, Inc.

Total Living	First Floor	Second Floor	Bonus	Bed	Bath	Width	Depth	Foundation	Price Category
3194 sq ft	2018 sq ft	1176 sq ft	N/A	4	3-1/2	66' 0"	61' 0"	Basement or Crawl Space	I

Design Features

- Multiple gathering spaces inside give families plenty of choices on where to spend their time.

- A keeping room just off the kitchen is a great place for reconnecting at the end of a busy day.

- The family room—complete with a fireplace and built-in cabinetry—makes a cozy retreat for evenings in front of the fire.

- An entertainment room upstairs gives children a designated place to spend their recreational time.

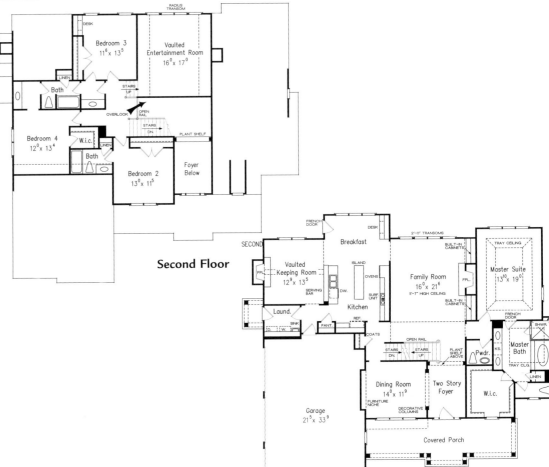

Second Floor

First Floor

Rear Elevation

Fitzgerald — TSPDG01-1018 — 1-866-525-9374

Total Living	First Floor	Second Floor	Bonus	Bed	Bath	Width	Depth	Foundation	Price Category
3196 sq ft	2215 sq ft	981 sq ft	402 sq ft	5	4	71' 11"	55' 10"	Crawl Space*	G

*Other options available. See page 347.

Design Features

- The floor plan features well-defined rooms, yet still remains open and family-friendly.

- Two-story ceilings highlight both the foyer and great room.

- A butler's pantry connects the kitchen to the dining room.

- Special elements include a bay window in the breakfast nook and a shower seat in the master bath.

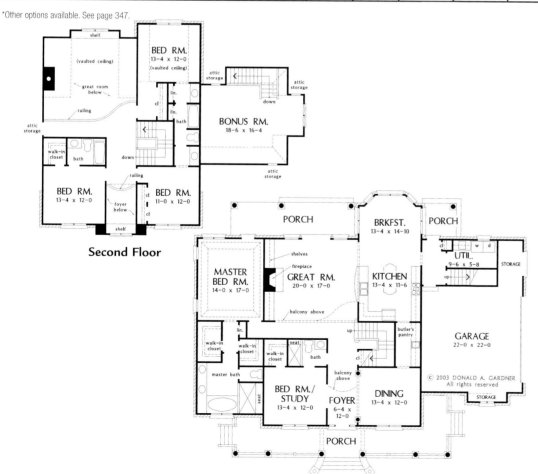

Second Floor

First Floor

Rear Elevation

Dogwood Ridge — TSPAL01-5005 — 1-866-525-9374

Total Living	First Floor	Basement	Bonus	Bed	Bath	Width	Depth	Foundation	Price Category
3201 sq ft	2090 sq ft	1111 sq ft	N/A	3	3-1/2	71' 1"	78' 6"	Hillside Walkout	O

Design Features

- This hillside walkout boasts a rear wall of glasswork on both floors.

- A patio, and rear and screened porches take living outdoors.

- The open floor plan distinguishes rooms by ceiling treatments and columns.

- For added convenience and future planning, an elevator makes living easier.

- Along with the master suite, the secondary bedrooms take advantage of the natural scenery.

Rear Elevation

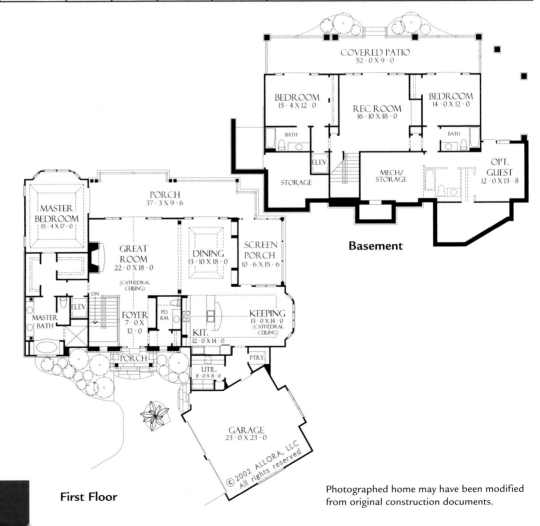

Basement

First Floor

Photographed home may have been modified from original construction documents.

Touchstone — TSPDG01-1099-D — 1-866-525-9374

Total Living	First Floor	Basement	Bonus	Bed	Bath	Width	Depth	Foundation	Price Category
3213 sq ft	2143 sq ft	1070 sq ft	N/A	4	4	73' 9"	60' 0"	Hillside Walkout	G

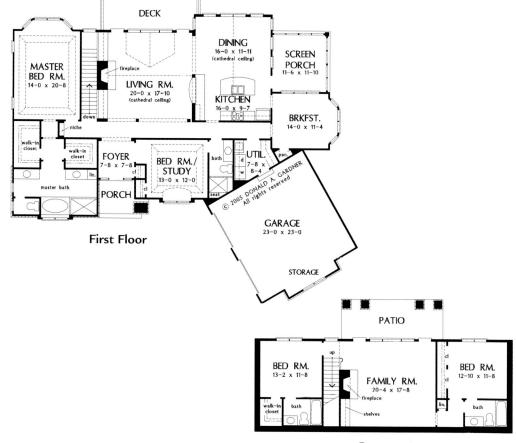

First Floor

Basement

Design Features

- Sweeping gables and a stone façade add drama to the exterior.
- The lofty living room overlooks a rear deck and gracefully flows into the dining room and kitchen.
- The breakfast room includes a walk-in pantry and looks out onto a screened porch.
- The lower-level family room opens onto a covered patio.

Rear Elevation

Donald A. Gardner Architects, Inc.

Tuscany — TSPDG01-877 — 1-866-525-9374

© 2000 Donald A. Gardner, Inc.

Total Living	First Floor	Second Floor	Bonus	Bed	Bath	Width	Depth	Foundation	Price Category
3219 sq ft	2477 sq ft	742 sq ft	419 sq ft	4	4	99' 10"	66' 2"	Crawl Space*	G

Design Features

- Built-ins in the great room and counter space in the utility/mud room add convenience.

- The family-efficient floor plan can be witnessed in the kitchen's handy pass-thru.

- The kitchen has rear-porch access for outdoor entertaining.

- Ceiling treatments highlight the master bedroom, bedroom/study, breakfast area and loft/study.

- The bonus room can be used as a home theatre, playroom or gym.

*Other options available. See page 347.

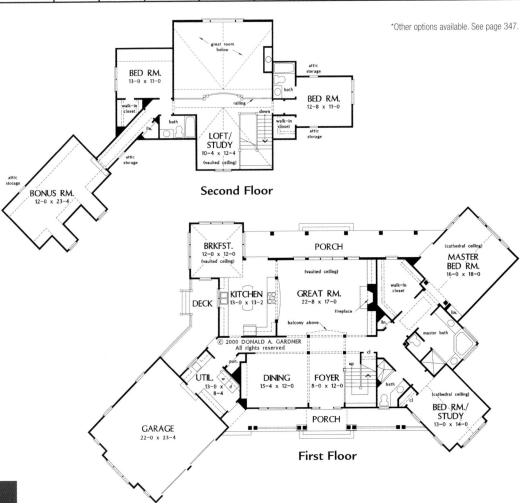

Second Floor

First Floor

Rear Elevation

© 2004 Frank Betz Associates, Inc.

McGinnis Ferry — TSPFB01-3879 — 1-866-525-9374

Total Living	First Floor	Second Floor	Bonus	Bed	Bath	Width	Depth	Foundation	Price Category
3254 sq ft	2224 sq ft	1030 sq ft	N/A	4	3	65' 4"	53' 8"	Basement or Crawl Space	H

First Floor

Second Floor

Design Features

- A vaulted keeping room connects to the kitchen and breakfast areas.

- Transom and radius windows illuminate the main floor with plenty of natural light.

- A mudroom and coat closet are placed just off the garage.

- A teen suite has been designed into the second floor near the bedrooms.

Rear Elevation

Carmichael — TSPFB01-770 — 1-866-525-9374

© 1994 Frank Betz Associates, Inc.

Total Living	First Floor	Second Floor	Bonus	Bed	Bath	Width	Depth	Foundation	Price Category
3262 sq ft	1418 sq ft	1844 sq ft	N/A	4	3-1/2	63' 0"	41' 0"	Basement, Crawl Space or Slab	I

Design Features

- A two-story breakfast area adjoins the sizeable kitchen.

- Entertaining is easy and convenient with a butler's pantry connecting the dining room to the kitchen.

- Decorative shelves are creatively tucked between the living room and family room.

- The master retreat is complete with private sitting area and two-sided fireplace.

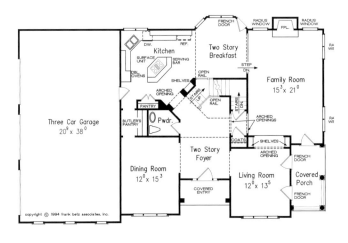

First Floor

Rear Elevation

Second Floor

3000 sq ft—4000 sq ft

Donald A. Gardner Architects, Inc.

© 1999 Donald A. Gardner, Inc.

Clairemont — TSPDG01-791-D — 1-866-525-9374

Total Living	First Floor	Basement	Bonus	Bed	Bath	Width	Depth	Foundation	Price Category
3272 sq ft	2122 sq ft	1150 sq ft	N/A	4	3	83' 0"	74' 4"	Hillside Walkout	G

First Floor

Basement

© 1999 DONALD A. GARDNER
All rights reserved

Design Features

■ A Craftsman combination of cedar shake and wood siding lends warmth to this custom home.

■ A stunning cathedral ceiling spans the open great room and spacious, island kitchen.

■ Two rear decks and a screened porch augment the home's ample living space.

■ Note the large utility room, basement-floor recreation room and plentiful storage.

Rear Elevation

Brookmere — TSPFB01-3452 — 1-866-525-9374

© 1999 Frank Betz Associates, Inc.

Total Living	First Floor	Second Floor	Bonus	Bed	Bath	Width	Depth	Foundation	Price Category
3274 sq ft	2332 sq ft	942 sq ft	305 sq ft	4	3-1/2	60' 0"	64' 4"	Basement, Crawl Space or Slab	H

Design Features

- The kitchen overlooks a large break-fast area and cozy keeping room.

- This area is gently divided from the grand room, creating two separate spaces to entertain guests.

- Decorative columns and a furniture niche give the dining room that extra dose of character.

- A sitting area surrounded by a wall of windows overlooks the backyard from the master suite.

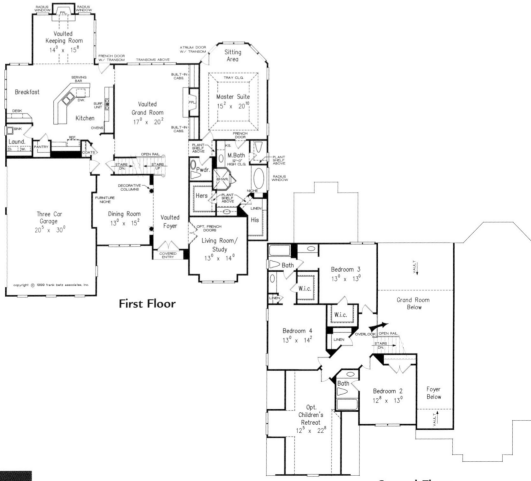

First Floor

Second Floor

Rear Elevation

Frank Betz Associates, Inc.

© 2003 Frank Betz Associates, Inc.

Graves Spring — TSPFB01-3852 — 1-866-525-9374

Total Living	First Floor	Second Floor	Opt. Bonus	Bed	Bath	Width	Depth	Foundation	Price Category
3280 sq ft	2096 sq ft	1184 sq ft	187 sq ft	5	4	67' 0"	54' 6"	Basement or Crawl Space	I

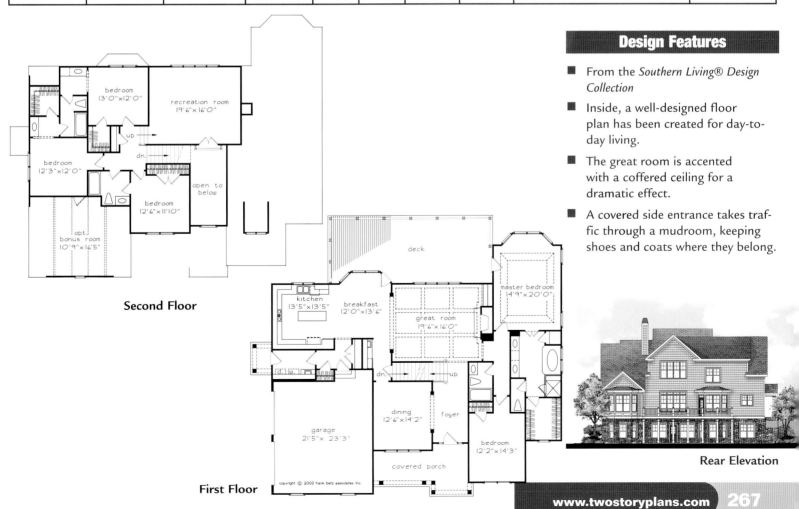

Second Floor

First Floor

Design Features

- From the *Southern Living® Design Collection*

- Inside, a well-designed floor plan has been created for day-to-day living.

- The great room is accented with a coffered ceiling for a dramatic effect.

- A covered side entrance takes traffic through a mudroom, keeping shoes and coats where they belong.

Rear Elevation

Berkshire — TSPDG01-748-D — 1-866-525-9374

Total Living	First Floor	Basement	Bonus	Bed	Bath	Width	Depth	Foundation	Price Category
3281 sq ft	2065 sq ft	1216 sq ft	N/A	4	3-1/2	82' 2"	43' 6"	Hillside Walkout	G

Design Features

- Stone, siding and multiple gables combine beautifully on this walkout basement home.
- Capped by a cathedral ceiling, the great room features a fireplace and built-in shelves.
- Twin walk-in closets and a private bath infuse the master suite with luxury.
- The nearby powder room offers an optional full bath, allowing the study to double as a bedroom.
- Downstairs, a large media/recreation room with wet bar and fireplace separates two bedrooms.

Rear Elevation

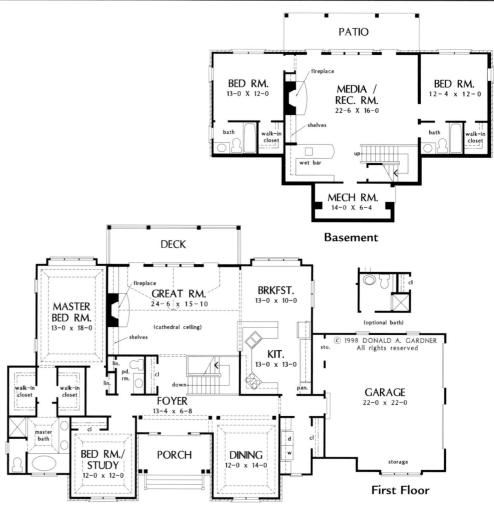

Basement

First Floor

© 1999 Frank Betz Associates, Inc.

Fenway — TSPFB01-3444 — 1-866-525-9374

Total Living	First Floor	Second Floor	Opt. Bonus	Bed	Bath	Width	Depth	Foundation	Price Category
3285 sq ft	2293 sq ft	992 sq ft	131 sq ft	4	3-1/2	71' 0"	62' 0"	Basement, Crawl Space or Slab	I

First Floor

Second Floor

Design Features

- A fire-lit keeping room borders the kitchen and breakfast area.

- Sunshine pours in through transom windows in the grand room.

- A buffet table or china cabinet will fit perfectly into the furniture niche in the dining room.

- Private lounging space has been incorporated into the master suite.

Rear Elevation

Les Anges — TSPDS01-6825 — 1-866-525-9374

© The Sater Design Collection, Inc.

Total Living	First Floor	Second Floor	Lower Level	Bed	Bath	Width	Depth	Foundation	Price Category
3285 sq ft	2146 sq ft	952 sq ft	187 sq ft	3	3-1/2	56' 0"	64' 0"	Island Basement	H

Design Features

- Built-in cabinetry and a massive fireplace anchor the central living space.

- The morning nook provides a bay window and fireplace view.

- A gourmet island kitchen serves the formal dining room.

- The secluded master wing enjoys a bumped-out window and twin walk-in closets.

- The upper level boasts a catwalk that connects two secondary suites.

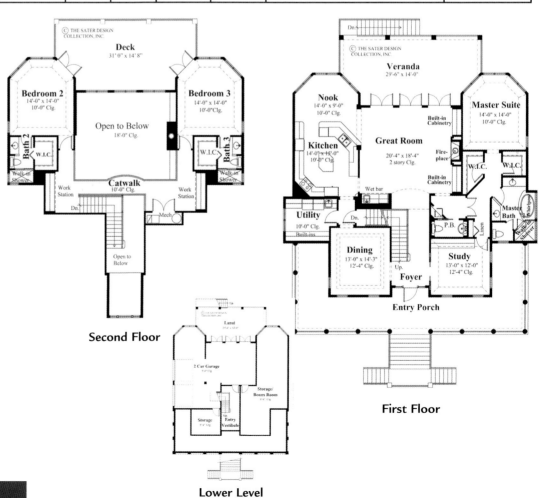

Second Floor

First Floor

Lower Level

Rear Elevation

© 2002 Frank Betz Associates, Inc.

Benedict — TSPFB01-3768 — 1-866-525-9374

Total Living	First Floor	Second Floor	Opt. Bonus	Bed	Bath	Width	Depth	Foundation	Price Category
3301 sq ft	2355 sq ft	946 sq ft	275 sq ft	4	3-1/2	60' 0"	64' 4"	Basement or Crawl Space	I

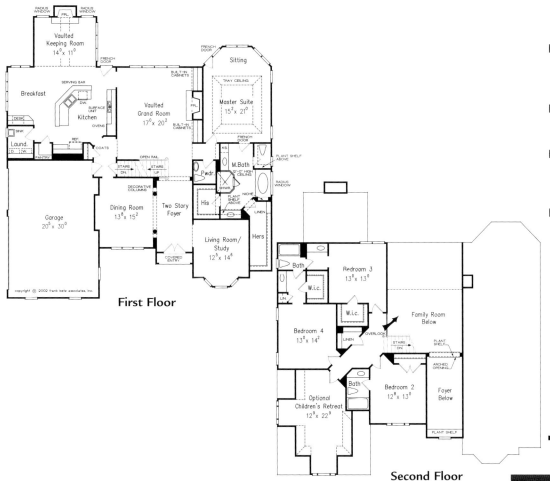

First Floor

Second Floor

Design Features

- Multiple gables, a cheery dormer and the classic combination of brick and siding make an impressive statement.

- The two-story foyer gives the perfect introduction to a vaulted grand room.

- Beyond the kitchen is a vaulted keeping room with radius windows and a fireplace.

- The master bedroom includes a secluded lounging area that overlooks the backyard.

Rear Elevation

Adelaide — TSPDG01-866-D — 1-866-525-9374

© 2000 Donald A. Gardner, Inc.

Total Living	First Floor	Basement	Bonus	Bed	Bath	Width	Depth	Foundation	Price Category
3301 sq ft	2151 sq ft	1150 sq ft	N/A	4	3	83' 0"	74' 4"	Hillside Walkout	G

Design Features

■ Twin dormers, board-and-batten siding and stone add curb appeal to this charming Craftsman design.

■ Two decks, a screened porch and spacious patio provide plenty of room for outdoor entertaining.

■ Counters in the kitchen and utility room offer ample workspace.

■ From decorative ceilings and columns to fireplaces and a shower seat, niceties are abundant.

First Floor

Basement

Rear Elevation

© 2001 Frank Betz Associates, Inc.

Sutcliffe — TSPFB01-3634 — 1-866-525-9374

Total Living	First Floor	Second Floor	Bonus	Bed	Bath	Width	Depth	Foundation	Price Category
3312 sq ft	1692 sq ft	1620 sq ft	N/A	5	4-1/2	60' 0"	56' 0"	Basement or Crawl Space	I

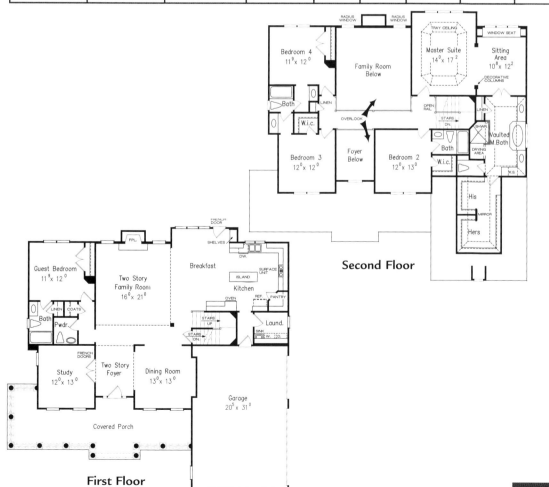

Second Floor

First Floor

copyright © 2001 frank betz associates, inc.

Design Features

- A sweeping front porch against a façade of cedar shake makes an attention-grabbing exterior.

- Decorative columns define this space to create a peaceful haven for the homeowner.

- A guest bedroom with its own private bath is designed into the main floor.

- The kitchen is an entertainers dream, with a large prep island, open layout and decorative shelving.

Rear Elevation

Laycrest — TSPDG01-995-D — 1-866-525-9374

© 2002 Donald A. Gardner, Inc.

Total Living	First Floor	Basement	Bonus	Bed	Bath	Width	Depth	Foundation	Price Category
3320 sq ft	1720 sq ft	1600 sq ft	N/A	4	3-1/2	59' 0"	59' 4"	Hillside Walkout	G

Design Features

- With Arts-n-Crafts charm, this hillside design starts with an exterior of siding and stone.

- A rear clerestory frames the sky and places it under the great room's cathedral ceiling.

- Built-in cabinetry and fireplaces enhance the great room and rec room.

- An angled counter separates the kitchen from the dining room and great room.

- A wet bar, bay windows and tray ceilings add custom style.

Rear Elevation

First Floor

Basement

Donald A. Gardner Architects, Inc.

Hollingbourne — TSPDG01-990 — 1-866-525-9374

Total Living	First Floor	Second Floor	Bonus	Bed	Bath	Width	Depth	Foundation	Price Category
3341 sq ft	2062 sq ft	1279 sq ft	386 sq ft	5	4-1/2	73' 8"	50' 0"	Crawl Space*	G

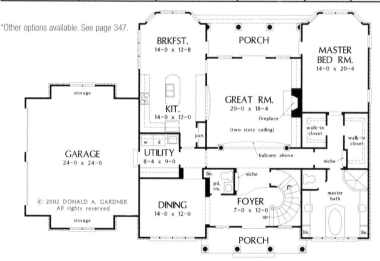

*Other options available. See page 347.

First Floor

Design Features

- Reminiscent of Old-World Manors, this stately home features a stone and stucco exterior.

- An elegant balustrade crowns the entryway, which is highlighted by columns and arches.

- A grand staircase leads to a balcony loft, which separates the two-story foyer and great room.

- Built-in cabinetry, a striking fireplace and French doors enhance the great room.

Second Floor

Rear Elevation

Vincent — TSPDS01-7038 — 1-866-525-9374

© The Sater Design Collection, Inc.

Total Living	First Floor	Second Floor	Bonus	Bed	Bath	Width	Depth	Foundation	Price Category
3342 sq ft	1865 sq ft	1477 sq ft	282 sq ft	4	2-1/2	79' 0"	79' 2"	Opt. Basement/Crawl Space	H

Design Features

- The glass entry is enhanced by a second-story bowed balcony.

- The leisure room boasts ample built-ins and French doors to the porch.

- A sweeping pass-thru counter connects the island kitchen and sunny nook.

- The coffered-ceiling living room is complete with fireplace.

- A uniquely shaped master suite hosts a private deck.

Porch View

Second Floor

First Floor

Photographed home may have been modified from original construction documents.

© 2004 Donald A. Gardner, Inc.

Innsbrook — TSPDG01-1076 — 1-866-525-9374

Total Living	First Floor	Second Floor	Bonus	Bed	Bath	Width	Depth	Foundation	Price Category
3356 sq ft	2113 sq ft	1243 sq ft	443 sq ft	4	4	68' 8"	51' 4"	Crawl Space*	G

*Other options available. See page 347.

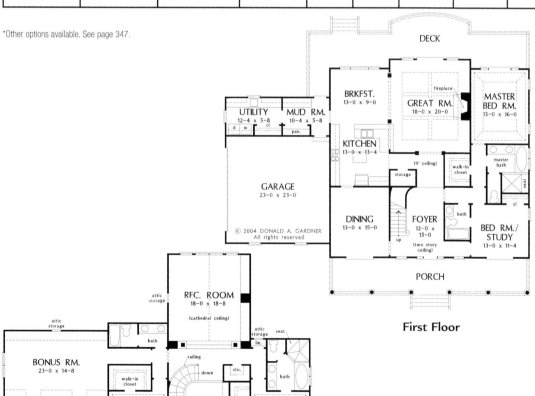

First Floor

© 2004 DONALD A. GARDNER
All rights reserved

Second Floor

Design Features

- Using columns, dormers and a prominent gable, the front elevation showcases architectural interest.

- A sweeping staircase makes a stunning focal point upon entering the foyer.

- With distinct room definition, the kitchen, breakfast nook and great room remain open to each other.

- The upstairs includes a recreation room with a cathedral ceiling and columns.

Rear Elevation

Donald A. Gardner Arohitects, Inc.

Oxfordshire — TSPDG01-1046 — 1-866-525-9374

© 2004 Donald A. Gardner, Inc.

Total Living	First Floor	Second Floor	Bonus	Bed	Bath	Width	Depth	Foundation	Price Category
3367 sq ft	2562 sq ft	805 sq ft	622 sq ft	4	4	87' 7"	59' 6"	Crawl Space*	G

*Other options available. See page 347.

Design Features

■ Majestic columns frame the metal-topped portico.

■ Columns and built-in cabinetry exude elegance.

■ Secondary bedrooms with balconies display attention to detail.

■ The bonus room fulfills expansion needs.

■ The master retreat pampers with a sitting bay, huge walk-in closets and a spacious bath.

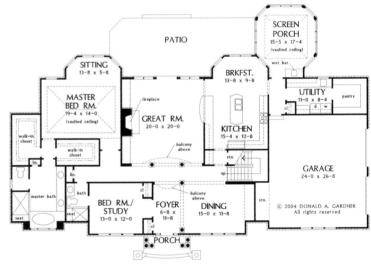

First Floor

Rear Elevation

Second Floor

© 2002 Frank Betz Associates, Inc.

Hamilton Place — TSPFB01-3723 — 1-866-525-9374

Total Living	First Floor	Second Floor	Bonus	Bed	Bath	Width	Depth	Foundation	Price Category
3379 sq ft	1765 sq ft	1614 sq ft	N/A	5	4	57' 0"	45' 4"	Basement or Crawl Space	I

First Floor

Second Floor

Design Features

- The combination of the kitchen, breakfast, keeping and dining rooms make entertaining a breeze.

- A butler's pantry was strategically placed next to the walk-in pantry.

- A two-story family room offers a bowed wall of windows.

- The laundry is located on the second level convenient to the spacious master bedroom.

Rear Elevation

McFarlin Park — TSPFB01-3808 — 1-866-525-9374

© 2003 Frank Betz Associates, Inc.

Total Living	First Floor	Second Floor	Bonus	Bed	Bath	Width	Depth	Foundation	Price Category
3397 sq ft	2434 sq ft	963 sq ft	N/A	5	4	64' 0"	62' 10"	Basement or Crawl Space	I

Design Features

- From the *Southern Living®* *Design Collection*

- The timeless façade of the *McFarlin Park* is a welcoming blend of stone and cedar shake.

- A vaulted keeping room adjoins the kitchen area, creating the perfect spot for lounging or entertaining guests.

- Unwind on the cozy screened porch, tucked away off the kitchen.

First Floor

Rear Elevation

Second Floor

© 2003 Donald A. Gardner, Inc.

Yesterview — TSPDG01-1002 — 1-866-525-9374

Total Living	First Floor	Second Floor	Bonus	Bed	Bath	Width	Depth	Foundation	Price Category
3419 sq ft	2237 sq ft	1182 sq ft	475 sq ft	4	3-1/2	85' 4"	56' 4"	Crawl Space*	G

*Other options available. See page 347.

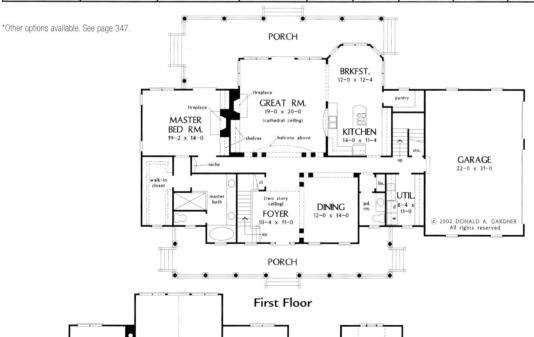

First Floor

Second Floor

Design Features

- An elegant exterior—with country and traditional flair—surrounds a modern floor plan.

- Two spacious porches promote outdoor entertaining.

- Built-in cabinetry and fireplaces complement the great room and master bedroom.

- Two staircases, a large laundry room with sink and a guest suite create enjoyable living.

Rear Elevation

Candace — TSPFB01-965 — 1-866-525-9374

© 1996 Frank Betz Associates, Inc.

Total Living	First Floor	Second Floor	Bonus	Bed	Bath	Width	Depth	Foundation	Price Category
3434 sq ft	2384 sq ft	1050 sq ft	228 sq ft	4	3-1/2	65' 8"	57' 0"	Basement, Crawl Space or Slab	I

Design Features

- The master suite has its own private sitting area.

- Decorative columns and arched openings serve as transitional points from various rooms.

- An optional bonus room upstairs has endless finishing possibilities.

- Plenty of volume in ceiling heights gives the entire first floor of this design a very spacious and roomy feeling.

Rear Elevation

First Floor

Second Floor

Concordia — TSPDG01-1083 — 1-866-525-9374

Total Living	First Floor	Second Floor	Bonus	Bed	Bath	Width	Depth	Foundation	Price Category
3449 sq ft	2468 sq ft	981 sq ft	393 sq ft	4	4	58' 7"	79' 6"	Crawl Space*	G

*Other options available. See page 347.

Design Features

■ Pillars, arches and banding lend a subtle sophistication to the brick exterior.

■ Opening to the dining room and bedroom/study, the foyer leads to a gallery with art niche.

■ Columns, built-in cabinetry and ceiling treatments enhance the open floor plan.

■ A butler's pantry, home office and spacious laundry provides convenience.

■ The balcony visually connects both floors, creating a sense of togetherness.

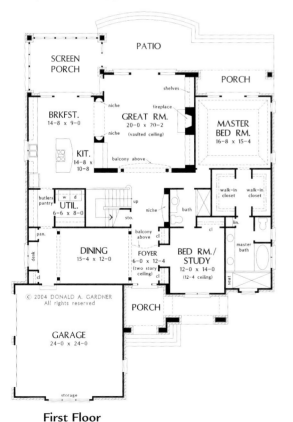

First Floor

Second Floor

Rear Elevation

Wentworth Trail — TSPDS01-6653 — 1-866-525-9374

© The Sater Design Collection, Inc.

Total Living	First Floor	Second Floor	Bonus	Bed	Bath	Width	Depth	Foundation	Price Category
3462 sq ft	2894 sq ft	568 sq ft	N/A	3	3-1/2	102' 0"	67' 0"	Slab	H

Design Features

- The foyer opens to a living room with corner pocketing doors.

- The kitchen has a walk-in pantry and cooktop island.

- The leisure room has a fireplace, bar and TV niche.

- The opulent master wing includes a study and sitting area.

- The loft has a wet bar and an observation deck.

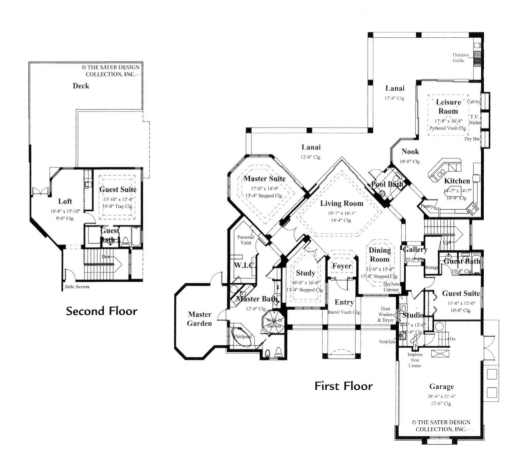

© THE SATER DESIGN COLLECTION, INC.—

Deck

Loft
18'-8" x 15'-10"
9'-0" Clg.

Guest Suite
13'-10" x 12'-4"
10'-0" Tray Clg.

Guest Bath

Attic Access

Second Floor

Outdoor Grille

Lanai
12'-8" Clg.

Leisure Room
17'-0" x 16'-8"
Pyramid Vault Clg.

Cab'try
T.V. Niche
Dry Bar

Lanai
12'-8" Clg.

Nook
10'-0" Clg.

Kitchen
14'-7" x 14'-7"
10'-0" Clg.

Master Suite
17'-0" x 14'-9"
13'-4" Stepped Clg.

Living Room
19'-7" x 16'-1"
14'-4" Clg.

Pool Bath

Pantry

Personal Valet

Dining Room
11'-6" x 15'-0"
15'-0" Stepped Clg.

Gallery
10'-0" Clg.

Guest Bath
10'-0" Clg.

Storage

W.I.C.

Study
10'-0" x 16'-6"
13'-4" Stepped Clg.

Foyer

DryAire Cabinet

Guest Suite
11'-0" x 12'-0"
10'-0" Clg.

Master Bath
12'-0" Clg.

Entry
Barrel Vault Clg.

Duet Washer & Dryer

Studio
13'-0" x 13'-6"
10'-0" Clg.

Walk-In Shower

Master Garden

SinkSpa

Whirlpool

Impress from Center

Dn

First Floor

Garage
28'-6" x 21'-6"
12'-6" Clg.

© THE SATER DESIGN COLLECTION, INC.—

Rear View

Photographed home may have been modified from original construction documents.

© The Sater Design Collection, Inc.

Salcito — TSPDS01-6787 — 1-866-525-9374

Total Living	First Floor	Second Floor	Bonus	Bed	Bath	Width	Depth	Foundation	Price Category
3473 sq ft	2357 sq ft	1116 sq ft	N/A	4	4-1/2	52' 2"	94' 0"	Slab	H

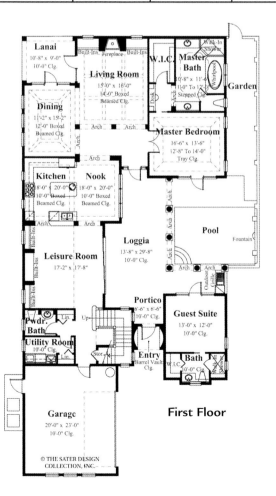

First Floor

Second Floor

© THE SATER DESIGN COLLECTION, INC.

Design Features

- This enchanting courtyard home features interesting spaces.

- Rooms open to a central loggia with fountain pool.

- The second level includes a study and two bedroom suites.

- Multiple upstairs decks have courtyard views.

- The main-floor leisure room has a two-story boxed-beamed ceiling.

Interior Courtyard

Kenthurst — TSPFB01-3755 — 1-866-525-9374

© 2002 Frank Betz Associates, Inc.

Total Living	First Floor	Second Floor	Bonus	Bed	Bath	Width	Depth	Foundation	Price Category
3482 sq ft	2461 sq ft	1021 sq ft	N/A	5	4-1/2	80' 0"	57' 0"	Basement or Crawl Space	I

Design Features

- A beautiful covered porch on the front of the home makes a wonderful entrance.

- A striking keeping room is connected to the kitchen, featuring a fireplace, built-in cabinetry and vaulted ceilings.

- Homeowners will appreciate the master bedroom's private sitting area when it comes time to rest for the evening.

- The main-floor fifth bedroom also makes an ideal location for a study.

Rear Elevation

First Floor

Second Floor

Donald A. Gardner Architects, Inc.

© 2000 Donald A. Gardner, Inc.

Churchdown — TSPDG01-867 — 1-866-525-9374

Total Living	First Floor	Second Floor	Bonus	Bed	Bath	Width	Depth	Foundation	Price Category
3499 sq ft	2755 sq ft	744 sq ft	481 sq ft	3	3-1/2	92' 6"	69' 10"	Crawl Space*	G

*Other options available. See page 347.

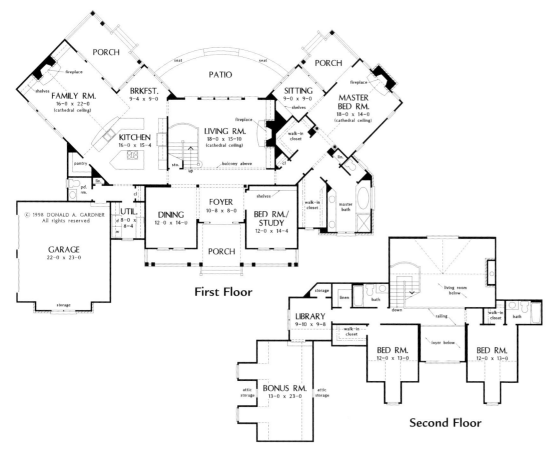

First Floor

Second Floor

Design Features

- The master suite features two walk-in closets, sitting area and a private porch.

- In addition to the living room, this plan also has a family room with a cathedral ceiling.

- Built-in shelves enhance the private study.

- Secondary bedrooms, two full bathrooms, a library and bonus room complete the second level.

Rear Elevation

New Brunswick — TSPDS01-8021 — 1-866-525-9374

© The Sater Design Collection, Inc.

Total Living	First Floor	Second Floor	Bonus	Bed	Bath	Width	Depth	Foundation	Price Category
3501 sq ft	2232 sq ft	1269 sq ft	N/A	4	4-1/2	80' 6"	63' 8"	Opt. Basement/Slab	I

Design Features

- This wonderful floor plan is centralized around a great room.

- The rear garage has a large bonus room/guest suite above.

- French doors open the great room to a wide veranda.

- The kitchen and nook open to a covered veranda with grille.

- The secluded master suite has private veranda access.

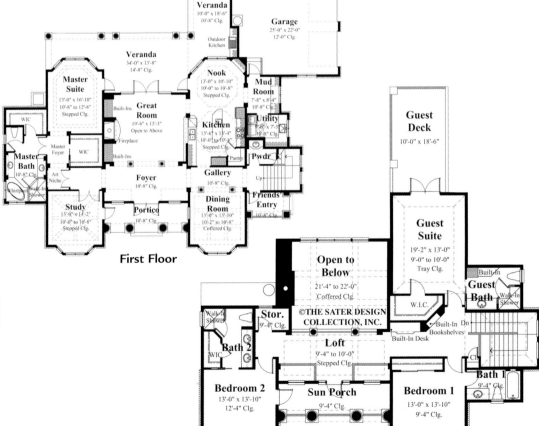

Rear Elevation

© 1996 Frank Betz Associates, Inc.

Greenwich — TSPFB01-1003 — 1-866-525-9374

Total Living	First Floor	Second Floor	Bonus	Bed	Bath	Width	Depth	Foundation	Price Category
3525 sq ft	1786 sq ft	1739 sq ft	N/A	5	4-1/2	59' 0"	53' 0"	Basement, Crawl Space or Slab	I

Design Features

- A massive archway over the double-door entrance makes a dramatic impression.

- A bright two-story family room sets the stage for a home that's upscale everywhere you look.

- The kitchen surrounds a large island with serving bar—big enough to use for casual dining.

- The master suite houses its own private sitting area, accented with an arched opening.

Second Floor

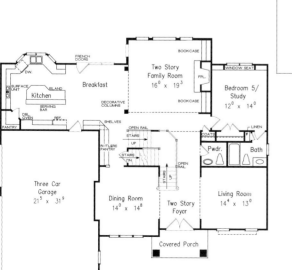

First Floor

Rear Elevation

Kingston Place — TSPFB01-3822 — 1-866-525-9374

© 2003 Frank Betz Associates, Inc.

Total Living	First Floor	Second Floor	Bonus	Bed	Bath	Width	Depth	Foundation	Price Category
3550 sq ft	1810 sq ft	1740 sq ft	N/A	5	4-1/2	59' 0"	53' 0"	Basement, Crawl Space or Slab	I

Design Features

- This home's stately brick façade with traditional white columns makes quite a statement from the street.
- A furniture niche is situated in the dining room, perfect for a buffet table or china cabinet.
- His-and-her closets in the master suite are divided by a mirror.
- The fifth bedroom features a window seat with built-in cabinets, making it easy to convert into a home office.

Rear Elevation

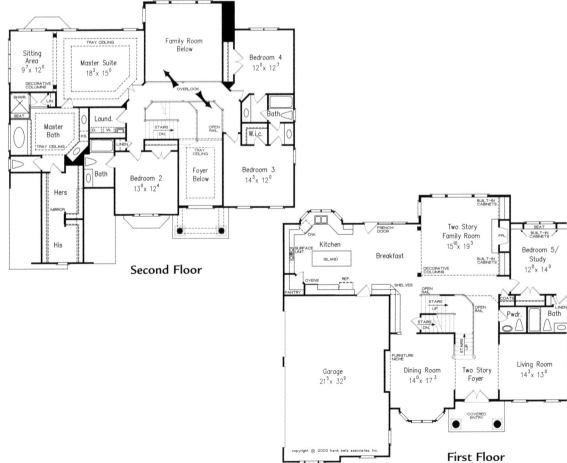

Second Floor

First Floor

Donald A. Gardner Architects, Inc.

Wellingley — TSPDG01-943 — 1-866-525-9374

Total Living	First Floor	Second Floor	Bonus	Bed	Bath	Width	Depth	Foundation	Price Category
3573 sq ft	2511 sq ft	1062 sq ft	465 sq ft	4	3-1/2	84' 11"	55' 11"	Crawl Space*	H

*Other options available. See page 347.

First Floor

Second Floor

Design Features

- Dormers echo the angles found in the gables, which provide a great contrast to the hip roof.

- A balcony overlooks, while fireplaces highlight, the two-story great room.

- The breakfast nook is expanded by a gentle curve.

- The sitting room off the master suite makes a perfect reading area or private study.

Rear Elevation

Corsini — TSPDS01-8049 — 1-866-525-9374

© The Sater Design Collection, Inc.

Total Living	First Floor	Second Floor	Bonus	Bed	Bath	Width	Depth	Foundation	Price Category
3578 sq ft	2163 sq ft	1415 sq ft	N/A	5	3-1/2	71' 0"	72' 0"	Opt. Basement/Slab	I

Design Features

- Triplets of French doors deck out the Spanish eclectic façade.

- The central gallery opens to the heart of the home.

- The kitchen, loggia, nook and dining room flow to make entertaining easy.

- The owners' wing includes a study and a powder bath.

- The upper level has a loft that connects four secondary bedrooms.

Rear Elevation

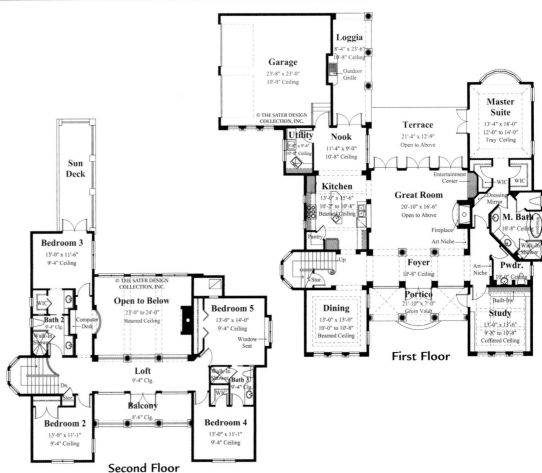

Second Floor

First Floor

Woodcliffe — TSPFB01-3757 — 1-866-525-9374

Total Living	First Floor	Second Floor	Opt. Bonus	Bed	Bath	Width	Depth	Foundation	Price Category
3585 sq ft	2225 sq ft	1360 sq ft	234 sq ft	4	3-1/2	68' 10"	60' 0"	Basement or Crawl Space	I

First Floor

Second Floor

Design Features

- Coffered ceilings create interesting dimension throughout the main level of this home.

- Art niches are designed at the top of the staircase, providing innovative decorating opportunities.

- The master suite has a quiet sitting area with views to the backyard.

- A large family recreation room is incorporated into the upper level, providing flexible options like a media room or exercise area.

Rear Elevation

Bellamy — TSPDS01-8018 — 1-866-525-9374

© The Sater Design Collection, Inc.

Total Living	First Floor	Second Floor	Bonus	Bed	Bath	Width	Depth	Foundation	Price Category
3610 sq ft	2483 sq ft	1127 sq ft	332 sq ft	4	3-1/2	83' 0"	71' 8"	Opt. Basement/Slab	I

Design Features

- The two-story foyer leads past columns and archways to the living and dining rooms.

- A fireplace is shared by the leisure room and study.

- The master suite enjoys a morning kitchen.

- An upstairs guest suite has an outdoor balcony.

- A second-floor common area offers a built-in computer desk.

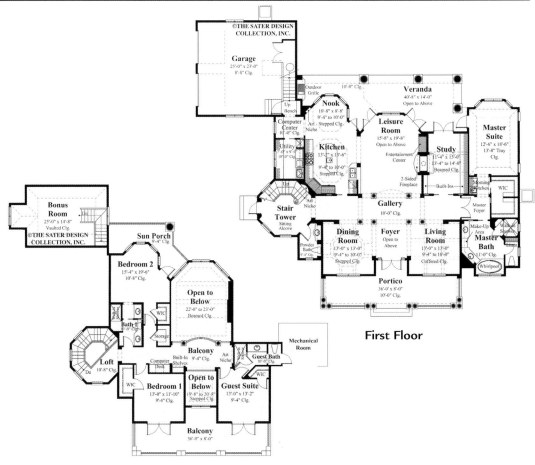

First Floor

Second Floor

Rear Elevation

Witherspoon — TSPFB01-3462 — 1-866-525-9374

Total Living	First Floor	Second Floor	Opt. Bonus	Bed	Bath	Width	Depth	Foundation	Price Category
3618 sq ft	2384 sq ft	1234 sq ft	344 sq ft	5	4-1/2	64' 6"	57' 10"	Basement, Crawl Space or Slab	I

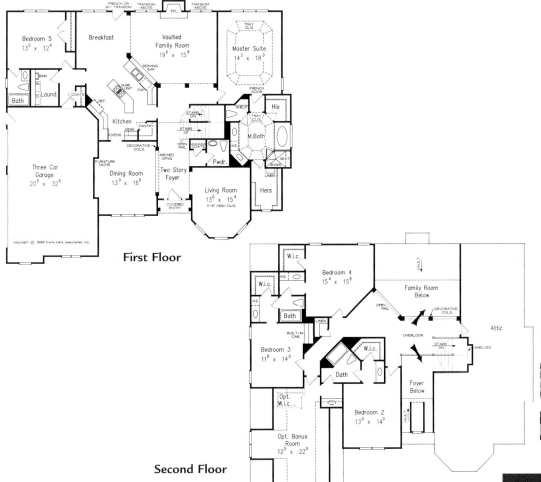

First Floor

Second Floor

Design Features

- A handy built-in message center and double ovens are included in the kitchen.

- Distinctive and elegant, tray ceilings enhance the master suite.

- A seated shower, double sinks and his-and-her closets make for an enviable master bath.

- Decorative columns and a furniture niche add charm to the formal dining room.

Rear Elevation

3000 sq ft—4000 sq ft

Frank Betz Associates, Inc.

Greywell — TSPFB01-3914 — 1-866-525-9374

© 2004 Frank Betz Associates, Inc.

Total Living	First Floor	Second Floor	Bonus	Bed	Bath	Width	Depth	Foundation	Price Category
3629 sq ft	2499 sq ft	1130 sq ft	N/A	5	4	67' 6"	69' 10"	Basement, Crawl Space or Slab	I

Design Features

- The kitchen overlooks a gracious breakfast area, as well as a fire-lit keeping room.

- A screened porch is accessed off this area as well, providing the perfect opportunity to take the party outside.

- Coffered ceilings and built-in cabinetry make the family room extra special.

- A children's retreat upstairs can be used as the perfect playroom, as a fourth bedroom or guest suite.

First Floor

Second Floor

Rear Elevation

© 2003 Frank Betz Associates, Inc.

Seagraves — TSPFB01-3843 — 1-866-525-9374

Total Living	First Floor	Second Floor	Bonus	Bed	Bath	Width	Depth	Foundation	Price Category
3688 sq ft	1935 sq ft	1753 sq ft	N/A	5	4-1/2	65' 6"	59' 0"	Basement or Crawl Space	I

First Floor

Second Floor

Design Features

■ A keeping room adjoins the kitchen.

■ The family room soars two stories high, giving a spacious and airy feel to the room.

■ Built-in cabinetry provides great decorating opportunities.

■ A main-floor bedroom is a convenient location for a guest room.

Rear Elevation

Greyhawk — TSPFB01-3479 — 1-866-525-9374

© 1999 Frank Betz Associates, Inc.

Total Living	First Floor	Second Floor	Bonus	Bed	Bath	Width	Depth	Foundation	Price Category
3728 sq ft	2495 sq ft	1233 sq ft	351 sq ft	4	3-1/2	66' 10"	57' 6"	Basement, Crawl Space or Slab	I

Design Features

- Stately brick and clapboard siding are a welcomed combination in neighborhoods new and old.
- The kitchen is situated at a unique angle, adding an interesting dimension to the main floor.
- The kitchen overlooks a breakfast area and keeping room that unite to create a casual, comfortable place for families to spend quality time.
- An arched opening and built-in cabinetry in the family room are special details that make this home unique.

First Floor

Second Floor

Rear Elevation

© 2003 Frank Betz Associates, Inc.

Baldwin Farm — TSPFB01-3831 — 1-866-525-9374

Total Living	First Floor	Second Floor	Bonus	Bed	Bath	Width	Depth	Foundation	Price Category
3733 sq ft	2503 sq ft	1230 sq ft	N/A	4	3-1/2	80' 0"	66' 0"	Basement, Crawl Space or Slab	I

Second Floor

First Floor

Design Features

- From the *Southern Living® Design Collection*
- The carriage doors and cupola on the garage go back in time to give a historic elegance to the façade of this home.

- The kitchen overlooks a cozy keeping room that adjoins a screened porch and deck.

- A mudroom is strategically situated inside the secondary entrance.

- The second level of the home features a vaulted family entertainment room.

Rear Elevation

Cotton Creek Trace — TSPDS01-6650 — 1-866-525-9374

© The Sater Design Collection, Inc.

Total Living	First Floor	Second Floor	Bonus	Bed	Bath	Width	Depth	Foundation	Price Category
3748 sq ft	3092 sq ft	656 sq ft	N/A	4	4	77' 4"	93' 10"	Slab	I

Design Features

- A wet bar easily serves the living and dining rooms.

- A large kitchen has a cooktop island and walk-in pantry.

- A bayed leisure room features large views to the rear yard.

- The owner's retreat showcases an extravagant oversized master suite.

- The second floor has a study and a guest suite with balconies.

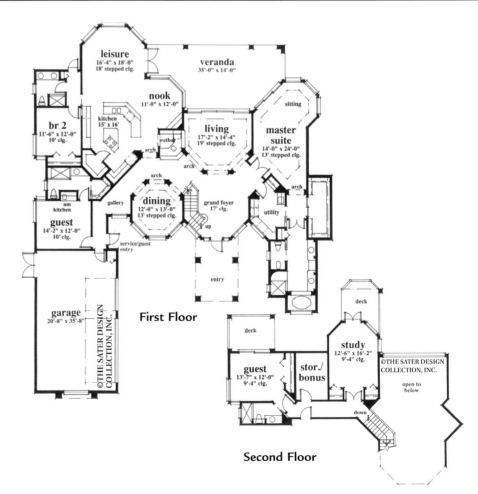

First Floor

Second Floor

Rear Elevation

Sunningdale Cove — TSPDS01-6660 — 1-866-525-9374

Total Living	First Floor	Second Floor	Bonus	Bed	Bath	Width	Depth	Foundation	Price Category
3792 sq ft	2853 sq ft	627 sq ft	312 sq ft	4	3-1/2	80' 0"	96' 0"	Slab	J

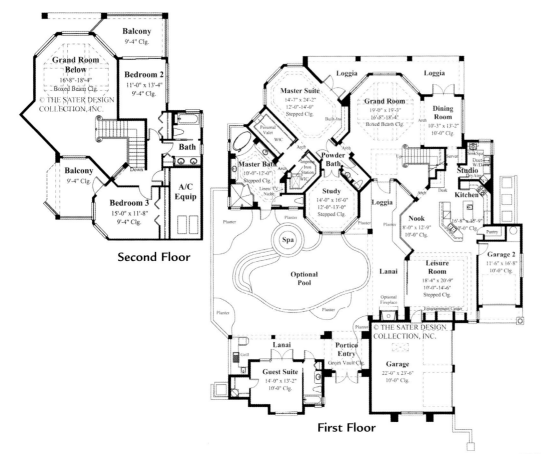

Second Floor

First Floor

Design Features

- The home features glass-walled rooms wrapped around a courtyard.

- The main-entry door opens to a two-story grand salon.

- The leisure room has an entertainment center and glass doors to the lanai and outdoor fireplace.

- The master wing has a bayed sitting area.

- Two upstairs bedrooms feature private decks.

Rear Elevation

3000 sq ft—4000 sq ft

The Sater Design Collection, Inc.

Griffith Parkway — TSPDS01-6721 — 1-866-525-9374

© The Sater Design Collection, Inc.

Total Living	First Floor	Second Floor	Bonus	Bed	Bath	Width	Depth	Foundation	Price Category
3872 sq ft	2924 sq ft	948 sq ft	N/A	4	3-1/2	65' 0"	91' 0"	Slab	I

Design Features

■ A high-pitched roof and dormers lend a Southern traditional feel.

■ Disappearing glass doors open the family spaces and master suite to the lanai.

■ The owner's wing includes a nearby study.

■ A private garden infuses the master bath with nature.

■ Two guest rooms, a loft and covered deck complete the second floor.

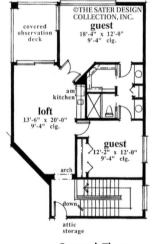

Second Floor

Rear Elevation

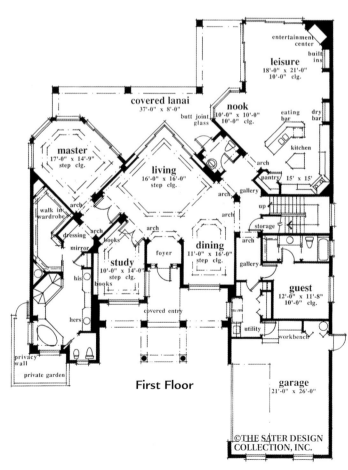

First Floor

3000 sq ft—4000 sq ft

Frank Betz Associates, Inc.

Flanagan — TSPFB01-1058 — 1-866-525-9374

Total Living	First Floor	Second Floor	Bonus	Bed	Bath	Width	Depth	Foundation	Price Category
3877 sq ft	2060 sq ft	1817 sq ft	N/A	5	4-1/2	54' 0"	78' 4"	Basement, Crawl Space or Slab	J

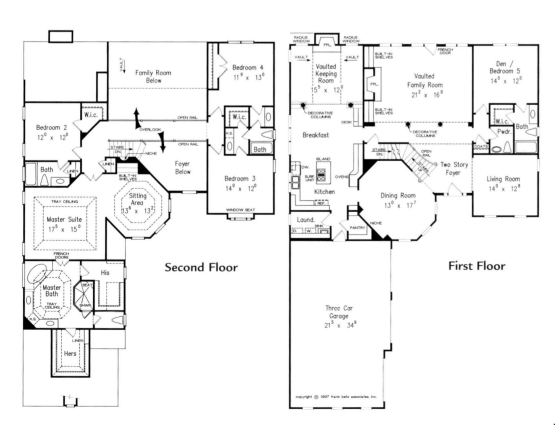

Second Floor

First Floor

Design Features

- A stunning stone turret gives this plan eye-catching curb appeal.

- A vaulted keeping room—accentuated by decorative columns—adjoins the kitchen area.

- The master suite earns its name, featuring a uniquely shaped sitting area

- A main-floor bedroom can be easily converted into a den or home office.

Rear Elevation

Fiddler's Creek — TSPDS01-6746 — 1-866-525-9374

© The Sater Design Collection, Inc.

Total Living	First Floor	Second Floor	Bonus	Bed	Bath	Width	Depth	Foundation	Price Category
3893 sq ft	2841 sq ft	1052 sq ft	N/A	4	3-1/2	85' 0"	76' 2"	Opt. Basement/Slab	I

Design Features

- Inside the foyer, columns frame a living room with a fireplace.

- A column-lined gallery hall leads down each wing.

- The kitchen has a walk-in pantry and counter to the veranda.

- The leisure room has a fireplace and French doors to the lanai.

- A balcony overlook leads to three upstairs bedrooms.

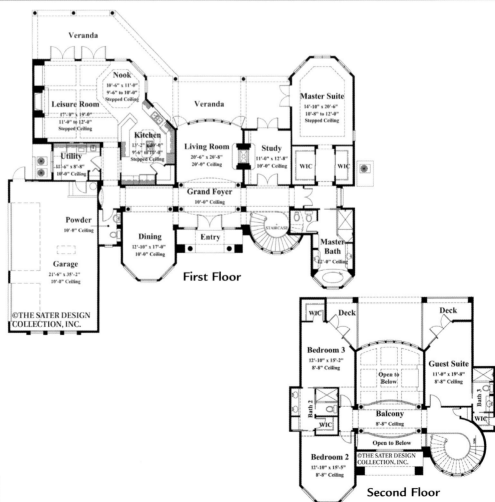

Rear Elevation

© 2003 Frank Betz Associates, Inc.

Hartford Springs — TSPFB01-3824 — 1-866-525-9374

Total Living	First Floor	Second Floor	Bonus	Bed	Bath	Width	Depth	Foundation	Price Category
3971 sq ft	2504 sq ft	1467 sq ft	N/A	4	3-1/2	73' 0"	66' 10"	Basement or Crawl Space	I

Design Features

■ From the *Southern Living® Design Collection*

■ Two decks are situated on the back of the home.

■ A crackling fire is the backdrop for family time spent in the vaulted keeping room.

■ The master suite includes a roomy sitting area with access to the deck.

Second Floor

First Floor

Rear Elevation

Royal Country Down — TSPDS01-8001 — 1-866-525-9374

© The Sater Design Collection, Inc.

Total Living	First Floor	Second Floor	Bonus	Bed	Bath	Width	Depth	Foundation	Price Category
3977 sq ft	2834 sq ft	1143 sq ft	N/A	4	3-1/2	85' 0"	76' 8"	Slab/Opt. Basement	I

Design Features

- Formal spaces radiate from a gallery of arches and columns.

- Built-in shelves frame an entertainment center in the leisure room.

- A gallery hallway leads to a spectacular master suite.

- Upstairs guest rooms feature private decks.

- A wrapping veranda opens all the rear rooms to the outdoors.

First Floor

Second Floor

Rear Elevation

© The Sater Design Collection, Inc.

Verrado — TSPDS01-6918 — 1-866-525-9374

Total Living	First Floor	Second Floor	Bonus	Bed	Bath	Width	Depth	Foundation	Price Category
3991 sq ft	3008 sq ft	983 sq ft	N/A	4	3-1/2	65' 0"	115' 0"	Slab	I

Second Floor

First Floor

Design Features

- Mediterranean façade details make this home ultra impressive.

- The interior impresses, too, with zero-corner pocket doors in the living room.

- The private master suite has an adjacent study and privacy garden.

- The Solana boasts an outdoor grill and fireplace.

- The second floor offers two guest rooms, bonus room and deck.

Rear Elevation

Capucina — TSPDS01-8010 — 1-866-525-9374

© The Sater Design Collection, Inc.

Total Living	First Floor	Second Floor	Bonus	Bed	Bath	Width	Depth	Foundation	Price Category
4005 sq ft	2850 sq ft	1155 sq ft	371 sq ft	4	4-1/2	71' 6"	83' 0"	Slab/Opt. Basement	I

Design Features

- A triple skylight overlooks the foyer.

- An open arrangement of the central living spaces creates great views.

- French doors open the leisure room to the outdoors.

- The master suite has a private sitting area and lanai access.

- Upstairs, a balcony hall connects the family bedrooms and guest quarters.

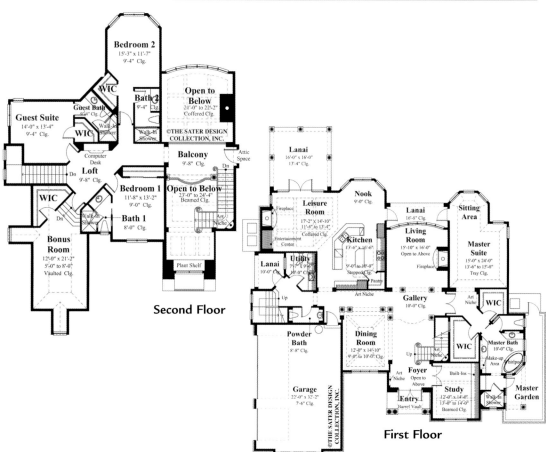

Second Floor

First Floor

Rear Elevation

© The Sater Design Collection, Inc.

Anna Belle — TSPDS01-6782 — 1-866-525-9374

Total Living	First Floor	Second Floor	Bonus	Bed	Bath	Width	Depth	Foundation	Price Category
4091 sq ft	2701 sq ft	1390 sq ft	N/A	4	4	98' 0"	60' 0"	Crawl Space/Slab	I

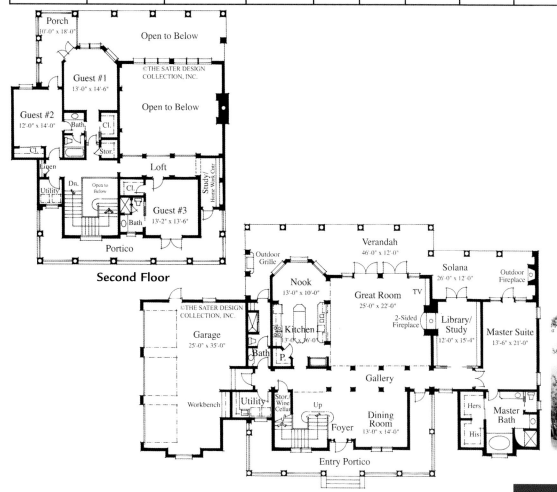

Second Floor

First Floor

Design Features

- The wide front-entry porch leads to a foyer with outdoor views.

- The master retreat begins with a library/study.

- The rear veranda has an outdoor grille, and a cozy Solana features an outdoor fireplace.

- The loft has desk space and leads to three guests suites.

- The kitchen features an oversized island and bayed nook.

Rear Elevation

The Westmoreland — TSPFB01-1025 — 1-866-525-9374

© 1997 Frank Betz Associates, Inc.

Total Living	First Floor	Second Floor	Bonus	Bed	Bath	Width	Depth	Foundation	Price Category
4135 sq ft	2037 sq ft	2098 sq ft	N/A	5	4.5	68' 6"	53' 0"	Basement, Crawl Space or Slab	X

- A truly remarkable design, the *Westmoreland* was designed for the homeowner who's looking for upscale living with a casual and comfortable feel.

- Its breathtaking two-story foyer leads you to the gallery with a curved staircase that makes a lasting impression.

- A fireplace, built-in cabinetry and bookshelves create a dramatic focal wall in the two-story family room.

Second Floor

Rear Elevation

First Floor

© The Sater Design Collection, Inc.

Stoney Creek Way — TSPDS01-6656 — 1-866-525-9374

Total Living	First Floor	Second Floor	Bonus	Bed	Bath	Width	Depth	Foundation	Price Category
4140 sq ft	3053 sq ft	1087 sq ft	N/A	4	3-1/2	87' 4"	80' 4"	Basement/Slab	L

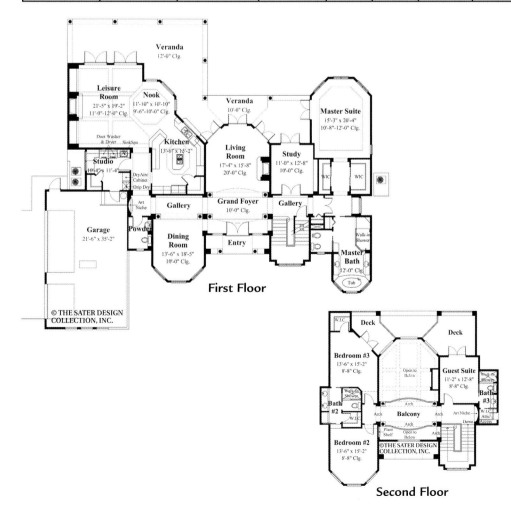

First Floor

Second Floor

Design Features

- The two-story living room features a fireplace and bayed glass doors.

- The open island kitchen easily serves the dining room.

- A relaxing leisure room has a fireplace and a television niche.

- The master wing is private and houses an opulent bath.

- Upstairs are three bedrooms linked by a gallery catwalk.

Rear Elevation

Cutlass Key — TSPDS01-6619 — 1-866-525-9374

© The Sater Design Collection, Inc.

Total Living	First Floor	Second Floor	Bonus	Bed	Bath	Width	Depth	Foundation	Price Category
4143 sq ft	2725 sq ft	1418 sq ft	N/A	4	5-1/2	61' 4"	62' 0"	Piling with slab	J

Design Features

- The two-story grand room offers a wet bar and see-through fireplace.

- Glass doors open to the screened veranda.

- The eat-in kitchen has a huge walk-in pantry.

- The large side porch features an outdoor grille.

- An opulent master suite gets privacy upstairs.

Second Floor

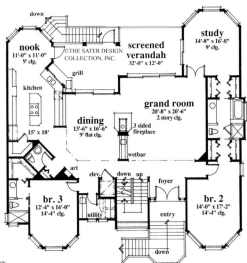

First Floor

Rear Elevation

Lower Level

© The Sater Design Collection, Inc.

La Reina — TSPDS01-8046 — 1-866-525-9374

Total Living	First Floor	Second Floor	Guest	Bed	Bath	Width	Depth	Foundation	Price Category
4151 sq ft	2852 sq ft	969 sq ft	330 sq ft	5	4-1/2	80' 0"	96' 6"	Slab	J

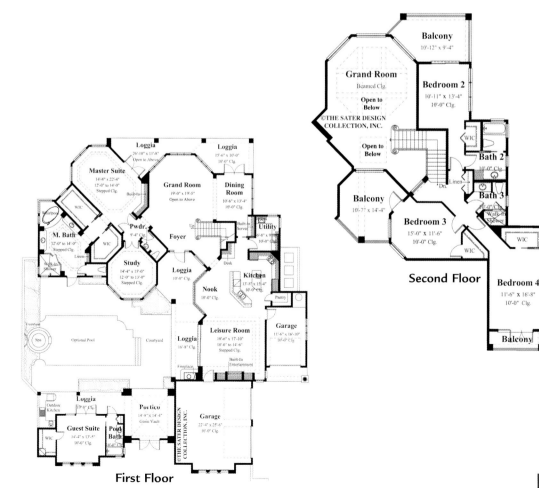

First Floor

Second Floor

Design Features

- A side balcony is spacious enough to serve as an outdoor room.

- The paneled portal opens to a portico and courtyard, and leads to the formal entry.

- To the front of the courtyard, a guest house can also serve as an office.

- The foyer opens directly to the grand room and formal dining room.

- The dining room leads to a loggia for *plein air* meals.

Rear Elevation

Vasari — TSPDS01-8025 — 1-866-525-9374

© The Sater Design Collection, Inc.

Total Living	First Floor	Second Floor	Bonus	Bed	Bath	Width	Depth	Foundation	Price Category
4160 sq ft	1995 sq ft	2165 sq ft	N/A	5	5-1/2	58' 0"	65' 0"	Slab/Opt. Basement	J

Design Features

- A gallery foyer and loft deepen the central living/dining room.

- A two-sided fireplace warms the central area and spacious study.

- Above the entry, a sun porch with a transom brings in light.

- The loft connects the family's sleeping quarters with a guest suite.

- The main level boasts a cabana-style guest suite.

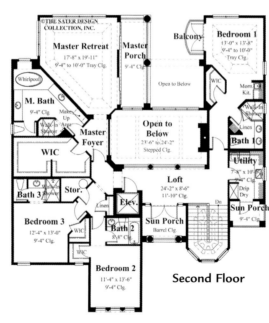

Second Floor

First Floor

Rear Elevation

© 2002 Frank Betz Associates, Inc.

The Clarendon — TSPFB01-3685 — 1-866-525-9374

Total Living	First Floor	Second Floor	Opt. Bonus	Bed	Bath	Width	Depth	Foundation	Price Category
4317 sq ft	2635 sq ft	1682 sq ft	114 sq ft	4	4.5	79' 0"	74' 5"	Basement or Crawl Space	J

First Floor

Second Floor

Design Features

- Soaring white columns against striking red brick create a time-honored façade that takes you back in time.

- An impressive foyer has a curved staircase and furniture niche. A library—with a massive wall of bookshelves—overlooks the entry.

- The well-appointed kitchen opens to a breakfast area, two-story keeping room and covered porch.

Rear Elevation

The Castlegate — TSPFB01-790 — 1-866-525-9374

© 1994 Frank Betz Associates, Inc.

Total Living	First Floor	Second Floor	Bonus	Bed	Bath	Width	Depth	Foundation	Price Category
4362 sq ft	2764 sq ft	1598 sq ft	N/A	4	3.5	74' 6"	65' 10"	Basement, Crawl Space or Slab	J

Design Features

- Your castle awaits.....A portico entry leads into a breathtaking two-story foyer with a curved staircase.

- The master suite is appropriately named, with its vaulted sitting room and luxurious bath.

- An art niche, built-in cabinetry and bookshelves, coffered ceilings and decorative columns are pleasant surprises as you wander the home.

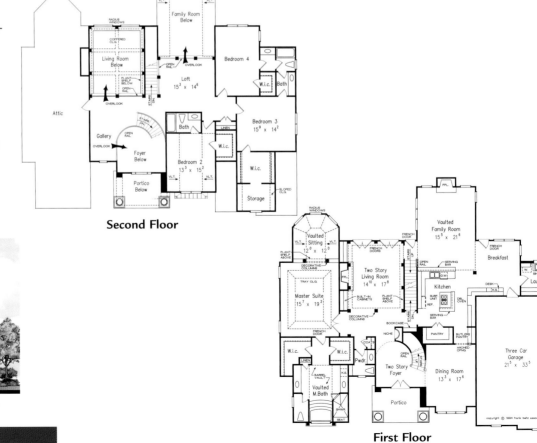

Second Floor

First Floor

Rear Elevation

© The Sater Design Collection, Inc.

San Filippo — TSPDS01-8055 — 1-866-525-9374

Total Living	First Floor	Second Floor	Bonus	Bed	Bath	Width	Depth	Foundation	Price Category
4391 sq ft	2913 sq ft	1478 sq ft	N/A	5	4-1/2	69' 4"	95' 4"	Slab/Opt. Basement	J

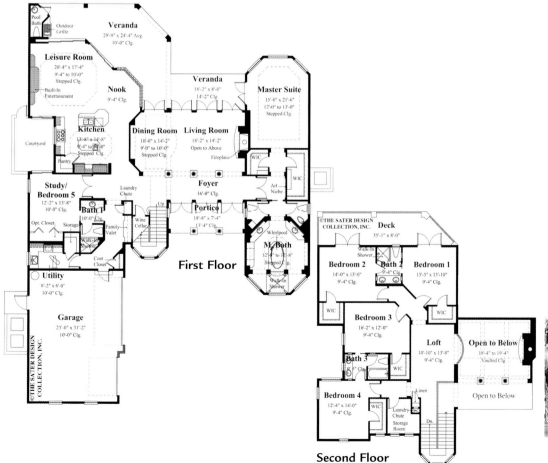

First Floor

Second Floor

Design Features

- Two bold turrets extend the foot-print of the home.

- Spacious, light-filled rooms allow unencumbered views.

- A massive hearth adds warmth to the living room.

- Retreating glass walls meld the out-doors with the nook and kitchen.

- Guests may step into a side court-yard from a flex room/study.

Rear Elevation

Collinwood — TSPFB01-1206 — 1-866-525-9374

© 1998 Frank Betz Associates, Inc.

Total Living	First Floor	Second Floor	Bonus	Bed	Bath	Width	Depth	Foundation	Price Category
4464 sq ft	2092 sq ft	2372 sq ft	N/A	5	4.5	75' 5"	64' 0"	Basement or Crawl Space	J

Design Features

- The *Collinwood's* wrap-around covered porch with a built-in gazebo is a true eye-catcher!

- A coffered ceiling and transom windows complement the family room.

- A second entrance on the side of the home gives an informal place for family to come and go.

- The master suite includes rare access to a covered porch that overlooks the backyard.

Rear Elevation

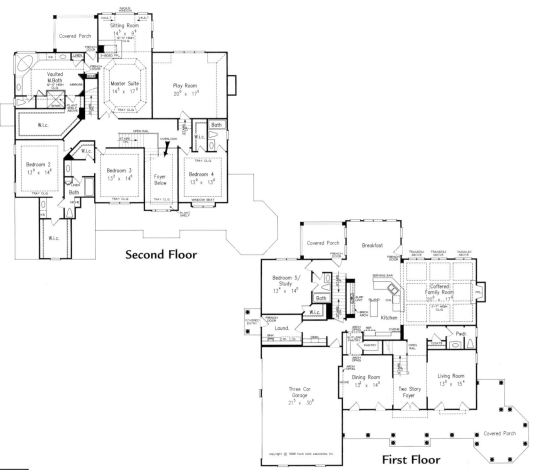

Second Floor

First Floor

The Sater Design Collection, Inc.

© The Sater Design Collection, Inc.

Stonehaven — TSPDS01-8032 — 1-866-525-9374

Total Living	First Floor	Second Floor	Bonus	Bed	Bath	Width	Depth	Foundation	Price Category
4465 sq ft	2163 sq ft	2302 sq ft	N/A	5	5-1/2	58' 0"	65' 0"	Slab/Opt. Basement	J

Second Floor

First Floor

Design Features

- This grand plan has a host of windows and outdoor places.

- Views fill the home through a wall of glass to the rear.

- A retreating glass wall in the leisure room brings in the outdoors.

- A wraparound loft leads to a master suite and child's suite.

- Two secondary suites, perfect for teenagers, are also upstairs.

Rear Elevation

Over 4000 sq ft

The Sater Design Collection, Inc.

Martinique — TSPDS01-6932 — 1-866-525-9374

© The Sater Design Collection, Inc.

Total Living	First Floor	Second Floor	Bonus	Bed	Bath	Width	Depth	Foundation	Price Category
4492 sq ft	3745 sq ft	747 sq ft	N/A	5	5-1/2	94' 10"	103' 5"	Slab	O

Design Features

- Radius glass brings views into the breakfast nook.

- The leisure room has pocketing glass doors to the veranda and outdoor kitchen.

- The master bathroom has a freestanding tub and private garden.

- Two ground-floor guest suites share a private garden.

- A third guest suite shares the upper level with a multipurpose loft.

Second Floor

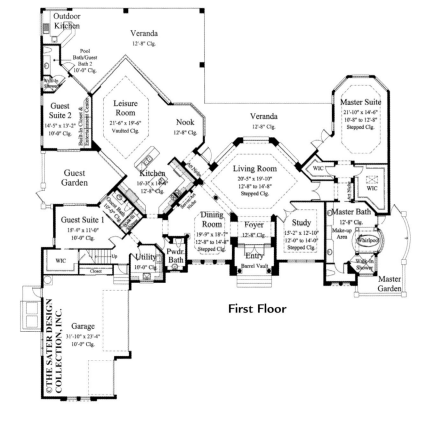

First Floor

Rear View

Photographed home may have been modified from original construction documents.
Not available for construction in Lee or Collier Counties, Florida.

© The Sater Design Collection, Inc.

Cantadora — TSPDS01-6949 — 1-866-525-9374

Total Living	First Floor	Second Floor	Bonus	Bed	Bath	Width	Depth	Foundation	Price Category
4532 sq ft	3623 sq ft	909 sq ft	N/A	4	4-1/2	91' 6"	122' 4"	Slab	O

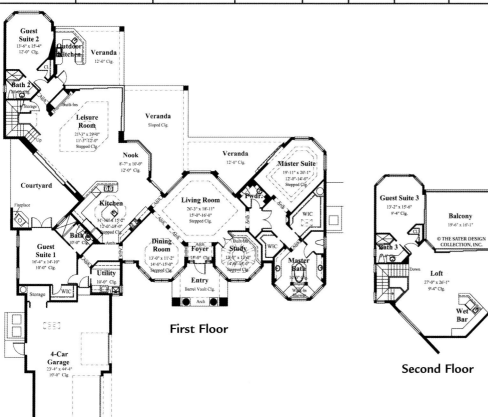

First Floor

Second Floor

Design Features

- A soaring stepped ceiling highlights an octagonal-shaped study.

- A 16-foot stepped ceiling gives spaciousness to the living room.

- The kitchen features a gleaming, carved-wood hood.

- The leisure room has disappearing glass doors to an outdoor kitchen.

- A regal master bath features an elevated marble spa tub.

Rear View

Photographed home may have been modified from original construction documents.
Not available for construction in Lee or Collier Counties, Florida.

The Oak Abbey — TSPAL01- 5003 — 1-866-525-9374

Total Living	First Floor	Basement	Bonus	Bed	Bath	Width	Depth	Foundation	Price Category
4547 sq ft	3006 sq ft	1541 sq ft	480 sq ft	4	4.5	93' 11"	80' 9"	Hillside Walkout	O

Design Features

- This hillside walkout features an open floor plan and plenty of outdoor living areas.

- Cathedral ceilings top the great room, master bedroom and screened porch.

- Three fireplaces, wet-bar, built-in cabinetry and an art niche add custom touches.

- Downstairs, each bedroom has its own walk-in closet and bath.

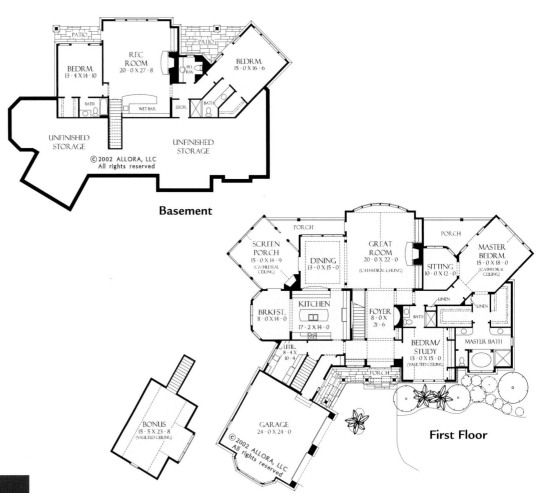

Basement

First Floor

Rear Elevation

© The Sater Design Collection, Inc.

Porto Velho — TSPDS01-6950 — 1-866-525-9374

Total Living	First Floor	Second Floor	Bonus	Bed	Bath	Width	Depth	Foundation	Price Category
4588 sq ft	4039 sq ft	549 sq ft	N/A	4	4-1/2	97' 2"	100' 9"	Slab	O

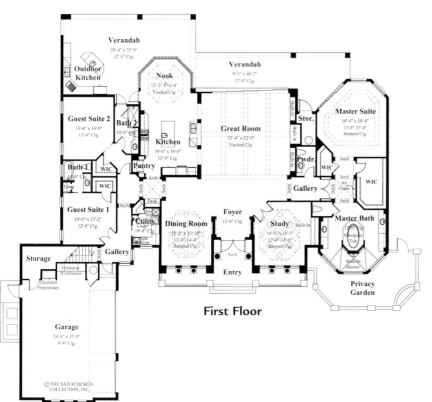

First Floor

Second Floor

Design Features

- Triple arches create pass-thus between the kitchen and great room.

- The great room is spacious, with a vaulted ceiling and wall of pocketing glass.

- Antiqued custom cabinetry and a furniture-style island enrich the kitchen.

- Floor-to-ceiling glass embraces the master sitting nook.

Rear View

Photographed home may have been modified from original construction documents.
Not available for construction in Lee or Collier Counties, Florida.

The Sater Design Collection, Inc.

Brendan Cove — TSPDS01-6740 — 1-866-525-9374

© The Sater Design Collection, Inc.

Total Living	First Floor	Second Floor	Bonus	Bed	Bath	Width	Depth	Foundation	Price Category
4633 sq ft	3670 sq ft	963 sq ft	N/A	4	5	76' 8"	113' 0"	Slab	J

Design Features

- The living room has views through a massive bowed window.

- An outdoor kitchen highlights the wrapping rear veranda.

- Two guest suites share views from a large upstairs deck.

- Two walls of pocketing glass doors connect the leisure room and outdoors.

- The master bath features an angled tub facing a private garden.

Rear Elevation

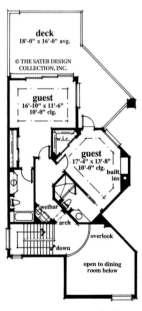

Second Floor

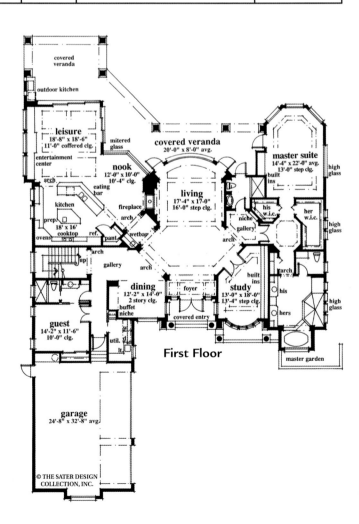

First Floor

© The Sater Design Collection, Inc.

Sherbrooke — TSPDS01-6742 — 1-866-525-9374

Total Living	First Floor	Second Floor	Bonus	Bed	Bath	Width	Depth	Foundation	Price Category
4652 sq ft	3933 sq ft	719 sq ft	N/A	4	4-1/2	89' 8"	104' 10"	Slab	O

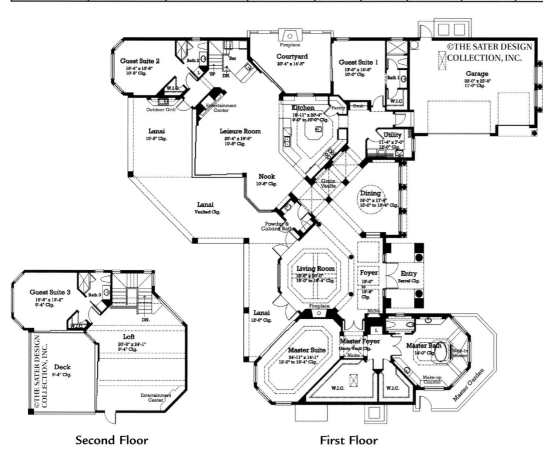

Second Floor

First Floor

Design Features

- The master suite has walk-in closets and a pampering bath.

- With an extended stonewall mantle and an art niche, the living room exudes detail.

- The patio is perfect for entertaining and includes an outdoor kitchen.

- The formal dining room features a high domed ceiling.

- Designed for convenience, the kitchen has plenty of counter space and double ovens.

Rear View

Coach Hill — TSPDS01-8013 — 1-866-525-9374

© The Sater Design Collection, Inc.

Total Living	First Floor	Second Floor	Bonus	Bed	Bath	Width	Depth	Foundation	Price Category
4664 sq ft	3018 sq ft	1646 sq ft	294 sq ft	4	4-1/2	70' 0"	100' 0"	Slab/Opt. Basement	J

Design Features

- The foyer surrounds a staircase enhanced with a dome ceiling.

- A stepped ceiling and arched columns define the living room.

- The great room features a two-story bow window and two-sided fireplace.

- French doors open the master suite to a private veranda.

- The owners' bath has access to an outdoor garden.

Second Floor

First Floor

Rear Elevation

© The Sater Design Collection, Inc.

Hillcrest Ridge — TSPDS01-6651 — 1-866-525-9374

Total Living	First Floor	Second Floor	Bonus	Bed	Bath	Width	Depth	Foundation	Price Category
4759 sq ft	3546 sq ft	1213 sq ft	N/A	4	3-1/2	95' 4"	83' 0"	Basement	L

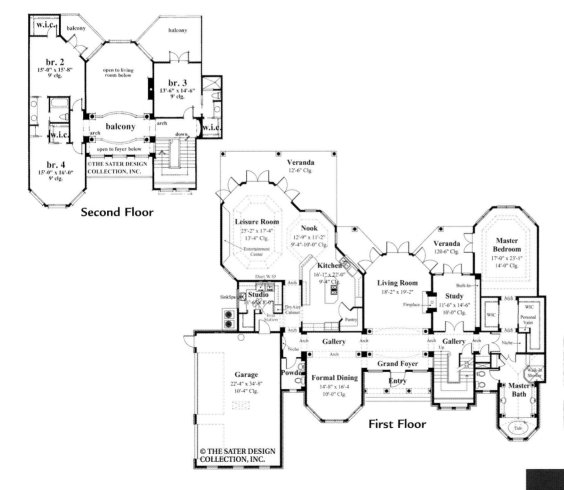

Second Floor

First Floor

Design Features

- An arched gallery hallway leads to the family areas.

- The expansive kitchen has a cooktop island and pass-thu to the veranda.

- Three sets of French doors open the leisure and living rooms to the rear.

- The owner's wing includes a study perfect for an office.

- Upstairs, two bedroom have private balconies, and a third has a bayed sitting space.

Rear Elevation

Flagstone Ridge — TSPDS01-6765 — 1-866-525-9374

© The Sater Design Collection, Inc.

Total Living	First Floor	Second Floor	Bonus	Bed	Bath	Width	Depth	Foundation	Price Category
4809 sq ft	3556 sq ft	1253 sq ft	N/A	4	4-1/2	95' 0"	84' 8"	Slab	J

Design Features

- Stacked stone and arch-top windows gives this home a warm façade.

- A front portico under triple arches leads to the foyer and living room.

- Three sets of French doors open the living room to the veranda.

- French doors lead outside from the study, master suite and leisure room.

- A second-floor loft connects to three upstairs bedrooms.

Second Floor

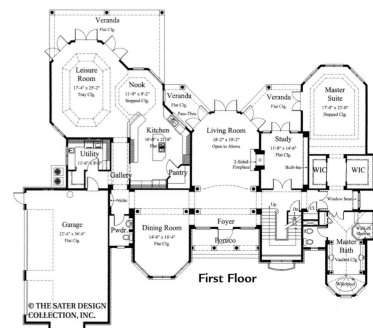

First Floor

Rear Elevation

© The Sater Design Collection, Inc.

Gambier Court — TSPDS01-6948 — 1-866-525-9374

Total Living	First Floor	Second Floor	Bonus	Bed	Bath	Width	Depth	Foundation	Price Category
4951 sq ft	3758 sq ft	1193 sq ft	N/A	4	4-1/2	93' 10"	113' 8"	Slab	O

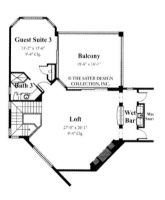

Second Floor

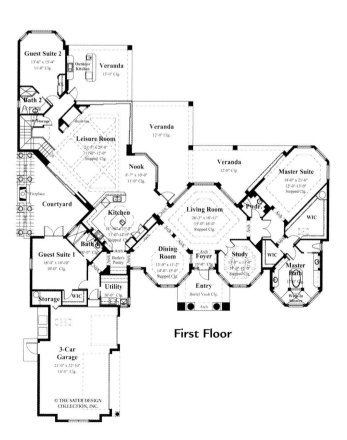

First Floor

Design Features

- A cozy courtyard features an extraordinary fireplace.

- The leisure room has a coffered ceiling and disappearing glass walls to the courtyard.

- The kitchen has an extended serving counter and center island.

- The large master retreat includes a cozy sitting area.

- A center garden tub and walk-in shower make the bath luxurious.

Rear View

Photographed home may have been modified from original construction documents.
Not available for construction in Lee or Collier Counties, Florida.

Saraceno — TSPDS01-6929 — 1-866-525-9374

© The Sater Design Collection, Inc.

Total Living	First Floor	Second Floor	Bonus	Bed	Bath	Width	Depth	Foundation	Price Category
5013 sq ft	4137 sq ft	876 sq ft	N/A	4	5	81' 10"	113' 0"	Slab	O

Design Features

- A thoughtful plan divides social areas and guest rooms from the master suite and study.

- A wet bar serves an open living and dining room.

- Zero-corner sliding doors connect the leisure room to the outdoors.

- A second-floor loft provides privacy for the fourth bedroom.

- The master bedroom has two tray ceilings and a sliding glass wall.

First Floor

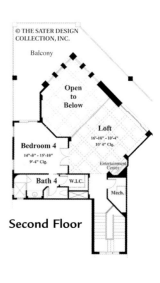

Second Floor

Rear View

Photographed home may have been modified from original construction documents.
Not available for construction in Lee or Collier Counties, Florida.

© The Sater Design Collection, Inc.

Sancho — TSPDS01-6947 — 1-866-525-9374

Total Living	First Floor	Second Floor	Bonus	Bed	Bath	Width	Depth	Foundation	Price Category
5292 sq ft	4666 sq ft	626 sq ft	N/A	4	5-1/2	95' 0"	134' 6"	Slab	O

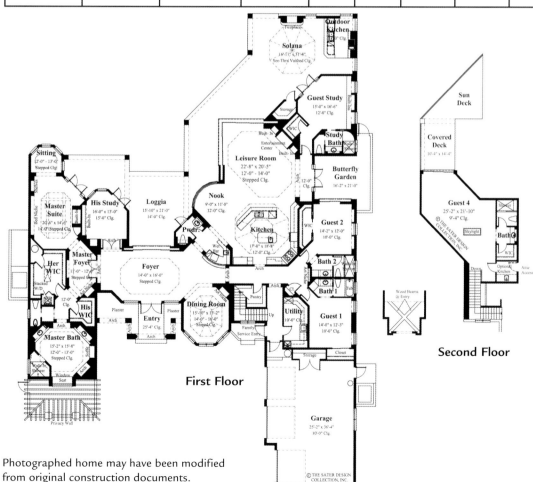

First Floor

Second Floor

Design Features

- Separate his-and-her studies feature custom built-ins.

- The foyer's stepped, exposed-beam ceiling towers sixteen-feet high.

- A charming chimenea-style fireplace warms the outdoor kitchen.

- An elegant art niche welcomes at the entrance of the guest suite.

- The master suite is incredibly private to one side of the home.

Solana

Photographed home may have been modified from original construction documents.
Not available for construction in Lee or Collier Counties, Florida.

Grayhawk Trail — TSPDS01-6748 — 1-866-525-9374

© The Sater Design Collection, Inc.

Total Living	First Floor	Second Floor	Studio/Guest	Bed	Bath	Width	Depth	Foundation	Price Category
5464 sq ft	4470 sq ft	680 sq ft	314 sq ft	4	5-1/2	102' 0"	131' 4"	Slab	O

Design Features

- Angled architecture permits sweeping views of the landscape.

- An arched gallery links the formal rooms to the kitchen and guest wing.

- The casual living zone opens through retreating doors to a wraparound porch.
- A rambling master wing features a grand bay window and luxe bath.

- Upstairs, a computer loft shares a sun deck with guest quarters.

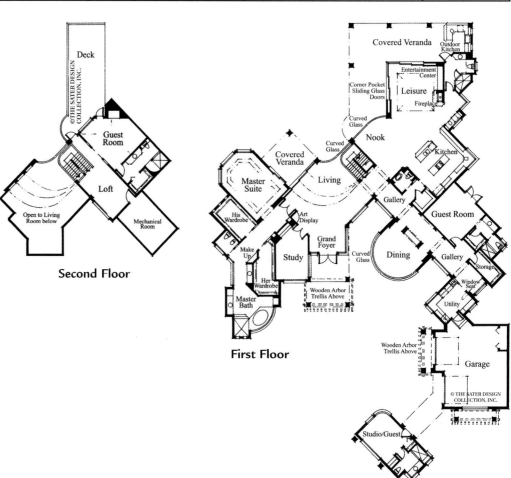

Second Floor

First Floor

Rear View

Photographed home may have been modified from original construction documents.

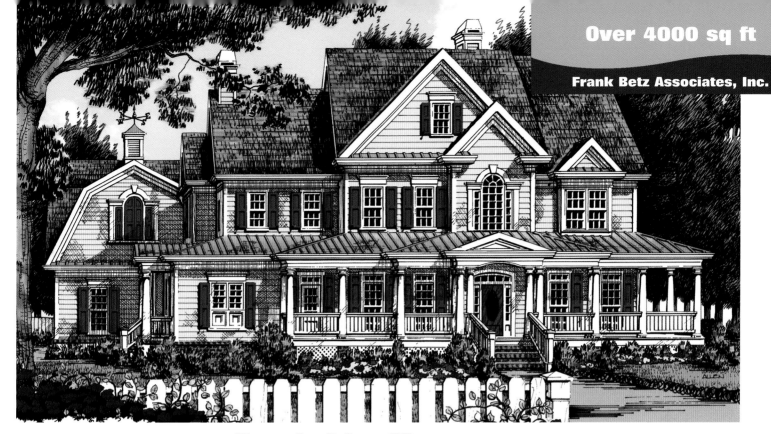

© Frank Betz Associates, Inc.

Brookshire Manor — TSPFB01-1184 — 1-866-525-9374

Total Living	First Floor	Second Floor	Bonus	Bed	Bath	Width	Depth	Foundation	Price Category
5466 sq ft	2732 sq ft	2734 sq ft	N/A	5	5.5	85' 0"	85' 6"	Basement, Crawl Space or Slab	J

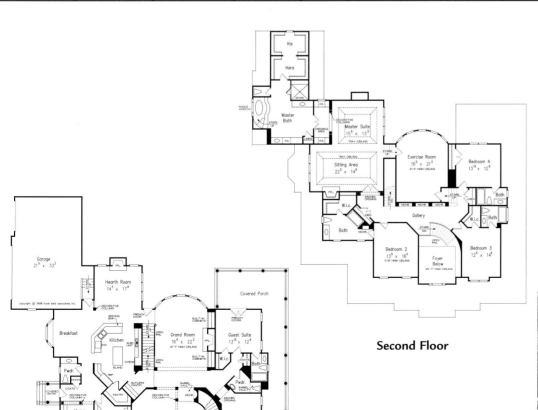

Second Floor

First Floor

Design Features

- Class, style, tradition and every creature comfort imaginable — the *Brookshire Manor* grants every wish!

- Relax and unwind by the fire in the cozy hearth room adjoining the kitchen.

- The master suite earns its name, featuring a personal lounging room with a fireplace and a lavish master bath with private dressing area and direct access to the exercise room.

Rear Elevation

Ristano — TSPDS01-6939 — 1-866-525-9374

© The Sater Design Collection, Inc.

Total Living	First Floor	Second Floor	Bonus	Bed	Bath	Width	Depth	Foundation	Price Category
5564 sq ft	4186 sq ft	1378 sq ft	N/A	5	6	96' 0"	111' 0"	Slab	O

Design Features

- The grand foyer leads to a vaulted living room.

- A graceful colonnade borders the formal dining room.

- A wet bar and butler's pantry bridges the formal and casual areas.

- The upper level features a loft with wet bar, two secondary suites and a deck.

- A portico overlooks the veranda with outdoor grill and kitchen.

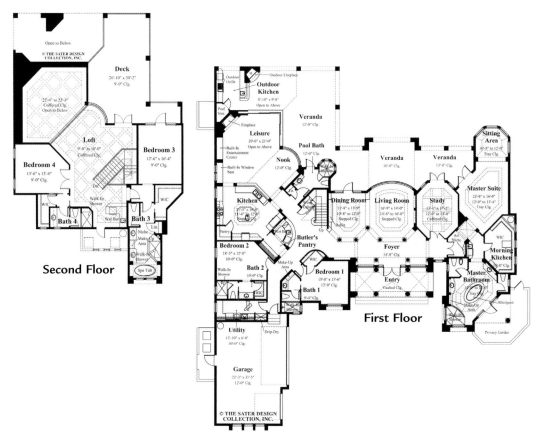

Rear View

Photographed home may have been modified from original construction documents.
Not available for construction in Lee or Collier Counties, Florida.

© The Sater Design Collection, Inc.

Sterling Oaks — TSPDS01-6914 — 1-866-525-9374

Total Living	First Floor	Second Floor	Bonus	Bed	Bath	Width	Depth	Foundation	Price Category
5816 sq ft	4385 sq ft	1431 sq ft	N/A	5	5-1/2	88' 0"	110' 1"	Slab	O

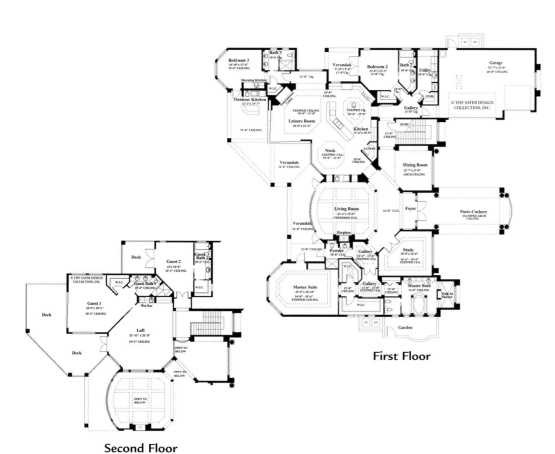

Second Floor

First Floor

Design Features

- A dramatic entry under a porte-cochere leads to a living room with a bowed window.

- Bedrooms are cozily sequestered and have full baths.

- Two gallery areas lead to a grand master bedroom.

- Multiple outdoor spaces include a winding verandah with kitchen.

- The second floor features a loft, two guests suites and two decks.

Rear View

Photographed home may have been modified from original construction documents.

Prestonwood — TSPDS01-6922 — 1-866-525-9374

© The Sater Design Collection, Inc.

Total Living	First Floor	Second Floor	Bonus	Bed	Bath	Width	Depth	Foundation	Price Category
5924 sq ft	4715 sq ft	1209 sq ft	N/A	3	3-1/2	117' 2"	131' 7"	Slab	O

Design Features

- A circular-shaped grand hall sits under a soaring 35-foot domed ceiling.

- A massive triangle-shaped veranda is flanked by the master suite and family living areas.

- The leisure room has an outdoor kitchen beyond zero-corner sliding glass doors.

- A regal staircase leads elegantly to a second-floor loft.

- Two upstairs guest suites share a deck overlooking the pool.

Rear View

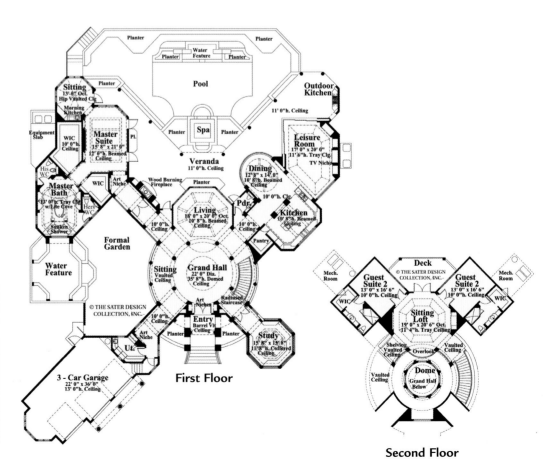

First Floor

Second Floor

Photographed home may have been modified from original construction documents.

© The Sater Design Collection, Inc.

Lindley — TSPDS01-6930 — 1-866-525-9374

Total Living	First Floor	Second Floor	Bonus	Bed	Bath	Width	Depth	Foundation	Price Category
6011 sq ft	5265 sq ft	746 sq ft	N/A	4	4-1/2	99' 4"	140' 0"	Slab	O

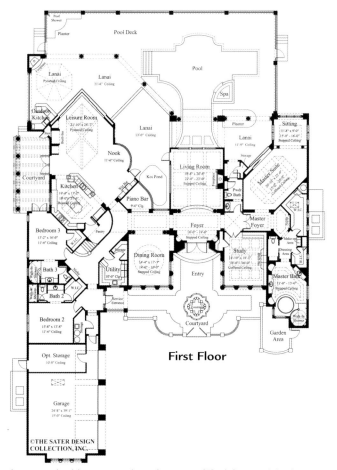

First Floor

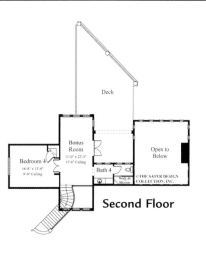

Second Floor

Design Features

- Stepped ceilings highlight the foyer, living room and master suite.

- Water features provide a pool and a koi pond.

- In the leisure room, arched niches nestle an entertainment center.

- French doors open the leisure room onto a private courtyard.

- The master bath showcases a gorgeous round tub.

Rear View

Photographed home may have been modified from original construction documents.
Not available for construction in Lee or Collier Counties, Florida.

Cataldi — TSPDS01-6946 — 1-866-525-9374

© The Sater Design Collection, Inc.

Total Living	First Floor	Second Floor	Bonus	Bed	Bath	Width	Depth	Foundation	Price Category
6011 sq ft	5265 sq ft	746 sq ft	N/A	4	4-1/2	99' 4"	140' 0"	Slab	O

Design Features

- The great room is elegant under an exposed wood beam ceiling.

- The kitchen boasts sleek granite countertops and scenic views.

- Volume ceilings create distinction and spaciousness in the dining room.

- A wraparound verandah forms the boundary between the indoor spaces and backyard pool.

Second Floor

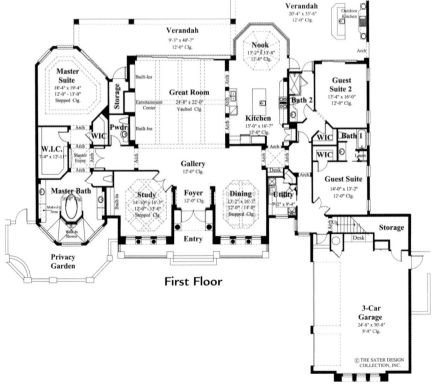

First Floor

Rear View

Photographed home may have been modified from original construction documents.
Not available for construction in Lee or Collier Counties, Florida.

© The Sater Design Collection, Inc.

Fiorentino — TSPDS01-6910 — 1-866-525-9374

Total Living	First Floor	Second Floor	Bonus	Bed	Bath	Width	Depth	Foundation	Price Category
6273 sq ft	4742 sq ft	1531 sq ft	N/A	4	4F/2H	96' 0"	134' 8"	Slab	O

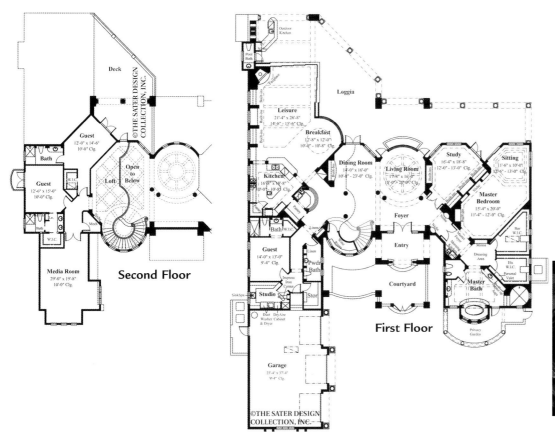

Second Floor

First Floor

Design Features

- The guest quarters and family spaces shares a wet bar with the formal rooms.

- The kitchen has a food-prep island and a serving counter.

- The leisure room has a fireplace and retreating glass walls to the loggia.

- The master suite is secluded, with a spa tub viewing a private garden.

- On the upper level, a computer loft overlooks the living room.

Rear View

Photographed home may have been modified from original construction documents.

Port Royal Way — TSPDS01-6635 — 1-866-525-9374

© The Sater Design Collection, Inc.

Total Living	First Floor	Second Floor	Bonus	Bed	Bath	Width	Depth	Foundation	Price Category
6312 sq ft	4760 sq ft	1552 sq ft	N/A	5	6-1/2	98' 0"	103' 8"	Slab	M

Design Features

- Columns and archways grace the view through the formal spaces.

- A two-story ceiling, fireplace and French doors add drama to the living room.

- The family room pleases with a media center and corner wet bar.

- An enormous master wing has a personal valet and center whirlpool tub.

- Multiple verandas and second-floor decks offer limitless outdoor living.

Rear Elevation

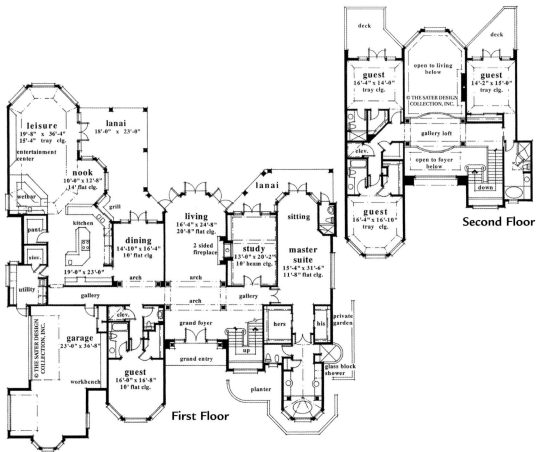

Second Floor

First Floor

© The Sater Design Collection, Inc.

Windsor Court — TSPDS01-6751 — 1-866-525-9374

Total Living	First Floor	Second Floor	Guest	Bed	Bath	Width	Depth	Foundation	Price Category
6457 sq ft	4715 sq ft	570 sq ft	1172 sq ft	3	3-1/2	137' 4"	103' 0"	Slab	O

Second Floor

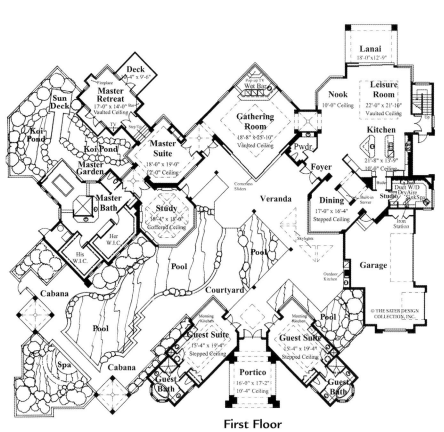

First Floor

Design Features

■ The nook has a floor-to-ceiling wall of curved glass.

■ The pool winds its way through the property, offering views from several rooms.

■ A sundeck is tucked in a private corner near a waterfall.

■ The gathering room has a glass-walled bar with huge views.

Courtyard View

Photographed home may have been modified from original construction documents.

Casa Bellisima — TSPDS01-6935 — 1-866-525-9374

© The Sater Design Collection, Inc.

Total Living	First Floor	Second Floor	Bonus	Bed	Bath	Width	Depth	Foundation	Price Category
6524 sq ft	5391 sq ft	1133 sq ft	N/A	4	5-1/2	104' 0"	140' 0"	Slab	O

Design Features

- A formal medallion-shaped fountain fills the courtyard entry.

- The formal living room features a fireplace flanked by art niches.

- A butler's pantry, wet bar and wine room are situated near the formal areas.

- The home's covered veranda is equipped with a fireplace.

- An elegant circular staircase curves upward to a game room and guest suite.

Second Floor

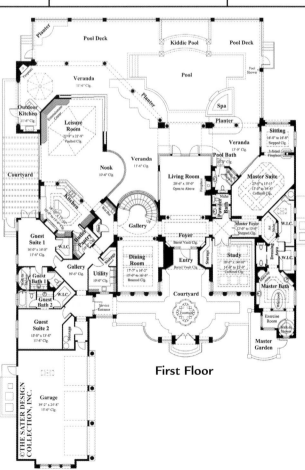

First Floor

Rear View

Photographed home may have been modified from original construction documents.
Not available for construction in Lee or Collier Counties, Florida.

Over 4000 sq ft

The Sater Design Collection, Inc.

© The Sater Design Collection, Inc.

Huntington Lakes — TSPDS01-6900 — 1-866-525-9374

Total Living	First Floor	Second Floor	Bonus	Bed	Bath	Width	Depth	Foundation	Price Category
6770 sq ft	5170 sq ft	1600 sq ft	N/A	3	4	140' 7"	118' 4"	Slab	O

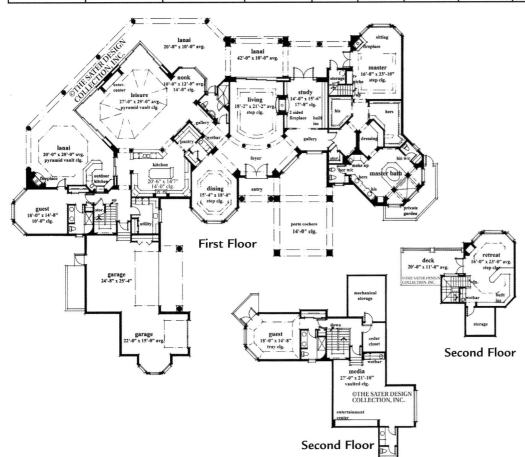

First Floor

Second Floor

Second Floor

Design Features

- Double doors open onto a bowed foyer and living room with two-sided fireplace.

- The lanai includes an outdoor kitchen and fireplace.

- Zero-corner sliding glass doors open the leisure room to the outdoors.

- A gallery leads to the private master suite, which enjoys a fireplace.

- A full guest suite and media room outfit the second floor.

Rear View

Photographed home may have been modified from original construction documents.
Not available for construction in Lee or Collier Counties, Florida.

Trissino — TSPDS01-6937 — 1-866-525-9374

© The Sater Design Collection, Inc.

Total Living	First Floor	Second Floor	Bonus	Bed	Bath	Width	Depth	Foundation	Price Category
7209 sq ft	6134 sq ft	1075 sq ft	N/A	4	4-1/2	142' 7"	118' 0"	Slab	O

Design Features

- A stepped ceiling defines the grand foyer.

- The vaulted living room has a wall of glass framed by French doors to the lanai.

- The dining room is crowned by a round coffer and harbored in a glass bay.

- A wine cellar offers a 1,700-bottle storage capacity near the kitchen.

- A master gallery leads to a secluded study and a morning kitchen.

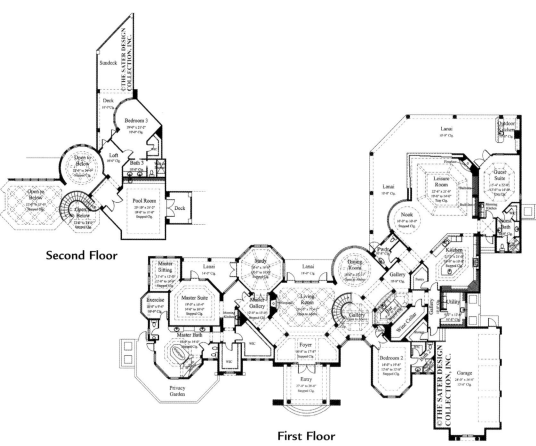

Second Floor

First Floor

Rear View

Photographed home may have been modified from original construction documents.
Not available for construction in Lee or Collier Counties, Florida.

© The Sater Design Collection, Inc.

Alamosa — TSPDS01-6940 — 1-866-525-9374

Total Living	First Floor	Second Floor	Bonus	Bed	Bath	Width	Depth	Foundation	Price Category
8088 sq ft	6122 sq ft	1966 sq ft	N/A	5	5F/2H	118' 0"	147' 10"	Slab	O

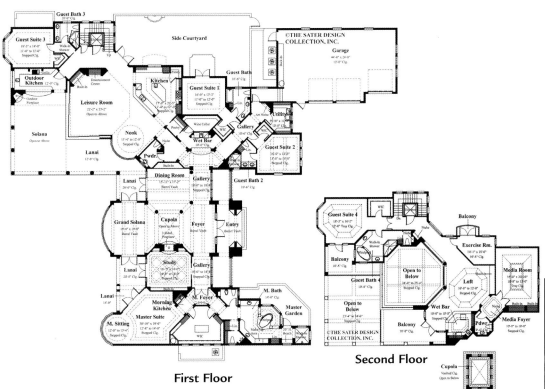

First Floor

Second Floor

Design Features

- Volume ceilings and stately columns define the living areas.

- An opulent master suite includes a morning kitchen and garden.

- A two-sided hearth warms the Grand Solana.

- The second floor houses a loft with wet bar, exercise room and media room.

- Outdoor areas include balconies, lanais, courtyards and a Solana.

Rear View

Photographed home may have been modified from original construction documents.
Not available for construction in Lee or Collier Counties, Florida.

BEFORE YOU ORDER
PLEASE READ THE FOLLOWING
HELPFUL INFORMATION

QUICK TURNAROUND

Because you are placing your order directly, we can ship plans to you quickly. If your order is placed before noon EST, we can usually have your plans to you the next business day. Some restrictions may apply. We cannot ship to a post office box; please provide a physical street address.

OUR EXCHANGE POLICY

Since our blueprints are printed especially for you at the time you place your order, we cannot accept any returns. If, for some reason, you find that the plan that you purchased does not meet your needs, then you may exchange that plan for another plan in our collection. We allow you sixty days from the time of purchase to make an exchange. At the time of the exchange, you will be charged a processing fee of 20% of the total amount of the original order, plus the difference in price between the plans (if applicable) and the cost to ship the new plans to you. Vellums cannot be exchanged. All sets must be approved and authorization given before the exchange can take place. Please call our customer service department if you have any questions.

LOCAL BUILDING CODES AND ZONING REQUIREMENTS

Our plans are designed to meet or exceed national building standards. Because of the great differences in geography and climate, each state, county and municipality has its own building codes and zoning requirements. Your plan may need to be modified to comply with local requirements regarding snow loads, energy codes, soil and seismic conditions and a wide range of other matters. Prior to using plans ordered from us, we strongly advise that you consult a local building official.

ARCHITECTURE AND ENGINEERING SEALS

Some cities and states are now requiring that a licensed architect or engineer review and approve any set of building documents prior to construction. This is due to concerns over energy costs, safety, structural integrity and other factors. Prior to applying for a building permit or the start of actual construction, we strongly advise that you consult your local building official who can tell you if such a review is required.

DISCLAIMER

We have put substantial care and effort into the creation of our blueprints. We authorize the use of our blueprints on the express condition that you strictly comply with all local building codes, zoning requirements and other applicable laws, regulations and ordinances. However, because we cannot provide on-site consultation, supervision or control over actual construction, and because of the great variance in local building requirements, building practices and soil, seismic, weather and other conditions, WE CANNOT MAKE ANY WARRANTY, EXPRESS OR IMPLIED, WITH RESPECT TO THE CONTENT OR USE OF OUR BLUEPRINTS OR VELLUMS, INCLUDING BUT NOT LIMITED TO ANY WARRANTY OF MERCHANTABILITY OR OF FITNESS FOR A PARTICULAR PURPOSE. Please Note: Floor plans in this book are not construction documents and are subject to change. Renderings are artist's concept only.

HOW MANY SETS OF PRINTS WILL YOU NEED?

We offer a single set of prints so that you can study and plan your dream home in detail. However, you cannot build from this package. One set of blueprints is marked "NOT FOR CONSTRUCTION." If you are planning to obtain estimates from a contractor or subcontractor, or if you are planning to build immediately, you will need more sets. Because additional sets are less expensive, make sure you order enough to satisfy all your requirements. Sometimes changes are needed to a plan; in that case, we offer vellums that are reproducible and erasable so changes can be made directly to the plans. Vellums are the only set that can be reproduced; it is illegal to copy blueprints. The checklist below will help you determine how many sets are needed.

PLAN CHECKLIST

_____ **Owner** (one for notes, one for file)

_____ **Builder** (generally requires at least three sets; one as a legal document, one for inspections and at least one to give subcontractors)

_____ **Local Building Department** (often requires two sets)

_____ **Mortgage Lender** (usually one set for a conventional loan; three sets for FHA or VA loans)

_____ **Total Number of Sets**

IGNORING COPYRIGHT LAWS CAN BE A
$1,000,000 *mistake!*

Recent changes in the US copyright laws allow for statutory penalties of up to $150,000 per incident for copyright infringement involving any of the copyrighted plans found in this publication. The law can be confusing. So, for your own protection, take the time to understand what you cannot do when it comes to home plans.

WHAT YOU CAN'T DO!

YOU CANNOT DUPLICATE HOME PLANS.

YOU CANNOT COPY ANY PART OF A HOME PLAN TO CREATE ANOTHER.

YOU CANNOT BUILD A HOME WITHOUT BUYING A BLUEPRINT OR LICENSE.

346 www.twostoryplans.com

HOW TO ORDER BY PHONE, MAIL OR ONLINE

1-866-525-9374

Select the option that corresponds to the designer of your home plan:

Frank Betz and Associates, dial 01
Donald A. Gardner Architects, dial 02
Sater Design Collection, dial 03

This puts you in DIRECT contact with the designer's office!

ADDITIONAL ITEMS**

Additional Blueprints (per set) $60.00
Reverse Mirror-Image Blueprints $125.00

MATERIALS LIST*

Plan Categories A — E $65.00
Plan Categories F — O $75.00

FOUNDATION OPTIONS*

(basement, crawl space or slab, if different from base plan)
(no charge for Frank Betz plans)

Plan Categories A — C $225.00
Plan Categories D — E $250.00
Plan Categories F — M $275.00
Specification Outline* $15.00

*Call for availability. Special orders may require additional fees.

SHIPPING AND HANDLING

Overnight . . . . $50.00   Ground $30.00
2nd Day . . . . $40.00   Saturday $70.00

For shipping international, please call for a quote.

**Products and prices vary for each designer. Call for specific availability and pricing.

BLUEPRINT PRICE SCHEDULE*

	1 STUDY SET	4 SETS	8 SETS	VELLUM
A	$465	$515	$565	$710
B	$510	$560	$610	$775
C	$555	$605	$655	$840
D	$600	$650	$700	$905
E	$645	$695	$745	$970
F	$690	$740	$790	$1035
G	$760	$810	$860	$1115
H	$835	$885	$935	$1195
I	$935	$985	$1035	$1295
J	$1035	$1085	$1135	$1395
K	$1135	$1185	$1235	$1495
L	$1235	$1285	$1335	$1595
M	$1335	$1385	$1435	$1695
N	$1435	$1485	$1535	$1795
O	Call for pricing			

* Prices subject to change without notice

Order Form

PLAN NUMBER _____

☐ 1-set [study only] . . . . $_____
☐ 4-set building package . . . . $_____
☐ 8-set building package . . . . $_____
☐ 1-set of reproducible vellums . . . . $_____

____ Additional Identical Blueprints @ $65 each $_____
____ Reverse Mirror-Image Blueprints @ $125 fee $_____

Foundation Options:
____ Crawl Space ____ Slab ____ Basement $_____
(no charge for Frank Betz plans)

Sub-Total $_____
Shipping and Handling $_____
Sales Tax *(will be determined upon placing order)* $_____

TOTAL $_____

Check one: ☐ Visa ☐ MasterCard

Credit Card Number _____

Expiration Date _____

Signature _____

Name _____

Company _____

Street _____

City _____ State____ Zip_____

Daytime Telephone Number (_____)_____

Check one:

☐ Consumer   ☐ Builder   ☐ Developer

PLAN INDEX

PLAN NAME	PLAN #	PAGE

PLAN INDEX

PLAN INDEX

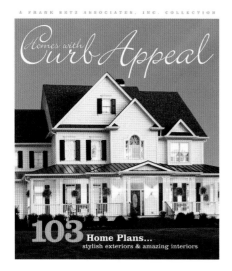

HOMES WITH CURB APPEAL

103 Home Plans...stylish exteriors & amazing interiors

More than ever, our homes need to work for us. Our living spaces need to reflect and support our daily activities. In this practical and beautiful book, Frank Betz & Associates offers the finest designs in four distinctive styles: Cottage, Craftsman, Country and Old World. Each style features its own design elements that set it apart from the next. But all share common goals: practicality, function, charm, and above all, curb appeal!

1,300 to over 5,400 sq ft

$12.95   160 full-color pages

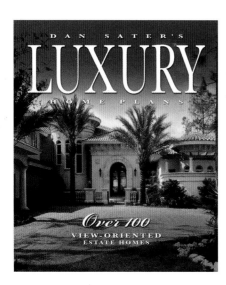

DAN SATER'S LUXURY HOME PLANS

over 100 view-oriented estate homes

A colorful and richly textured collection of more than 100 exquisite floor plans. This stunning display of unique Sater homes features more than 220 pages of exciting interior and exterior photography, with unique design ideas for the most gracious living. Whether you're seeking plans for a 6,000-square-foot estate or a 3,000-square-foot villa, you can find them in this truly inspirational portfolio of Dan's best luxury home plans.

2,700 to over 8,000 sq ft

$16.95   256 full-color pages

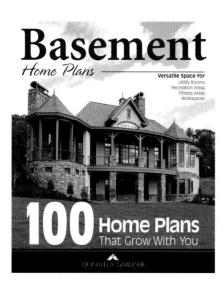

BASEMENT HOME PLANS

100 Home Plans that Grow with You

Created by the nation's leading architectural firm, *Basement Home Plans* presents an exciting compilation of home designs from Donald A. Gardner Architects. Featuring basement plans for every home, this stunning book showcases a variety of homes for every region. If you're looking to retire in the mountains, settle down in the suburbs or live on several acres, these plans with comfortable interiors and visually stunning exteriors fit any locale.

1,400 to over 4,700 sq ft

$12.95   176 full-color pages